A MANUAL OF PRACTICAL LABORATORY AND FIELD TECHNIQUES IN PALAEOBIOLOGY

A Manual of Practical Laboratory and Field Techniques in Palaeobiology

by

Owen R. Green
Department of Earth Sciences,
University of Oxford, United Kingdom

KLUWER ACADEMIC PUBLISHERS

DORDRECHT / BOSTON / LONDON

Library of Congress Cataloging-in-Publication Data

Green, Owen R.
A manual of practical laboratory and field techniques in palaeobiology / by Owen R. Green.
p. cm.
Includes bibliographical references and index.
ISBN 0-412-58980-X (hardbound : alk. paper)
1. Paleobiology--Methodology--Handbooks, manuals, etc. I. Title.

QE719.8 .G74 2001
560'.78--dc21

2001023212

ISBN 0-412-58980-X

Published by Kluwer Academic Publishers,
P.O. Box 17, 3300 AH Dordrecht, The Netherlands.

Sold and distributed in North, Central and South America
by Kluwer Academic Publishers,
101 Philip Drive, Norwell, MA 02061, U.S.A.

In all other countries, sold and distributed
by Kluwer Academic Publishers,
P.O. Box 322, 3300 AH Dordrecht, The Netherlands.

Printed on acid-free paper

Cover caption: Polyester resin impregnated and prepared surface of a piston core sediment containing scattered bivalve molluscan fragments and seagrass roots supported in a micritic carbonate mud, Holocene, Tavernier mud mound, Florida Keys, U.S.A.

Printed in the Netherlands.

For Jane, Alex, Caitlin and Leonie.

To the memory of
S.T.R.

A *technique* may require one or more *procedures*,
that are performed using a variety of *methods*
comprising numerous *stages*.

Acknowledgements

This manual has been influenced by my association, initially as a student and latterly as a member of staff, with undergraduate courses at South London College and the Universities of London and Oxford. To the lecturing, research and technical staffs, undergraduate and post graduate students who have tutored, discussed and influenced methodology with me, I am eternally grateful.

I am particularly indebted to friends and colleagues at South London College, Tony Grindrod, Brian Hunt, Lindzi and Richard Marsh, Steve Rye and Alec Quarterman for enthusiastically introducing me to geological specimen preparation techniques. While in the Department of Geology at Goldsmiths' College (University of London) Dan Bosence, Derek Briggs and Andy Johnstone encouraged the demonstration and application of palaeontological preparation procedures in undergraduate courses. For the opportunity to practice established and develop new techniques I am thankful. This opportunity continued in the Department of Earth Sciences at Oxford University in establishing the Palaeobiology Laboratories and assisting Martin Brasier and Bob Spicer. The latter to whom I am particularly grateful for encouraging me in pursuing the publication of this work, following its early draft as a laboratory manual. I am also indebted to Steve Packer for discussions on numerous preparation techniques, and their application in the commercial sector. Numerous commercial technical experts from industrial chemical and laboratory equipment suppliers have freely provided data and information relating to the use of their products.

The editorial staff, initially at Chapman and Hall and latterly at Kluwer Academic Publishers, have been supportive, encouraging, helpful and sympathetic in allowing me to develop and pursue this project, particularly as it has grown in complexity. In particular I am grateful to the editorial contributions of Ruth Cripwell, Ian Francis, Jane Plowman, Petra van Steenbergen, Manja Fredriksz and Donna Lynch for providing the final encouragement in the conclusion of this project.

Finally, I would like to thank my wife Jane for continued emotional support and encouragement, and in undertaking proof reading of the entire manuscript, and finally to my children, for allowing me a few quiet minutes during the evening and at weekends, and posing the most searching question - why?

FOREWORD

The user

This manual is designed for the use of geo-scientists with an interest and need in developing palaeobiological materials as a potential source of data. To meet this objective practical procedures have been formatted for use by both professional and semi-professional students with an initial understanding of palaeobiological research aims as a primary source of scientific data.

I have attempted to provide an explanation and understanding of practical procedures which may be required by students undertaking palaeobiological projects as part of a degree course. The layout of this manual should be particularly beneficial in the instruction and training of geotechnologists and museum preparators. Graduate students and scientists requiring an outline of a preparation procedure will also be able to use the manual as a reference from which to assess the suitability of a procedure.

This manual is also intended for use by the "committed amateur". Many of the techniques described in this manual have been devised by non-palaeontologists, and developed from methods used in archaeology, zoology and botany, as well as other areas of geology. A considerable number of the methods can be undertaken by the amateur, and in the case of many of the field procedures, should be used. This will ensure that specimens and samples can be conserved in such a manner as to facilitate any later research, and not invalidate the results of subsequent geochemical analytical techniques which might be employed.

Despite this manual covering a wide range of field and laboratory techniques used by palaeobiologists, it is by no means exhaustive. Although a large number of *modifications* to *standard* techniques are included, most *local customising* of methods has been omitted. A large part of this omission I readily accept as ignorance on my behalf of such modifications. The control and availability of chemicals, equipment and products within local markets, or the implementation of health and safety regulations and by-laws may also effect how procedures are undertaken in different countries.

The manual's scope

The systematic approach to procedures adopted in this manual is biased towards my own personal interests. Methods are, however, not mutually exclusive to the headings under which they described or assigned. In adopting this approach I have attempted to provide the user with the optimum pathway in determining a suitable technique. It is hoped that this format and numerous cross-referencing to related methods will assist the novice, while more experienced preparators will have attention focused to a specific procedure.

Following a short introductory section outlining practical procedures, emphasis is placed on documentation and record keeping. The bulk of this manual, the practical procedures employed by preparators and geotechnologists, are broadly divisible into one of two sections, (i) field techniques, and (ii) laboratory techniques. Within the first section, field collecting methodology and aspects of site and specimen conservation are outlined, prior to more detailed laboratory preparation. To assist in recent approaches to morphological and taphonomical studies, procedures for the collection and preservation of live zoological material are also included. Semi-quantitative determinations of the two principle biominerals (calcite and phosphate) are also described, although as stated above, these techniques are not exclusive to field use. Procedures used in specimen stabilization and replication, prior to packaging and removal to the laboratory are also discussed. The third section of the manual covers fundamental laboratory procedures. Success within this area will be considerably enhanced in a well designed laboratory, implementing the highest safety considerations. Prior to preparation the stabilization methods, including those used in the preparation of recent (living) specimens is described. The bulk of this manual is then devoted to describing in detail the wide array of physical and chemical techniques commonly used in palaeontology.

Following the successful preparation of specimens and materials, some consideration is given to selected analytical procedures, in particular electron microscopy and x-radiography. Geochemical analytical procedures are not discussed, as to do this sub-discipline justice a companion volume would be required. In conclusion I thought it fitting to leave prepared specimens suitable for display. The final chapters of this manual are devoted to replication, conventional photographic techniques and a brief outline of procedures used in the display and illustration of prepared specimens.

How to use the manual

Following the initial stages of collecting and stabilization, the preparator is required to identify those techniques specific to preparing the fossil group of interest. Methods outlined in this manual provide a means of fulfilling this objective. Two complimentary approaches have been adopted in detailing the methods within the manual. Methods described as *physical procedures* have been grouped according to *subject area* or the *fossil group* to which they are most applicable, and where chemical methods are supplementary or of secondary importance. Within this part, methods have been detailed by *subject*, from which it will be appreciated that applications can be adapted for work on any fossil group supported within a matrix of known composition or degree of induration.

The second part, concentrating on chemical procedures is more specific in application to fossils or matrices of a known composition, reflecting preservational differences. This approach is particularly critical in processing for microfossils, where the small specimen size may inhibit direct, constant monitoring of chemical effects on the organisms shell so readily evident in macrofossil preparation. Consequently many microfossil extraction techniques are protracted and more complex than those employed in macrofossil preparation. I make no apology for an apparent bias towards the processing of samples for microfossils. I have always found it more convenient to modify a technique towards the extraction of macrofossils than vice versa.

Many of the techniques described in these sections (*e.g.* palynological, palaeobotanical and acid insoluble procedures) may appear superficially similar to one another. However, subtle differences are detailed at critical parts of each the procedure. This, inevitably, results in some repetition, but has been purposely retained to ensure each section provides a *stand-alone* description and explanation of the technique. However, where the repetition of a procedure is considered superfluous to the understanding and continuity of the technique, it has been omitted. If a specialized procedure has been detailed elsewhere, cross reference is made. Experienced preparators will be familiar with these procedures, and understand the significance of the stage. Once preparators are familiar with the methods and procedures outlined, reference need only be made to the accompanying flow charts. These have been designed to show the relationships between the methods and stages comprising the technique.

The manuals layout is designed to answer two questions posed by potential users:

- *how can I do this.......?*
- *I want to examine this fossil* (of a known composition) *from this rock.*

The two main parts of this manual (Physical and Chemical Laboratory Techniques) attempt to direct users in answering these questions. Furthermore, some element of thought has been given to describing the consequences of the method's actions, both directly and indirectly, during specimen development. Technological advances in other scientific disciplines, and subsequently applied to palaeobiology will doubtless improve methods of examining and recording palaeobiological data. Furthermore, to fully appreciate our palaeontological heritage, in an age when it is increasingly difficult to collect a new data set from the field, consideration must be given to the correct preparation of specimens within existing collections. I sincerely hope that this manual will provide some assistance to preparators in performing this task in the coming years.

Holt, Norfolk
April, 2000

O.R.G.

A MANUAL OF PRACTICAL LABORATORY AND FIELD TECHNIQUES IN PALAEOBIOLOGY

CONTENTS

Acknowledgements vii
Foreword ix

I. INTRODUCTION

1 Palaeontological Techniques - an introduction to practical procedures 2
2 Documentation in the Laboratory 12

II. FIELD TECHNIQUES

3 Field collecting procedures 20
4 Collecting techniques for microfossil and live foraminifera 27
5 Consolidation, stabilization and replication techniques 43
6 Field staining techniques for determining calcite, dolomite and phosphate 55
7 Field documentation, sample packing and transportation 59

III. LABORATORY TECHNIQUES

A. INTRODUCTION

8 Laboratory design and layout 64
9 Safety in the laboratory 68
10 Preservation, consolidation and repair of unstable specimens 82
11 Preparation of recent material for comparative studies 98

B. PHYSICAL PROCEDURES

12 Mechanical methods for preparing fossil specimens 110
13 Fossil extraction techniques by thermal disintegration 120
14 Disaggregation and dispersal of partially consolidated and unconsolidated sediments 127
15 Microfossil sample contamination and reliability problems 138
16 Washing and sieving techniques used in micropalaeontology 146
17 Centrifuge techniques used in micropalaeontology 154
18 Mechanical separation of microfossil residues 160
19 Flotation and liquid separation techniques 168
20 Thin section and slide preparation techniques of macro and microfossil specimens 182
21 Staining techniques used in micropalaeontology 211
22 Preparation of carbonate stained acetate peels and thin sections 219
23 Preparation of amber specimens containing fossils 234
24 Preparation and conservation of vertebrate fossils 242

C. CHEMICAL PROCEDURES

25 Extraction techniques for palaeobotanical and palynological material 256
26 Extraction techniques for acid insoluble microfossils 288
27 Extraction techniques for phosphatic fossils 318
28 Extraction techniques for uncrushed graptolites 331

29	Extraction techniques for calcareous microfossils from argillaceous sediments	334
30	Extraction techniques for calcareous microfossils from carbonate sediments	342
31	Extraction techniques for agglutinated foraminifera from calcareous sediments	350
32	Specialist techniques used in the preparation of individual microfossil specimens	355
33	Extraction techniques for calcareous nannofossils	366
D. ANALYTICAL PROCEDURES		
34	Electron microscopy techniques	376
35	X-radiography techniques	391
E. EXHIBITION OF FOSSIL MATERIAL		
36	Fossil replication techniques	400
37	Photomacrography and photomicrography techniques	418
38	Illustrating and exhibiting for display and publication	441
REFERENCES		450
APPENDIX		
1	Equipment and chemical suppliers	490
2	Describing sedimentary rocks and fossils	498
3	Centrifuge times for bromoform separation and water/acetone residue washing	503
4	Hazard symbols	506
5	Respirator and filter cartridge colour coding	507
6	Fire extinguisher classification	507
7	Laboratory glove resistance properties	508
8	Care and handling of hydrofluoric acid	511
9	Spillage and disposal of unwanted chemicals	514
10	Conversion data	515
	(i) Temperature	
	(ii) Weight	
	(iii) Volume	
	(iv) Length	
11	Formulae for the preparation of standard solutions	516
	(i) Weight % (w/w)	
	(ii) Volume % (v/v)	
	(iii) Weight/Volume % (w/v)	
	(iv) Mole Fraction	
	(v) Standard solutions	
INDEXS		517
Author Index		517
Subject Index		528
Chemical Index		534

PART I

INTRODUCTION

1. Palaeontological Techniques - an introduction to practical procedures

2. Documentation in the Laboratory

1. PALAEONTOLOGICAL TECHNIQUES - AN INTRODUCTION TO PRACTICAL PROCEDURES

1.1. INTRODUCTION

The Palaeontological technician, conservator, preparator and geo-technologist is required to perform and understand a variety techniques. These may range from relatively simple cleaning procedures to complex multi-stage techniques requiring great dexterity, concentration and many hours of patience to successfully complete. The type and mode of preservation of specimens encountered during normal working practices is vast, with fossil remains of different primary and secondary mineralogical compositions appearing within the laboratory. Seldom will two specimens respond in the same manner, and an element of trial-and-error has to be practised by the preparator to achieve the optimum results and gain experience. The ultimate objective of the specimen preparation is a combination of enhancing the appearance, revealing more information, and ensuring its longevity.

This brief introduction provides a broad chronological classification of procedures and techniques applicable in four areas of a specimen's history, and the principal working environment in which they are performed (Table 1.1). Although the majority of preparation and conservation work is undertaken within the laboratory, consideration must be given to sample or specimen field collection, its initial stabilization and transportation back to the laboratory.

TECHNIQUES/STAGE	WORKING ENVIRONMENT
1. COLLECTION	FIELD
2. CONSERVATION	FIELD & LABORATORY
3. PRESERVATION	LABORATORY
4. CURATION	LABORATORY

Table 1.1. Many techniques are applicable in both the field and laboratory, and can be used on all specimens regardless of quality (*e.g.* museum and research). For simplicity, techniques have been placed in one of the four sub-divisions.

1.2. COLLECTION OF SAMPLES

Sample and specimen collecting techniques are some of the most neglected procedures by geologists. Badly prepared field excursions or ill thought out collecting strategies may result in disappointment on returning to the laboratory. This is often compounded by the time, effort and expense that has gone into collecting, transporting and carefully unpacking samples, only to find broken specimens, damaged beyond repair, or displaced from labels or collecting data. These problems are examined in more detail in section 5 COLLECTING TECHNIQUES.

Ethical considerations in sample collecting must also be emphasised, particularly to new or inexperienced collectors. Collecting techniques should reflect the nature of the material and the purpose for which it is collected, while collectors must consider the importance or value of specimens to the objectives of their work. Summaries and discussions on the many aspects and considerations of collecting geological specimens have been outlined Bassett (1979), Brunton *et al.*, (1985), Compton (1962) and Goldring (1991). Strategies should be thoroughly discussed with colleagues before embarking on field work. It may also prove pertinent to discuss collecting objectives with site or locality owners, from whom permission for access should always be gained.

Intimately connected with sample collecting is site conservation. Monitoring the state and accessibility of sites has been a major priority for many local and national geological groups and societies. Numerous high profile reports (*e.g.* Benton and Wimbledon 1985, Crowther and Wimbledon 1989, Nature Conservancy Council 1990) have been produced reflecting the importance of site conservation.

The history of a specimen, once collected, can become a complex affair, with various sub-samples studied by a variety of invasive (destructive) and non-evasive analytical techniques. A methodical documentation procedure is required for each stage and technique, diagramatically illustrated in figures 1.1a and 1.1b.

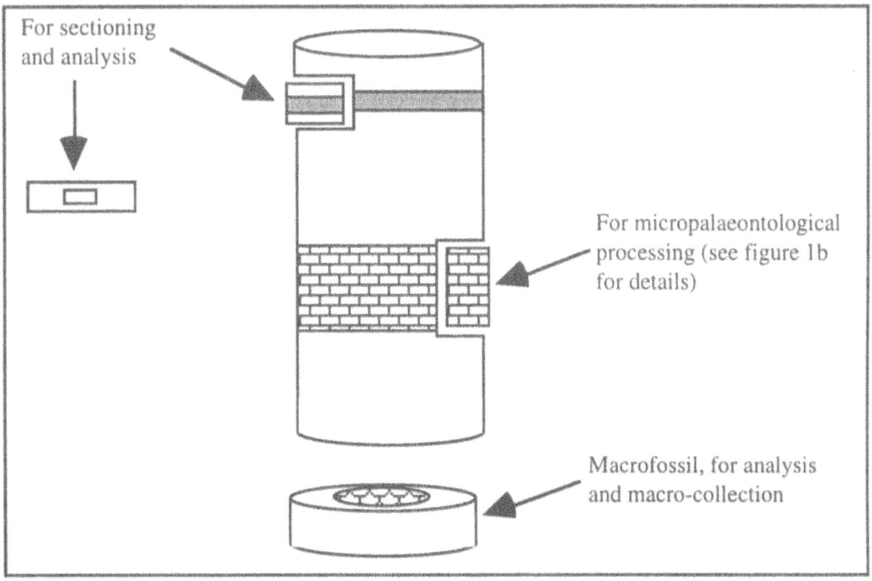

Figure 1.1a. A potted history of a fictitious sample illustrates the possible scenario regarding the stages of preparation together with many of the techniques that may be applied. At the base macrofossil material, and to the right sub-samples for micropalaeontological processing detailed in figure 1.1b. (Re-drawn from Rushton 1979).

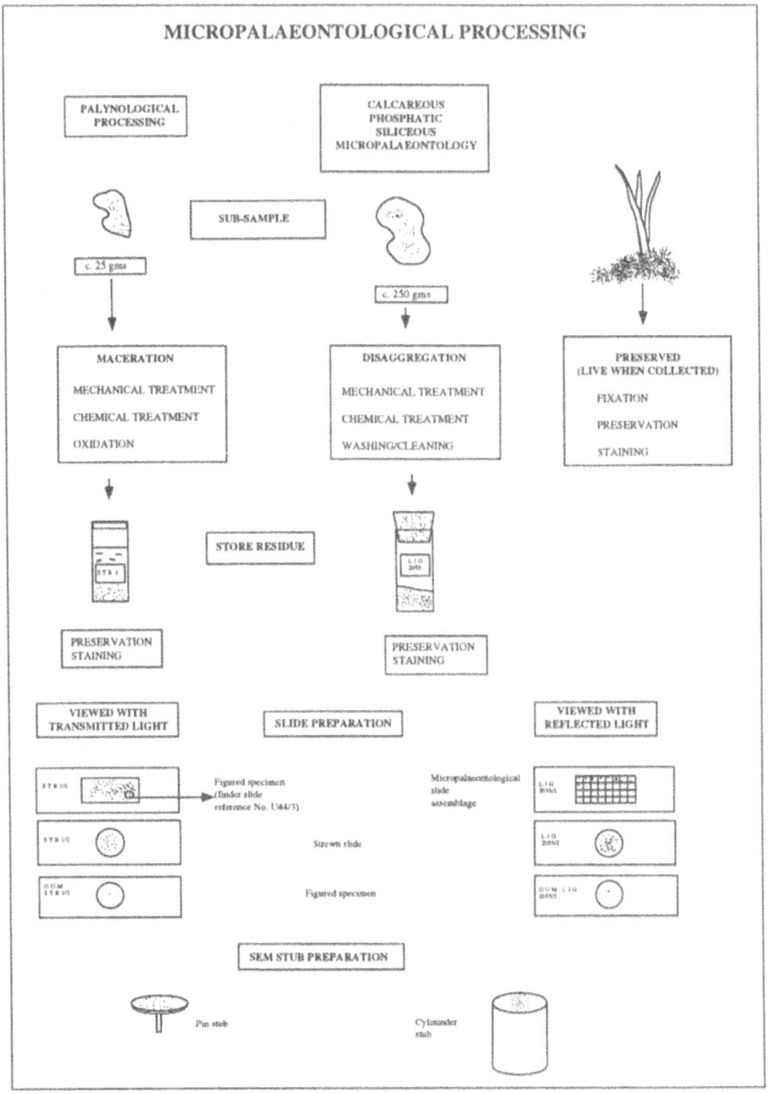

Figure 1.1b. Diagrammatic illustration of micropalaeontological processing (calcareous/siliceous/phosphatic) showing the stages from lithified sample, specimen dissociation, cleaned residue and microscopical examination (from Green 1995b).

Together these diagrams illustrate the care required to ensure that no information is lost during any process during the sample's laboratory life. There may be many years between the initial collecting of a sample and its final use. Each preparation stage may be undertaken by a different worker, it is therefore vital to ensure that all procedures are correctly documented (see also figure 1.4).

1.2.1. Sample labelling

As a precaution against any accident it is wise to place a label within the sample bag, as well as clearly labelling the outside of the sample bag. A small receipt book (with a duplicate copy) allows the top sheet to be slipped into the collecting bag, and the duplicate to be retained as a separate independent record. It is also essential to record locality data in a field notebook.

Labelling the outside of the sample bag allows for rapid identification of material without the need to open it. Tie-on labels also have this advantage, but are susceptible to becoming detached from the bag during transit. Specimens wrapped in newspaper should also be taped to avoid the danger of them becoming accidentally unwrapped. On fresh hard rock surfaces adhesive plaster can be stuck and written on, in ball-point or fibre pen (Rixon 1976).

1.2.2. Packaging

1.2.2a. Sample bags

A selection of bag sizes made of a durable material should always be taken into the field. These can be of polythene, cloth, canvas or hessian material. However, some thought must be made as to the objectives of collecting, thus ensuring that the most suitable bag is used for the type of sample required. Avoid placing unconsolidated fine grained sediments in sample bags that might leak. More serious problems may be encountered by cross contamination of samples in which the biostratigraphically important information is primarily determined by analysis of a microfossil group (see section 15 MICROFOSSIL SAMPLE CONTAMINATION AND RELIABILITY PROBLEMS).

For ease of use in the field it can often help if bags are easily recognisable when placed on the ground, and marked with the specimen or locality data.

(i) Polythene bags

Polygrip bags have the advantage of a sealable top and usually an area on the front on which the sample and locality details can be written with either a felt or biro pen. This type of bag should only be used once as angular specimens can make holes in them. Some good quality freezer bags will serve as suitable sample bags, and often have the added advantage of coloured indicator strips to show when they are closed.

(ii) Cloth, canvas and hessian bags

These often have a drawstring or tape method of closing, and a stitched label at the base. Although more expensive than polythene bags cloth bags are more durable and can be used for a number of years. The stitched label usually deteriorates first. This type of sample bag is more suited to the collection of consolidated rocks rather than sands and gravel's, or samples that are going to be used for micropalaeontological studies.

1.2.2b. Boxes and containers

For fragile and small samples it may be desirable to pack them immediately in boxes or tins directly in the field. Line the container first with paper towel, soft tissue paper or cotton wool. Cover the specimen with more packing to avoid damage on its journey back to the laboratory. Plastic food containers make ideal alternatives, and can often be purchased locally to the field area.

1.2.2c. Newspaper

This is ideal for wrapping all hard rock and fossil specimens before placing them in a collecting bag. Double wrapping of specimens is advisable, particularly if fragile fossil material occurs on an exposed surface (*e.g.* compression fossils). Collecting part and counter-part of samples, packed so that the fossil is sandwiched between the two halves prior to wrapping is preferable.

1.2.3. Transportation and shipping of large collections

Strong wooden or cardboard boxes and metal tins or drums are useful for the transportation of material back to the laboratory. Containers should not be over packed or too heavy, particularly if they are being sent through an independent carrier. If possible thoroughly dry samples before packing, this will help to reduce weight and fungal or mould development on the specimen surface that might potentially destroy fragile fossil material or disintegrate matrix supporting a fossil. Ensure that sufficient packing material is in the box (shredded newspaper or polystyrene chips) to protect specimens and reduce movement and damage during transit. Remember to keep a record of the number and date cartons were dispatched, and the contact details of the carrier for material shipped from overseas. Additional hints on packaging and transportation can be found in Compton (1962). It is also important to remember that packages may be inspected by customs, and that the correct documentation should be obtained from the authorities before attempting to ship specimens.

A summary of the main stages outlined in relation to collecting is diagramatically illustrated in the flow chart illustrated in figure 1.2. This also lists some of the specialized conservation techniques that might be employed in the field before the removal of specimens back to the laboratory. These techniques are usually applied to palaeontological and sedimentological specimens, and will outlined in detail in later sections (see part II, FIELD TECHNIQUES).

1.3. CONSERVATION

A summary of specimen conservation techniques is outlined in the flow chart shown in figure 1.3. Initial sample cleaning of both macro- and micropalaeontological specimens is similar. However, processing techniques for microfossils involve long, complex and repetitious stages to obtain a cleaned faunal rich residue. Recent and sub-fossil material may require special treatment, particularly if it has been stored in a preservative. This involves removal and cleaning before analytical examination can commence. These techniques are detailed in later sections of the manual (see part III LABORATORY

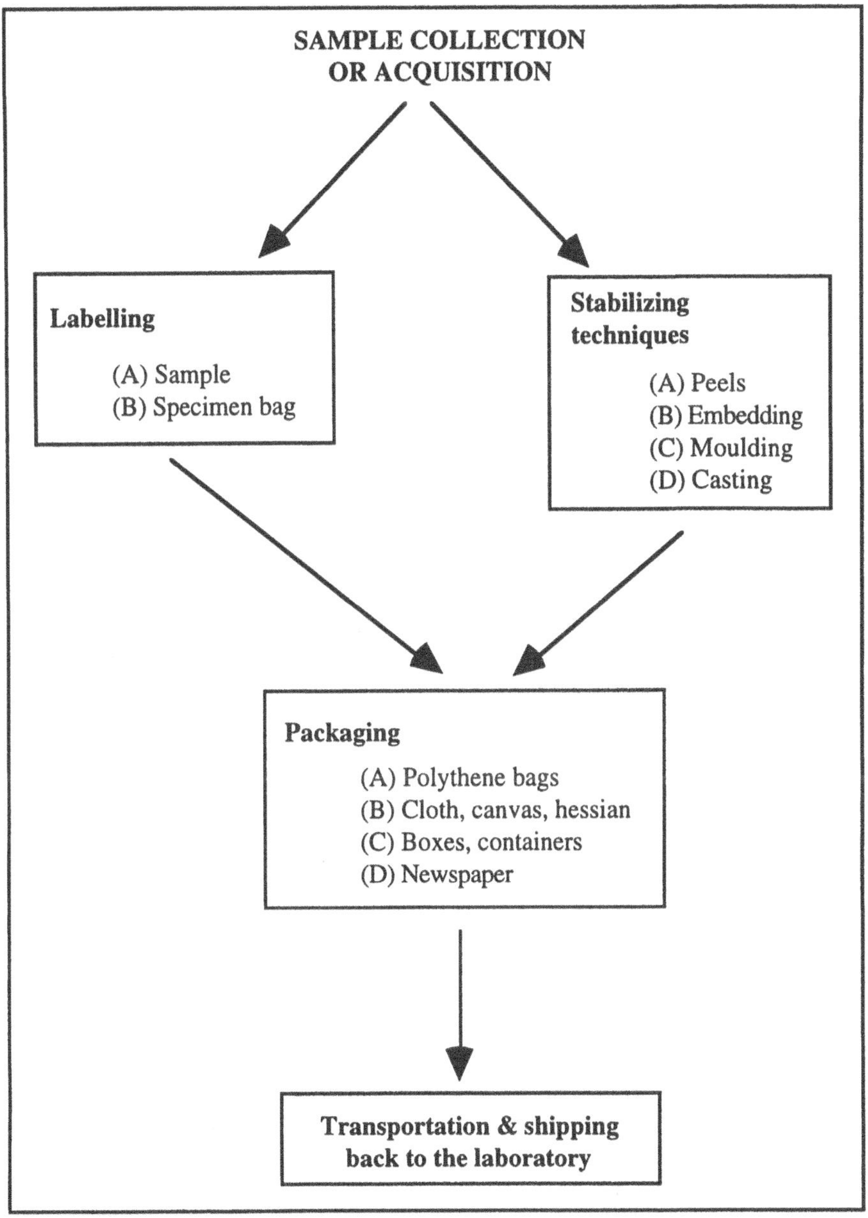

Figure 1.2. Summary flow chart showing stages a sample may pass through from acquisition to arrival in the laboratory.

TECHNIQUES). Mechanical cleaning of macrofossil material involves the use of specialist equipment and considerable patience when working on fragile material. These techniques are outlined in detail in section 12 MECHANICAL METHODS OF PREPARING FOSSIL SPECIMENS.

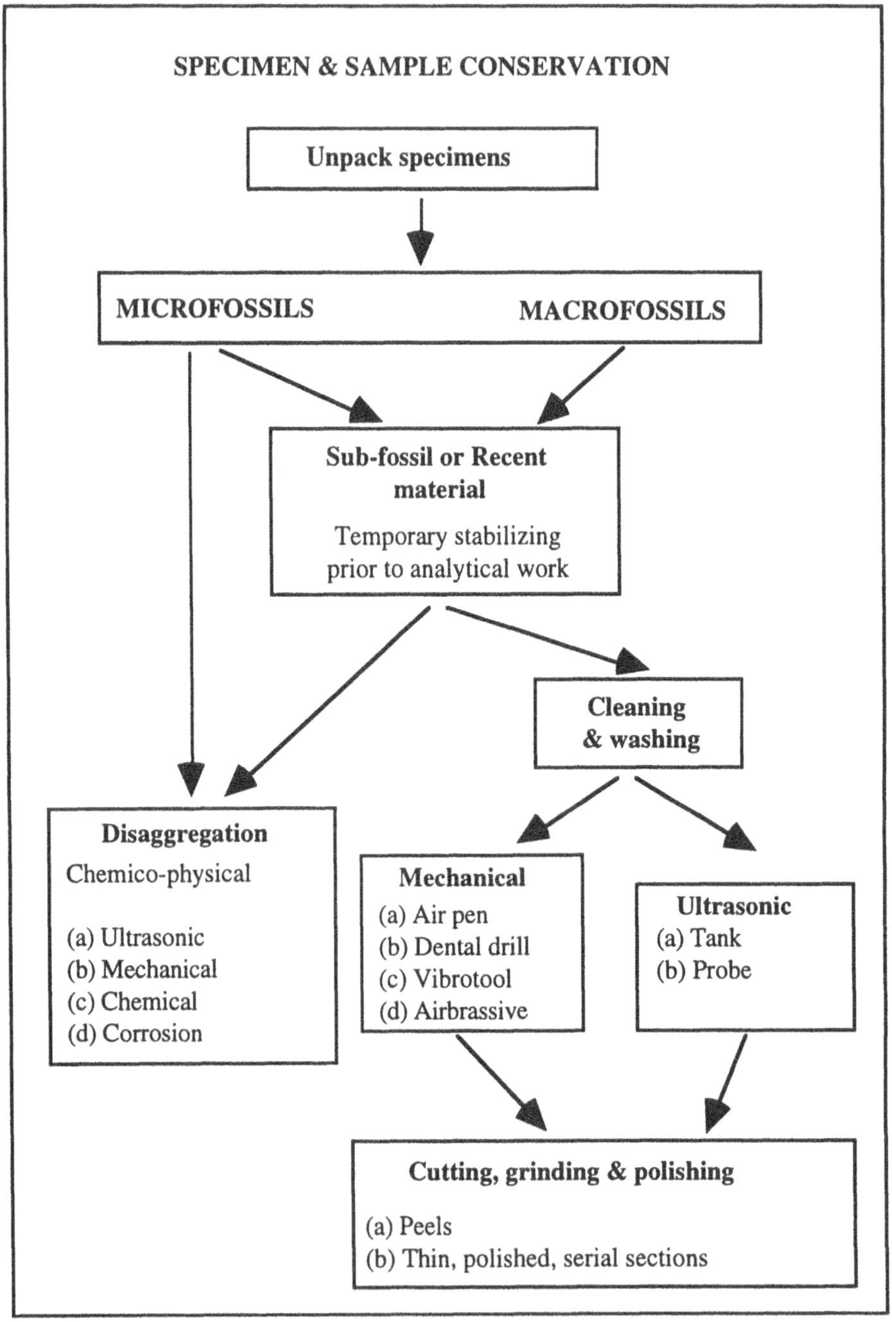

Figure 1.3. Flow chart illustrating processing procedures for palaeobiological specimens entering a research laboratory. Initial emphasis is placed on preparation and conservation of material for analytical work before preservation and curation (see figure 1.4). (Modified from Green 1995b).

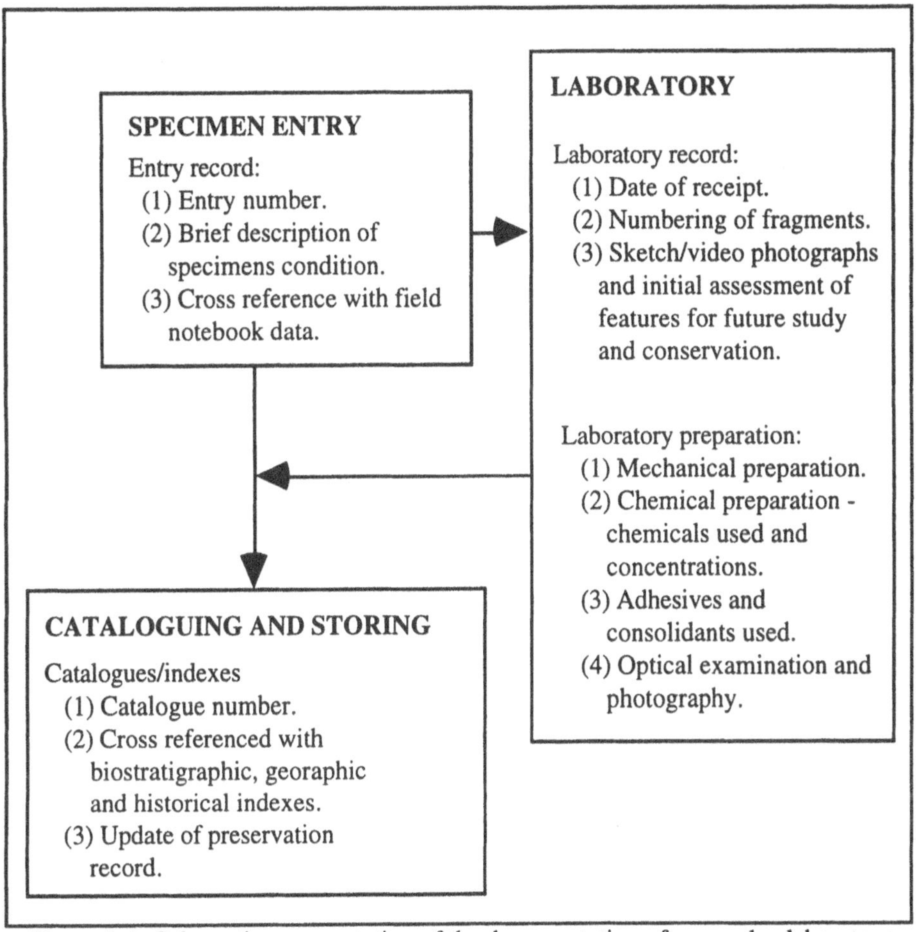

Figure 1.4. Schematic representation of the documentation of a complex laboratory operation. A laboratory record number is initiated for the sample. Individual fragments can be assigned sub-codes, which should be permanently marked on each of the specimens. All elements should be illustrated or photographed before reconstruction work begins. Copies of all original notes **SHOULD NOT** be stored with the specimens, but filed separately in case of accident (fire or water damage). (Data from Brunton *et al.*, 1985).

1.4. PRESERVATION AND CURATION

The immediate treatment carried out on specimens, either in the field or back in the laboratory, is undertaken to arrest decay, stabilize the specimen and render it suitable for transportation back to the laboratory and immediate research work. After initial preparation and conservation techniques have been performed, it may be necessary to carry out long term preservation techniques, culminating in specimen curation and eternal storing.

There are a large number of techniques available to the geo-technologist, conservator and preparator (Table 1.2) to stabilize and arrest specimen decay.

PRESERVATION & CURATION PROCEDURES	
SURFACE PROTECTION	(A) SURFACE PAINTING (i) PVA/Butvar/Paraloid (ii) Varnish (B) EMBEDDING (C) IMPREGNATING (i) Polyester resins (ii) Epoxy resins
SPECIMEN REPLICATION	(A) MOULDING (i) Vinamould (ii) Silastic (iii) Latex (iv) RTV silicone rubbers (B) CASTING (i) Plaster of Paris (ii) Polyester resin (iii) Epoxy resin (iv) Fibreglass
IDENTIFICATION **LABELLING** **STORAGE**	Documentation and photography of associated thin sections and specimens Publication, citation and figuring

Table 1.2. A summary of the practices and techniques used during the long term preservation and curation of specimens.

These are of varying complexity, depending on the value and importance of the specimen, and the ultimate use to which it is to be put. Most of the techniques require specialist equipment or chemicals, and must be undertaken within a fume cupboard or well-ventilated area. Many specimens may require only superficial surface protection, while more valuable specimens can be replicated to prevent damage to an irreplaceable original. The quality of many replications is sufficient for comparative morphological research work to be undertaken. Some analytical research procedures, however, result in the destruction of all or part of a fossil. For these high quality photographs and detailed line drawings must be obtained at predetermined stages during specimen destruction.

All curated specimens, particularly type and figured, must be stored in a controlled environment (controlled humidity, temperature and relative humidity) preferably a museum repository, ensuring their availability to all genuine research workers. Even under these conditions rock stores and collections must be subject to monthly or periodic visual inspections, and any decay monitored and corrected.

The entire range of actions encountered and followed from a specimen's arrival in a laboratory to storing are diagramatically illustrated in figure 1.4. The physical and chemical procedures employed during the stages of specimen preparation form the main text of this manual.

REFERENCES

BASSETT, M. G. (ed) 1979. Curation of palaeontological collections. *Special Papers in Palaeontology*. **22**, 1-279.

BENTON, M. J and WIMBLEDON, W. A. 1985. The conservation and use of fossil vertebrate sites: British fossil reptile sites. *Proceedings of the Geologists Association*. **96**, (1), 1-6.

BRUNTON, C. H. C, BESTERMAN, T. P and COOPER, J. A. 1985. Guidelines for the Curation of Geological Materials. *Miscellaneous Paper of the Geological Society (London)*. No. 17.

COMPTON, R. R. 1962. *Manual of field geology*. John Wiley & Sons, Inc. (New York). xi + 378p.

CROWTHER, P. R and WIMBLEDON, W. A. (eds) 1988. The use and conservation of palaeontological sites. *Special Papers in Palaeontology*. **40**, 1-200.

GOLDRING, R. 1991. *Fossils in the Field*. Longman Scientific (Harlow, U. K.). xiv + 218p.

GREEN, O. R. 1995b. Pitfalls, problems and procedures in micropalaeontological preparation and conservation. *The Geological Curator*. **6**, (4), 157-166.

NATURE CONSERVANCY COUNCIL. 1990. *Earth science conservation in Britain - A strategy*. Nature Conservancy Council, (Peterborough). 1-84 + Appendices - a handbook of earth science conservation techniques (compiled by A. P. McKirdy).

RIXON, A. E. 1976. *Fossil animal remains: their preparation and conservation*. Athlone Press, (London). viii + 304p.

RUSHTON, A. W. A. 1979. The fossil collections of the Institute of Geological Sciences, pp 57-66; In: M. G. Bassett (ed), *Curation of palaeontological collections*. Special Papers in Palaeontology. **22**, 1-279.

2. DOCUMENTATION IN THE LABORATORY

2.1. INTRODUCTION

One of the most neglected practices within the laboratory is record keeping. This discipline has to be acquired and exercised by all laboratory personnel. Within the laboratory it is essential that all stages of every technique are correctly documented, and that personnel maintain logical concise records detailing the history of specimen and sample development (Table 2.1). This provides an invaluable data base for later workers wishing to re-examine, analyse or under take additional conservation or preservation techniques, particularly in relation to irreplaceable material.

Recording procedures begin during field sampling. These include cross referencing photographs and sketches with map locations, logging notes detailing sediment type and structure with fossil orientation and mode of preservation (see Appendix 2). Additional notes on fossil specimens may be required if conservation and stabilization procedures have to be performed before transportation back to the laboratory. In some instances it may not prove possible or viable to extract the specimen, and a plastic, plaster or rubber replica can be taken. In such cases it is important to record all chemicals and procedures used. It is also advisable to indicate on the wrapped specimen any hazard or other safety implication in handling and moving the package. Always ensure specimens are well labelled with all relevant data, and that duplicate information is recorded within a separate notebook, particularly if they are to be transported back to the laboratory by a second party.

AREA REQUIRING LABORATORY DOCUMENTATION	NOTES
LABORATORY MANAGEMENT	• Document chemical stocks and supplies. • Provide clear and concise instructions for the use of equipment and laboratory facilities. • Users of chemicals must indicate they have read appropriate COSHH forms. • Label containers with chemicals in use. • Record annual tests on equipment. • Record all accidents.
PREPARATION & CONSERVATION	• Continue records initiated from field collecting. • Document mechanical equipment and chemicals used in preparing specimens.
SPECIMEN CATALOGUING, INDEXING & CURATING	• Record all repairs and adhesives used. • Cross index related thin sections, photographs (field and laboratory) and publications. • Record movement and loans of specimen.

Table 2.1. Prominent laboratory and sample areas requiring documentation.

In addition to documentary details of individual specimens and samples, record keeping within the laboratory involves a wide variety of documentation (Table 2.1). Documentary accounts of chemicals used (re-ordering and stock rotation), and the recording of equipment and machine servicing and testing are essential records in a well-run laboratory. Statutory requirements mean that much of this documentation has to be undertaken by qualified specialists (*e.g.* electrical testing, testing of pressure tanks, testing the capability of fume extraction systems). The purpose of this testing is to ensure the safety of the equipment and laboratory personnel. In practice it results in a general awareness among laboratory users that the interaction of all personnel and procedures undertaken result in a safe and secure working environment.

2.2. GOOD LABORATORY PRACTICE (GLP)

Criteria and codes of practice used to reflect the professional behaviour and approaches to work of laboratory personnel manifest themselves under the general heading of Good Laboratory Practice (GLP). In essence this is a process of "self regulation" within the work place, primarily motivated by a "self preservation" common sense mentality to one's self and other laboratory users. Good Laboratory Practice is not a legal requirement, although many aspects may result from regulations and legislation imposed upon employers and employees (see section 9 SAFETY IN THE LABORATORY).

In some laboratory's specialist areas, or separate rooms, may exist for procedures that have to be undertaken within controlled environments, or areas free from external contamination. Within these areas limited access of personnel can considerably reduce, amongst others, cross sample contamination problems. However, the majority of palaeobiology preparation laboratories are multipurpose with multi-user personnel. The smooth running of such a laboratory requires a code of equipment use and laboratory procedures. New or inexperienced users in the laboratory require guidance not only in a specific technique or procedure, but also in the way the laboratory functions, the limitations of equipment and personnel, and the location of safety equipment (*e.g.* first aid box, chemical spillage kit, fire extinguishers, escape routes and assembly areas). The initial introduction a new colleague to the laboratory should start with an informal tour of the laboratory, introducing aspects outlined above. A more detailed discussion of the work they intend to do should include the following points:

- Discuss with colleagues the procedures, equipment and chemicals intended for use. Ensure sufficient quantities and grades of all chemicals and consumable materials are present within the laboratory before commencing the procedure.
- Determine how long the procedure will take to complete, and ensure enough time has been allocated. Avoid multi-use of bench areas and equipment, particularly if more than one technique is being undertaken. This will reduce the risk of cross sample contamination, and multiple use of fume cupboards.
- Avoid hurrying colleagues into completing techniques. Chemical and physical procedures should not be rushed; 'short-cuts' invariably result in longer techniques.
- Ensure all laboratory workers mark containers and beakers with sample numbers, chemicals in use, their concentration and the appropriate hazard warning label. Chemical stock bottles must be correctly labelled. All laboratory personnel must be aware of any dangerous chemicals used in the fume cupboard.

2.3. SAFETY REQUIREMENTS

Legislation covering safety requirements within the laboratory are detailed in section 9 SAFETY IN THE LABORATORY. However, it is good practice to ensure that all relevant information (self adhesive safety hazard warning labels, chemicals in use and concentrations) is indicated on all bottles and containers. It is also vital to see that samples and specimens in preparation are clearly labelled, particularly if chemical reactions require them to stand in an acid (*e.g.* acetic acid preparations) or oxidizing solution (*e.g.* sodium hypochlorite cleaning) for many hours or even days.

Sufficient personal protective equipment (*e.g.* gloves, eye protection, face masks) should be readily available. Stocks should be periodically checked, and replaced if signs of wear are in evidence (*e.g.* check gloves, paying particular attention to the areas between fingers or marked colour deterioration's; respirator cartridges degenerate if not sealed; ear protector foam distorts and looses shape), and discarded accordingly. When performing procedures ensure that there is sufficient protective equipment for all laboratory users. It is no good providing face masks or ear protectors to the user of noisy equipment, if it is not available to other laboratory users while the equipment is in use.

2.4. PRACTICAL PROCEDURES

Recording details of techniques employed in specimen preparation is an increasingly important requirement of geo-technologists, preparators and conservators work. Documenting all practical procedures and the history of development of a specimen is essential in providing future workers with a data base for deciding what course of action to follow in later preparation or stabilization (*e.g.* Green 1995b, Hodgkinson 1991). Previous technical data may be critical in determining which mechanical preparation and analytical techniques are employed (Gaffey *et al.*, 1991, Whybrow 1982). Before beginning preparation work on the holotype specimen of *Archaeopteryx lithographica* stored at the Natural History Museum of London, Whybrow (1982) evaluated previous development of the fossil. This had been initiated over 60 years before, and provided the starting point for the preparation that followed.

Recent studies evaluating micropalaeontological processing procedures testify to the importance of recording the effects laboratory chemicals used can have (Gaffey and Bronnimann 1993, Kontrovitz *et al.*, 1991, Pingitore *et al.*, 1993). Such studies clearly reflect the importance in maintaining accurate records of preparation procedures and chemicals used within a palaeobiological laboratory.

Similar data has been amassed by numerous museum conservators and preparators from all disciplines in relation to materials used in conservation (*e.g.* consolidants and adhesives, Horie 1987), helping to build up a comprehensive data base on the long term effect resins and bonding agents have, and solvents capable of dissolving them.

2.5. SPECIMEN COLLECTIONS

Documentation of material and specimens entering a museum or laboratory form an important aspect of curators and conservators work. Ethical considerations and standardized guidelines for documenting new material entering a collection have been extensively debated, and examples from museum institutions liberally illustrated in the literature (*e.g.* Anon 1992, Brunton *et al.*, 1985, Converse 1989, Cooper 1987,

Doughty 1981, Rixon 1976). A specimens "value" is considerably enhanced if both collection and preparation data are recorded. Some points to note in relation to specimen documentation are as follows:

- Maintain entry records for all specimens deposited, specimen enquiries, loans and potential acquisitions.
- Keep a bound register or catalogue of all accessions, containing at minimum, an inventory number and sufficient information for collection management purposes. At maximum a cataloguing system should bring together all primary information, including cross references. Loans must also be recorded in the same or a parallel register. Fully automated systems must be backed up by a bound hard-copy on archival-quality paper.
- Each specimen must be marked or labelled with a unique inventory number. Care must be taken to ensure that the number does not obscure details of taxonomic or morphological importance. Old numbers (field numbers and redundant accession numbers), and labels must be preserved. An automated system should ensure that links between individual specimens (sections/peels, residue mounts, photographs/negatives, electron microscope stubs) and assemblages are maintained.
- Maintain back-up copies of all records. At least one copy should be kept in a separate room.
- Record the movement and loan of specimens, both within and outside the repository.

On establishing a suitable recording and documentation system, it is essential to ensure that the optimum storage conditions are available. Conditions for specimen storage are adequately detailed in specialist publications (*e.g.* Brunton *et al.*, 1985, Thompson *et al.*, 1992). Storage of records and documentation is just as important, even in a working laboratory. Fortunately, guidelines and storage conditions are well established for museums (Anon 1992, Converse 1989). A summary of the physical factors required for the long term preservation of documentary media and related materials are provided in table 2.2.

Cooper (1987) discusses the lack of success of implementing, and universal acceptance of any formal scheme (*e.g.* Anon 1977, Brunton 1979) in documenting conservation and preparation stages. He surmised that more simplified formats (*e.g.* Brunton *et al.*, 1985, Converse 1989, Rixon 1976) fulfilled the needs of the curator conserving and the conservator curating. Few would dispute that a documentary record of all preparation and conservation procedures is necessary, and indeed mandatory (Cooper 1987). However, further thought and discussion by concerned organisations are required to standardize a documentation format not only applicable to small and large institutions, but comprehensible too professional and semi-professional alike.

Materials	Ambient Temperature (°C)	Ambient Relative Humidity (%)	Notes
Paper Documents	13-18	55-65	
Black & white prints	15-20	30-50	Avoid condensation
Black & white negatives	<20	15-40	Avoid condensation
Magnetic tapes	18-22	35-45	
Optical & Laser discs	18-22	35-45	
Microfilm	<20	15-40	
Colour slides & negatives	≤2	25-30	High RH values accelerate deterioration
Colour prints	≤2	30-50	High RH values accelerate deterioration
Microslides	18	60	Fungus can form in water based mounting media at high RH's

Table 2.2 Suggested temperature and relative humidity range for the storage of documents, photographic materials, microscope slides and magnetic and disc storage. (Data from Anon 1992).

REFERENCES

ANONYMOUS. 1977. Proposals for the documentation of conservation in museums. *Museum Documentation Association. Occasional Paper.* **1**, 1-40.
ANONYMOUS. 1992. 2. *Standards in the Museum Care of Biological Collections 1992.* Museum and Galleries Commission (London). 55p.
BRUNTON, C. H. C. 1979. The development of a computer-based curitorial system for palaeontology at the British Museum (Natural History), pp 159-173; In: M. G. Bassett (ed) *Curation of palaeontological collections. Special Papers in Palaeontology.* **22**, 1-279.
BRUNTON, C. H. C, BESTERMAN, T. P and COOPER, J. A. 1985. Guidelines for the Curation of Geological Materials. *Miscellaneous Paper of the Geological Society, (London).* No. 17.
CONVERSE, H. H. 1989. *Handbook of paleo-preparation techniques* (2nd edition, revised by R. W. McCarty). Florida Paleontological Society. v + 109.
COOPER, J. A. 1987. Conservation documentation, pp 407-412; In: P. R. Crowther and C. J. Collins (eds), The Conservation of Geological Material. *Geological Curator.* **4**, (7), 375-474.

DOUGHTY, P. S. 1981. The state and status of Geology in U. K. Museums. *Miscellaneous Paper of the Geological Society, (London).* No. 13, 118p.
GAFFEY, S. J and BRONNIMANN, C. E. 1993. Effects of bleaching on organic and mineral phases in biogenic carbonates. *Journal of Sedimentary Research.* **63**, (4), 752-754.
GAFFEY, S. J, KOLAK, J. J and BRONNIMANN, C. E. 1991. Effects of drying, heating, annealing and roasting on carbonate skeletal material, with geochemical and diagenetic implications. *Geochimica et Cosmochimica Acta.* **55**, 1627-1640.
GREEN, O. R. 1995b. Pitfalls, problems and procedures in micropalaeontological preparation and conservation. *The Geological Curator.* **6**, (4), 157-166.
HODGKINSON, R. L. 1991. Microfossil processing: a damage report. *Micropaleontology.* **37**, (3), 320-326.
HORIE, C. V. 1987. *Materials for conservation. Organic consolidants, adhesives and coatings.* Butterworth-Heinemann (Oxford). xi + 281.
KONTROVITZ, M, SLACK, J. M and YUHONG, Z. 1991. On the use of some phosphates in the preparation of ostracod shells. *Journal of Micropalaeontology.* **10**, (2), 121-126.
PINGITORE, Jr, N. E, FRETZDORFF, S. B, SEITZ, B. P, ESTRADA, L. Y, BORREGO, P. M, CRAWFORD, G. M and LOVE, K. M. 1993. Dissolution kinetics of $CaCO_3$ in common laboratory solvents. *Journal of Sedimentary Research.* **63**, (4), 641-645.
RIXON, A. E. 1976. *Fossil animal remains: their preparation and conservation.* Athlone Press, (London). viii + 304p.
THOMPSON, J. M. A. (eds). 1992. *Manual of Curatorship* (2nd Edition). Butterworth-Heinemann/Museums Association. 756p.
WHYBROW, P. J. 1982. Preparation of the cranium of the holotype of *Archaeopteryx lithographica* from the collections of the British Museum (Natural History). *Neues Jahrbuch für Mineralogie, Geologie und Paläontologie, Monatshefte.* H3, 184-192.

PART II

FIELD TECHNIQUES

3. Field collecting procedures

4. Collecting techniques for microfossil and live foraminifera

5. Consolidation, stabilization and replication techniques

6. Field staining techniques for determining calcite, dolomite and phosphate

7. Field documentation, sample packaging and transportation

3. FIELD COLLECTING PROCEDURES

3.1. INTRODUCTION

The current awareness through media coverage of environmental and "green" issues have increased public knowledge and perception of conservation related topics. Unfortunately, the term conservation has, in its widest sense, become regarded as a synonym with a "look but do not touch" or "hands-off" attitude (Young 1993). However, in an eloquent discussion Besterman (1988) examines definitions of conservation with particular reference to palaeontological sites and specimens. He recognizes the apparent contradictions arising from (i) preservation with no-change, to (ii) a flourishing continued existence, and hints that this elevates collecting and the removal of specimens to a privilege not a right. Furthermore, at many important sites this has been supplemented by legislation, national and international moratoriums and celebrity endorsement of conservation issues. Both collectors, amateur and professional, and the sites of collection are under increased scrutiny. Although nationally important sites and areas can be controlled and monitored by governmental organisations and local authorities, with legal powers to impose restrictions on access and collecting, the legal status at other sites is less well defined. Taylor and Harte (1988) outline some of the legal problems pertinent to palaeontological site conservation within the UK, examining governmental control (English Nature, formerly the Nature Conservancy Council and Local Councils), site occupier's powers and liability, and the ownership of fossils both *in-situ* and in scree deposits. They conclude (p. 35) that the legal questions must be discussed by palaeontological community, and that resources spent on public education provide a better opportunity of conserving sites and individual fossils than compulsory legislation.

Most localities and collecting areas are situated on private or public lands, and restrictions against access usually exist. Public areas situated within national or state controlled parks and reserves require permits to be obtained before collecting can be undertaken. Access to private sites can only be undertaken with the permission of the land owner.

Privately owned mines and quarries are usually fenced and posted with "Private - no trespassing" signs. These are designed to both warn and protect (Converse 1989). Disused quarries and mines are extremely dangerous places, frequently flooded and subject to cave-ins. Working quarries are subject to legislation, requiring people who enter the premises to wear appropriate safety hats, and where heavy machinery operates, fluorescent or brightly coloured "day-glow" over jackets. Permission to enter a quarry (operational or closed) must be obtained from the owner, and on the day of collecting, liaison with the quarry manager or representative.

3.2. THE ROLE OF THE COLLECTOR

Amongst the natural sciences geology, is unique in both encouraging and regarding the collecting specimens as acceptable (Brunton *et al.*, 1985). However, over zealous collecting and site abuse has resulted in restricted access and collecting from many privately owned sites. This coupled with over collecting from type and "classic" exposures put additional pressure on the collector to show restraint during sampling, and greater care in preparing a strategy for collecting. Collectors must accept an increasing burden of responsibility as conservationists (Young 1993).

3.3. FOLLOWING A COLLECTING STRATEGY

Although comprehensive descriptions of samples can be recorded in field note books and on logs, rock and fossil samples must be collected for a number of reasons. Compton (1962) and Brunton *et al.*, (1985) collectively provide a rationale for bringing samples back to the laboratory. These can be summarised as follows:

- To maintain or further the best interests of geological science, even if the specimens may not in themselves show great scientific interest.
- A more reliable and exact identification plus detailed lithological or fossil descriptions can be made in the laboratory, particularly following the preparation and examination of samples under a microscope. Accurate determination of porosity, permeability and mineral ratios in lithologic samples, and the recognition of biostratigraphic and palaeoenvironmental factors from fossil material.

Before field sampling, formulate objectives, and collect with a clear purpose and in a logical manner. Design a sampling strategy, particularly if working in a team, containing standardized names and descriptions. A working objective should be one of collecting samples providing a complete stratigraphic coverage that encompasses different lithologies to allow for intra and inter facies variations to be determined. Ideally samples must be representative of the unit studied, preferably obtained after several outcrops have been examined.

If fossils form part of the collecting objective, it is useful to acquaint ones self with their appearance and potential modes of preservation. This can be achieved by studying collections and photographs, consulting with experts and undertaking a literature search. Take copies of the most useful publications into the field, and prepare a field guide with illustrations and descriptions of the fossils and associated lithologies.

3.4. GEOLOGICAL FIELDWORK CODE

Within the United Kingdom a self regulatory code for geological fieldwork has been widely circulated to both professional and amateur geologists for over 20 years (Geologists' Association 1975). Outlined within this fact sheet is a code of conduct that should be followed by individuals on visiting geological sites and collecting specimens. Important health and safety information are also provided, together with brief guidelines to the role of field parties from societies and educational institutions, and research workers in collecting. A second edition of this information sheet (Geologists' Association 1995) includes information on a code for coring, originally outlined in a separate fact sheet (Geologists' Association 1989). As with the first edition it includes an address for landowners requiring information about parties visiting sites, or who wish to register a complaint against a party or individual misusing a site.

The Institution of Geologists, now amalgamated with the Geological Society of London, published a detailed code of practice for geological group visits to quarries, mines and cave (Institute of Geologists 1985). Detailed advice for under graduate students, and research students just beginning a project have been published by the Natural Environment Research Council (1990) a government grant funding institution in the UK, and the UK Committee of Heads of University Geoscience Departments (1995). More than ever the conduct of individuals, parties and professional geologists requires the co-operation of local authorities and landowners to maintain free access to geological sites. Only by strictly adhering to a code of practice can access for future generations of geologists be guaranteed. The following points should be considered to ensure this remains the case:

1 A visit to a geological site, especially a working quarry or mine, should be regarded as a privilege granted by the owner.

2 The introduction of the Health and Safety at Work Act and the Mines and Quarries Act ensure that site owners and managers are responsible for the safety of employees and visitors. Protective clothing (*e.g.* safety helmets) must be worn by all employees and visitors at all times. Other precautions may also be required of visitors on entering a site (see 3.4.2 below).

3.4.1. Preparing for a visit

1 Seek written permission from the landowner or manager to visit the site at least 3-6 weeks in advance of the proposed date. Provide information on:
- the interests of the site,
- intentions (*e.g.* collection and removal of specimens from the site),
- proposed date and time of arrival,
- expected length of visit,
- number in the party (age ranges may be required for entering working quarries).

2 When written permission has been obtained, inform other party members of visiting conditions and restrictions agreed with the owner. A prior inspection of the site by the leader, particularly if leading an educational visit, can assist in familiarizing with the potential hazards and the geology. If permission is withheld, under no circumstances trespass.

3.4.2. Insurance

Most companies possess Public Liability Insurance cover. However, insurance companies may be unwilling to provide indemnity covering visiting parties without payment of an additional premium. Furthermore, even if a visiting party provide their own indemnity cover it will not provide protection for the operator. The responsibility of the operator and site manager cannot be delegated. Some companies have arrangements with their own insurers, requiring the leader and or members of a visiting party to sign an indemnity before a visit. Party leaders are advised to ensure that proper and adequate insurance cover arrangements exist for group members.

3.4.3. Equipment

1 Wear suitable field equipment for the locality. Safety helmets are mandatory in working mines and quarries. Eye protection must be worn when hammering. A first aid kit should be available amongst the group.

2 Equipment restrictions may apply when entering working mines. Matches, lighters, torches and camera flash equipment are not permitted. Electrical equipment (*e.g.* geophysical, CB, radio, audio and video) may be permissible with the owners consent, as their use may interfere with sensitive process control equipment or electronic detonation systems.

3.4.4. Arriving at the locality

1 Obey the country code, and observe local bylaws. Shut gates and leave no litter.

2 Do not interfere with machinery.

3 Report to the site owner, manager or his representative.

4 Avoid leaving rock fragments when sampling. These may cause injury to livestock, or be hazardous to vehicles or pedestrians.

5 Avoid any unnecessary disturbance of wildlife. Flora and fauna may inadvertently be displaced or destroyed by careless actions, particularly during winter seasons when dormant or in hibernation.

6 Keep away from unstable cliff faces, particularly with overhangs liable to collapse without warning, scree slopes and gravel pit mounds. Avoid flooded areas where there is uncertainty as to the depth of water, and areas of semi-liquid material covered by a thin, potentially unstable, crust (sludge lagoons).

7 When visiting coastal sections obtain tidal information and data on local hazards such as the access and stability of cliffs. If in doubt contact the local coast guard service.

8 When working in remote areas, either at altitude in mountainous areas or underground, consult specialized guides (*e.g.* Barry and Jepson 1988) or recognised authorities. Seek advice and work in the company of experienced personnel. Inform someone of the intended route and location of the work. Never go alone. When necessary inform of your return from the field. Never take risks on insecure cliffs or rock faces. Consider those who may visit the exposure at a later date, and never leave sites in a dangerous condition.

3.4.5. Collecting and educational field parties

1 Avoid indiscriminately hammering at exposures. Encourage students to observe, record and photograph *in-situ* material and examine scree deposits and collecting float specimens.

2 Keep collecting to a minimum. Avoid the removal of *in-situ* fossils, minerals and rocks unless required for genuine research.

3 For teaching and demonstration purposes use float specimens. Collect a good representative fossil sample that can have replicas made from it.

4 Never sample directly from walls and buildings. Do not remove specimens that might undermine fences, walls' bridges, or other structures.

3.4.6. Collecting for research

1 No researcher has the right to "dig out" a site.

2 Avoid disfiguring rock faces with symbols and numbers, and never use permanent markers or paints on rock surfaces.

3 In fill pits and excavations for both safety reasons and to protect the outcrop.

3.4.7. Field coring

The results of small diameter cores taken from outcrops, using hand-held power tools are becoming increasingly evident at some localities. This technique provides a quick alternative to hammering, and produces a clean sample with only a small weathered surface. However, it can cause unsightly scarring of an exposure, particularly when taken from faces in open view. This form of collecting is without doubt the most aggravating to landowners and members of the public. Permission must be obtained before core samples can be collected.

1 Take cores from the least exposed faces. Never sample those visible from the site entrance, or exhibiting a structure.

2 Take only the minimum number of samples and avoid closely spaced patterns.

3 Try to plug holes with debris similar in colour to the core material.

3.4.8. After departing from site

Following fieldwork and collecting from a site, send a letter of appreciation to the landowner or site manger. Comment on any preliminary geological results or observations of particular interest.

The collector is responsible for the condition of the samples until they reach the laboratory. Specimens and records should never be entrusted to a person who does not appreciate their meaning or value (Camp and Hanna 1937). Research material and notebooks should eventually become available to others by depositing them with an appropriate institution. A representative collection of student research material, including specimens illustrated within a thesis or any publication, should be lodged either at the institute of study or a recognised repository (Tunnicliff 1983).

3.5. SITE CONSERVATION

Many aspects of geological site conservation have been documented, with an emphasis on palaeobiological and mineralogical sites of interest, both of which have received high profile consideration within the United Kingdom and world-wide (*e.g.* Agnew *et al.*, 1989, Clemens 1988, Ferguson 1988, O'Halloran *et al.*, 1993, Wild 1988, Wimbledon 1988).

Many palaeobiological sites within the UK have been extensively examined and fossil fines documented. The need for conservation of these sites and their potential contribution to further scientific research has been evaluated (*e.g.* Benton 1988, Benton and Wimbledon 1985, Cleal 1988, Cope 1988, Kermack 1988). Wimbledon (1988, p. 41) states that "the only valid justification for conservation of palaeontological sites is their scientific value and the potential they have to be *used* for future research." However, this fails to recognise the multiple use to which sites can attain. Successful site conservation lies in beneficial exploitation by a wide range of users (Besterman 1988). Abandoned mines and spoil heaps may also provide unique microbiological and botanical habitats (*e.g.* Jenkins and Johnson 1993, Purvis 1993).

Approximately 2,200 Sites of Special Scientific Interest (SSSI) are conserved, by statutory obligation, by the site owners and occupiers, planning authorities and English Nature. Sites designated as SSSIs are maintained and managed, countering any threats, because of research work and visitations by groups and individuals providing improved and updated documentation. Non-statutory sites may be included in the RIGS (Regionally Important Geological/Geomorphological Sites) network. Such sites may be receiving active conservation from local voluntary groups. Site documentation for both SSSIs and RIGS sites is provided by The National Scheme for Geological Site Documentation (NSGSD). This scheme was initiated by the Geological Curators Group in 1977, and the 100,000+ records are now co-ordinated and maintained by the British Geological Survey. A proposal by English Nature (Nature Conservancy Council 1990) outlines a conservation strategy of six themes in improving site knowledge, documentation and public awareness. Geological site conservation continues to face problems, although the approach outlined by English Nature focuses attention and outlines a procedure to accommodate professional and recreational geoscience needs. Many of the problems are shared, or derived from, other parts of the conservation movement (Black 1988), and can only be addressed when considered jointly.

REFERENCES

AGNEW, N, GRIFFIN, H, WADE, M, TEBBLE, T and OXNAM, W. 1989. Strategies and techniques for the preservation of fossil tracksites: an Australian example, pp 397-407; In: D. G. Gillette and M. G. Lockley (eds), *Dinosaur tracks and traces*. Cambridge University Press, (Cambridge). xvii + 454p.

BARRY, J and JEPSON, T. 1988. *Safety on Mountains*. British Mountaineering Council, (Manchester). 37p.

BENTON, M. J. 1988. British fossil reptile sites, pp 73-84; In: P. R. Crowther and W. A. Wimbledon (eds), *The use and conservation of palaeontological sites*. Special Papers in Palaeontology. **40**, 1-200.

BENTON, M. J and WIMBLEDON, W. A. 1985. The conservation and use of fossil vertebrate sites: British fossil reptile sites. *Proceedings of the Geologists' Association*. **96**, (1), 1-6.

BESTERMAN, T. P. 1988. The meaning and purpose of palaeontological site conservation, pp 9-19; In: P. R. Crowther and W. A. Wimbledon (eds), *The use and conservation of palaeontological sites*. Special Papers in Palaeontology. **40**, 1-200.

BLACK, G. P. 1988. Geological conservation: a review of past problems and future promise, 105-111; In: P. R. Crowther and W. A. Wimbledon (eds) *The use and conservation of palaeontological sites*. Special Papers in Palaeontology. **40**, 1-200.

BRUNTON, C. H. C, BESTERMAN, T. P and COOPER, J. A. 1985. Guidelines for the Curation of Geological Materials. *Miscellaneous Paper of the Geological Society (London)*. No. 17.

CAMP, C. L and HANNA, G. D. 1937. *Methods in Paleontology*. University of California Press, (Berkeley, California). xxiii + 153p.

CLEAL, C. J. 1988. British palaeobotanical site, pp 57-71; In: P. R. Crowther and W. A. Wimbledon (eds), *The use and conservation of palaeontological sites*. Special Papers in Palaeontology. **40**, 1-200.

CLEMENS, W. A. 1988. Challenges of management of palaeontological site resources in the United States, pp 173-180; In: P. R. Crowther and W. A. Wimbledon (eds), *The use and conservation of palaeontological sites*. Special Papers in Palaeontology. **40**, 1-200.

COMMITTEE OF HEADS OF UNIVERSITY GEOSCIENCE DEPARTMENTS and THE GEOLOGICAL SOCIETY, LONDON. 1995. *Safety in geoscience fieldwork: precautions, procedures and documentation*. (London). 18p.

COMPTON, R. R. 1962. *Manual of field geology*. John Wiley and Sons, Inc. (New York). xi + 378p.

CONVERSE, H. H. 1989. *Handbook of paleo-preparation techniques* (2nd edition, revised by R. W. McCarty). Florida Paleontological Society. v + 109p.

COPE, J. C. W. 1988. The conservation of biostratigraphical sites, pp 95-104; In: P. R. Crowther and W. A. Wimbledon (eds), *The use and conservation of palaeontological sites*. Special Papers in Palaeontology. **40**, 1-200.

CROWTHER, P. R and WIMBLEDON, W. A. (eds). 1988. The use and conservation of palaeontological sites. *Special Papers in Palaeontology*. **40**, 1-200.

FERGUSON, L. 1988. The 'fossil cliffs' at Joggins, Nova Scotia: a Canadian case-study, pp 191-200; In: P. R. Crowther and W. A. Wimbledon (eds), *The use and conservation of palaeontological sites*. Special Papers in Palaeontology. **40**, 1-200.

GEOLOGISTS' ASSOCIATION. 1975. *A code for geological fieldwork*. (London).

GEOLOGISTS' ASSOCIATION. 1989. *Take care when coring*. (London).

GEOLOGISTS' ASSOCIATION. 1995. *A code for geological fieldwork*. (London).

INSTITUTE OF GEOLOGISTS. 1985. *Code of practice for geological visits to quarries, mines and caves*. (London).

JENKINS, D. A and JOHNSON, D. B. 1993. Abandoned metal mines: a unique mineralogical and microbiological resource. *Journal of the Russell Society*. **5**, (1), 40-44.

KERMACK, K. A. 1988. British Mesozoic mammal sites, pp 85-93; In: P. R. Crowther and W. A. Wimbledon (eds), *The use and conservation of palaeontological sites*. Special Papers in Palaeontology. **40**, 1-200.

NATURAL ENVIRONMENT RESEARCH COUNCIL. 1990. *Guidance note safety in fieldwork*. (Swindon). 32p.

NATURE CONSERVANCY COUNCIL. 1990. *Earth science conservation in Britain - A strategy*. Nature Conservancy Council, (Peterborough). 1-84 + Appendices - a handbook of earth science conservation techniques (compiled by A. P. McKirdy).

O'HALLORAN, D. A, MOAT, T, MATHEWS, R. E and McKIRDY, A. P. 1993. Mineralogical conservation in Great Britain - a governmental conservation agency approach. *Journal of the Russell Society*. **5**, (1), 61-63.

PURVIS, O. W. 1993. The botanical interest of mine spoil heaps - the lichen story. *Journal of the Russell Society*. **5**, (1), 45-48.

TAYLOR, M A and HARTE, J. D. C. 1988. Palaeontological site conservation and the law in Britain, pp 21-39; In: P. R. Crowther and W. A. Wimbledon (eds), *The use and conservation of palaeontological sites*. Special Papers in Palaeontology. **40**, 1-200.

TUNNICLIFF, S. P. 1983. *I am beginning my research: What shall I do with my specimens? A note of advice*. Natural Environmental Research Council (Swindon). 16p.

WILD, R. 1988. The protection of fossils and palaeontological sites in the Federal Republic of Germany, pp 181-189; In: P. R. Crowther and W. A. Wimbledon (eds), *The use and conservation of palaeontological sites*. Special Papers in Palaeontology. **40**, 1-200.

WIMBLEDON, W. A. 1988. Palaeontological site conservation in Britain: facts form, function, and efficacy, pp 41-55; In: P. R. Crowther and W. A. Wimbledon (eds), *The use and conservation of palaeontological sites*. Special Papers in Palaeontology. **40**, 1-200.

YOUNG, B. 1993. The mineral collector as a conservationist - the right to collect? *Journal of the Russell Society*. **5**, (1), 35-36.

4. COLLECTING TECHNIQUES FOR MICROFOSSIL AND LIVE FORAMINIFERA

4.1. INTRODUCTION

Attempting biostratigraphical, palaeoecological, ecological or taxonomic work using microfauna as the primary data source requires methodical and systematic field collecting. In this section, sampling techniques used for the collection of fossil and live material is outlined. Fossil sample collecting, from outcrop (macro and micro) and borehole, will be described first followed by marine sediment sampling techniques, and the collection of live material. Methods associated with marine protozoan culture is not outlined as these are adequately detailed in Hemleben *et al.*, (1989).

4.2. FIELD SAMPLING

Microfossils are commonly found in rocks containing abundant macrofossils. However, some lithological types will yield more and better preserved specimens than others. Table 4.1 lists potentially microfossil rich sedimentary rocks that should be examined and sampled in the field.

CLAYSTONES	In macro-fossiliferous limestones, the beds may be only a few mm thick.
LIMESTONES & DOLOMITES	Calcareous, agglutinated and siliceous forms common.
CALCAREOUS SHALES	Can be rich in microfossils when interbedded with limestones.
SANDSTONES & SILTSTONES	In fine grained sediments palynomorphs can be common, while coarser clastics are poor or barren.
GLAUCONITE RICH SANDSTONES	Foraminifera and ostracods abundant in fine to medium grained deposits.
BLACK SHALES	Fissile deposits may be associated with cyclotherms, can be rich in microfossils and palynomorphs.
MARINE CHERTS	May contain agglutinated foraminifera, diatoms and radiolaria.

Table 4.1. Main sedimentary rock types and associated microfossil groups (after Jones 1965).

Successful micropalaeontological sampling requires careful thought and the design of a sampling strategy to include the full stratigraphic coverage of the field area, incorporating all lithologies and facies variations. The collection of samples (preferably unweathered) should be accompanied by notes, logs and photographs of lithological, sedimentological and palaeoecological variations.

4.3. FIELD EQUIPMENT

Before undertaking field work careful consideration must be given to the sampling of different lithologies. This will assist in determining what specialized equipment is required to ensure that the most useful samples are obtained. An extensive array of hand held tools are available to the collector. These range from standard hammers and chisels (see section 12 MECHANICAL METHODS OF PREPARING FOSSIL SPECIMENS) to pneumatic drills (Jackhammers) and grinders, more suitable for field use by having petrol driven engines. Operating in either a rotary or percussion manner, these are ideal for the removal of large quantities of matrix and overburden, exposing fossiliferous beds. Experienced operators only should use the equipment, and wear suitable ear and eye protection.

4.3.1. Pneumatic drills

Use of these can generate large amounts of dust and particulate matter, and are extremely noisy when in use. A variety of bits for digging, breaking or drilling can be used in association with wedges and pins. Chaney (1989) outlines a method to expose a fossiliferous bed in a quarry or cutting:

1 Drill a series of holes in the overburden (using a rotary bit).

2 Insert pins and hammer in wedges.

3 Repeat the process. The positioning and insertion of successive wedges generates a stress line, and will eventually fracture the rock.

4.3.2. Rock saws

Rock saws with carborundum blades are operated dry (unlike diamond blades). Consequently they generate large amounts of dust and are extremely noisy. Saws have a considerable advantage over percussion equipment in that they eliminate vibration caused by the prolonged use of impact tools (Chaney 1989). However, two disadvantages reinforce the need for experienced operators only to use this type of machinery. Firstly, macrofossils not evident on the surface are more liable to damage as the saw blade penetrates any line of weakness. Secondly, carborundum saw blades are susceptible to violent explosive disintegration if inadvertently twisted during operation. The operator must wear full face mask or high impact eye protection when using a rock saw in the field.

4.4. SPECIMEN SAMPLING

There are no fixed rules or universally accepted methods governing the collecting of rock samples. As a standard procedure, horizons containing macrofossils should also be sampled for microfossils. Specific sampling techniques are dependent on the type of work undertaken, and whether an initial reconnaissance survey is undertaken, or a more detailed stratigraphic or taxonomic follow up. Once both scope and purpose of the project are established, an appropriate sample size and interval can be determined. Over large stratigraphic intervals, widely spaced samples (3m, 10m or 15m apart) of approximately 1kg may suffice, while detailed stratigraphic work may require larger bulk samples, or samples spaced only a few centimetres apart. Todd *et al.*, (1965) suggest a well-reasoned approach in determining sample size and interval. Collect samples from each lithology at a number of locations, and then examine each to ascertain faunal uniformity and abundance, determining which lithologies to re-sample, at what interval, and the size required. However, this approach may require some modification if collecting is restricted to a single field season.

Three types of microfossil sampling are used: (1) spot sampling, (2) continuous or channel sampling and (3) core sampling. Spot sampling through apparently uniform beds of shale or sandstone, even at regular intervals, may fail to locate horizons only a few millimetres thick containing microfossils.

None of the methods outlined below, however, reflect a true picture of the microfaunal assemblage present at a specific time. This is because the absolute time of the stratigraphic interval sampled is not known. Spot samples will, however, represent a considerably shorter time than continuous samples. However, small environmental changes and many generations of microfossils are reflected in only a few millimetres of sediment (Todd *et al.*, 1965).

4.4.1. Spot sampling

Sampling at regular intervals, or in a regular pattern, is most effective where outcrops of uniform lithology, such as thick shales, siltstones or limestones occur (Fig. 4.1). It can also be employed on thin shales or clay breaks inter-bedded in thicker limestones, sandstones, or shale outcrops containing limestone lenses. Contamination is minimised because of the small area from which the sample is taken, while faunas present can be accurately placed at specific stratigraphic horizons. However, as the intervening lithology remains unsampled, some of the faunal elements may be missed using this method. Varker (1967), collecting for conodonts from limestones and shales of the Yoredale Series of the Askrigg and Alston Blocks of northern England, used sampling intervals ranging from 30cm to a maximum of 1.5m. Spot sampling may be used during a reconnaissance survey, or for comparative morphological studies, and for supplementing or establishing reference collections.

Targeting the richest horizons for microfossils may require the establishment of a "field laboratory" to undertake initial processing to determine the best preserved material. Acid etching, washing and a method of observing results can be used to guide the collector to the best lithologies. This procedure may be of particular use to radiolarian workers where preservation in Mesozoic sequences varies widely within a single lithological section (Baumgartner *et al.*, 1981), although it should not be considered exclusive to this microfossil group (for details see below).

4.4.2. Continuous (channel) sampling

Continuous or near continuous sampling is the preferred method when taxonomic or biostratigraphic work is performed (Collinson 1965). In some cases the outcrop is first trenched, to reveal fresh surfaces, and then samples are collected from all

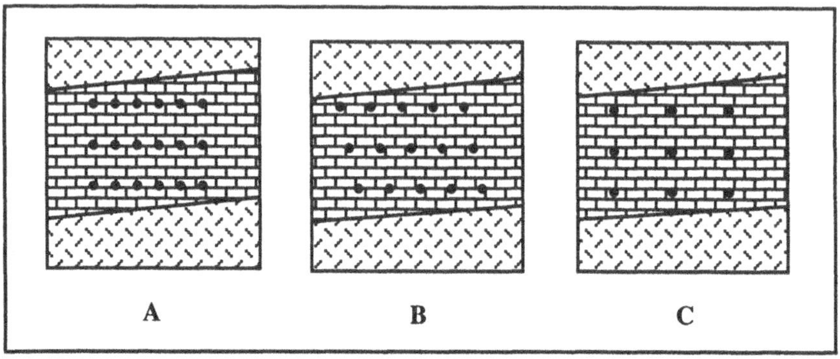

Figure 4.1. Diagram illustrating the disposition of spot sampling sites. (a) by a square spacing pattern, (b) by a rhombic spacing pattern, and (c) by a rectangular spacing pattern. (After Maximov *et al.*, 1973).

levels within pre-determined stratigraphic intervals (Fig. 4.2). In uniform lithologies, 30cm long samples every metre can be taken (essentially these represent large spot samples). This type of sampling is preferable in smaller outcrops, or where a rapid alternation or intercalation of lithologies occurs, as it ensures the inclusion of all fossils present. However, the danger of stratigraphic contamination is increased. In outlining a sampling procedure for conodont collecting, Collinson (1965, pp 98-100) follows a three stage approach:

1 **Reconnaissance collecting** - continuous channel sampling of large exposures and all stratigraphic units represented. An adequate representation of the facies, members or formations is more desirable than the uniform geographic spacing of samples.

2 **Selective sampling** - concentrating on zones abundant with organisms to gain a comprehensive knowledge of the fauna. Collecting organised on a grid system (Fig. 4.1a), or spot sample spacing very close.

3 **Bulk collecting** - large samples (up to 230kg) from beds where faunas are sparse but of special significance (*e.g.* where a stratigraphic boundary problem must be resolved).

The durability of conodonts may result in problems of contamination (Collinson 1965). This is particularly likely where conodonts are concentrated in weathered material, at the base of argillaceous deposits where slumping and re-working are evident. Avoid collecting from unstable outcrops or where recent slumping has occurred, and discard samples suspected of containing re-worked assemblages (Collinson 1965), as recycled material may be very difficult to recognise (see section 15 MICROFOSSIL SAMPLE CONTAMINATION AND RELIABILITY PROBLEMS).

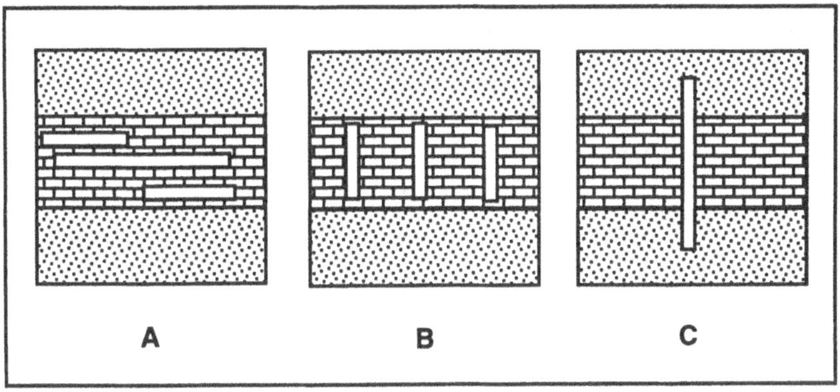

Figure 4.2. Diagram illustrating the likely position of channel samples (unshaded). (a) from ditches parallel to bedding. (b) from ditches at right angles to bedding. (c) sectional channel sampling across lithological boundaries. (After Maximov *et al.*, 1973).

4.4.3. Core sampling

Core samples are frequently used in oil exploration for determining the stratigraphic age and position of strata (Hills 1949). Consequently characteristics of both the sediment and the microfauna must be accurately identified and described to fulfil these aims. By its nature core material is often irreplaceable, and should therefore be subdivided before undertaking any destructive processing. Small sub-samples (250g) provide sufficient material for most microfossil processing. However, rarely are cores continuous rock samples, but more commonly drill or well cuttings, of about 500g that represents approximately a 3 metre sub-surface interval. The cuttings are generated by the rotary action of a drilling bit as it penetrates the rock. A closed circulatory system of drilling fluid ("mud") lubricates and cools the bit and returns the well cuttings to the surface. At the head of the drilling rig, well cuttings are collected on "mud screens" or "shale shakers", and at regular intervals sampled by the mud logger or rig geologist, bagged and labelled up. Great care must be taken during this stage to ensure that samples are not lost or mislabelled. Accurate record keeping and documentation is an essential part of this type of sampling.

Before the material can be processed or examined it has to be cleaned of dried mud contamination and sub-sampled. An immense amount of skill is required by the examining geologist to differentiate between the microfauna and sediments from the sub-surface rock, and additives included within the mud. Furthermore, skill, judgement and experience are required to differentiate well chipping's that might be either stratigraphically higher or lower than the interval recorded on the sample bag.

Continuous rock cores provide better samples for foraminiferal analysis than chipping's. Unfortunately they are much more expensive and time consuming to collect. However, once cleaned of drilling mud the stratigraphical resolution and control offered by cores is far superior to that of the well cuttings. Depth is more accurately recorded and lithological changes and contacts are readily distinguished as are structural features (*e.g.* faults). Furthermore, there is less chance of cross sample contamination. Once back in the laboratory, cores can be sub-sampled (see section 1 PALAEONTOLOGICAL TECHNIQUES: AN INTRODUCTION TO PRACTICAL PROCEDURES).

4.5. ON-SITE FIELD OBSERVATIONS

4.5.1. Preliminary optical observations

Before taking a sample always inspect the bedding surfaces with a hand lens. Although not seen in detail, calcareous microfossils exceeding 1mm in size such as ostracods, benthic foraminifera and calcium phosphatic conodonts, often stand proud on bedding planes of fissile shaley rocks or the weathered surfaces of limestones. Such careful observations can save time in collecting vast amounts of rock. Smaller forms such as diatoms, radiolaria, spores and pollen are rarely seen, their presence confirmed by enhancing prepared surfaces, laboratory processing or preliminary field laboratory observations (Baumgartner *et al.*, 1981). A basic field laboratory should include a work table suitable to accommodate a stereozoom binocular microscope, and facilities to immerse specimens in acid, wash and dry them. Suitable protective equipment (*e.g.* gloves, safety glasses) should be used. A guide for the reconnaissance method is as follows:

1 Establish the field laboratory as close as possible to the section.

2 Sample every 20 to 50cm collecting 50 to 100g from the main lithologies exposed.

3 While wearing acid resistant gloves and safety goggles etch carbonate-free lithologies for several minutes in concentrated hydrofluoric (HF) acid. Calcareous cherts and siliceous limestones must be etched in concentrated hydrochloric (HCl) acid first. When effervescence has ceased specimens can be etched in HF. Preliminary HCl treatment removes carbonate and prevents the formation of calcium fluoride, insoluble to HF.

> **Safety note:** When using concentrated acids in the field, ensure appropriate precautions are taken (*e.g.* acid resistant gloves and safety goggles must be worn during preparation). Avoid breathing acidic fumes, and working in adverse weather. For more detailed safety information see **Appendix 8 - Care, handling and disposal of Hydrofluoric (HF) acid.**

4 After etching, totally immerse samples in water and thoroughly wash for several minutes.

5 Allow the samples to dry (camping stoves can assist in this process).

6 Examine the etched surface, viewing under a stereozoom binocular microscope (with magnifications of x30 to x100), or a high powered achromatic hand lens. Abundant radiolarian tests will appear in relief on the etched surface.

The rapidly obtained results can then be used to guide collecting to specific faunal rich horizons, and in particular eliminate lateral sampling of the horizon that frequently indicates that good preservation is generally of patchy occurrence (Baumgartner *et al.*, 1981).

4.5.2. Surface enhancement

The recognition of biogenic structures (trace fossils) and the presence of organisms within fine-grained pelagic sediments can be difficult, even in laboratory conditions. The apparent homogeneity between the structures and matrix results in poor contrast effects. A technique of oil immersion, developed and modified for the observation of structures and trace fossils present in light coloured chalk lithologies (Bushinsky 1947, Bromley 1980), has been adapted for friable diatomaceous rocks (Savrda *et al.*, 1985). The procedure is as follows:

1. Prepare a smoothed surface (cut with a spade, knife or trowel). Clean loose debris with a fine brush and compressed-air dust gun.

2. Spray surface with a light lubricating oil (*e.g. WD-40*). In the laboratory, oil can be applied to more lithified surfaces with a brush or roller (Bromley 1980). Some poor or non-porous lithologies may require soaking for a short while in an oil bath (Bushinsky 1947).

3. Several minutes after spraying or removal from the oil bath, contrast between structures and matrix will be significantly improved, allowing features to be photographed or sampling directed to burrow openings or termination's.

Surfaces prepared in this manner provide comparable results, quicker and cheaper, than those achieved by X-radiography (Savrda *et al.*, 1985, see section 35 X-RADIOGRAPHY TECHNIQUES).

4.5.3. Work ethic

During collection ensure that all possible precautions are taken to minimise the risk of sample contamination. Todd *et al.*, (1965) stressed this as a major problem during sample collection, and will be discussed further in section 15 MICROFOSSIL SAMPLE CONTAMINATION PROBLEMS AND RELIABILITY PROBLEMS. Samples sent through a third party back to the laboratory must be properly packed to avoid sample disruption and the possibility of contamination (see part II FIELD DOCUMENTATION, section 7 FIELD DOCUMENTATION, SAMPLE PACKING AND TRANSPORTATION).

Clean hammers and spades between sampling, and collect only from fresh, clean and preferably unweathered surfaces. At an outcrop, the presence of gypsum ($CaSO_42H_2O$), jarosite ($KFe_3((OH)_6(SO_4)_2)$) or limonite ($FeOOHnH_2O$) on the weathered surfaces of rocks should be noted. The formation of these minerals generates acids that dissolve calcareous foraminifera and ostracods (Jones 1965). Contaminated surfaces should be removed before unweathered rock is examined and assessed for its suitability for collection. Remove all recent plant debris before sampling, and lightly brush the surface with a fine soft one centimetre wide paint brush. It may be prudent to photograph bagged samples in position on the outcrop, or draw a rough sketch in a notebook to stratigraphically place them when back in the laboratory.

4.6. MARINE SEDIMENT SAMPLES

The problems and specialized equipment required for collecting Recent marine sands and muds is briefly outlined by Haynes (1981). A wide variety of ship based "gravity" corers, "box" corers and 'grab' samplers have been cited in the literature (*e.g.* Emery and Broussard 1954, Fuller and Meisburger 1982, Hovorslev and Stetson 1946, Kullenburg 1947, Phleger 1960, Smith and McIntyre 1954, Weaver and Schultheiss 1983). Manual piston coring devices (*e.g.* Benson and Kaesler 1954, Burnham 1988, Ginsburg and Lloyd 1956, Hargis and Twilley 1994) have also been successfully used for obtaining undisturbed foraminiferal rich sediments. Sub-samples should be taken immediately after the sample has been collected, and preserved with a fixative to prevent decay of the microfauna. Mechanical percussion corers must also be considered in obtaining samples, although some thought must be given to the logistics of moving equipment to the field site and terrain of operation (Gilbert and Glew 1985, Imperato 1987, Lanesky *et al.*, 1979, Parker and McCann 1988, Perillo *et al.*, 1984, Smith 1984, Smith 1987).

4.7. COLLECTING LIVE MATERIAL

Arnold (1974) provides an excellent summary on all aspects of studying live foraminifera. Recent foraminifera can be collected from intertidal areas along shallow water coastlines. Although widely distributed, their abundance varies greatly geographically, seasonally and diurnally.

Quartz rich sand beaches of temperate and high latitude areas do not provide samples with large assemblages. Sand grade sized particles usually indicate vigorous wave or current action - environmental conditions that quickly lead to foraminiferal test destruction. However, at certain times of the year strand-line deposits rich in foraminifera may be found on these beaches. Abundant foraminiferal assemblages are found in carbonate environments of warmer climates. They most frequently occur, like their fossil ancestors, in fine grained sediments.

Live, or recently living material may often be intermingled with fossil forms, making the study of the living component of the fauna difficult. In slightly deeper water, away from the surf line, simple dredge samples can be obtained. More elaborate equipment is required for deeper water sampling (Cushman 1948, Bothner and Valentine 1982). This may involve the use or construction of temporary platforms (*e.g.* Hoyt and Demarest 1981, Irwin *et al.*, 1983, Stone and Morgan 1992), or SCUBA divers directly coring underwater (*e.g.* Anima 1981, Bonem and Pershouse 1981, Jones *et al.*, 1992, Martin and Miller 1982, Nichols and Eichenlaub 1991). Generally most equipment and procedures are variations on the gravity corer principle.

Live foraminiferan studies involve observing the effects of differing environmental parameters on growth, development, and shell morphology (vital effects), under carefully controlled field or laboratory conditions. Laboratory cultures ensure that specimens can be studied in an environment similar to their natural habitat. However, mortality rates are high and continuous cultures are difficult to establish (Hemleben *et al.*, 1989). Thus the continued collection of live material is required, employing techniques that minimise the amount of trauma subjected to specimens. Material that is to be examined several weeks after collection must follow through the stages defined by Maybury and Ap Gwynn (1993): (i) fixation, (ii) preservation, (iii) staining, (iv) sorting. The first two stages are critical if material is to remain in pristine condition, and are described in more detail below.

Free living planktonic foraminifera can be collected by using a series of plankton nets. The nets usually have an open neck of between 0.5 to 1 metre in diameter (avoids the capture of large numbers of fish and mammals), and a mesh diameter of no less than 75μm for the collection of mature specimens, or 10μm for juveniles. Nets are deployed at varying depths throughout the water column, and allowed to drift with the boat or towed at specified angles relative to the surface. Opening and closing mechanisms allow for quantitative studies from a variety of environments from pre-determined depths. The collecting time should be kept to a minimum, usually between 3 to 5 minutes, to prevent excessive organism trauma.

In shallower waters SCUBA diving and snorkelling can be employed to collect samples, further reducing the effects of specimen trauma if they are required in ecological studies. Material can be carefully scooped into a jar with a screw top lid, ensuring water and organisms are from only one part of the water column.

Collecting procedures for other invertebrate groups are summarised in Lincoln and Sheals (1979). In general the collecting of intertidal and shallow water species can be achieved without the aid of specialized equipment. Greater care is required in storage and transportation if living specimens are required for study. Collecting specimens in deeper waters requires diving and dredging.

4.8. FIXATION, PRESERVING AND STORING OF LIVE MATERIAL

Protein constituents of tissues can be stabilized after death by the process of fixation. This enables the assumption that during later studies' tissues and structures retain in some small degree the form and position they possessed in life. Furthermore, fixation can aid light microscopical studies by raising the refractive index of the cell contents and rendering tissues more susceptible to staining (Lincoln and Sheals 1979). Most fixatives act by coagulating protein, while others (*e.g.* formaldehyde) increase viscosity of the protein colloidal suspensions or convert them to gels. No one fixative can be regarded as "all-purpose", the choice depending on the material and the purpose it is required for (Lincoln and Sheals 1979). Chemicals commonly used include formaldehyde (H.CHO), ethanol (ethyl alcohol, C_2H_5OH), acetic acid ($CH_3.COOH$), picric acid ($(NO_2)_3C_8H_2.OH$), mercury (II) chloride (mercuric chloride, $HgCl_2$), osmic acid (OsO_4), potassium dichromate ($K_2Cr_2O_7$) and chromium (VI) oxide (chromium trioxide or chromic acid, CrO_3). Although some of these chemicals can be used in isolation, mixtures are usually made (Table 4.2) combining the best properties (Lincoln and Sheals 1979). If the prime objectives of study involve histological or cytological examination, specialist texts should be consulted (*e.g.* Pantin 1969).

Preservatives are solutions in which material can be stored indefinitely, without seriously distorting or destroying tissues, and which inhibit cell autolysis (cell breakdown by enzymes produced in the cell, a self-digestion) (Lincoln and Sheals 1979). The arresting of any autolytic action differentiates a preservative from a fixative. Although most commonly used preservatives have some fixative action, the "phenoxetols" (Table 4.3) do not inhibit autolysis, and must be regarded as 'incomplete preservatives' or as *post-fixation* preservatives (Lincoln and Sheals 1979).

If material is required for abundance and distribution studies, it should be immediately fixed to prevent the loss of spines that often occurs if organisms are allowed to remain in suspension for long periods of time prior to fixation. Live material can be stored in alcohol, or fixed in a 4% formalin solution buffered with hexamine (Hexamethylenetetramine) buffered to pH 8.2. This technique has a disadvantage in that the buffer must be replenished at intervals beginning at 4 weeks, then 6 months, and then every 2 years (Hemleben *et al.*, 1989). Replenishing ensures the correct pH is maintained, and prevents test dissolution. Similarly never use concentrated formaldehyde as this too will destroy tests. Walker *et al.*, (1974) uses a fixative solution consisting of 50ml of 40% formaldehyde and 2g of calcium chloride mixed in 1 litre of sea water. Buffer the solution to pH 8.3 using a buffer salt mixture. A 5 to 10% formalin solution is recommended by Haynes (1981), buffered with a solution of borax (10g/litre) to reduce the acidity and prevent dissolution.

Alternatively samples can be washed in fresh water and dried immediately after collection. However, they do tend to become clumped together, and test damage may result when attempting to separate them. Washing in fresh water prior to fixation causes the protoplasm to rupture and be lost. Fixation acts not only as a protoplasm preservative, but also coagulates and hardens it, retaining the essential feature for staining (see section 21 STAINING TECHNIQUES USED IN MICROPALAEONTOLOGY).

Two methods used in the wet processing of Recent material have been described (Maybury and Ap Gwynn 1993). Although both methods reduce the effects of shell dissolution, they both require the use of hazardous carcinogenic chemicals and solutions. Sediments and specimens must be treated immediately after collection. This procedure is not suitable for material that is in transit for long periods of time.

Safety Note: Always undertake these procedures in a fume cupboard, using PPE (nitrile gloves, safety glasses).

4.8.1. Glutaraldehyde ($CH_2(CH_2.CHO)_2$)

The Glutaraldehyde ($CH_2(CH_2.CHO)_2$) method is suitable for fixing most pelagic and smaller benthic foraminifera (maximum 1mm diameter), collected in plastic screw top containers that might require further detailed analysis (*e.g.* critical point drying, mass spectrometry, thin sectioning). Solutions penetrate and fix most soft tissues to a depth of about 0.5mm.

(i) Fixation

1. Dissolve 1g analytical reagent quality calcium chloride 6-hydrate in a mixture of 8ml of electron microscopy grade 25% glutaraldehyde and 92ml of filtered seawater. Filter the solution. At 20°C the pH will be 6.75. Dissolution effects are minimal as the solution is not in contact with specimens for very long (Maybury and Ap Gwynn 1993).

2. Top up sediments with a 2% glutaraldehyde solution. Use equal volumes of sediment and glutaraldehyde solution.

3 Gently agitate the sediment in solution to saturate the organisms. NB do not shake.

4 Leave for 1 hour.

5 Filter the residue. Excess glutaraldehyde will polymerize in the sediment if all traces are not removed. The sediment should remain wet at all times, continuously wash with seawater.

(ii) Preservation

1 Dissolve 21g of laboratory grade sodium cacodylate [$(CH_3)_2AsO_2Na.3H_2O$] in 1 litre of filtered seawater. Filter to remove residual undissolved chemicals. At 20°C the solution has a pH 8.93). Add equal quantities of solution to residue and gently agitate.

Specimens and residues can be stored indefinitely in this solution.

(iii) Staining

1 Dissolve 4g of analytical grade ammonium hydrogen carbonate (NH_4HCO_3) in 1 litre of distilled water (pH 8.6 at 20°C).

2 Dissolve 1g of rose bengal and 4g of analytical grade ammonium hydrogen carbonate in 1 litre of distilled water. Immerse specimens in the solution and leave overnight to stain.

3 Wash of excess stain using a solution of ammonium hydrogen carbonate in distilled water.

4.8.2 Paraformaldehyde (HCHO)n

The second method detailed by Maybury and Ap Gwynn (1993) is relatively straight forward, although the process should again be carried out in a fume cupboard. Two stock solutions are required.

(i) Fixation

1 Dissolve 40g analytical grade sodium hydroxide pellets (NaOH) in 1 litre of distilled water.

2 Dissolve 2g of laboratory grade paraformaldehyde in 100ml of filtered seawater. Dissolution can be enhanced by heating at 60°C (undertaken in a fume cupboard) and constantly stirring and adding drops of 1M sodium hydroxide solution. Buffer the solution with 0.8g analytical grade *di*-sodium tetraborate ($Na_2B_4O_7.10H_2O$) in 100ml 2% formaldehyde in filtered seawater (pH 9.22 at 20°C). The solution should be stored at 5-15°C.

3 Add the fixative to the residue in a 9:1 ratio.

(ii) Preservation

The fixative also functions as a preservative, but the pH must be monitored every month, and renewed when the value falls below 8.6. The staining procedure follows that outlined under the glutaraldehyde procedure.

Debate continues to surround optimum pH values required for the safe storage of residues and specimens (Maybury and Ap Gwynn 1993, Green 1995b). Optimum values appear to be related to biomineralization and shell chemistry of the microfossil group studied (*e.g.* Hay 1977, Hodgkinson 1991, Lewin 1961, Pfannkuche and Thiel 1988). Constant monitoring of the preservative, and its immediate replacement when the pH deteriorates is essential for the long term safe storage of specimens.

Procedures of fixation and preservation used for zoological specimens are too numerous to be detailed in this volume. However, tables summarising the chemicals and solutions used are included (Fixation - table 4.2; and Preservation table 4.3), as these may provide suitable alternatives in some studies. Invertebrate specimen anaesthetization procedures are summarised in section 11 PREPARATION OF RECENT MATERIAL FOR COMPARATIVE STUDIES.

The requirement to find a suitable mounting medium for live material prompted Maybury *et al.*, (1991) to examine the availability of products and undertake a long term study of the effects mounting media have on an organism's tests. Glycerol jelly, composed of 50g edible gelatine, 300ml of distilled water and 350ml of glycerol, mixed together and gently heated until the gelatine dissolves, was recommended. When the mixture has cooled but not set, 1ml of a saturated solution of phenol is added to prevent bacterial and fungal growth. The mixture must be stored in an air-tight container. Slide making is a simple procedure.

1 Gently heat but do not boil the jelly in a 50°C water bath.

2 Transfer 2-3 drops of jelly to a warm slide, position specimen and cover with a warm cover glass.

3 Allow the slide to cool and seal the edges with clear nail varnish. (*Glyceel* recommended by Maybury *et al.*, (1991) is no longer available, and no suitable alternative has been found).

REFERENCES

ANIMA, R. J. 1981. A diver operated reverse corer to collect samples of unconsolidated coarse sand. *Journal of Sedimentary Petrology*. **51**, (2), 653-654.

ARNOLD, Z. M. 1974. Field and laboratory techniques for the study of living foraminifera, pp 153-206: In; R. H. Hedley and C. G. Adams (eds), *Foraminifera*, Volume 1. Academic Press (London). iv + 276p.

BAUMGARTNER, P. O, BJØRKLUND, K. R, CAULET, J-P, DE WEVER, P, KELLOGG, D, LABRACHERIE, M, NAKASEKO, K, NISHIMURA, A, SCHAAF, A, SCHMIDT-EFFING, R and YAO, A. 1981. EURORAD II, 1980 - Second European meeting of radiolarian paleontologists: current research on Cenozoic and Mesozoic radiolarians. *Eclogae geologicae Helvetiae*. **74**, (3), 1027-1061.

BENSON, R. H and KAESLER, R. L. 1965. The Benson-Kaesler bolapipe dredge. *Micropaleontology*. **11**, (3), 396-372.

BONEM, R. M and PERSHOUSE, J. R. 1981. Inexpensive, portable underwater coring device. *Journal of Sedimentary Petrology*. **51**, (2), 654-656.

BOTHNER, M. H and VALENTINE, P. C. 1982. A new instrument for sampling flocculent material at the sediment/water interface. *Journal of Sedimentary Petrology*. **52**, (2), 639-640.

BROMLEY, R. G. 1980. Enhancement of visibility of structures in marly chalk: modification of the Bushinsky oil technique. *Bulletin of the Geological Society of Denmark*. **29**, 111-118.

FIXATION SOLUTION	CONCENTRATION, METHOD & ORGANISMS EFFECTED
Formaldhyde & Formalin	Formalin is commercially available as a 40% aqueous solution. Diluted to 10% to form a cytoplasmic fixative. Marine animals subject to distortion following osmotic changes. Prevented by making solution with sea water. Acidic - unsuitable for animals with calcareous skeletons. Can be buffered: (i) addition of 4% NaOH, added by dropper until formalin is neutral to phenol red; (ii) addition of hexamine; 200g/l of 40% formalin: forms a penetrating fixative. Fixed in 48 hours. N.B. To make 10% formaldehyde: 3.5 parts water to 1 part commercial formalin (40% formaldehyde). To make 10% formalin solution: 9 parts water to 1 part commercial formalin.
Steedman's solution	For general fixation and preservation of marine zooplankton. Combines fixative properties of formalin (5ml) with the preservative and softening action of propylene phenoxetol (0.5ml) and propylene glycol (4.5ml) (forming a strong stock solution), and water (90ml - fresh/marine).
Bouin's fluid (Picro-formol)	Suitable for marine micro-invertebrates. Combines picric acid* (75ml sat aqueous sol), formalin (25ml of 40%) and acetic acid† (5ml). Fixed in 12+ hours.
Alcoholic Bouin (Dubosq-Brasil fluid)	More penetrating than Bouin's, suitable for animals with tough exoskeletons (e.g. arthropods). Combines picric acid* (1g), acetic acid † (15ml), formalin (60ml of 40%) and 80% alcohol (150ml). Fixed in 2 hours (longer for larger animals, although prolonged contact time results in brittle material).
Heidenhain's Susa mixture	Good fixative for material requiring sectioning. Combines mercury (II) chloride• (45g), sodium chloride (5g), water (dist/marine 800ml), trichloracetic acid (20ml), acetic acid† (40ml) and formalin (200ml of 40%). Fixed in 3-24 hours. Transfer to 90% iodized alcohol (see below) to remove mercuric precipitates from tissues.
Viets' solution	Combines acetic acid† (3 parts), glycerol (11 parts) and distilled water (6 parts). Preserves and fixes water mites.
Oudemans' fluid	Combines acetic acid† (8 parts), glycerol (5 parts) and 70% alcohol (87 parts). Preserves and fixes terrestrial mites.
Zenker's fluid	Good fixative for micro-anatomical work. Combines mercury (II) chloride• (5g), sodium chloride (5g), acetic acid† (5ml), potassium dichromate (2g), sodium sulphate (1g) and distilled water (100ml). Fixed in 3-12 hours. Wash in running water and transfer to 50% alcohol to remove mercuric precipitates from tissues.

Table 4.2. For details and additional notes see over.

Table 4.2 (continued)

FIXATION SOLUTION	CONCENTRATION, METHOD & ORGANISMS EFFECTED
Flemming's solution	Combines 1% chromium (VI) oxide (150ml), 2% osmic acid° (40ml) and acetic acid† (10ml). Wash in running water to remove traces of osmic acid which might blacken material.
Chromic/Osmic acid mixture	Combines 1 % chromium (VI) oxide (100ml) and 1% osmic acid° (2ml).
Chromic/Acetic acid mixture	Combines 1% chromium (VI) oxide (100ml) and acetic acid† (5ml).
Corrosive sublimate (mercuric chloride•)	Usually used as a saturated solution mixed in either fresh or sea water. All traces of mercuric precipitates must be removed after fixation. Wash for several hours in running water, or in numerous changes of 70% alcohol. Most effective method is to leave overnight in iodized alcohol (see below).
Corrosive acetic	The addition of a small quantity of acetic acid † (<10% to vol) to corrosive sublimate may reduce tissue shrinkage.
Iodized alcohol	Used to remove traces of mercury (II) chloride• from fixed material. Combines iodine (3g), potassium iodide (6g) and 70% alcohol (300ml). Add this solution (known as tincture of iodine) to 70% alcohol in a sufficient quantity to give a brown sherry colour. Wash specimens until brown colour ceases to disappear. Transfer to fresh 70% alcohol to remove traces of iodine.
Schaudinn's solution	An alcoholic solution of corrosive sublimate. Combines saturated mercury (II) chloride• (2 parts) with 90% alcohol (1 part). A few drops of acetic acid† can be added to reduce tissue shrinkage.
T.A.F.	Combines commercial formalin (14ml), triethanolamine (4ml) and distilled water (82ml).

Table 4.2. Commonly used solutions and mixtures for the fixation of zoological marine and fresh water specimens.

*Picric acid: an explosive substance, which readily detonates when in contact with certain metals. **Never store in a metal container**. Store in a glass vessel under water.

†Acetic acid: cytoplasmic detail is lost in fixatives containing acetic acid.

•Mercury (II) chloride: **extremely poisonous,** with no odour and looks like water. Avoid contact with steel instruments since it has a strong corrosive reaction.

°Osmic acid: the solution and vapour are **highly toxic**.

Data from Lincoln and Sheals (1979).

PRESERVING SOLUTION	CONCENTRATION, METHOD & ORGANISMS EFFECTED
Formaldhyde & Formalin	A 5-10% aqueous solution is a good general preservative (see table 4.2 for method). Formaldehyde tends to stiffen and harden animals making them brittle.
1-(3-chlorallyl)5,7-triaza-1-azoniaadamantane	A yellow/white solid, highly soluble in water (fresh and sea). As a 10% aqueous solution it can be used as a fixative as well as a preservative. Slightly acidic, but can be buffered with calcium carbonate. Less toxic than formalin, non-flammable and non-volatile. Does not cause distortion or brittleness in specimens.
Ethanol (ethyl alcohol)	Dilute with water to a 70% concentration. Regarded as the best general preservative. N.B. alcohol extracts water from specimens placed in it. If the bulk of the alcohol does not exceed that of the fresh specimens, the strength of the alcohol will eventually be reduced by at least one half. Soft-bodied specimens placed directly in strong alcohols are liable to shrinkage. This can be avoided by placing material in weaker solutions and gradually increasing the strength. Many animals become hard if stored in concentrations stronger than 80%.
Propylene phenoxetol	A successful post-fixation preservative for a wide range of animals as a 1-2% aqueous solution, although vigorous stirring is required as it dissolves with difficulty. Specimens retain natural colour and remain pliable. Non-flammable and non-volatile. Should only be used after adequate fixation. Fresh unfixed material will decompose.
Phenoxetol BPC	Used as a post-fixation preservative as a 1-2% aqueous solution. Less efficient than propylene phenoxetol as a bactericide and fungicide.
Ethylene glycol	A useful preservative for marine organisms when used as a 50% aqueous solution. Non-volatile, non-flammable and does not precipitate when mixed with sea or fresh water with high mineral content. Best results for marine plankton when fixed with formalin, decanted, washed in sea water and placed in ethylene glycol. Causes little or no shrinkage.

Table 4.3. Commonly used solutions for preserving zoological marine and fresh water specimens. Data from Lincoln and Sheals (1979).

BURNHAM, R. J. 1988. A large-diameter coring device for use in shallow water. *Journal of Paleontology*. **24**, (4), 479-484.

BUSHINSKY, G. I. 1947. Textures and structures of Cretaceous chalk and flints (Russian with English summary). *Bulletin of Society of Naturalists, Moscow*. New Series 52, (Geology). **22**, 37-43.

CHANEY, D. S. 1989. Hand-held mechanical preparation tools, pp 186-203; In: R. M. Feldman, R. E. Chapman and J. T. Hannibal (eds), *Paleotechniques*. The Paleontological Society Special Publication. **4**. iv + 358.

COLLINSON, C. 1965. Conodonts, pp 94-102: In; B. Kummel and D. Raup (eds), *Handbook of Paleontological Techniques*. W. H. Freeman and Co., (San Francisco). xiii + 852p.

CUSHMAN, J. A. 1948. Chapter 3, Collecting and preparing material, pp 25-28: In; *Foraminifera, their classification and economic use*, (4th Edition), revised and enlarged, with an illustrated key to the genera. Harvard University Press (Cambridge, Massachusetts). 605p.
EMERY, G. R and BROUSSARD, D. E. 1954. A modified Kullenberg piston corer. *Journal of Sedimentary Petrology*. **24**, (3), 207-211.
FULLER, J. A and MEISBURGER, E. P. 1982. A simple ship-based vibratory corer. *Journal of Sedimentary Petrology*. **52**, (2), 642-644.
GILBERT, R and GLEW, J. 1985. A portable percussion coring device for lacustrine and marine sediments. *Journal of Sedimentary Petrology*. **55**, (4), 607-608.
GINSBURG, R. N and LLOYD, R. M. 1956. A manual piston coring device for use in shallow water. *Journal of Sedimentary Petrology*. **26**, (1), 64-66.
GREEN, O. R. 1995b. Pitfalls, problems and procedures in micropalaeontological preparation and conservation. *The Geological Curator*. **6**, (4), 157-166.
HARGIS, T. G and TWILLEY, R. R. 1994. Improved coring device for measuring soil bulk density in a Louisiana deltaic marsh. *Journal of Sedimentary Research*. **A64**, (3), 681-683.
HAY, W. W. 1977. Calcareous nannofossils, pp 1055-1200 (techniques 1056-1069), In: A. T. S. Ramsay (ed), *Oceanic Micropalaeontology* (2 Volumes). Academic Press (London). xii + 1453.
HAYNES, J. R. 1981. *Foraminifera*. Macmillan Publishers Ltd., (London). x + 433p.
HEMLEBEN, CH, SPINDLER, M and ANDERSON, O. R. 1989. Chapter 3, Collecting and Culture Methods, pp 33-55: In; *Modern Planktonic Foraminifera*. Springer-Verlag, (New York). xiii + 363p.
HILLS, J. M. 1949. Sampling and examination of well cuttings. *Bulletin of the American Association of Petroleum Geologists*. **33**, (1), 73-91.
HODGKINSON, R. L. 1991. Microfossil processing: a damage report. *Micropaleontology*. **37**, (3), 320-326.
HOVORSLEV, M. J and STETSON, H. C. 1946. Free-fall coring tube: a new type of gravity bottom sampler. *Bulletin of the Geological Society of America*. **57**, (10), 935-950.
HOYT, W. H and DEMAREST, II, J. M. 1981. A versatile twin-hull barge for shallow-water vibracoring. *Journal of Sedimentary Petrology*. **51**, (2), 656-657.
IMPERATO, D. P. 1987. A modification to the vibracoring technique for sandy sediment. *Journal of Sedimentary Petrology*. **57**, (4), 788-789.
IRWIN, J, PICKRILL, R. A and deL. MAIN, W. 1983. An outrigger for sampling from small boats. *Journal of Sedimentary Petrology*. **53**, (2), 675-676.
JONES, B, PHIMESTER, K. F, HUNTER, I. G and BLANCHON, P. 1992. A quick, inexpensive, self-contained sediment coring system for use underwater. *Journal of Sedimentary Petrology*. **62**, (2), 725-728.
JONES, D. J. 1965. Chapter 2, Collection, preparation, and Preservation of Microfossils, pp 7-18: In; *Introduction to Microfossils*. Harper, (New York).
KULLENBURG, B. 1947. The piston core sampler. *Svenska hydrogr-biol. Komm, Skr. (3,Hydr)*. **1**, (2), 1-46.
LANESKY, D. E, LOGAN, B. W, BROWN, R. G and HINE, A. C. 1979. A new approach to portable vibracoring underwater and on land. *Journal of Sedimentary Petrology*. **49**, (2), 654-657.
LEWIN, J. C. 1961. The dissolution of silica from diatom walls. *Geochim Cosmochim Acta*. **21**, 182-198.
LINCOLN, R. J and SHEALS, J. G. 1979. *Invertebrate animals. Collection and preservation*. British Museum (Natural History), (London and Cambridge University Press). vii + 150p.
MARTIN, E. A and MILLER, R. J. 1982. A simple diver-operated coring device for collecting undisturbed shallow cores. *Journal of Sedimentary Petrology*. **52**, (2), 641-642.
MAYBURY, C, MORRISON, L and STEWART, V. 1991. The search for a reliable mounting medium for Recent "live" foraminifera. *Journal of Micropalaeontology*. **10**, (1), 16.
MAYBURY, C and AP GWYNN, I. 1993. Wet processing of Recent calcareous foraminifera: methods for preventing dissolution. *Journal of Micropalaeontology*. **12**, (1), 67-69.
MAXIMOV, A, MILOSERDINA, G and EREMIN, N. 1973. *Short course of Geological Prospecting and Exploration*. MIR Publishers, (Moscow). 1-256.
NICHOLS, J. A and EICHENLAUB, Jr, C. M. 1991. A simple, diver-operated pneumatic corer. *Journal of Sedimentary Petrology*. **61**, (4), 634-635.
PARKER, A and McCANN, C. 1988. Percussion corer and extension device for unconsolidated sediments. *Journal of Sedimentary Petrology*. **58**, (4), 752-753.
PERILLO, G. M. E, ALBERO, M. C, ANGIOLINI, F and CODIGNOTTO, J. O. 1984. An inexpensive, portable coring device for intertidal sediments. *Journal of Sedimentary Petrology*. **54**, (2), 654-655.
PFANNKUCHE, O and THIEL, H. 1988. Sample Processing, pp 134-145: In, R. P. Higgins and H. Thiel (eds), *Introduction to the study of meiofauna*. Smithsonian Institution Press (Washington).
PHLEGER, F. B. 1960. *Ecology and Distribution of Recent Foraminifera*. John Hopkins Press, (Baltimore). vii + 297.
PANTIN, C. F. A. 1969. *Notes on microscopical techniques for zoologists*. Cambridge University Press. 77p.
SAVRDA, C. E, BOTTJER, D. J and GORSLINE, D. S. 1985. An image-enhancing oil technique for friable, diatomaceous rocks. *Journal of Sedimentary Petrology*. **55**, (4), 604-605.
SMITH, D. G. 1984. Vibracoring fluvial and deltaic sediments: tips on improving penetration and recovery. *Journal of Sedimentary Petrology*. **54**, (2), 660-663.
SMITH, D. G. 1987. A mini-vibracoring system. *Journal of Sedimentary Petrology*. **57**, (4), 757-758.
SMITH, W and McINTYRE, A. D. 1954. A spring loaded bottom sampler. *Journal of the Marine Biological Association (UK)*. **33**, 257-264.
STONE, G. W and MORGAN, J. P. 1992. Jack-up pontoon barge for vibrocoring in shallow water. *Journal of Sedimentary Petrology*. 62, (4), 739-741.
TODD, R, LOW, D and MELLO, J. F. 1965. Smaller foraminifera, pp 14-20: In; B. Kummel and D. Raup (eds), *Handbook of Paleontological Techniques*. W. H. Freeman and Co., (San Francisco). xiii + 852p.
VARKER, W. J. 1967. Conodonts of the genus *Apatognathus* Branson and Mehl from the Yoredale Series of the North of England. *Palaeontology*. **10**, (1), 124-141.
WALKER, D. A, LINETON, A. E and SCHAFER, C. T. 1974. Sudan Black B: a superior stain to rose Bengal for distinguishing living from non-living foraminifera. *Journal of Foraminiferal Research*. **4**, (4), 205-215.
WEAVER, P. P. E and SCHULTHEISS, P. J. 1983. Detection of repenetration and sediment disturbance in open-barrel gravity cores. *Journal of Sedimentary Petrology*. **53**, (2), 649-654.

5. CONSOLIDATION, STABILIZATION AND REPLICATION TECHNIQUES

5.1. INTRODUCTION

The removal of fossil specimens from the field to the laboratory is the responsibility of the collector. During the stages of removal, packing and transportation, ensure the minimum amount of damage and information loss from the specimen. Preliminary conservation procedures, specimen consolidation and stabilization, may have to be performed in the field to achieve this. Occasionally it may be neither practical nor possible to remove specimens back to the laboratory (*e.g.* moulds and large trace fossils). Fossil replication procedures, using compounds such as moulding clays, plaster, silicone rubbers, reinforced rubbers (latex) and glassfibre plastics, may be undertaken in such circumstance's.

Consolidation and stabilization techniques have been used by palaeontologists for over one hundred and thirty years (for reviews *see* Camp and Hanna 1937, Carreck and Adams 1969 and Greenwald 1989). Procedures used on large specimens (*e.g.* vertebrate bones) have changed little during this time (*e.g.* Hotton 1965, Rixon 1976, Converse 1989, Greenwald 1989), although new compounds, chemical processes and technological advances have been incorporated in procedures, and provide useful alternatives for consideration by field palaeontologists.

5.2. FIELD SPECIMEN CONSOLIDATION AND STABILIZATION

Extra time spent in preparing material in the field before removal, may save several weeks work on return to the laboratory (Camp and Hanna 1937). Only when all preliminary note taking and photography is complete should removal of the fossil begin. Site examination and evidence obtained from fossils collected as float samples from scree slopes and spoil heaps can be used to evaluate the fragility of *in-situ* specimens and their successful removal from supporting and enclosing matrix. Any immediate field preparation undertaken on material may well result in extensive damage, irreparable back in the laboratory. To ensure the successful removal of material back to the laboratory three criteria have to be satisfied. The specimen must be (i) stabilized, (ii) supported, and (iii) protected. All of these stages assist in consolidating the specimen for removal to the laboratory. Initial procedures must stabilize and re-attach all fossil bearing fragments. Hardeners, adhesives and emulsion solutions are used for this purpose, but only where necessary (Rixon 1976). All other fragments should be carefully packed separately.

Consolidants and hardeners such as poly(vinyl butyral) (dissolved in isopropyl alcohol or acetone), and less commonly poly(vinyl acetate) (dissolved in 10% acetone to form an adhesive, and in toluene to form a consolidant), or poly(vinyl acetate) emulsions (dissolved in water) are commonly used. Both can be mixed to varying thicknesses, depending on how thick a coat or depth of surface penetration is required. For field procedures pre-mixed solutions are best transported in unbreakable wide neck containers with leak proof tight fitting screw-top caps. The solutions can be transferred to a plastic squeeze bottle and liberally applied to the fragile surface. Alternatively, if applied with a brush, take care not to contaminate the solution with fragments of matrix. To be effective the solutions must soak into the specimen (re-apply a second coat if necessary), and thoroughly dry before continuing with the excavation.

Wet specimens should if possible, be removed (with a surrounding margin of matrix), allowed to dry and immediately treated with a consolidating poly(vinyl acetate) emulsion. Where the specimen is in danger of collapse on drying, it must be coated with a poly(vinyl butyral) solution. The process of wet specimen decay may be further deferred by collecting from the locality during a dry season. However, time and resources dictate the action taken when collecting from water logged environments (Baird 1980, Rixon 1976). If field time is short, then carefully remove any matrix surrounding the fossil, and pack with multiple layers of heavy duty aluminium foil moulded around the specimen for transportation (Converse 1989). When working on a site for a few days, cover the area with a tarpaulin or heavy duty polythene sheet to protect specimens before treatment. If solvent based solutions are used in consolidating specimens, ensure adequate ventilation, reducing vapour build-up in an enclosed area.

Standard vertebrate fossil consolidation techniques cannot be employed in waterlogged and intertidal areas. Fossils are rarely dry enough, or access is limited by tidal activity, for conventional collecting, consolidating and stabilizing procedures to be used. A rapid procedure (doping) using 2-Ethoxyethyl acetate ($CH_3.COO.CH_2.CH_2.OC_2H_5$, a cellulose adhesive) and burning off the solvent, has been outlined by Baird (1980). The only proviso is that the dissolved celluloid must have a sufficiently low viscosity that it will rapidly soak into the specimen. The method is as follows:

1 With the aid of a dropper bottle, liberally apply the solution to the specimen.

2 Standing upwind, carefully ignite the solvent. Protect the blaze until it burns itself out. Burning will assist to dry the specimen surface, although care must be taken with areas of projecting bone, as these are liable to char in the flames if not adequately protected. Moisture present within the specimen will make the cellulose turn white. This can be reversed in the laboratory fume cupboard by applying acetone under a heat lamp.

3 Repeat the procedure until the entire specimen is consolidated with a skin of cement. Baird (1980) suggests that 2 to 3 applications may be required.

4 On completion of stabilization, trench and extract the specimen in the conventional manner described below.

5.3. JACKETING SPECIMENS

Supporting and protecting stabilized specimens is a demanding and time consuming process, requiring a great deal of skill, patience, ingenuity and occasionally physical strength. In many respects it can be considered "a science in itself" (Greenwald 1989). The most well known and widely described method is that of constructing a plaster jacket ("jacketing") around the specimen (Fig. 5.1). The method can be adapted for specimens found in unconsolidated gravels and lithified limestones and sandstones. Specimens occurring in clays can be protected by wrapping and moulding a layer of clay around the fossil (Converse 1989). The stickiness of the clay should prevent the fossil from fracturing, and it can be kept moist and from drying by wrapping in damp newspaper. There is no size limit on the specimens that can be jacketed, although for larger specimens (and complete skeletons) Auffenberg (1967) considers this procedure, the palaeontologists most valuable tool. Converse (1989) describes three variations of the jacketing procedure using widely available materials:

i Burlap or hessian cloth strips saturated in a mixed solution of plaster of Paris. This is the conventional procedure, almost as old as the science of palaeontology, which has been in use for over 130 years (Davies 1865, Woodward 1865).

ii Plaster impregnated medical bandages. These provide a faster and more convenient method of transporting material to a field site, and simpler field application. However, bandage size may restrict the versatility of the method when jacketing large specimens (Converse 1989).

iii *3M Scotchcast* Casting tape. This is a polyurethane resin impregnated knitted fibreglass matting that is also used in the medical profession. As with the plaster impregnated bandages the curing reaction is initiated by immersing the fabric in water for 10-15 seconds. Within 6 minutes of placing in position it becomes rigid. Converse (1989) suggests this material is particularly useful when working in adverse field conditions.

All three variations of the procedure rely on an adequate supply of water at the field site. Carreck and Adams (1969) examined alternative methods, and detailed a successful procedure using expanded polyurethane (see below). Whatever compounds are used in jacketing, site preparation procedures are essentially the same.

5.3.1. Specimen preparation

1 Carefully remove all loose fragments of matrix. Excavate all around the specimen, revealing a complete outline of the specimen but exposing as little of the fossil as possible. This will ensure a protective layer of sediment remains around the fossil. Excavation will leave the specimen on a pedestal of rock. Continue to excavate beneath the suspected depth of the fossil, to ensure the jacket will support it (Fig. 5.1a).

2 When the specimen is thoroughly dry, stabilize the surface with a coat of poly(vinyl butyral). This has now superseded the natural resin shellac used by Camp and Hanna (1937), which can prove difficult to remove from the specimen surface.

3 Once safely below the suspected depth of the specimen, carefully start to undercut to a distance of 2-3cm. This will give a "mushroom"

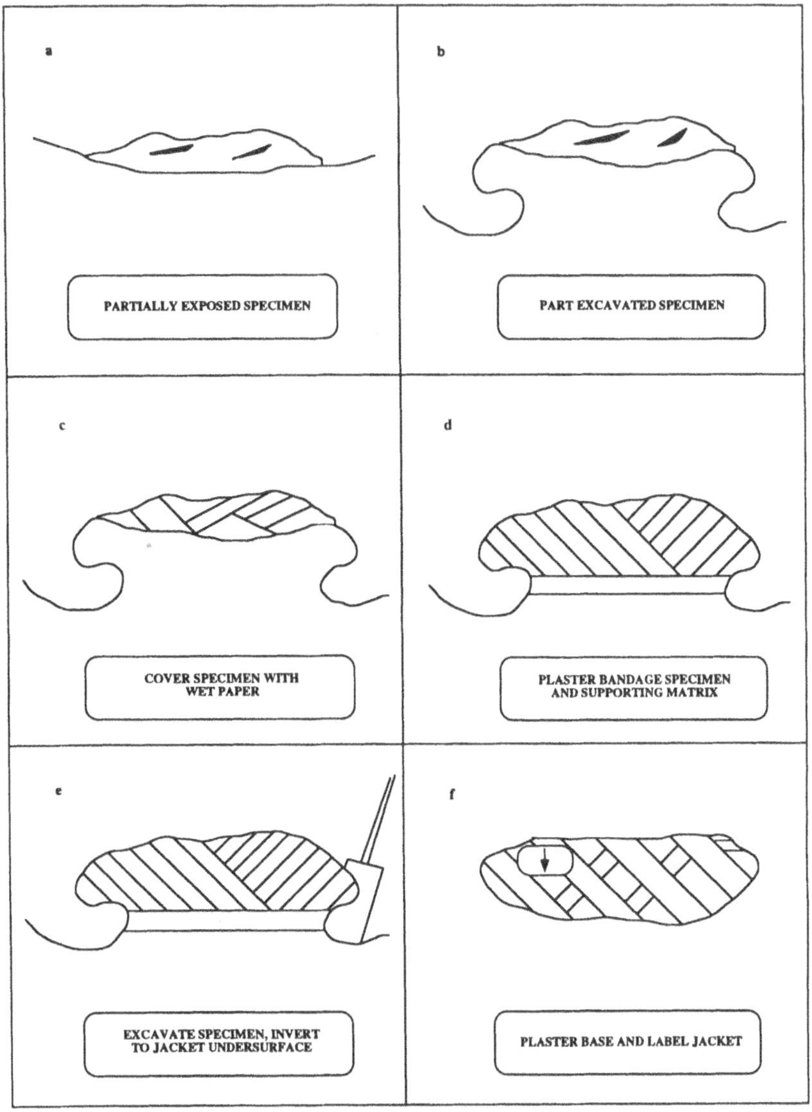

Figure 5.1. Diagrammatic illustration of the stages of jacketing a specimen. (a) Specimen partly exposed in the ground. (b) Excavate around the base of the specimen, and slightly undercut to leave it on a pedestal of sediment. (c) Apply a layer of wet paper, tissue or towels over the exposed fossil material. (d) Apply plaster bandages to the damp paper covered surface. (e) Excavate the specimen by cutting through the sediment pedestal. (f) Turn the specimen over, repeat the plaster bandage stage, completely encasing the specimen. (Diagram re-drawn from Converse 1989).

shaped profile to the specimen raised on its pedestal (Converse 1989, Fig. 5.1b). This allows for the jacketing material to be tucked well under the specimen, providing maximum support and ensuring the hardened jacket will not slip off when the specimen is released from the ground. Allow a wide enough trench so that working conditions all around the specimen are comfortable (Converse 1989).

4 Cover the specimen with wet newspaper, tissue, towels or any water absorbent fabric (Fig. 5.1c). This provides a protective layer around the fossil, preventing it from becoming attached to the plaster by direct contact. The absorbent layer must be well worked into all depressions and crevices (dampening with a mist spray dispenser ensures it will conform to the specimen contours), to a minimum thickness of 5mm.

5.3.2. Plaster jacketing

5 Cut an open weave material capable of absorbing plaster into strips approximately 1m in length and 7-15cms in width. Loosely roll and immerse in a bucket of water.

6 Mix the plaster of Paris by gently sprinkling powder into a bucket of water until about 3mm of liquid stands above the level of the settled powder. Hand stir it to a creamy consistency, removing any undissolved lumps. If the plaster is added quickly to the water, much of it will remain undissolved, and it will not evenly adhere to the matting fabric.

7 Dampen the paper covered fossil surface as this aids in allowing the plaster soaked strips to bond to the surface.

8 Wring out the matting strips, and immerse them in the plaster mixture. Ensure the strips are thoroughly soaked, but remove all excess plaster. Use a slow setting plaster, as this will give a longer working time and will set harder than rapid setting plaster.

9 Start at one end of the specimen, applying the saturated bandage to the top of the specimen, working down to any undercut areas, and smoothing lightly by hand. Repeat the process with a second bandage, slightly overlapping the first. Continue this process along the length of the fossil (Fig. 5.1d).

10 Apply a second layer of bandages at right angles to the first, making sure undercut area is generously covered.

11 Very large vertebrate specimens will require additional support. Either metal or timber splints can be incorporated in the jacket to provide reinforcement and carrying handles. Splints should always be placed between two layers of bandages or fixed to the outside of the package.

Before entering the field ensure an adequate supply of materials, and the means to cut them to appropriate lengths are available.

12 The thickness of the plaster jacket is dependent on specimen size, but is critical in providing support and protection, thus preventing collapse during removal and transportation back to the laboratory (Andrews and Cooper 1928). The plaster thickness should range from 5mm (small specimens) up to several centimetres for large specimens (Converse 1989). When the upper surface is completely bandaged, wind a plaster "collar" around the base using a long strip of material. This will assist in securing the block and preventing loose material from falling out when it is turned over.

13 Allow the top of the jacket to set and thoroughly dry. If possible leave for 24 hours. Label the outside of the jacket, indicating all relevant data such as orientation. Situate the label so that it can be read when the specimen is in its position of transportation and storage.

14 When ready to remove the specimen, start by gently excavating below the level of the specimen (Fig. 5.1e). In softer sediments (clays and gravels) this can be achieved by pushing a shovel under the jacket until it breaks free from the ground. Lithified rocks may require the use of hammers and chisels.

15 When the pedestal has been cut and the jacket freed, quickly and gently turn the specimen over.

16 Remove excess matrix and apply additional plaster strips following stages' 4-10 above, until the fossil is completely sealed (Fig. 5.1f). Allow the jacket to dry before transporting to the laboratory. In some instances, where transportation from the field is not far, and laboratory work will commence immediately, this stage may not be required (Rixon 1976).

5.3.3. Expanded polystyrene

Carreck and Adams (1969) could not recommend the use of expanded polystyrene as it required the application of steam, and the use of apparatus thought to be costly and cumbersome for the field, particularly for small scale use. Furthermore, results from preliminary field tests were not encouraging. However, the arrival of styrofoam in spray-cans provides a handy simple alternative (Converse 1989). Following the preparation of the specimen surface described above (stages 1-3), the method is as follows:

1 Cover the specimen in a layer of clear plastic food wrap, moulding it to conform with the specimen contours.

2 Spray the foam sealant around the specimen, smoothing it with a spatula or by a rubber glove coated hand.

3 After approximately 45 minutes the foam jacket will set, forming a strong, lightweight protective shell.

5.3.4. Expanded polyurethane

Initial fears that the exothermic reaction resulting from the mixing of compounds would, on curing, result in fracturing of the specimen, proved ill founded (Carreck and Adams 1969). Moreover, fears were further outweighed by the advantages of the jacket's strength, its lightweight and simplicity and ease in field application. Two compounds are required:

i a three-part semi-pre-polymer which may require heating (although higher temperatures produce shorter reaction times and shorter working life), composed of an ether-glycol resin, an isocyanate and an emulsifier;

ii an ether-glycol resin, a catalyst and a halogenated hydrogen blowing agent.

The two compounds are mixed in equal parts by weight or volume (depending on the manufacturers recommendation). The exothermic reaction causes the mixture to expand by vaporisation of the freon, and become a foam. Continuation of the reaction causes the gas cell walls increase in viscosity and become solid, eventually resulting in the formation of a stable and rigid foam.

SAFETY NOTE: Isocyanates are toxic. Avoid contact with skin, eyes or clothing and vapour inhalation. When mixing in a sheltered field position, avoiding rain and wind etc., the risk of inhaling fumes may be greater.

Other two stage compounds are available, although they may have slightly different bases, catalysts and blowing agents, but the principle behind the system is always similar. The reaction for these systems is always: a polyurethane is produced by the reaction of a diisocyanate and polyol. The diisocyanate reacts with substances containing active hydrogen atoms (Carreck and Adams 1969). The field application is as follows:

1 Expose as much of the bone as possible, leaving the base on a pedestal All loose matrix must be removed.

2 Cover the specimen with aluminium foil (0.018mm thick). If required the surface of the fossil can be strengthend with a surface application of poly(vinyl butyral).

3 Construct a retaining wall around the fossil using wood and cardboard, providing approximately 20cm clearance around the side and above the height of the specimen. Tape the sides of the wall to avoid excess loss of foam during initial pouring.

4 Around large specimens, pouring must be done in stages. This will avoid excessive viscosity increases and rise in temperature generated from mixing large quantities, and thus reduce wastage. Mix a maximum of 1 litre quantities at a time (Carreck and Adams 1969).

5 Mix compounds thoroughly by hand using a large spatula, ceasing to stir when the mixture begins to look whiter than its original syrup-yellow. A creamy consistency results in about 30 seconds, and a "full-cream" after a further 15 seconds Excessive mixing will form a very white mixture and produce below strength jackets. Better results are produced when the compounds are under mixed (Carreck and Adams 1969).

6 Pour the mixture as quick and evenly as possible.

The working life of the mixed compounds is very short (approximately 140 seconds), while the foam ceases to be adhesive after 160 seconds. The ambient working temperature for mixing polyurethane compounds is 18-21°C. This may impose climatic restrictions on field environments in which the products are used, and must be assessed before putting into practice. The success of this technique will be greatly enhanced by practice in the laboratory **before** attempting in the field.

7 After pouring, leave for 20 minutes to allow for complete curing. Excess foam can be trimmed using a wire cutter or hot knife.

8 Undercut the pedestal, and quickly and gently turn the block over (as in the plaster jacket method), and repeat the process on the open surface.

The expense of using polyurethane products in jacketing can be significantly reduced by plaster jacketing the underside, revealed when the specimen is turned over (Rixon 1976). This has the advantage of allowing the specimen too dry-out through the porous plastered side.

As a safety precaution Carreck and Adams (1969) suggest remnants of isocyanate-based products present in empty containers should be deactivated. This is best achieved by filling the container with a dilute ammonia solution (5vols laboratory grade (25%) NH_3), plus methylated spirits (55vols) and water (40vols). A 13.5% w/v aqueous solution of sodium carbonate *10-hydrate* ($Na_2CO_3.10H_2O$) is also effective, but takes longer to react. Containers should be left in the open as deactivation liberates carbon dioxide (CO_2).

Carreck and Adams (1969) noted that the polyurethane method had limitations, particularly when used on damp ground and in damp atmospheres. Excess water vapour impairs surface quality at the foam:sediment interface, resulting in the formation of a layer of friable foam. This can be compensated for when preparing the excavation, by allowing a greater jacket thickness and then removing the inferior outer foam layer.

Once back in the laboratory the foam support is removed from the specimen by cutting with a blade or hacksaw. This action will generate a large amount of dust, so undertake the work in a well-ventilated area and frequently vacuum. Alternatively a heated wire or knife can be used to remove the foam, but this action has the disadvantage of generating toxic fumes, and there is always the possibility of the foam igniting (Carreck and Adams 1969). Thinner surface layers of foam will be removed with the foil when it is gently peeled off.

5.4. FIELD SPECIMEN REPLICATION

When a fossil is weathered out or dissolved from the enclosing sediment, all that remains is a natural mould. Any size fossil can be dissolved to leave a mould, and as with jacketing, replication procedures are essentially similar, differing in the compounds used by the preparator. It may not be possible or practical to remove the mould from the field to the laboratory. Following photography, moulding and casting may be the only option available for obtaining a three dimensional reproduction of a fossil. Techniques have been successfully used on large vertebrate tracks (*e.g.* Maceo and Riskind 1989, Obata *et al.*, 1989), and small invertebrate remains (*e.g.* Johnson and McKerrow 1995, Parsley 1989). However, the best results are usually obtained from lithified sediments that have been little weathered.

5.4.1. Latex

Methods using pre-vulcanised natural rubber latex have to a large part been superseded by techniques using silicone rubber. Latex moulds have the disadvantage of suffering shrinkage with age, while the time required to build up a mould may limit the methods application in the field. A further disadvantage is that the mixtures usually have to be made up before going into the field. Consequently on site application will be delayed as the latex is allowed to settle and entrapped bubbles dispersed. Mixtures are water based, with a minimum of three consistencies usually used (one thinned to the consistency of light oil, one to a heavy oil, one neat latex, Parsley 1989). Mixtures are naturally white, and can be coloured to enhance contrast, with either a drop of Indian ink or graphite powder. The method of application is as follows:

1 Clean the mould of any loose material and ensure it is dry.

2 Paint a thinned (light oil consistency) latex mixture into the mould and allow it to dry. Wash the brush immediately in a weak solution of ammonia.

3 Apply subsequent coats of thinned latex until all fine detail is obscured. **N.B.** always allow coats to dry between applications, **and** wash brushes. Drying will be accelerated if the temperature is raised to between 70-80°C.

4 Apply a more viscous coat of latex (heavy oil consistency). At this stage the mould can be reinforced by the addition of a layer of hard-weave, fine-mesh cheese cloth (Parsley 1989), which will also reduce shrinkage. Additional layers can be applied as the latex is setting, however bonding will be stronger with a layer that is already solid.

5 The final layer can, if required, be thickened by a latex gel coat. Use a maximum 5% (weight or volume) latex thickener (Ammonium

Polymethacrylate) with undiluted latex. **N.B. the latex thickener is an irritant. Avoid skin and eye contact. Should contact occur, wash the effected area with plenty of water and seek medical advice.**

6 When the final layer has dried, gently peel the latex cast from the specimen, starting at the edges and working to the centre.

As a rule of thumb, large specimens require more layers than small specimens. A minimum of three layers is about average, although moulds of approximately 10cm diameter will require at least one layer of cloth reinforcement. For smaller moulds the procedure may only require two layers. A full discussion of the latex moulding and casting will be provided in section 36 FOSSIL REPLICATION TECHNIQUES.

A redeeming feature of latex is that it can be diluted with additional ammonia solution (to a maximum ratio of 1:8) and a small amount of liquid detergent for optimum surface contact. This makes it a particularly useful medium for penetrating fine cavities (*e.g.* burrows) when the surrounding matrix can be chemically removed leaving an internal cast of the structure (Garner 1953).

5.4.2. Silicone rubber

On site casting procedures of dinosaur tracks using RTV silicone rubbers have been outlined by Maceo and Riskind (1989) and Obata *et al.*, (1989). The procedure described by Obata *et al.*, (1989) reveals the versatility of the silicone rubber method in that it was undertaken on a natural cliff surface dipping at 70°, and in winter working conditions with temperatures as low as -10°C. Furthermore, the size of the mould (6x9m) was larger than average. This provides another example of the specialized techniques used in the collection and study of larger vertebrate remains.

When successfully performed in the field, silicone rubber replication procedures reduce field time, mitigate the expense of quarrying and laboratory preparation of large slabs. However, the quality of the palaeontological evidence preserved is intricately dependent on the care taken during the preparation of the specimen. The method is as follows:

1 Clean the rock surface of all loose material. Reinforce the fossil surface with a coat of poly(vinyl butyral). For protection of partly destroyed areas of the footprint, Obata *et al.*, (1989) covered the area with tin foil (0.002mm thick). Corral the sides with cardboard, plasticine or clay to prevent excessive loss of rubber. On inclined surfaces construct a support and working platform (Obata *et al.*, 1989).

2 Apply a generous coat of modelling wax with a bristle brush. This will act as a release agent. Ensure that all undercut areas are adequately coated. Allow 20 minutes for this to dry. If the matrix is porous, apply a second coat. A second coat should also be applied if the working temperature is in excess of 30°C, conditions in which the wax melts and soaks into the matrix (Maceo and Riskind 1989).

3 Mix the RTV silicone rubber base with catalyst, (approximately 20 drops of catalyst per 500ml of rubber, depending on the manufacturers' instructions). Large quantity's rubber and catalyst (*e.g.* 180kg:1.8kg as used by Obata *et al.*, 1989) must be thoroughly mixed to ensure complete curing.

4 Pour the catalysed rubber into the mould, ensuring all undercuts are filled. Evenly spread the rubber with a palette knife or wooden spatula over the entire surface. For smaller moulds pre-activated rubber from a cartridge can be applied using an application gun (Maceo and Riskind 1989). This provides an excellent method of transporting materials into the field. A wide variety of *Dow Corning* adhesive sealants are available in this form, and can be experimented with in the laboratory before embarking on field work, as the quality of reproduction varies with each product (Maceo and Riskind 1989).

5 For additional strength the rubber can be reinforced with a fibreglass matting or gauze sheet backing. This must be done as the rubber is being applied, before curing begins (Maceo and Riskind 1989).

6 Allow the rubber to cure (from a few hours to 24 hours depending on the compounds used).

7 Excess rubber can be trimmed off when the rubber has cured, with a sharp knife, and the mould carefully peeled off the surface.

Following curing Maceo and Riskind (1989) proceeded to jacket the silicone rubber mould with plaster to ensure it retained its natural shape. Laboratory replication of the mould produced by Obata *et al.*, (1989) could only be achieved by measuring the curvature of the cliff face, and constructing a support to display the cast. Because of the large size of the mould, it had to be cut into 5x1.8m rolls for transportation. Once in the laboratory a two part (2x6m, 2x4m) replica composed of glassfibre reinforced plastic (GRP) was constructed (see section 36 FOSSIL REPLICATION TECHNIQUES section for details). By far the greatest disadvantage in using silicone rubbers is the expense of the compounds.

REFERENCES

ANDREWS, C. W and COOPER, C. F. 1928. *On a specimen of Elephas antiquus from Upnor.* British Museum (Natural History), London.

AUFFENBERG, 1967. What is a plaster jacket? *The Plaster Jacket.* **6**, 1-7.

BAIRD, D. 1980. The burnt dope technique and other intertidal ploys from America. *Newsletter of the Geological Curators Group.* **2**, (8), 519-520.

CAMP, C. L and HANNA, G. D. 1937. *Methods in Paleontology.* University of California Press, (Berkeley, California). xxiii + 153p.

CARRECK, J. N and ADAMS, S. J. 1969. Field extraction and laboratory preparation of fossil bones and teeth, using expanded polyurethane. *Proceedings of the Geologists' Association.* **80**, (1), 81-89.

CONVERSE, H. H. 1989. *Handbook of paleo-preparation techniques* (2nd edition, revised by R. W. McCarty). Florida Paleontological Society. v + 109p.

DAVIES, W. 1865. On the preservation of fossil mammalian remains found in Tertiary deposits. *Geological Magazine.* **2**, 239-240.

GARNER, M. R. 1953. The preparation of latex casts of soil cavities for the study of tunnelling activities of animals. *Science.* **118**, 380-381.

GREEWALD, M. T. 1989. Techniques for collecting large vertebrate fossils, pp 264-274; In: R. M. Feldman, R. E. Chapman and J. T. Hannibal (eds), *Paleotechniques.* The Paleontological Society Special Publication. **4**. iv + 358p.

HOTTON, N. 1965. Tetrapods, pp 119-125: In; B. Kummel and D. Raup (eds), *Handbook of Paleontological Techniques*. W. H. Freeman and Co., (San Francisco). xiii + 852p.
JOHNSON, M. E and McKERROW, W. S. 1995. The Sutton Stone: an early Jurassic rocky shore deposit in South Wales. *Palaeontology*. **38**, (3), 529-541.
MACEO, J and RISKIND, D. H. 1989. Field and laboratory moldmaking and casting of dinosaur tracks, pp 419-420; In: D. G. Gillette and M. G. Lockley (eds), *Dinosaur tracks and traces*. Cambridge University Press, (Cambridge). xvii + 454p.
OBATA, I, MARUO, H, TERAKADO, H, MURAKAMI, T, TANAKA, T and MATSUKAWA, M. 1989. Replicas of dinosaur tracks, using silicone rubber and fibreglass-reinforced plastics, pp 413-415; In: D. G. Gillette and M. G. Lockley (eds), *Dinosaur tracks and traces*. Cambridge University Press, (Cambridge). xvii + 454p.
PARSELY, R. L. 1989. Latex casting of microinvertebrate fossils, pp 275-281; In: R. M. Feldman, R. E. Chapman and J. T. Hannibal (eds), *Paleotechniques*. The Paleontological Society Special Publication. **4**. iv + 358p.
RIXON, A. E. 1976. *Fossil animal remains: their preparation and conservation*. Athlone Press, (London). viii + 304p.
WOODWARD, H. 1865. How the skull of a mammoth was got out of the Brickearth at Ilford. *Geological Magazine*. **2**, 93-94.

6. FIELD STAINING TECHNIQUES FOR DETERMINING CALCITE, DOLOMITE AND PHOSPHATE

6.1. INTRODUCTION

Quick and simple field techniques are described for determining and distinguishing the carbonate mineral's calcite and dolomite. A further procedure for determining the presence of phosphate is also described. All of these field staining procedures can be modified for, or have been adapted from techniques used in the laboratory. A review of the field applications of geological techniques employing stains and dyes can be found in Green (1993). The methods are described below:

6.2. CARBONATE (CALCITE AND DOLOMITE) DETERMINATION

6.2.1. Dolomite (Mann 1955)

1 On a clean rock surface add a few drops of dilute (10%) hydrochloric acid HCl to etch the surface. This releases the Ca and Mg ions into solution. Continue with the procedure even if no reaction is visible (*i.e.* high Mg content). The etching stage may help separate a pure dolomite from a limestone (but not a low Mg limestone, MgO $<10\%$).

2 When any vigorous reaction has ceased, add a drop of alkaline solution (0.002g *p*-Nitrobenzene-azoresourcinol in 100ml of 2N sodium hydroxide).

3 Within the first minute of adding the solution observe the results of any reaction.

RESULTS

HIGH-Mg CALCITE	Blue precipitate is evident (indicates a carbonate high in available MgO). The intensity of the blue colour will decrease with MgO content.
LOW-Mg CALCITE	Violet colour indicates a carbonate with no MgO.
DOLOMITE	Yellow colour (seen when no reaction occurs after etching and alkali solution is applied).

As a blue coloration will be evident on evaporation of the alkali solution, observations must be noted within the first minute of application. Shield the test area from direct sunlight while undertaking this technique to avoid the loss of staining solution through evaporation, and detect light colour hues of the precipitates.

Following three month's field and laboratory testing, Mann (1955) was able to semi-quantify the presence of magnesium into one of four categories:

(i) >30% MgO
(ii) 10-30% MgO
(iii) 1-10% MgO
(iv) <1% MgO

However, it is unfortunate that the intensity of the blue coloration is not indicated for each category. This technique is of particular use in areas where contacts between formations are poorly exposed, or where lithologic boundaries are not visible (Mann 1955).

6.2.2. Calcite from dolomite (Warne 1962)

Warne (1962) modified the staining scheme of Friedman (1959) for field and laboratory use. However, the scheme requires the use of seven solutions to determine Calcite, Dolomite and High-Mg Calcite. Three of the stages require boiling - facilities that may not be readily available at every field station. Initial determination is achieved by etching the specimen, followed by immersion in Alizarin Red-S, an acidic solution (dye + HCl), and then boiled in an alkali solution (dye + NaOH). Warne's staining scheme, like those suggested by Friedman (1959), are best performed in the laboratory (see section 22 PREPARATION OF STAINED ACETATE PEELS AND SECTIONS).

A simple procedure, although non-quantitative, which can be undertaken in adverse weather, uses Alizarin Red-S to differentiate calcite (and aragonite) from dolomite.

1 Etch a clean rock surface with 10% HCl.

2 Place a few drops of Alizarin Red-S (0.1g dye in 50ml 1.5% HCl) on the area. Calcite will stain within 2-3 minutes.

RESULTS

CALCITE (ARAGONITE)	Pink or red (depending on crystal orientation).
DOLOMITE	Unstained.

The tests for determining iron in carbonate, using Potassium Ferricyanide either by itself, or in combination with Alizarin Red-S, and the determination of Mg-calcite using Titan Yellow, is not readily adaptable for field use. This is mainly due to the unstable nature of the freshly mixed stains. Potassium Ferricyanide is light sensitive, and must be used within a few hours of being mixed. Titan Yellow is non-permanent unless correctly "fixed" within a strong alkali solution, a procedure suitable for laboratory staining schemes only (for details see section 22 PREPARATION OF CARBONATE STAINED ACETATE PEELS AND THIN SECTIONS).

6.3. PHOSPHATE DETERMINATION

A quick and simple field test for the determination of phosphate in rocks has been outlined by Mann (1950) and Shapiro (1952) using an acidified vanadate-molybdate solution. Add the solution to a small amount of powdered sample. A sufficient concentration of phosphate ions will give a positive reaction, indicated by a yellow precipitate of ammonium phosphomolybdate.

The acidic vanadate-molybdate solution is made up as follows:

(i) *Ammonium molybdate*

1 Dissolve 10g of ammonium molybdate ($(NH_4)_6MO_7O_{24}.4H_2O$) in 100mls of distilled water.

The solution, if stored in a dark bottle, has an extended shelf life. However, to ensure its sensitivity, test before use, on a specimen of known phosphatic composition.

(ii) *Vanadomolybdate*

2 Dissolve 0.3g of ammonium metavanadate (NH_4VO_3) in 60ml of distilled water, and add 40ml of concentrated nitric acid (HNO_3).

3 Mix the two solutions.

The combined solution is applied as follows:

4 Scrape a clean surface on the sample (or in the laboratory crush a small amount of sample).

5 Place a small amount of the acidic vanadate-molybdate solution on the specimen surface.

6 While immersed, check the reaction (this may require the use of a handlens or, in the laboratory, a binocular microscope).

RESULTS

The presence of phosphate is indicated by a yellow precipitate within the solution. By comparing with prepared standard solutions of 5, 10 and 15%, it may be possible to determine the percentage P_2O_5 (Shapiro 1952). A more precise method of determining total P_2O_5 has been outlined by Sandon, Peachey and Vickers (1989), although a broad indication is as follows:

Colour of solution	Precipitate present	Estimated P_2O_5%
Colourless or pale yellow	No	<3
Yellow	Yes	3-10
Orange	Yes	>13

Flooding the sample with distilled water inhibits the reaction and precipitate formation, while the rock or grains tested remain unstained (Hoskings 1957, Reid 1969).

Swanson (1981) describes a very rapid field method, confirming the presence of phosphatic shells or fragments within a sample. The method is as follows:

1. Place a small crystal of pure ammonium molybdate on the suspected mineral.

2. Using a pipette, administer a couple of drops of dilute nitric acid onto the crystal. The presence of phosphate will be confirmed by the bright yellow colour of the precipitate.

An alternative, the strychnine-molybdate method, is restricted to laboratory use, as it involves more elaborate procedures in solution preparation and equipment (Kittrick and Hope 1967).

REFERENCES

FRIEDMAN, G. M. 1959. Identification of carbonate minerals by staining methods. *Journal of Sedimentary Petrology.* **29**, (1), 29-38.

GREEN, O. R. 1993. The use and applications of stains and dyes in geological techniques. *Science Technology.* **28**, 2-26.

HOSKINGS, K. F. G. 1957. The identification - essentially by staining techniques - of white and near white mineral grains in composite samples. *The Camborne School of Mines Magazine.* **57**, 5-16.

KITTRICK, J. A and HOPE, E. W. 1967. Organic dyes and strychnine-molybdate reagent as aids in isolating phosphate mineral grains. *American Mineralogist.* **52**, (1/2), 263-272.

MANN, V. I. 1950. A spot test for phosphorus in rocks. *Journal of Sedimentary Petrology.* **20**, (2), 116-117.

MANN, V. I. 1955. A spot test for dolomite limestones. *Journal of Sedimentary Petrology.* **25**, (1), 58-59.

REID, W. P. 1969. Mineral staining tests. *Mineral Industries Bulletin, Colorado School of Mines.* **12**, (3), 1-20.

SANDON, P. T. S, PEACHEY, D and VICKERS, B. P. 1989. Field Methods Project 10. Assessment of methods for the determination of the chemical reactivity, and hence the agronomic potential, of African phosphate rocks. *British Geological Survey, Technical Report WI/89/5.* Analytical and Applied Geochemistry Report Series. 22p.

SHAPIRO, L. 1952. Simple field method for the determination of phosphate in phosphatic rocks. *American Mineralogist.* **37**, (3/4), 341-342.

SWANSON, R. G. 1981. *Sample examination manual.* American Association of Petroleum Geologists, Methods in Exploration Series, No. 1, 117p.

WARNE, S. St. J. 1962. A quick field or laboratory staining scheme for the differentiation of the major carbonate minerals. *Journal of Sedimentary Petrology.* **32**, (1), 29-38.

7. FIELD DOCUMENTATION, SAMPLE PACKING AND TRANSPORTATION

7.1 INTRODUCTION

Collecting geological specimens can be a painstaking and occasionally hazardous task. Furthermore, the collecting of palaeobiological material may be compounded by the need to preserve recent (biological and ecological) material or undertake conservation procedures to stabilize material for the journey back to the laboratory. For the collector it is frustrating to arrive back to the laboratory to find specimens beyond repair or use through damage caused while in transit. It would be disastrous, however, to receive specimens with no or incomplete documentation. Although there may be some way for the collector to salvage something from this scenario, a catastrophic situation can result from a complete loss of all samples and documentation. Clearly if transportation of material is entrusted in the hands of a third party it may be difficult to obtain compensation, no matter how careful the collector has been, an element of good fortune is necessary.

It is essential that the collector documents and records all material and procedures employed in the field. On return to the laboratory it may be many months or even years before the material are unpacked and worked on. Furthermore the preparator may be a different person from the collector. The success and future reliability assigned to quantitative and qualitative data from specimens, requires analysts and preparators to have confidence in the documentation and information provided by the collector. In this section field documentation, packing and sample transportation to the laboratory is discussed.

7.2. DOCUMENTATION

Self discipline and a logical approach by the collector are an essential first step in documenting samples collected in the field. The conventional practice is for collectors to assign numbers in the field, while museums and repositories designate numbers on receipt of samples or when laboratory preparation and analysis commence. This practice has been going on for many years in institutions (*e.g.* Camp and Hanna 1937, Rushton 1979), and doubtless will continue to do so.

Some national or government supported institutions may have a policy designating specimen numbers to an employed or designated collector undertaking field work. Any specimen or numbering system used by a collector must be matched with that used in field note books, on maps and field and logging slips or aerial photographs (Compton 1962). Furthermore, the numbering system must be individual to a locality, outcrop or section where more than one specimen is likely to be collected. Incorporating the date or year of collection in the number may also be useful in cross referencing later. As a general guideline record the following in a field notebook:

- the date of collection and field number of the specimen/sample with a brief description (see Appendix 2).
- indicate on any sketch the position of the sample, and its relationship to the horizon or feature from which it came plus any associated structures, fabrics or orientations. Indicate if collected *in-situ* or a float

(loose) sample. Record photographic frame numbers (and weather conditions at the locality at the time of collection).

- indicate the position on any sketch map, aerial photograph, or log sheet. If a national grid system covers the area, record it in full.

In short, record and document as much information as possible to allow for other unforeseen uses of the material (Brunton *et al.*, 1985).

7.2.1. Rocks and consolidated specimens

Most rock samples can be marked directly with a felt-tip pen. Some field workers may prefer to write on adhesive tape that is attached to the specimen. Take care with the initial label to ensure it does not obscure any feature or detail. Where directional data (*e.g.* dip and strike, way-up, younging) is recorded it must be indicated on the specimen. Paper labels can also be packed with specimens, although these are subject to rapid degradation if the sample is damp or they are left un-opened for any time.

7.2.2. Wet specimens

Indelible pens or wax pencils can be used on wet or damp specimens. Paper labels are best avoided, although strips of drafting film (water repellent tracing paper) are an effective substitute. If these cannot be taped to specimens, it may be possible to use wire or twine (very effective on Holocene scleractinian coral specimens).

7.2.3. Live specimens and unconsolidated materials

This type of material is usually collected directly into an open topped jar or bottle containing preservative that can be sealed for transportation, or a sample bag with a secure top.

Labelling and numbering should be such that it can be understood by others, avoiding unnecessary coding (Brunton *et al.*, 1985). It should also be evident on the specimen bag so it is clearly recognisable without unpacking. Before undertaking any fieldwork it is well worth spending time thinking about the process of collecting, the types of samples to be collected, the manner in which they are to be packed, and transported back to the laboratory. Ensure adequate supplies are taken into the field, as it may not always be possible to obtain suitable bags, containers, labels and other equipment required for specimen collecting locally.

7.3. SAMPLE PACKING

Individually wrapping specimens and parts of specimens is recommended to prevent damage (Brunton *et al.*, 1985). Rock and consolidated samples should be put in a suitable bag. A wide variety are currently available on the market, and the most common types are described below.

7.3.1. Cloth bags

Linen, cotton-poly fibre and canvas bags are commonly used to pack rock specimens, and are constructed in a variety of different gauge threads. Heavy cotton drill and canvas bags provide sturdy bags for heavy duty work. Canvas bags usually have tie tapes to secure the top, while cotton bags may have drawstrings. The bags usually have a printed water, insect and mildew resistant tag stitched into the side seam.

7.3.2. Plastic bags

Can now be woven like cloth bags, and in some cases may be as strong, although in bulk much lighter, consequently well worth considering if weight restrictions are a problem. Standard and heavy gauge polypropylene and polyethylene sample bags are more commonly used. These are usually available in a wider range of sizes, with self-seal tops and in some cases write-on panels.

7.3.3. Paper bags

Unconsolidated sand and geochemical soil sample bags are usually made of paper with water resistant adhesives. This allows for moisture to evaporate while the paper and adhesive remain dry. The sample is prevented from escaping by a double fold or wire tie closure.

Double wrapping of specimens may be required for fossil specimens. Delicate fossil material may have to be packed in a strong rigid box, padded with polystyrene chips or bubble wrap to absorb vibrations. Cotton wool is not recommended because of the difficulty in removing loose fibres and the potential to damage a delicate specimen (Brunton *et al.*, 1985).

7.4. REMOVAL AND TRANSPORTATION

Removing, transporting and shipping specimens back to the laboratory requires some thought before proceeding. Small numbers of light specimens can be packed and shipped in strong cardboard boxes. For large quantities or heavy specimens, wooden crates or metal boxes should be used, particularly if transportation is over any great geographical distance. Always include an address label in the box in case it has to be opened or re-addressed.

Before shipping, rock samples should be dried and if required re-bagged and wrapped in newspaper. Try to remove sharp edges which may pierce bags or do damage to other specimens. Samples which are stored in preservatives, and cannot be dried, may require special arrangements for shipping. Boxes should not be over filled with rocks, or under filled, as most damage to specimens is caused by loose samples abrading against each other. Boxes have to be strong enough to withstand rough handling.

REFERENCES

BRUNTON, C. H. C, BESTERMAN, T. P and COOPER, J. A. 1985. Guidelines for the Curation of Geological Materials. *Miscellaneous Paper of the Geological Society (London).* No. 17.

CAMP, C. L and HANNA, G. D. 1937. *Methods in Paleontology.* University of California Press, (Berkeley, California). xxiii + 153p.

COMPTON, R. R. 1962. *Manual of field geology.* John Wiley and Sons, Inc. (New York). xi + 378p.

RUSHTON, A. W. A. 1979. The fossil collections of the Institute of Geological Sciences, pp 57-66; In: M. G. Bassett (ed), *Curation of palaeontological collections.* Special Papers in Palaeontology. **22**, 1-279.

PART III

LABORATORY TECHNIQUES

A. INTRODUCTION

8. Laboratory design and layout

9. Safety in the laboratory

10. Preservation, consolidation and repair of unstable specimens

11. Preparation of recent material for comparative studies

8. LABORATORY DESIGN AND LAYOUT

8.1. INTRODUCTION

A range of work can be requested from a Palaeobiology Laboratory. The nature of the work frequently results in the use of techniques involving multiple procedures and the repetition of chemical and mechanical stages. Few of the techniques and procedures can be completed within a single working day, and many require many days, weeks and even months from initiation to completion. The success and efficiency of techniques employed in sample and specimen preparation is intricately connected to the laboratory management and its initial design.

Irrespective of the subject area and work undertaken, an efficient safe laboratory can only be successful when certain factors have been satisfied. Everett (1992) recognised eight factors requiring consideration for the operational success of the laboratory:

- (i) a suitable laboratory management system;
- (ii) clearly defined and stated policies and targets;
- (iii) an internal communication system integrating all staff into a coherent unit;
- (iv) effective information and training systems;
- (v) selection and appointment of suitably qualified staff appropriate for the work of the laboratory;
- (vi) installation of equipment suitable and relevant for the work of the laboratory;
- (vii) well designed premises, conducive to efficient and effective working;
- (viii) an efficient system of maintenance and support services.

Clearly however, any institution or company establishing, modifying or running a laboratory complex has additionally the budget requirement to consider. In turn, this may be allied to staffing levels, services offered and medium or long term laboratory development and expansion (Johnson 1983). Until recently, many of these considerations may have been thought of as problems for "commercial" and industrial laboratories. However, the volatility within today's market economies of countries situated in both the east and western worlds has resulted in a significant emphasis placed on the financial justification in developing and running a laboratory.

8.2. LABORATORY DESIGN

The luxury of designing a laboratory for construction in new premises is experienced by few. More commonly establishments are modifying existing laboratories or adapting rooms within existing buildings. However, in both cases it is vital that the specification for the laboratory includes requirements for the immediate need and possible developments within the establishment or company. If a laboratory is at the forefront of science, it is essential that a certain degree of flexibility is incorporated. Short term personal fads will not encourage investment in the laboratory required for future development

To avoid this scenario, a project officer must be appointed. This person should preferably be someone who is familiar with the new laboratory's function, aims, and staff, and may even be required to work within it. Furthermore, the project officer must have the full support of his or her immediate manager. Preferably the appointment should be from project inception to completion, liaison initially with the design team, architect and contractors.

A considerable number of equipment suppliers and specialist companies provide help in laboratory design and planning (see APPENDIX 1, CHEMICAL AND EQUIPMENT SUPPLIERS). However, such expertise is usually in general laboratory design, making it essential that the project officer emphasises the specific function and objectives of a specialized or commercial laboratory. Questions have to be carefully thought out, clearly asked and understood and thoroughly answered. Determine the main work of the laboratory, what materials and equipment will be used, and what risks are associated with the procedures. Can related work and equipment be confined to one area of the laboratory? Plan a "route" for material/specimens and procedures through the laboratory as they are being worked on. This is a pre-requisite for large laboratories, used by more than one person, or where more than one procedure may occur at the same time. It is essential to plan and design an effective working laboratory covering all aspects of the procedures to be undertaken, even if compromises have to be made because of limitations of the size of the premises (Johnson 1983). Route planning for procedures and equipment is essential where a "conveyor-belt" production line is required. Thought must also be given to the optimum positioning of the operator, ensuring maximum time-management during multi-stage procedures (Hodgkinson and Margerum 1986). Mistakes are expensive, frequently hard to rectify, and invariably result in complaints from the personnel who work within the laboratory. For example, wet sieving of dark residues, particularly palynomorph rich residues, is best undertaken over a light coloured sink, as this allows the operator to monitor how clean the residue is (L. Ross *pers. com*).

8.3. SAFETY CONSIDERATIONS

Considerable thought must be given to the safety requirements of the laboratory. In most cases these will have to comply with Local Authority, National, Federal and International laws and agreements, besides standard company or institution policies. Within the United Kingdom employers must comply with the Health and Safety at Work Act (1974) in providing employees with clearly written and unambiguous statements of their current health and safety policy, and intended implementation by both management and employees (Everett 1992). This has been amplified by subsequent European Economic Community Directives (*e.g.* 76/90/EEC illustrating hazard warning symbols) and legislation (*e.g.* UK Packaging of Dangerous Substances Regulations, 1978; Control of Substances Hazardous to Health Regulations 1988, and 1994 (COSHH); Management of Health and Safety at Work Regulations 1992; Chemical (Hazardous Information and Packaging) Regulations 1993 and 1994 (CHIP 1 and 2)).

A bewildering array of legislative regulations, codes of practice and directives now exists (Moore 1993). Although these do not have to be personally read by every laboratory worker, it is essential to know of legislative changes, and how they affect procedures and working practices within the laboratory. It is in every laboratory worker's interest to receive safety information from within their establishment, or if possible by attending conferences where experts, familiar with changes in regulations, can provide synopses on how working practices should change (Green 1995a). For more detailed aspects of safety refer to section 9 SAFETY IN THE LABORATORY.

8.4. SERVICES TO THE PREMISES

Services (natural gas, water and electricity) are standard essentials required by any multipurpose laboratory. The positioning of service points can often pre-determine where equipment, benching and furniture must be situated. In turn this may apply limitations to the planning and layout of the laboratory. Direct line water supplies are required for many distilled and deionising water stills. It is important to ensure that the water pressure is sufficient to effectively run this type of apparatus. Many small rock saws and grinding wheels have integral water re-circulatory tanks. Consequently positioning within the laboratory is easier, but it is important to remember that these machines can generate large amounts of "dirty" fine spray. Sighting of such equipment should accordingly be well away from delicate equipment such as microscopes and computers.

Additional *specialized* services may be required for certain procedures within a palaeobiology laboratory. A compressed air line is essential if pneumatic and air abrasive equipment is present (see section 12 MECHANICAL METHODS OF PREPARING FOSSIL SAMPLES). Although this type of work can generate large amounts of dust, this should be controlled and limited by working in a dust box connected by a flexible hose system to a dust extractor. However, it is worth noting the positioning of smoke detectors, some types of which can be activated by excessive dust in the atmosphere. Air compressors supplying these machines are excessively noisy, and continually on, although usually only intermittently running, while the hand tools are in use. In most laboratories, attempts are made to situate compressors away from the immediate work bench. However, there in no denying the machines are "anti-social" for other laboratory workers. Continuous noise levels above 85 decibels can induce deafness (Everett 1992), and should be monitored and measured by recognised professionally competent persons. In the United Kingdom refer to The Noise at Work Regulations (1989), supported by the 1989 Health and Safety Executive booklet, *Noise at Work: Guidance on the Regulations*, published by HMSO, London.

Compressed and liquefied gas bottles may also be required for particular equipment in certain areas of the laboratory (*e.g.* argon gas for sputter coating specimens in an electron microscopy laboratory). Many of these are situated close to the associated equipment, clamped in place to the bench or wall. It is essential to ensure that sufficient room is allowed for the safe operation of the machine and access to replace the cylinder. The external sighting of tanks and bottles is preferable in terms of refilling or replacing, although this may be constrained by safety considerations in installing flexible delivery pipes, and budget.

8.5. INTERNAL CONSIDERATIONS AND LAYOUT

Some problems related to the internal planning and layout of the laboratory has already been alluded to (*e.g.* machines generating noise or requiring compressed gas supplies). Planning and positioning of equipment is not always as easy as it would appear, particularly if instruments are purchased as "add-ons" to existing equipment. The efficient use of equipment and space is only achievable when the laboratory is completed, successfully processing samples and generating data. Minor problems can be rectified with few modifications to satisfy individual operator preferences (Johnson 1983).

8.6. MISCELLANEOUS CONSIDERATIONS

Probably the most important consideration here, for non-basement laboratories, is the floor loading. Excess heat generated from equipment which is continually running (*e.g.* refrigerators, freezers, compressors, furnaces *etc.*) can be an annoying problem, particularly if it affects other heat sensitive equipment or chemicals. Carefully consider the site and situation of specialist equipment within the laboratory, and plan well in advance the range of procedures to be undertaken. Removal of heat, via ventilation systems and toxic fumes through fume cupboards can result in small eddies and dominant air-flow routes through the laboratory, essential for the efficient working of this type of equipment. However, within a specialized palaeobiology laboratory (*e.g.* palynology), this can result in sample contamination. Palynological preparations require special techniques and procedures, using considerable amounts of highly toxic chemicals. Storage and disposal must be given special consideration (Costa 1983).

Within the standard palaeobiology laboratory, adequate rock and specimen storage is a necessity, that has to be provided for. As with standard laboratory furniture, specialized space saving storage systems can be installed (Upton 1987), and are particularly effective in large rock stores (Bishop and Stanley 1987, Howe 1987, Powell 1987). Essentially two types of storage areas are required:

- Short-term storage - an area to hold unprocessed recently collected samples and core material;
- Long-term storage - a system to accommodate the permanent storage of processed residues, hand specimens and core material.

Live material may be required temporarily, and appropriate storage facilities for specimens and preservatives should be considered (Lincoln and Sheals 1979). Care should be taken in the disposal of these materials, which may have been stored in a biological fixative (*e.g.* Bouin's solution, Flemming's solution), governed by COSHH (1994), which implements the 1990 and 1993 Biological Agents Directive, expanding previous provisions regarding the control of harmful micro-organisms.

An optional area, usually a separate clean room or office, containing light microscopes and slide cabinets may be required. Type material and figured specimens and collections should ultimately be lodged within a recognised museum or repository.

REFERENCES

BISHOP, M. J and STANLEY, M. F. 1987, (for 1986). Bruynzeel systems in use: 2, Buxton Museum. *The Geological Curator*. **4**, (9), 548-550.
COSTA, L. I. 1983. Safety in palynology laboratories. *Norwegian Petroleum Directorate-Bulletin 2 (Oljedirektoratet)*. 75-84.
EVERETT, K. 1992. Safety planning and laboratory design, pp 18-29; In: S. G. Luxon (ed), *Hazards in the Chemical Laboratory* (5th edition). Royal Society of Chemistry (London). xix + 675p.
GREEN, O. R. 1995a. Risk assessment in Science and Technology, conference report. *Science Technology*. **35**, 13-15.
HODGKINSON, R. L and MARGERUM, R. 1986. On the preparation of oriented sections of larger foraminifera occurring in hard limestones. *Micropaleontology*. **32**, (2), 158-162.
HOWE, S. R. 1987, (for 1986). Bruynzeel systems in use: 1, National Museum of Wales. *The Geological Curator*. **4**, (9), 545-547.
JOHNSON, B. 1983. Gearhart Geodata's stratigraphic laboratories, Aberdeen - the design concept. *Norwegian Petroleum Directorate-Bulletin 2 (Oljedirektoratet)*. 1-11.
LINCOLN, R. J and SHEALS, J. G. 1979. *Invertebrate animals. Collection and preservation*. British Museum (Natural History), (London and Cambridge University Press). vii + 150p.
MOORE, D. C. (Ed). 1993. *Laboratory handbook*. Institute of Science Technology (Lichfield). 68p.
POWELL, H. P. 1987, (for 1986). Bruynzeel systems in use: 3, Oxford University Museum. *The Geological Curator*. **4**, (9), 551.
UPTON, J. 1987, (for 1986). The Bruynzeel approach to geological storage. *The Geological Curator*. **4**, (9), 541-544.

9. SAFETY PROCEDURES IN THE LABORATORY

9.1. INTRODUCTION

In many countries compulsory legislation compels employers and employees to enforce health and safety guidelines. Chemical suppliers now include the relevant health and safety data sheets when products are purchased. The information provided in these data sheets identifies the product and manufacturer/supplier, and categorises the chemical composition. The hazards are summarised together with handling, transportation, spillage disposal and first aid measures. This information must be brought to the attention of users of chemicals, and readily available for consultation. Within the United Kingdom all chemicals are transported and packaged under the **Carriage of Dangerous Goods by Road, Rail - Classification, Packaging and Labelling** (CDG-CPL) legislation, effective from July 1995. This legislation incorporates the use of a world-wide packaging standard (UN approved) of hazardous chemicals for transportation, providing protection of both the general public and carriers from accidental exposure. Although approved packaging must meet the requirement of remaining leak proof from 1.8m drop tests, unpacking of chemicals within the laboratory should always be undertaken with care. Good laboratory practice provides an element of self discipline within the laboratory environment. Furthermore, it is a valuable aid to the successful running of a laboratory, and essential within a multi-user research environment generating quantitative and qualitative data.

All chemical procedures must be undertaken in a well-ventilated area. Commercially available extraction systems, fume hoods and cupboards should be installed and regularly checked by qualified personnel. Many techniques and procedures require continuous monitoring, particularly where rapid chemical reactions have to be terminated before damage to the specimen occurs, or toxic fumes are generated. Where particular techniques generate acidic or toxic fumes, ensure all work is undertaken using a fume hood. Commercially available extract systems have superseded "in-house" systems (*e.g.* Sass 1963). Commercial laboratories, or those dedicated to processing large volumes of material, should give special consideration to the extract system during the design stage of the laboratory (*e.g.* Costa 1983, Everett 1992, Johnson 1983).

Within a scientific laboratory the most common hazards can be attributed to one of three areas: (i) chemicals, (ii) equipment, (iii) the nature of the material in preparation. Well defined laboratory ground rules used by laboratory personnel reduce the risk of an accident occurring:

- All chemicals, solutions and powders must be treated with respect.
- Always work in a well ventilated and illuminated area. Use a fume cupboard or extraction system when working with chemicals and powders.
- Ensure all containers are well labelled, and have the correct hazard information label.
- Stock bottles, containing modified solutions, should have new concentrations clearly indicated, and the appropriate hazard information label.
- Inform fellow workers of the chemicals and procedures in use, and the relevant location of apparatus and equipment.

- On completion of work, ensure all chemicals are returned to stock cupboards, and any equipment used is thoroughly cleaned and returned to store. Inform the appropriate laboratory personnel about malfunctioning or damaged equipment.
- Safely dispose of all unwanted and out-of-date chemicals. Broken glass and unwanted sharps should be disposed of in the appropriate container.
- Ensure all mechanical and electrical equipment if regularly cleaned, serviced and tested. Moving parts must be covered by guards.
- Avoid having food and drink around the laboratory.

Such rules supplement legislation governing procedures' employers and employees are required to observe. Radical changes in effecting safety legislation are continually coming into effect within the single European Union. British Standards (BS) are superseded by new European (EN) Standards. Many of these standards have been adopted as Approved Standards by the Health and Safety Executive under the **Control of Substances Hazardous to Health (COSHH) Regulations (1999)**. Accident prevention should be a fundamental aim of all laboratory personnel (Thomas 1989).

9.2. RISK ASSESSMENT AND HAZARDS

A core consideration in laboratory work is defining hazards and assessing risks while undertaking a procedure, and reducing any danger to the operator (i.e. the person performing the task or using equipment), and personnel in the vicinity of the work area to a minimum. Defining hazards and risks in the context of legislation is not a straight forward procedure, and even generates discussion amongst safety experts (Green 1995a). However, in general a **hazard** is defined as having the potential to cause harm, while the **risk** is the probability of the hazard occurring. The hazard itself cannot be eliminated, and the potential to cause harm is always present (*e.g.* acidic solutions will always burn, moving parts of equipment can trap body and clothes). However, by performing risk assessments, the potential for causing harm is considerably reduced.

Within the laboratory environment hazards can be assigned to one of two general areas:

- Chemicals: the direct corrosive effects of coming into contact with or inhalation of fumes from strong acidic or alkali solutions, toxic solvents and compressed gases. Ingestion and inhalation of fine and potentially toxic powder. Combustibility of solutions and powders.
- Equipment: physical contact with moving parts (cutting, grinding, drilling, crushing), and large heavy objects. Electrical connections, particularly where water cooling systems are in proximity. Radiation. Fire hazards.

9.2.1. Chemicals

(i) Handling chemicals

By following the same routine procedure the handling of chemicals (solutions and powders) in the laboratory can be made considerably safer. A suggested routine is as follows:

- Read the label and relevant safety data sheets before opening the container, and identify all products. If the label is missing treat the substance as dangerous.
- Know the storage requirements of the chemicals. Check the chemical is the correct grade, quality and strength required for the procedure.
- Consider the hazards of the chemicals used (see hazard warning symbols, appendix 5), including and violent reactions resulting from the mixing of chemicals. Use suitable Personal Protective Equipment (PPE). Work in an appropriate well-ventilated area within the laboratory. If inadequate, defective or unavailable PPE, or a safe working environment is not available, **DO NOT PROCEED**.
- Open containers, and mix chemicals carefully. Identify the hazards in using the chemicals, and employ methods reducing the risk of inhalation, ingestion and direct contact.
- Avoid direct heat and flame coming into contact with chemicals and powders.
- Avoid eating, drinking or smoking while handling and using chemicals.
- After use, ensure chemical containers are correctly sealed and returned to store. All contaminated glassware and protective clothing must be thoroughly cleaned before storing.

Hazards associated with dangerous laboratory chemicals (flammable, toxic and miscellaneous substances) are summarised below in table 9.1. Much debate and misunderstanding surround the toxicity and carcinogenic activity of many chemicals used by palaeontological preparators and conservators (*e.g.* Howie 1987, Perera and Petito 1982). Treat all chemicals as potentially hazardous, and avoid direct skin contact.

(ii) Storing chemicals

The safe storage of chemicals is essential (and a legal requirement) in ensuring that risks are kept to a minimum. Points to remember in the storage of chemicals are as follows:

- Store chemicals in well ventilated areas, away from temperature extremes and sources of ignition.
- Ensure chemicals are secured against unauthorised use.
- Segregate stocks to reduce hazards (*e.g.* do not store oxidizing chemicals with flammable liquids). Check the correct hazard label (see appendix 4) is visible on the container.
- Monitor chemical stocks, and rotate as required. Dispose of out of date reagents, and those with a short shelf life that may have deteriorated.
- Enforce a no smoking policy in the presence of flammable chemicals.
- Take care in handling hazardous chemicals.

(iii) Corrosive, toxic and flammable substances

Ensure that all acidic chemicals (except hydrofluoric acid) are stored in plastic coated glass bottles, thus providing the user with two advantages. Firstly, the glass is inert. Secondly, should the bottle become accidentally broken then the plastic coating reduces the danger of leakage and splashing. Chemicals capable of dissolving silica (*e.g.* hydrofluoric acid) can **only** be stored in plastic bottles. Other chemicals (*e. g.* oxidants and alkalis) should be stored in polythene bottles (coloured black to prevent light deterioration). Laboratory powders are supplied in glass or plastic screw top bottles. Chemicals must be stored following manufacturers' guidelines, and in containers recommended by the supplier.

In handling any chemical within the laboratory it is essential to identify all hazards (Table 9.1), and use the correct and suitable PPE. Laboratory coats and aprons (rubber and disposable) are essential items, as are gloves. However, some knowledge on glove resistance (see appendix 7 for details) to chemicals is essential, and a variety of gloves should always be available in a wet chemical laboratory. Respiratory and eye protection is discussed in detail below.

Following the Montreal Protocol and Copenhagen Agreement ozone depleting solvents (Table 9.1), many of which have been used in palaeobiological preparation and conservation procedures, are now being phased out. As a consequence, suppliers of these products may have difficulty in fulfilling customer demands as manufacturers switch production to replacement products. Preparators and conservators must proceed with care in using replacement products, as these may not possess the same properties as the original substance.

(iv) Dealing with chemical spillage's

Spilt chemicals must be regarded as hazardous when:

- they are flammable or potentially explosive;
- they promote burning of combustible materials;
- they are poisonous by ingestion, skin contact or have a harmful vapour or dust which may affect lungs or eyes;
- they are corrosive and liable to oxidize skin or damage clothing, equipment, furniture or flooring with which they come into contact.

Spillage kits must be available, and clearly identified, within the laboratory. They should contain suitable absorbent materials to prevent spills from spreading. A plastic bag should be used to hold all contaminated materials. Spillage kits should be used in conjunction with standard PPE such as laboratory coats, aprons, gloves, face shields and eye protection and respirators. Spillage's should be dealt with using precautions and methods appropriate to the hazards of the chemicals. A standard check list in dealing with a hazardous spill follows:

- After the spill has occurred or been detected, evacuate the area, move away and get help by notifying other workers.
- Seal off the area by means of barrier tape and warning signs.
- Look for any injuries on those working near the spillage. Treat appropriately.
- Identify the hazards resulting from the spill. This will enable the clean up to be effectively undertaken.
- Based on the hazards recognised, devise a plan of action for proceeding with the clean-up.
- Ensure appropriate personal protective clothing for dealing with the spill is available.

SUBSTANCE	TIMETABLE		
	Montreal Protocol	**EC Regulations 594/91 & 3952/92**	**Copenhagen agreement**
Chlorofluorocarbons	50% cut by 1.1.95 85% cut by 1.1.97 Phase out by 1.1.2000	85% cut by 1.1.94 Phase out by 1.1.95	75% cut by 1.1.94 Phase out by 1.1.96
Halons	50% cut by 1.1.95 Phase out by 1.1.2000	Phase out by 1.1.94	Phase out by 1.1.94
Carbon tetrachloride	85% cut by 1.1.95 Phase out by 1.1.2000	85% cut by 1.1.94 Phase out by 1.1.95	85% cut by 1.1.95 Phase out by 1.1.96
1,1,1-Tricloroethane	70% cut by 1.1.2000 Phase out by 1.1.2005	50% cut by 1.1.94 Phase out by 1.1.96	50% cut by 1.1.94 Phase out by 1.1.96
Hydrochloro-fluorocarbons			35% cut by 2004 65% cut by 2010 90% cut by 2015 99.5% cut by 2020 Phase out by 2030
Methyl bromide			Freeze at 1991 levels from 1.1.95

Table 9.1. The phasing out timetable of ozone depleting solvents.

- Use the correct absorbent/neutralizing materials.
- Contain the spill, and if possible reduce the contaminated area.
- Clean up the spill and place all contaminated material for disposal in a suitable container. Thoroughly wash the contaminated area with soap and water.
- Seal the container/bag containing contaminated materials, and clearly label with suitable hazard and tactile warning symbols. Dispose of according to **The Chemicals (Hazard Information and Packaging for Supply) (CHIP) Regulations 1994**.

Further details of dealing with spillage's and the disposal of unwanted chemicals are presented in appendix 9. Hydrofluoric acid presents unique problems in handling, dealing with spills and disposal, and is dealt with separately in appendix 8.

9.2.2. Equipment

(i) General consideration

All machinery within a laboratory should only be operated by qualified personnel, or those that have undergone training, with a qualified operator in attendance. **Never** use electrical or other laboratory equipment without informing the person in charge of the facility. Any problems encountered with using an item of equipment, should be reported. **Never** leave equipment in a dangerous state, put a warning notice on it stating the problem, and inform regular laboratory users so that repairs can be initiated. Statutory legislation now requires all electrical equipment to be inspected and tested annually. Equipment not conforming to the required standards should be disabled by removing the plug, and not be used until repaired.

(ii) Moving parts and dust generation

Within the palaeobiology laboratory the hazards of various items of equipment must be assessed to ensure a safe working environment for both the operator and other laboratory personnel. The hazards from the moving parts of cutting and grinding (rock saws and lapping wheels) equipment is readily evident, particularly where specimen preparation requires objects to be hand held against wheels. What may not be readily appreciated are hazards to eyes generated by small particles, and the respiratory system by a fine mist of dust particles and coolant, briefly held in suspension around the working area. Care must also be employed when working with hazardous materials, particularly when working with specimens embedded or impregnated with epoxy resins.

Dust may also be generated around air-abrasive units (see section 12 MECHANICAL METHODS OF PREPARING FOSSIL SPECIMENS), from hand-held percussion and powder abrasive tools. This type of equipment should be operated in a dust box, or an area where dust extraction ensures an air-flow away from the operator. Crushing and grinding machines generate similar hazards, although a greater danger exists from contact with the moving parts of such machines (Crelling and Dutcher 1980). When in use, ensure the manufacturer's recommended protective safety guards are in place over fly-wheels and movable jaws.

The hazards associated with ultrasonic tanks are detailed in section 12 MECHANICAL METHODS OF PREPARING FOSSIL SPECIMENS. It is important to ensure that the machine is switched off before the removal of specimens from the bath. **Never allow fingers or hands to come into contact with the water while the machine is operational, as permanent soft tissue damage may result.** It should also be remembered that ultrasonic tanks generate heat. Water within the bath should be regularly changed (for cleaning and cooling purposes), and during operation situated in an adequately ventilated position (Howie 1987).

(iii) Radiation

Exposure to ionising radiation is increased by the sheer volume of x-ray equipment currently in use in palaeobiological laboratories. The risk of exposure is substantially increased if the equipment if not regularly maintained, and personal monitoring is not undertaken. The primary risk is not from high dose radiation accidents, which are extremely rare, but from constant radiation exposure. Symptoms can range from skin irritation and nausea to cancer, septic shock and death (Thomas 1989). Individuals working with x-ray generating equipment should be issued with a thurmoluminescent dosimeter to monitor local exposure.

The **absorbed dose** is the term given to irradiated material. The unit measuring the absorbed dose is the gray (Gy), (where 100 rad = 1 Gy). However, the biological effect of a given dose depends on factors such as total exposure time, irradiation rate and type of radiation. Thus for protection purposes the term **dose equivalent** is used. The unit measuring the dose equivalent is the sievert (Sv), (where

100 rem = 1 Sv). Any absorbed dose (Gy) is numerically equal to the corresponding dose equivalent (Sv), in which dose limits are expressed (Luxon 1992).

9.3. SAFETY PRECAUTIONS

Once a hazard has been identified, an assessment of the risk to the operator and others within the working vicinity will identify what personal protective safety equipment is required. The use of safety equipment, coupled with correct working procedures reduces the potential risk to the operator.

9.3.1. Safety Equipment

Personal protective safety equipment must be made available to all operators and laboratory users. A general supply of such equipment should be readily accessible in clearly labelled drawers and cupboards within the laboratory. Specific equipment (*e.g.* ultra violet lamps) may require specialized safety equipment when in use (*e.g.* UV protective glasses) which should be stored nearby. In a general purpose preparation laboratory personal protective equipment for the following areas must be available:

- **Exposed skin and clothes**

  Laboratory coats are essential in general laboratory use. Plastic sleeve protectors and rubber aprons offer additional protection when solutions may come into contact with skin or clothes. The chemical resistance of gloves must be understood to ensure the correct type is used (see chart in appendix 7, LABORATORY GLOVE RESISTANCE). This type of equipment must be regularly inspected, particularly after coming into contact with strong acids or alkalis, and replaced if found defective or sub-standard.

- **Eye protection**

  Eye protection is essential in handling chemicals and using equipment (bright or dangerous light sources, lasers and VDU screens). A wide variety of protective eye wear is available, particularly concerning impact resistance. These are usually available as either goggles or spectacles. When using a UV source (short wave 254-312 and long wave 365nm), the operator must wear appropriate eye protection (glasses or face shields). The dangers produced by short waves (254nm = UV-C), which readily propagate bacteria, and medium waves (312nm = UV-B), which can cause erythema, are medically well established, while long waves (365nm = UV-A), can be dangerous to sensitive people or those under medical treatment. Both types of radiation are dangerous to the eyes and skin.

  New European Community directives and Health and Safety (Display Screen Equipment) Regulations 1992, have now addressed problems arising from the environmental setting of visual display screens and workstations. Although this is generally not high risk to users, prolonged use can lead too muscular and other physical problems, eye fatigue and mental stress. The regulations cover equipment used and the working environment. Essential to the user is the lighting of the working environment (artificial and natural), and the reflections and glare on the screen are reduced. VDU spectacles and anti-glare, anti-static and anti-radiation screen filters should be used to improve screen clarity.

Eye protectors must be used to absorb and filter harmful laser light of particular wavelengths. Filters reduce or eliminate the harmful wavelengths, while light of other wavelengths is not obstructed. It is essential for the operator to ensure that the laser type and medium are known, and that the correct eye protection is used.

- **Noise**

In the United Kingdom, the **Noise at Work Regulations** was implemented in 1990. Legal obligations require employers to protect their workforce from exposure to damaging noise levels by providing suitable ear protection. Three action levels have been recognised:

(i) **85dB(A) and above** (decibels measured on an "A" weighted network sound meter) - suitable protective equipment must be provided.

(ii) **90dB(A) and above** - ear protection compulsory. The upper level at which ear protectors can be considered to provide temporary protection. If noise levels of 90dB(A) and above are sustained over an 8 hour day progressive hearing loss, which may not become apparent until sometime after exposure, may result (Brunton *et al.*, 1985).

(iii) **140dB(A) and above** - hearing loss induced from these noise levels may be instantaneous, and stringent control methods must be used.

A large amount of mechanical equipment used in the preparation of palaeobiological material generates noise and vibration. Hazards from noise fall into two categories (Howie 1987): (i) sustained noise, (ii) impulse noise. The most hazardous is impulse noise, where a rapid percussion noise can result in exposure to levels of up to150dB(A) or more, and lead to rapid hearing loss (Table 9.2).

Equipment operators in the vicinity of excessively noisy machinery, or equipment generating a persistent pitch should wear ear defenders. A large variety of ear defenders and earplugs are available. To ensure the quality of these products, check they conform to a national standard (in European Community Countries, equipment is manufactured and tested to EN352 Parts 1 and 2: 1993). Good quality ear defenders are suitable for noise reduction in the 10-30dB(A) range, while ear plugs provide a lower level of protection within the range 5-10dB(A) (Howie 1987). For some machines it may be possible to enclose them in acoustic hoods (Howie 1987), although personal ear defenders provide greater versatility for the equipment operator. Work areas containing noisy machinery should be clearly identifiable to other workers with warning signs and posters.

dB(A)	Noise Intensity	Activity	Effects	
			Short tern exposure	**Long term exposure**
30	Very quiet	Traffic through double glazing	None	None
60	Quiet	Normal conversation	None	None
75		Balanced extract fan	None	None
80-85	Loud	Electric drill or airbrassive	None known at present	Sustained exposure at 85dB(A) may cause gradual hearing loss over many years
90	**Present legal limit**	Compressed air & woodworking equipment	Temporary deafness for a few seconds to hours (may cause tinnitus) especially at levels of 95-105dB(A)	Sustained exposure at 90-105dB(A) may cause permanent hearing loss over many years. Social noise may also enhance the risk of hearing loss. 8 hours of 90dB(A) is equivalent to 15 mins of 105dB(A) noise
90-100	Very loud	Band saw, vibro tools		
95-105	Conversation difficult	Grinding casts		
100-110	Uncomfortably loud	Rock saw	Prolonged temporary deafness	Rapid loss of hearing likely
135	Painfully loud	Hammering hard rock	Repeated exposure causes severe damage. Pain causes withdrawl	
150-160			Eardrum ruptures plus instant deafness caused by destruction of sensory mechanism of middle/inner ear	

Table 9.2. Expected laboratory and workshop noise levels (Data from Howie 1987).

Pumps and continuously run equipment should have noise dampening inserts at points of contact with benches or ridged surfaces.

- **Vibration**

  Hand held tools can generate considerable vibration in association with noise. Prolonged use can result in the formation of cysts in the bones of the wrists, although these are relatively harmless (Howie 1987). More serious is vibration induced white finger (VWF), which can result in partial disablement (Howie 1987). Both hazards can be reduced considerably by ensuring the hand piece and the object worked is sufficiently dampened.

At present there are no official standards for ultrasonic equipment, but high-frequency sound can be reduced by encasing equipment in a shield of 6mm perspex (Brunton *et al.*, 1985). The duration of daily working with vibratory hand tools should be restricted to a maximum of 4 hours (Howie 1987).

- **Respirators**

  Five respiratory hazards have been recognised:

  (i) Dusts

  Particulate materials resulting from solid break down. The smallest particles remain airborne longer, and are thus easier to inhale.

  (ii) Mists

  Minute liquid droplets (<1μm), usually colloidal suspensions, form a liquid in a gas by atomization and condensation. Mists are often combinations of several hazardous chemicals.

  (iii) Metal fumes

  The vaporization of metals at extremely high temperatures' results in fumes laden with metallic particulate material. Rapid cooling results in condensation of the vapour containing the particles.

  (iv) Gases

  A state of matter in which molecules move freely, remaining airborne at room temperature. Gases very quickly occupy the total volume that contains them.

  (v) Vapours

  Substances maintained in a gaseous state by storing at a temperature below its critical temperature. At room temperature they revert to a liquid or solid state.

  Harmonised European Standards have been introduced in relation to respiratory protection (EN 149, filtering face piece dust respirators; EN140, conventional re-usable half masks; EN141, for gas and combination gas particle filters; EN143, for particle filters; EN405, for valved filtering, half mask respirators for gases or gases and particles). Respirators with interchangeable filters are particularly useful. The versatility of the cartridge system coupled with the colour coding of filter cartridges (see appendix 5 RESPIRATOR AND FILTER

CARTRIDGE COLOUR CODING), provides a laboratory scope for storing a wide selection to cover all types of work.

9.3.2. Occupational hygiene

A laboratory hygiene policy is essential for the occupational health and safety of laboratory personnel. The handling of some materials can put laboratory personnel at some risk if basic procedures are ignored.

- **Controlled disposal of unwanted products**

  A wide range of containers is available for the safe disposal of sharps, aerosols, spilled liquid waste and broken glass. Clearly marked provide for the safe disposal of products that **must not** be placed in the general waste disposal system. Many of these waste products, although not aerosols, require controlled disposal by incineration, and should only be dealt with by specialized waste disposal firms.

  Biohazard material, which may be pathogenic, must be similarly dealt with, through consultation with local authorities and environmental waste agencies. The specific disposal of this type of material is beyond the scope of this manual.

- **Toxic minerals**

  Brunton *et al.*, (1985) listed the toxic properties of a number of minerals, expanded on by Howie (1987) listing 94 minerals with toxic properties (ranging from moderate too high), although the list is not considered exhaustive. Generally, however, it is only through prolonged exposure to uraninite and other uranium minerals, that hazardous conditions might result. A greater danger may exist from the chemical effect of uranium on the kidneys. Guidelines for personal hygiene in handling radioactive minerals have been outlined by Howie (1986), and are summarised as follows:

  (i) Clothing

  Use disposable protective clothing and filter respirators.

  (ii) Washing

  Wash hands after handling radioactive minerals, particularly if specimens are dusty or friable. Do not handle material when hands are cut or skin is broken.

  (iii) Food hygiene

  Do not eat, drink or smoke in areas where radioactive minerals are stored or handled.

  Particular care must also be taken with asbestos minerals and minerals containing arsenic and thallium compounds.

- **Handling and lifting of heavy loads**

  The **Manual Handling Regulations (1992)** have been designed to reduce the annual number of accidents and working days lost attributed to lifting and handling operations. Prior to their introduction, in 1993, over a quarter of all occupational accidents in the UK were through injuries caused by manual handling operations. As such, the Regulations **do not** ban manual handling, but require employers to take all reasonably practical precautions. A four stage handling assessment provides an evaluation of:-

  (i) **the task**
  how is the load to be manipulated?
  the distance the load is moved
  the number of similar tasks to be carried out
  how many people are involved?
  the posture of the individual, is stooping involved?

  (ii) **the load**
  how heavy?
  how bulky?
  how unstable?
  how sharp or difficult to grasp?

  (iii) **the environment**
  the amount of space in which the operation is to be performed
  the type of work surface or flooring
  the lighting and ventilation

  (iv) **individual capability**
  strength of the person
  man or woman and age
  any existing health problems

Information gained from the assessment can then decide how best to reduce the risk of injury by redesigning the task (sharing the load, reducing the weight or changing the shape of individual items, improving the layout/floor of the environment) or providing mechanical assistance or adoption of strain-free lifting techniques (using the body more effectively).

9.3.3. Safety signs

Laboratory safety signs can be divided into one of four categories:

Prohibitory: Signs with a white background and black symbols. Text is white on a red background, with the prohibition indicated by a red circular band and cross bar over the symbol.

Mandatory: White background with white symbols in a blue background circle, and white text in a blue box.

Warning: White background, with a black triangle containing a black symbol on a yellow background, and black text in a yellow box.

Safe condition: Green background with white symbols and text.

Each of these is a combination of colour and design within which standard symbols are illustrated. In the United Kingdom these signs must comply with British Standard 5378, and BS 5499, although the process of harmonisation with Europe will see changes in the legislation covering safety signs. Two areas where these changes will be readily seen are with tactile warning signs and luminous safety signs. The **Chemicals (Hazard Information and Packaging for Supply) Regulations 1994** (CHIP 2) require the use of Tactile Danger Warnings on substances sold to the public containing toxic, very toxic, harmful, corrosive, extremely flammable of highly flammable products. The raised triangular signs are designed to alert the blind or visually impaired that they are handling a dangerous product.

Photoluminescence signs are designed to help indicate directional escape routes and safety equipment in low visibility situations, particularly when all external light sources have failed.

9.3.4. First aid

Minor accidents can be treated at the workplace by a competent qualified first aid college. However, if in any doubt casualties should be transferred to hospital or consult their doctor. Chemical casualties should be treated as follows:

- **Skin contact**
  - (1) Remove chemically contaminated clothing, unless it is stuck to the skin.
  - (2) Drench the affected area with running water until the chemical is removed (minimum of 10 minutes).
  - (3) Check for burns and general symptoms of poisoning. Arrange transfer to hospital.
- **Eye contact**
  - (1) Flush the chemical out of the eye with running water. Arrange transfer to hospital.
- **Ingestion**

  Do not induce vomiting

  If unconscious
  - (1) Administer nothing by mouth.
  - (2) Check for breathing and pulse (heartbeat). Give artificial ventilation and external chest compression if necessary. If breathing and pulse are normal, place in the recovery position.

  If conscious
  - (1) Give 1 pint of water to drink immediately.
  - (2) Give 1 pint of milk If chemical is corrosive, **unless** it contains **phosphorus**, **chlorinated hydrocarbons**, or **degassing solvents**. In these cases give repeated drinks of water.
  - (3) Transfer the casualty to hospital.
- **Inhalation**

Safety Note: Safeguard yourself by wearing appropriate protective clothing.

  - (1) Remove the casualty from the area of exposure

  If unconscious

(1) Administer nothing by mouth.
(2) Check for breathing and pulse (heartbeat). Give artificial ventilation and external chest compression if necessary. If breathing and pulse are normal, place in the recovery position.

If conscious

(1) Make the casualty lie or sit down quietly. If breathing becomes a problem (rapid or bubbly), place casualty in the sitting position and give oxygen if available.
(2) Transfer the casualty to hospital.

9.3.5. Security

Personal responsibility for laboratory security is essential for equipment operators and other laboratory users well being. Any laboratory that undertakes mechanical and chemical preparation procedures operates with an increased number of hazards. Access to any working environment that might inadvertently expose people to hazardous procedures, should be limited to those requiring use of the facilities. Most specimens passing through the laboratory during their processing and preparation are the property of others. Responsibility for their security during preparation resides with the preparator. Type and figured material requires particular care, as on completion of work this material must be clearly labelled and securely stored in a recognised repository.

Care must also be taken to ensure toxic and corrosive chemicals are securely stored. Laboratory personnel, particularly temporary or visiting workers, should not feel restricted in undertaking their work, but feel secure in the knowledge that they will not inadvertently encounter dangerous chemicals or equipment. Security within the laboratory is intricately linked to safety, with many factors controlled by legislation.

REFERENCES

BRUNTON, C. H. C, BESTERMAN, T. P and COOPER, J. A. 1985. Guidelines for the Curation of Geological Materials. *Miscellaneous Paper of the Geological Society*. No. 17.

COSTA, L. I. 1983. Safety in palynology laboratories. *Norwegian Petroleum Directorate-Bulletin 2 (Oljedirektoratet)*. 75-84.

CRELLING, J. C and DUTCHER, R. R. 1980. *Principles and applications of coal petrology*. Society of Economic Paleontologists and Mineralogists - Geological Society of America meeting, Indiana. Short Course No. **8** (Tulsa, Oklahoma). 1-127.

EVERETT, K. 1992. Safety planning and laboratory design, pp 18-29; In: S. G. Luxon (ed), *Hazards in the Chemical Laboratory* (5th edition). Royal Society of Chemistry (London). xix + 675p.

GREEN, O. R. 1995a. Risk assessment in Science and Technology, conference report. *Science Technology*. **35**, 13-15.

HOWIE, F. M. 1987. Safety considerations for the geological conservator, pp 379-401; In: P. R. Crowther and C. J. Collins (eds), The Conservation of Geological Material. *The Geological Curator*. **4**, (7), 375-474.

JOHNSON, B. 1983. Gearhart Geodata's stratigraphic laboratories, Aberdeen - the design concept. *Norwegian Petroleum Directorate-Bulletin 2 (Oljedirektoratet)*. 1-11.

LUXON, S. G. 1992. *Hazards in the Chemical Laboratory* (5th Edition). Royal Society of Chemistry (London). xix + 675p.

MOORE, D. C. (ed). 1993. *Laboratory handbook*. Institute of Science Technology (Lichfield). 68p.

PERERA, F and PETITO, C. 1982. Formaldehyde: a question of cancer policy? *Science*. **216**, 1285-1291.

SASS, D. B. 1963. Controlled acid preparation of fossils. *Journal of Paleontology*. **37**, (5), 1129.

THOMAS, K. M. 1989. Health hazards in the Paleontology laboratory, pp 30-36; In: R. M. Feldman, R. E. Chapman and J. T. Hannibal (eds), *Paleotechniques*. The Paleontological Society Special Publication. **4**. iv + 358.

10. PRESERVATION, CONSOLIDATION AND REPAIR OF UNSTABLE SPECIMENS

10.1. INTRODUCTION

Stabilizing fragile geological material involves an understanding of chemical reactions between the specimen and the chemicals and compounds used in preservation and conservation. Furthermore, a specimen's stability is effected by the environment in which it is stored. Through a number of key publications (*e.g.* Brunton *et al.*, (eds) 1985, Crowther and Collins (eds) 1987, Collins (ed) 1995, Howie (ed) 1992) conservators and preparators understand the importance of these aspects. Continued monitoring of specimen storage and their local environment adds to the data base, influencing decisions on what materials to use (*e.g.* Ashley Smith 1987, Doyle 1983, 1987, Howie 1979a & b, 1984), and reinforces the importance of documenting conservation procedures.

A comprehensive review of adhesives and consolidants suitable for conservation is by Horie (1987), in a reference that also covers polymer science (the long chain molecules forming most adhesives), genetically identifying each polymer group. Elder *et al.*, (1997) present an extremely useful guide to the range of use adhesives and consolidants are applicable to geological and palaeontological conservation. Relevant data, such as the chemical family and its composition, trade and manufacturer's name, thermoplastic operating temperatures, reversibility and solvents, plus notes on their principle use in geological conservation are presented. This extremely useful document is adaptable as a laboratory wall chart.

The following review approaches the subject from a practical perspective. Temporary and permanent repairs must be completed before undertaking long term specimen preservation. Where possible, repair and impregnation (surface or thorough) techniques should be performed within a controlled laboratory environment rather than the difficult conditions encountered in the field. However, many of the techniques outlined in this section can, and have to be executed at the site of the find. For example, large vertebrate bones usually require stabilizing before removal from the field back to the laboratory. Details of specific field techniques are presented in the part II FIELD TECHNIQUES, section 5, CONSOLIDATION, STABILIZATION AND REPLICATION TECHNIQUES.

Stability of the local environment within which the specimen is stored is a fundamental part of conservation. It provides a cornerstone of a strategic policy for the long term preservation of specimens. Factors affecting specimen stability are examined, and procedures to arrest decay are described with particular reference to fragile and unstable specimens (*e.g.* pyritized specimen, sub-fossil bone, amber).

10.2. THE REPAIR OF SPECIMENS

Deciding which adhesive or resin is suitable for a specimen is not a simple task (Wolberg 1989b). A bewildering range of adhesives, consolidants and resins is commercially available, although few are exclusive to palaeobiological preparation. However, a significant advantage is that many adhesives are used as consolidants, through adjustments to the viscosity during preparation. An understanding of the chemical interactions between adhesives, consolidants and the specimen is essential. Damage caused from using an inappropriate conservation material is second only to the environmental storage of the specimen (Keene 1987). Consequently, a basic knowledge of these interactions is advantageous, particularly during the salvage of specimens where the initial conservation materials used are now considered inappropriate (Howell 1977).

Keene (1987) recognised that suitable adhesives and consolidants should possess the following features:

- non-toxic to the user, work in well ventilated areas if harmful vapours are generated.
- easy to apply to the specimen, and with a consistency to stay in the area of application, setting quickly or slowly as required.
- harmless to and chemically stable with the object, not likely to expand, contract or undergo colour alteration and deteriorate with age.
- strong enough to support the repair or consolidate the surface, but weak enough to fail if excess stress is applied to the specimen.
- chemically compatible with the object, some adhesives will not bond to certain materials because of their chemical nature.

As a general rule-of-thumb, the repair of material should be undertaken using non-permanent adhesives. Permanent epoxy resin adhesives (see below) or cyanoacrylates bonding agents ("superglues"), should only be applied to specimens where the re-examination of broken surface is not required. Epoxy resins are prone to yellow with age. The long-term stability of cyanoacrylates is poorly understood (Whybrow and Lindsay 1990), although many degrade under alkaline conditions and through prolonged exposure to ultraviolet light. Until more data is available on their physical degradation and reversibility regarding palaeontological use, they should only be employed in exceptional circumstances where a strong bond is necessary. Broken specimens should be set in a sand box before applying adhesive. Voids can be filled by mixing coloured powdered fillers or crushed matrix with the adhesive. Pigmented glassfibre reinforced plastics (GRP's) can be used to add strength. The most commonly used adhesives and consolidants in palaeontology are the acrylic polymers (*e.g. Paraloid* B72, B67) and poly(vinyl butyral) (*e.g. Butvar* B98, B76).

10.3. SURFACE IMPREGNATION OF DRY SPECIMENS

Particular care must be taken to ensure that all specimen surfaces are thoroughly dried and free from loose debris before applying an adhesive. Similarly specimens that require impregnating must be thoroughly dry before proceeding. Although thin impregnating solutions penetrate deeper than viscous solutions, on porous surfaces multiple applications are required.

The use of acrylic polymer and poly(vinyl butyral) solutions has replaced the use of *Alvar* 1570 mixed with toluene, or mixed with amyl acetate and methylated spirit, a poly(vinyl acetal) solution suggested by Rixon (1976), unfortunately no longer manufactured. Acrylic polymer and poly(vinyl butyral) solutions have also superseded poly(vinyl acetate) solutions, although these are commonly used as an adhesive. Two other surface consolidants used in palaeontology, gum tragacanth mixed in water and the natural resin shellac dissolved in ethyl alcohol, are now rarely used in conservation. Shellac imparts a shine and brown colour to the surface (depending on the grade used) besides being brittle.

An alternative procedure, applied in archaeology, fine art sculpture and commercial stone conservation, employs silanes as deep penetrating consolidants (Bradley 1987). The two common types used in stone conservation are: (i) alkoxy silanes - form water repellent polymers, although water vapour can penetrate; and (ii) silicone esters - form silica based polymers not water repellent. Application is by brush or pipette applying successive coats until it no longer penetrates the surface. However, excess pools must be removed as they dry and produce an unsightly glassy appearance. A comprehensive introduction to their chemistry, use and application in geology is provided by Bradley (1987).

Safety Note: Silanes are irritants, avoid skin and eye contact. Fume inhalation can affect respiration. Exclude others from the work area, or provide suitable PPE.

10.4. SURFACE IMPREGNATION OF WET SPECIMENS

Reduce the possibility of wet specimens from rapidly drying. In the field, wrap specimens in damp newspaper, and place in plastic bags to retain moisture. Thoroughly wash marine specimens saturated with salts before applying one of the solutions listed below (test before use):

- **Acrylic polymer resins and emulsions:** are soluble in acetone to a workable viscosity. Emulsions are useful adhesives and consolidants for bone and some damp fossil material.
- **Poly(vinyl butyral)** (PVB): soluble in alcohol as a consolidant, or soluble in acetone as an adhesive.
- **Poly(ethylene glycol)** (PEG) (see below).
- **Polyvinyl acetate** (PVA): emulsions can be thinned with either distilled water or with toluene plus 25% volume industrial detergent. The use of these has, to a large extent been superseded by the consolidants listed above. Palaeontological use is not recommended (Elder *et al.*, 1997).

Where a large number of specimens require impregnating, application of the emulsions can be made more efficient by use of an aerosol type dispenser. Frequently turn specimen to obtain an even coating of emulsion. **Undertake this operation in a well-ventilated area.**

10.5. THOROUGH IMPREGNATION USING EPOXY AND POLYESTER RESIN

The impregnation of permeable samples is extensively documented in Allman and Lawrence (1972), Bouma (1969), Carver (1971), and Hutchison (1974), with later modifications provided by Jim (1985) and Awadallah (1991). However, fine grained samples (with poor permeability and low porosity) frequently present problems. A two stage vacuum and pressure impregnation procedure has been successfully used on Holocene carbonate rich sediments collected as cores (Ginsburg *et al.* 1966, Crevello *et al.* 1981, Green 1986).

Either thermo-setting epoxy resins or room temperature curing polyester resins can be used in the procedure. The latter have an advantage in that the resin is usually pre-accelerated, although a more significant advantage is that a "flexible" resin can be mixed with an embedding resin when impregnating large or irregularly shaped specimens. The addition of flexible resin with its elastic properties assists in dispersing stresses induced by the exothermic reaction during resin curing, reducing the risk of specimen distortion or fractures appearing in the resin block.

An outline of the procedure is as follows:

10.5.1. Specimen preparation

It is essential for the success of this procedure that the specimen is thoroughly dry. Failure to do so will result in an incomplete impregnation.

1 Oven dry the specimen at temperatures of between 30-50°C for a minimum of 12 hours. Saturated specimens must be allowed to drain for approximately 36 hours before beginning oven drying.

Alcohol and solvent dehydration procedures tend to have adverse effects on samples, reducing the bonding strength between the specimen and resin, and lowering the vapour pressure of the resin mix when under vacuum. However, the presence of any solvent extends the soaking time of the specimen in the resin before gelling and polymerization occurs. Crevello *et al.*, (1981) freeze-dried samples, but even this failed to eliminate shrinkage and the formation of micro-desiccation cracks. It appears that no matter what drying procedure is employed, some disruption of the original internal structure may result.

In cases where polyester resins are used, specimens must be placed in a container coated with a PVA release agent. This will prevent the resin from bonding to the container.

10.5.2. Resin system

All work must be undertaken in a well-ventilated area, with the preparator wearing suitable eye and hand protective equipment. Resins and additives (accelerator, pigment, etc.) should be weighed and thoroughly mixed in disposable cups, before the introduction of catalyst or hardener.

10.5.2a. Polyester resin

Quantities are dependent on the number and size of specimens. In the following method the order in which components are mixed should be noted, while the percentages indicated are by weight of the total resin mixture.

1 Mix pre-accelerated clear polyester embedding resin with flexible resin (maximum of 45% by weight of flexible to embedding resin).

2 Embedding and flexible resin mixtures require additional Cobalt Napthenate accelerator. However, this reduces the cure time, and thus the soaking time. A successful resin system enables the maximum time for both the vacuum and pressure stages to be effective before gelling and polymerization occur.

3 Add 1-2% Methyl Ethyl Ketone Peroxide (catalyst). The exact quantity required will depend on the addition of styrene (see 4).

4 Additional styrene (0.75%) with wax can be added. This has a two-fold purpose. Firstly, styrene aids in reducing the viscosity of the resin. However, the addition of styrene must be off-set by reducing the amount of catalyst used. Failure to do this can result in the formation of cracks in the cured block (Crevello *et al.*, 1981). Secondly, the wax forms a layer at the air:resin interface, reducing (although not eliminating) surface tackiness in the cured block.

10.5.2b. Epoxy resin

CIBA-GEIGY Araldite and Hardener systems are extensively used in impregnating geological specimens (Allman and Lawrence 1972), with medium viscosity mixes (AY 18:HZ 18, AY 105:HY 953F) the most common, although the latter mixes and cures to an amber colour. However, MY 778 is a colourless liquid resin that also contains a reactive diluent. When it is mixed with hardener HY 905 and accelerator DY 063, it forms a low viscosity resin, suitable for surface impregnation (Allman and Lawrence 1972). The addition of acetone reduces the viscosity further, making it suitable for thorough impregnation. As with polyester resin systems, quantities used depend on the number and size of specimens requiring impregnation. Percentages indicated represent volume of the total resin mixture.

1 Mix equal quantities of Araldite MY 778 and Hardener HY 905.

2 Add 1% of accelerator DY 063 and thoroughly mix.

3 Add 9% acetone to reduce the viscosity. To an AY 18:HZ 18 mixture Lister (1978) added an equal volume of acetone, and kept the specimen in the uncured resin for up to a week before slowly evaporating the acetone and oven curing at 100°C. Alternatively, if solvents are not used, resins can be heated (to a temperature below their cure point) to reduce their viscosity.

10.5.3. Vacuum treatment

Epoxy resins are particularly susceptible to boiling (frothing) when placed under vacuum. Excess frothing can be avoided by evacuation of the resin and specimen in separate vessels (*e.g.* Lister 1978, figure 1 Green 1986). The resin must be maintained at a higher pressure (lower vacuum) than the specimen. When the evacuated resin is introduced into the specimen chamber, the pressure differential assists in forcing the resin into the sample voids. An idealised vacuum and pressure system is illustrated in figure 10.1, and the use of the components described in the following method.

1 Evacuate the specimen within the chamber (**A**). The valve on the extract line (**B**) must be open. The valve on the intake line (**C**), and the stopcock at the base of separating funnel (**D**) must be closed.

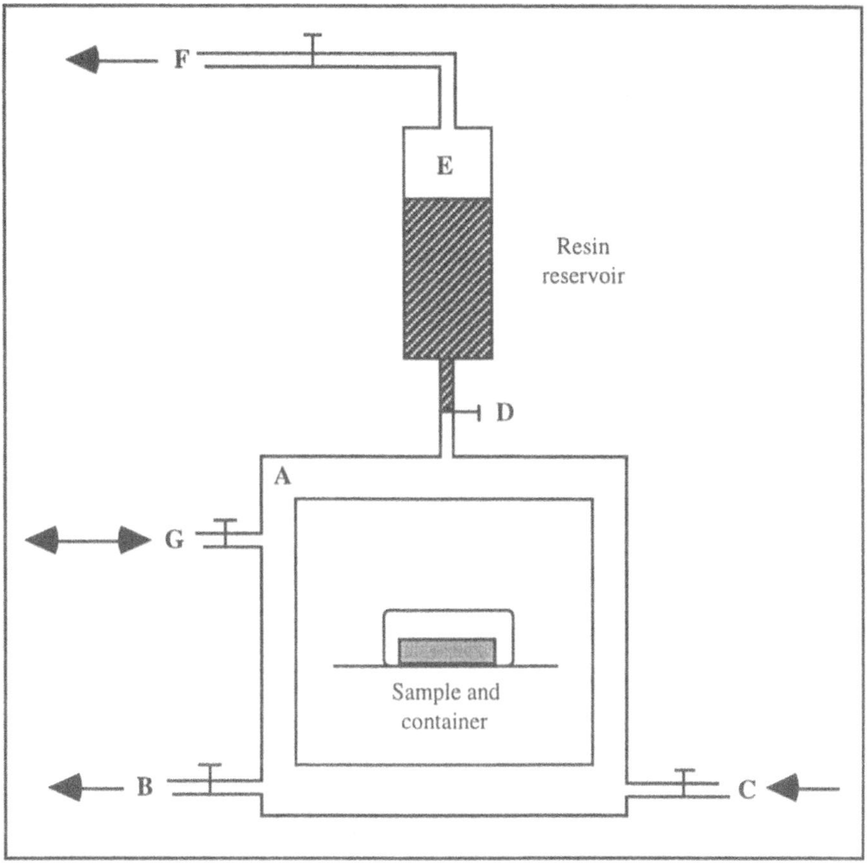

Figure 10.1. Diagrammatic illustration of a vacuum-pressure impregnating system. A - specimen chamber; B - chamber extract line valve; C - chamber intake line valve; D - stopcock at base of resin reservoir chamber; E - Resin reservoir; F - resin chamber extract line valve; G - specimen chamber leak valve (based on Green 1986).

2 Pour the freshly mixed resin into the resin reservoir (**E**), open extract line valve (**F**) and evacuate independently of the specimen chamber. As the resin is now in a catalysed state it is essential to work swiftly, before gelling and polymerization occurs (critical with polyester resins).

3 Evacuation should be maintained until a vacuum of 650mm Hg (28ins) is attained. Close valve (**B**).

4 Carefully open the stopcock (**D**) and slowly release the resin onto the specimen. The speed of resin release is controlled by use of the stopcock. The specimen should be covered by at least 2cm of resin.

Additional resin can be added to the reservoir by closing the extract line valve (**F**) and stopcock (**D**). A fresh quantity of resin can be evacuated and introduced onto the specimen. Alternatively, the specimen can be partly immersed in resin, and evacuated, although there is a risk of frothing. Regardless of the method employed, a small amount of resin **must** remain in the reservoir to maintain the pressure differential.

5 Close the extract line valve (**B**) before switching the vacuum pump off. This will avoid air and oil from the pump being sucked back, and contaminating the chamber. Once immersed in resin the vacuum should be maintained for a 10 to 15 minute period. Slowly pressurise the specimen chamber by opening the chamber leak valve (**G**), noting that as pressure increases, the resin level covering the specimen will drop (hence it is critical to ensure that coverage was sufficient during the initial vacuum stage). On equalizing the pressure ensure that the resin level remains approximately 1cm above the top of the specimen. On completion of this stage, pressure treatment can continue.

10.5.4. Pressure treatment

Ideally a combined vacuum-pressure chamber enables both stages to be undertaken without the problem of physically moving specimens soaking in uncured resin.

1 Ensure the extract line valves to the vacuum pump are closed. Slowly open the inlet valve (**C**), and pressurise the chamber to 7 bars (100psi). Close valve **C** and maintain the pressure for a minimum of four hours. Polyester resins will begin to polymerize within this time. Epoxy resins cure at elevated temperatures, which can be initiated at any time if the vacuum-pressure chamber possesses a heating element.

2 Reduce the pressure by slowly opening the leak valve (**G**). Release the door and examine the specimen. Polyester resins are best cured at room temperature, but additional oven curing of epoxy resin blocks can be undertaken if required.

10.6. ENVIRONMENTAL CONSIDERATIONS

A number of important environmental conditions must be considered when planning the long term preservation of unstable geological materials. These factors apply equally to petrological and mineralogical as well as palaeontological (micro and macro) specimens. The main factors are:

(i) Relative humidity (rh)
(ii) Bacterial activity
(iii) Temperature
(iv) Atmospheric pollutants
(v) Light
(vi) Vibration

Ashley Smith (1987) recognises four ways of "effecting a rescue" of a specimen in threat from its environmental conditions:

- remove the object,
- remove the source of danger,
- create a barrier,
- control the source.

To some extent, the first three provide short-term immediate solutions, while the fourth requires the implementation and maintenance of control mechanisms, while in practice combinations of all four ways are used to provide the optimum environment for storing material. The following account addresses these problems and their effects on unstable material, and the solutions tried, and in many examples failed in long and short term preservation.

10.6.1. Relative humidity

The ratio between the actual water vapour pressure in a given volume of air and the saturated vapour pressure concerning water at the same temperature provides a measure of the moisture content or relative dampness within the atmosphere. Values are expressed as percentages. Extreme variations in the relative humidity (rh) can result in chemical and physical changes in certain minerals. If this remains unchecked complete disintegration of the specimen may result. The early recognition of these changes is critical in specimen preservation. Evidence for reactions induced by high and low relative humidity variations is presented in table 10.1.

10.6.1a. Oxidation

Oxidation commonly affects three groups of metallic minerals: the sulphides (pyrite, chalcopyrite, pyrrhotite, sphalerite and chalcocite), rarer arsenides, and sulphosalts. Generally it results from high or fluctuating rh, coupled with grain size and the state of aggregation within the sample. Specimens containing porous pyrite become reactive when soluble salts in the water vapour are absorbed and transported through the matrices. Specimens undergo a change in colour or crystalline form, (Ashley Smith 1987). In a wide range of specimens this occurs at about 60% rh at normal temperatures. The cause of pyrite oxidation for specimens in air is probably wholly chemical in nature. Decay begins in the presence of moisture and oxygen within the air, and quickly becomes self-sustaining with the formation of hydrogen sulphide, sulphur dioxide, various iron sulphates, sulphur and sulphuric acid (Wolberg 1989a).

- **Iron Sulphides**

Decay of iron sulphide fossil specimens is more common than that of mineral specimens. Treatment is essential if long term preservation is required, particularly on porous specimens, such as pyritized plant tissues, which are especially difficult to stabilize. Pyrite decay can be recognised by:

- Loss of surface shine, powdering of the surface with the development of white or yellow crystals and the presence of a sulphurous smell.

REACTION	EVIDENCE
Oxidation/corrosion (High humidities)	Surface tarnishing, followed by specimen corrosion from atmospheric vapours. Can result in complete specimen disintegration.
Hydration/dehydration (Low humidities)	Destructive dimensional changes in specimens, including efflorescence, deliquescence and hydrolysis.

Table10.1. Specimen evidence from the reactions resulting from extremes in relative humidity (rh).

- The paper label below the specimen turns brown and crumbles, while the specimen exhibits disintegration, commonly from the base upwards.

The visual decay can be confirmed by moistening a small amount of the white powder with ammonium hydroxide. On drying a brick red colour (iron hydroxide) will be imparted to the powder, and occasionally yellow flecks of sulphur may be evident (Rixon 1976).

Prevention of sulphide oxidation can be achieved by the long term storing material at low rh. Short periods of storage at rh levels of 60% or above result in rapid specimen deterioration. The stabilization of material can be summarised in three key stages:

(i) initial cleaning and drying specimens.
(ii) prevention of immediate decomposition, especially during the initial stages of drying i.e. keep specimens damp.
(iii) long term preservation i.e. specimens dry.

(i) Initial Cleaning

1 Remove loose dry dirt with a sable hair brush.

2 Specimens collected from damp or wet environments (*e.g.* river banks, waterlogged quarries, mine shafts and beach foreshores) should be thoroughly washed to remove all salts in deionised water **before** they dry out.

(ii) Prevention of immediate decomposition

Safety Note: All pyrite preservation techniques must be undertaken in a well-ventilated area, fume hood or cupboard, with the operator wearing a face mask.

• **Ammonia vapour, ammonium hydroxide** method (Rixon 1976, Howie 1977a, Wolberg 1989a).

In this procedure specimen decomposition is related to the formation of sulphuric acid and is neutralized in the presence of strong base.

1 Consolidated, non-porous specimens can be placed directly in a solution. Unconsolidated porous specimens should not be immersed, but instead placed in a sealed inflated plastic bag with an open container of a 5% ammonia solution.

2 Leave specimen in the fume filled environment for approximately 1 week. However, check daily to ensure the solution has not evaporated, and top up as necessary during this time.

Alternatively employ the following method:

1 Paint the specimen with a solution of ethyl alcohol through which ammonia gas has been bubbled. Proceed to stage 3.

Following the ammonia treatment, dry the specimen as follows:

3 Remove the specimen from the ammonia saturated environment and immediately rinse in distilled water, followed by solvent (alcohol or acetone) dehydration.

4 Coat the dried specimen with a sealant (*e.g.* Poly(vinyl acetate) in toluene or acetone, see long term preservation, below).

• **Poly(ethylene glycol)** (PEG) 4000 in water.

This technique hardens the specimen and reduces shrinkage, and is particularly effective on friable material such as bone and wood, even when the specimens are damp.

1 Poly(ethylene glycol) 4000 melts at 50°C, and is miscible with water. Make up a 50% solution of PEG and water.

2 Paint the surface, or immerse the specimen in a bath of the solution.

3 The under surface of the specimen can also be treated with PEG, which can be strengthened by the addition of surgical gauze (Whybrow and Lindsay 1990).

(iii) Long term preservation

A number of solutions have been experimented with over the past thirty years. It is important to remember that constant handling of specimens will remove any protective coating, so continually examine and re-apply if signs of decomposition are evident. Any of the following methods will aid in long term preservation. Application is by total immersion or painting on the surface.

• **Emulsions and solutions** (Rixon 1976, Howie 1977a).

A variety of emulsions and solutions have been suggested and used in short term conservation. Product ranges continually change, and data on the long term effectiveness of these compounds is limited (Horie 1987). This must be carefully assessed before proceeding with any of the combinations listed below:

- **Poly(butyl methacrylate)** - acrylic polymer resins and suspensions are some of the most effective consolidants available (Rixon 1976) which can be used to conserve fragile specimens. Their application can result in shiny surfaces, but this can be diminished by the addition of toluene ($C_6H_5.CH_3$) and a small amount of aerogel silica added to the solution. Thick applications should be avoided, as pools may dry with a white appearance. Apply with brush or pipette.
- **Poly(vinyl acetate)** - mixed as a solution with either toluene, acetone or industrial methylated spirit (IMS). Leaves the surface with a very high gloss, while excessive handling causes the specimen to become tacky.
- **Poly(vinyl butyral) and poly(vinyl acetal)** - paint with a mixture of 1 part poly(vinyl butyral) or poly(vinyl acetate), the quantity determines viscosity, to 1 part methylated spirit and 4 part's amyl acetate. Alternatively, a slower drying solution of 75% toluene with 25% PVB or PVacetal can be applied.
- **Morpholine** - paint with a 5-10% solution of morpholine (C_4H_9ON) in ethanol (C_2H_5OH).
- **Polyurethane varnish** - surfaces may appear shiny, and may degrade if stored at excessive temperatures or light levels.
- **Ethanolamine thioglycollate** - (Cornish and Doyle 1983, 1984; Cornish 1987).

The ethanolamine thioglycollate technique has, to a large extent, replaced the ammonia and morpholine procedures (Rixon 1976, Howie 1979a). The main advantages for using an ethanolamine thioglycollate solution, in preference to other techniques outlined, are that it effectively neutralizes acidic pyrite oxidation by removing sulphate bi-products, and arresting immediate decay thus stabilizing the specimen. Moreover, the solution is alkaline and water-free, and thus there is little risk of further pyrite damage. Depending on the preservational state of specimens, one of two variations of the technique can be employed. However, there are disadvantages, and the treatments are fairly complex (see below), and expensive in comparison to the ammonia procedure (Wolberg 1989a).

• **Well preserved, consolidated specimens**

1 Immerse, to a depth of 5-6cms in a 2-5% ethanolamine thioglycollate solution mixed with 95% IMS for 4 hours.

2 During neutralization the solution will change from being clear to a violet colour as iron ferrothioglycollate is produced.

3 Repeat the treatment until all neutralization has ceased.

4 After each immersion the specimen should be washed in IMS and air dried.

• **Friable, unconsolidated, porous or large specimens**

1 Apply a paste to the decayed area of 3-5% ethanolamine thioglycollate solution mixed with IMS and added to sepiolite in equal portions.

2 Prevent rapid evaporation by covering the specimen with polythene or aluminium. Leave for up to 3 hours to dry, during which time the decayed products are drawn up and form a hardened paste.

3 Once dry brush off, and repeat if required until all decay products have been removed.

Cornish (1987) describes a complex procedure using ethanolamine thioglycollate for the stabilization of palaeobotanical seeds and fruits. Immersion of the specimens in an ethanolamine thioglycollate solution is followed by washing in IMS. Specimens are then dehydrated over a six day period in a series of different solvent mixtures, and finally stored in silicone fluid. This is advisable because the silicone fluid is immiscible with the alcohol used in the initial wash. Solvent mixtures of acetone and toluene followed by toluene and silicone is successful (see Cornish 1987, p452 for recommended solvent concentrations and times).

All techniques outlined within this section, require periodic monitoring and repeating to ensure the long term preservation of specimens. **The storage of material at a suitable rh (between 40-50%) is the critical factor in controlling pyrite decay**. If the rh exceeds 60% then decay will readily occur even on specimens that have been treated.

• **Sub-fossil bone, ivory, enamel and carbonised plants**

The alternating swelling and shrinkage of hygroscopic minerals forming clays and shales frequently results in cracking, distortion and spalling off specimen surfaces. Ultimately the complete breakdown of the matrix and any fragile fossils supported within can occur. Fossils affected when rh values fall below 40% include sub-fossil bone and lignite rich plant fossils.

Specimens collected from coastal areas are particularly susceptible to fluctuating rh levels, and the development of salt crystals. These contaminants must be washed off using deionised water before specimen preservation continues. With very fragile specimens it is advisable to impregnate them with a dilute coat of a PVA emulsion containing a fungicide. Doyle (1983, 1987) constructed an apparatus to hold vertebrate bones vertically while a dilute PVA solution was allowed to percolate through. The bottom end of the specimen was sealed with a concentrated PVA solution, and once dry (approximately 2 hours), a second emulsion mix (ratio of 1:10 with water), was applied until the specimen was thoroughly saturated, *i.e.* when the specimen becomes pliable. To avoid further specimen splitting during drying, it is advisable to dry the specimen slowly in a controlled environment with an initial rh of 85%. Reduce rh slowly over a number of days until the specimen is thoroughly dry. Excess PVA on the specimen surface can be removed by wiping with a soft cloth soaked in acetone. **N.B. Do not allow specimens to dry out, store them damp until they can be stabilized within the laboratory.**

• Glycerol

1 Small valuable specimens such as seeds and microfossils can be preserved by placing them directly in a tube of glycerol.

10.6.1b. Corrosion

Corrosion is commonly evident in natural metals such as silver, copper, antimony or the metallic iron in some meteorites. Corrosion is the result of a combination of factors, but in the main, high relative humidity (rh) is the most destructive component. Characteristically it is seen as surface tarnishing, but when hygroscopic salts are present it may result in the formation of pits on the specimen surface. Corrosion may also occur on specimens stored within a contaminated environment, in which the air contains acidic vapour traces. **N.B. Never use acidic materials for the cleaning of native metals.**

10.6.1c. Hydration and hydrolysis

The uptake of water from air to form a higher hydrate is known as hydration. Hydrated minerals undergo a constitutional change if rh levels vary. This occurs because water soluble minerals, such as halite (sodium chloride, NaCl), melanterite (hydrated iron sulphate, ($FeSO_4.7H_2O$) and trona (hydrous acid sodium carbonate, $Na_3H(CO_3)_2.2H_2O$) absorb moisture (deliquesce) when rh levels are at a higher water vapour pressure than that exerted by a saturated solution of the particular mineral. If these minerals remain within this unstable environment for prolonged periods of time they will dissolve.

A less common and rare occurrence is hydrolysis. This results from chemical reactions between minerals and atmospheric water vapour, resulting in the formation of new products. If halides and sulphates are involved then acidic gases may be liberated, which may effect other minerals in close proximity.

10.6.1d. Dehydration

Efflorescence results from a loss of water from a hydrated mineral. It is caused by a decrease in the rh vapour pressure to a level below its equilibrium. If this occurs it may result in mineral decomposition or a change in the crystal structure as new products form on the surface. Unfortunately, the formation of these new products, as with those resulting from hydrolysis, is also known as efflorescence.

10.6.2. Bacterial activity

Until the work of Howie (1977a & b), the main cause of pyrite decay had been attributed to activity of bacteria from the genera *Thiobacillus* and *Ferrobacillus* (Booth and Sefton 1970). This resulted from studies on acid mine waters in the late 1940's, in which it was observed that bacteria could oxidize ferrous iron too ferric in aqueous environments and produce sulphuric acid. The treatment suggested for specimen suffering from bacterial decay was the application of a low surface tension antiseptic detergent such as aqueous *Cetavlon* or *Savlon*, or an alcoholic solution of *Cetrimide* B.P.

However, experimental work (Howie 1977b) indicated, that high concentrations of sulphuric acid would inhibit the activity of these bacteria, as would high sulphate levels, while the bacteria that could withstand the high acidic environments would not oxidize metallic sulphides. Furthermore, all materials tested had been sterilised by washing in acetone and oven dried before oxidation. Continued work on this problem from Australia, relating to the various thiobacilli bacteria, has shown that on several actively decaying samples iron oxidizing *Thiobacillus ferro-oxidans* was not present, and that the sulphur oxidizer *Thiobacillus thio-oxidans* was present in one sample, but not in sufficient quantities to cause damage. Although the role of bacteria cannot be discounted in pyrite decay, clearly the main cause and in many cases the only cause, is chemical.

10.6.3. Temperature

Temperature is closely associated with rh. Variations in temperature alone are unlikely to cause much damage to specimens, although a rise in temperature may increase chemical reaction rates (Ashley Smith 1987), affecting mineral oxidation, efflorescence and deliquescence. Phase transitions of some minerals (*e.g.* tin) can result from temperature changes. However, a more serious problem will result from objects consisting of materials with differing coefficients of expansion (Ashley Smith 1987). An uncontrolled drop in temperature may result in a high rh.

10.6.4. Atmospheric pollutants

Urban atmospheres contain sulphur dioxide, carbon monoxide, ammonia and traces of mineral acids, all with the potential of causing considerable external damage to specimens. Unexpected local sources may also present a problem (*e.g.* acetic acid vapour from vinegar used in catering, Ashley Smith 1987). Dust can also affect both the appearance and value of specimens. The removal by brushing with a nylon brush, or cleaning within an ultrasonic tank may result in permanent damage.

A major problem in some older museums has been the build up of Organic Acid Vapours (OAV). Carbonate specimens are susceptible to the acetate, formate and nitrate vapours derived from adhesives used in the construction of display cabinets. The vapours condense forming unsightly mineral alteration products. Oak and birch ply are the most reactive woods to use. Both fresh and seasoned timber will generate OAV's.

- **Byne's Disease**

High temperatures and relative humidities will accelerate the hydrolysis of hemicellulose in wood to form acetic acid. If this occurs in the presence of carbonates or calcareous rich specimens, it may result in the development of thread-like white crystals (calclacite) on the surface. The crystals should be removed by gently brushing, and the specimen impregnated with a dilute consolidant (*e.g.* acrylic polymer, PVB).

Other OAV's include formaldehyde vapours (an irritant), which may result from urea formaldehyde glues used in the manufacture of some chipboards and blockboards. Some plastics may also generate OAV's and compounds that may undergo reactions in air to form corrosive substances, *e.g.* the rapid auto-oxidation of aldehydes to organic acids in air.

10.6.5. Light

Approximately 90 minerals are known to alter or decompose in the presence of light (Brunton *et al.* 1985). Some of the more common examples include sulphides, halides and chromates, which are also affected by oxygen rich environments. Minerals are affected to varying degrees. For example, fluorite and native sulphur are liable to cracking or parting along cleavage surfaces, realgar (AsS) suffers a permanent colour loss, while proustite (Ag_3AsS_3) suffers only a temporary colour loss. Some of the changes can occur at low prolonged light levels, and can only be avoided by keeping specimens in dark environments. No records of fossils decomposing as a result of being stored in direct light have been found, although damage to labels and other documentation is known.

10.6.6. Vibration

All specimens are susceptible to inadvertent shock or sustained exposure to vibration (Howie 1979b). Damage to specimens may be increased by poor storage in metal trays or cabinets in which resonance is readily amplified. Fragile fossils and minerals can be protected by being placed in trays lined with cotton wool, and large specimens can be placed on "bubble-wrap" paper, or supported on a bed of polystyrene chips.

- **Amber and gums**

Amber ($C_{12}H_{20}O$) or Succinite (because it contains succinic acid), is a water soluble coniferous tree resin. A related gum product is Copal, but this does not contain succinic acid. By polymerization and oxidation, resulting in a loss of volatile components, both become semi-stable. However, amber melts at temperatures between 300-375°C and copal at temperatures between 187-232°C.

The soft nature of these products, coupled with its instability, makes it difficult to work with. Cutting and grinding using diamond based wheels and plates may result in the amber melting, or a rippled effect on the surface that may feel sticky. Friction between the blade and the specimen may result in strain, a build up of a negative electric charge, and of static electricity generating a mild shock to the operator. However, more damaging is the possibility of specimen shattering. Broken specimens should not be repaired with solvent based adhesives as these will corrode amber. Epoxy resin adhesives can be used to repair fractured specimens. Further details can be found in the section 23 PREPARATION OF AMBER SPECIMENS CONTAINING FOSSILS.

REFERENCES

ALLMAN, M and LAWRENCE, D. F. 1972. *Geological Laboratory Techniques*. Blandford, (London). 1-335p.
ASHLEY SMITH, J. 1987. Environmental considerations, pp 403-405; In: P. R. Crowther and C. J. Collins (eds), The Conservation of Geological Material. *The Geological Curator*. **4**, (7), 375-474.
AWADALLAH, S. A. 1991. A simple technique for vacuum impregnation of unconsolidated, fine-grained sediments. *Journal of Sedimentary Petrology*. **61**, (4), 632-633.
BOOTH, G. H and SEFTON, G. V. 1970. Vapour phases inhibition of *thiobacilli* and *ferrobacilli*. A potential preservative for pyritic museum specimens. *Nature*. **226**, 185-186.
BOUMA, A. H. 1969. *Methods for the Study of Sedimentary Structures*. Wiley-Interscience, (New York). xvi + 458p.
BRADLEY, S. M. 1987. An introduction to the use of silanes in stone conservation, pp 427-432; In: P. R. Crowther and C. J. Collins (eds), The Conservation of Geological Material. *The Geological Curator*. **4**, (7), 375-474.
BRUNTON, C. H. C, BESTERMAN, T. P and COOPER, J. A. 1985. *Guidelines for the Curation of Geological Materials*. Geological Society (London). Miscellaneous Paper No. 17.
CARVER, R. E. 1971. *Procedures in sedimentary petrology*. Wiley-Interscience, (New York). xiii + 653p.
COLLINS, C. (Ed). 1995. *The Care and Conservation of Palaeontological Material*. Butterworth-Heinemann, (Oxford). xi + 139p.
CORNISH, L. 1987. The treatment of decaying pyritiferous fossil material using ethanolamine thioglycollate, pp 451-454; In: P. R. Crowther and C. J. Collins (eds), The Conservation of Geological Material. *The Geological Curator*. **4**, (7), 375-474.
CORNISH, L and DOYLE, A. M. 1983. Ethanolamine Thioglycollate as a chemical agent for the neutralization and removal of oxidised pyrite. *The Geological Curator*. **3**, (8), 512-513.
CORNISH, L and DOYLE, A. M. 1984. Use of Ethanolamine Thioglycollate in the conservation of pyritized fossils. *Palaeontology*. **27**, (2), 269-280.
CREVELLO, P. D, RINE, J. M and LANESKEY, D. M. 1981. A method for impregnating unconsolidated cores and slabs of calcareous and terrigenous muds. *Journal of Sedimentary Petrology*. **51**, (2), 658-660.
CROWTHER, P. R and COLLINS, C. J. (Eds). 1987. The Conservation of Geological Material. *The Geological Curator*. **4**, (7), 375-474.
DOYLE, A. M. 1983. The conservation of sub-fossil bird bone. *The Geological Curator*. **3**, (7), 447-451.
DOYLE, A. M. 1987. The conservation of sub-fossil bone, pp 463-465; In: P. R. Crowther and C. J. Collins (eds), The Conservation of Geological Material. *The Geological Curator*. **4**, (7), 375-474.
ELDER, A, MADSEN, S, BROWN, G, HERBEL, C, COLLINS, C, WHELAN, S, WENZ, C, ALDERSON, S and KRONTHAL, L. 1997. Adhesives and consolidants in geological and paleontological conservation: a wall chart. *Society for the Preservation of Natural History Collections Leaflets*. **1**, (2), 1-4.
GINSBURG, R. N, BERNARD, H. A, MOODY, R. A and DAIGLE, E. E. 1966. The Shell method of impregnating cores of unconsolidated sediments. *Journal of Sedimentary Petrology*. **36**, (4), 1118-1125.
GREEN, O. R. 1986. Vacuum and pressure impregnation of fine grained sediments. Proceedings of Geo-Techniques '85. Part 1. *Science Technology*. **15**, 21-24.
HORIE, C. V. 1987. *Materials for conservation. Organic consolidants, adhesives and coatings*. Butterworth-Heinemann (Oxford). xi + 281p.
HOWELL, A. C. 1977. Problems with resin at Bolton. *Newsletter of the Geological Curators Group*. **1**, (9), 466.
HOWIE, F. M. P. 1977a. Pyrite and conservation Part 1: historical aspects. *Newsletter of the Geological Curators Group*. **1**, (9), 457-465.
HOWIE, F. M. P. 1977b. Pyrite and conservation Part 2. *Newsletter of the Geological Curators Group*. **1**, (10), 497-512.
HOWIE, F. M. P. 1979a. Physical conservation of fossils in existing collections. *Newsletter of the Geological Curators Group*. **2** (5), 269-280.
HOWIE, F. M. P. 1979b. Museum climatology and the conservation of palaeontological material. *Special Papers in Palaeontology*. **22**, 103-125.
HOWIE, F. M. P. 1984. The conservation and storage of geological materials, pp 308-322: In; J. M. A. Thompson (ed), *Manual of curatorship. A guide to Museum practice*. Museums Association/Butterworth and Co. xiv + 553p.
HOWIE, F. M. P. (Ed). 1992. *The Care and Conservation of Geological Material*. Butterworth-Heinemann, (Oxford). xv + 138p.
HUTCHISON, C. S. 1974. *Laboratory handbook of Petrographic Techniques*. Wiley-Interscience (New York). xiii + 527p.
JIM, C. Y. 1985.Impregnation of moist and dry unconsolidated clay samples using Spurr resin for microstructural studies. *Journal of Sedimentary Petrology*. **55**, (4), 597-599.
KEENE, S. 1987. Some adhesives and consolidants used in conservation, pp 421-425; In: P. R. Crowther and C. J. Collins (eds), The Conservation of Geological Material. *The Geological Curator*. **4**, (7), 375-474.
LISTER, B. 1978. The preparation of polished sections. *Report of the Institute of Geological Sciences*, **78/27**. 20p.
RIXON, A. E. 1976. *Fossil animal remains: their preparation and conservation*. Athlone Press, (London). 304p.
WHYBROW, P. J and LINDSAY, W. 1990. Practical Techniques, pp 499-502: In; D. E. G. Briggs and P. R. Crowther (eds), *Palaeobiology - A synthesis*. Blackwell Scientific Publications (Oxford). xiii + 583p.
WOLBERG, D. L. 1989a. Stabilizing pyritic material , pp 244-248; In: R. M. Feldman, R. E. Chapman and J. T. Hannibal (eds), *Paleotechniques*. The Paleontological Society Special Publication. **4**. iv + 358p.
WOLBERG, D. L. 1989b. Glues and other stick'ums and patch'ums stabilizing compounds for strengthening fossils, pp 249-259; In: R. M. Feldman, R. E. Chapman and J. T. Hannibal (eds), *Paleotechniques*. The Paleontological Society Special Publication. **4**. iv + 358p.

11. PREPARATION OF RECENT MATERIAL FOR COMPARATIVE STUDIES

11.1. INTRODUCTION

Palaeobiological investigations utilizing comparative morphological and anatomical (form and structure) studies between recent and fossil organisms are increasing in importance. The techniques and procedures employed generate data more complex than that deduced from a comparison of fossil material with its living counterpart (Norris 1989). Observations, particularly those concerned with comparative ultrastructure, must be undertaken in conjunction with studies of soft tissue decay, compaction, transportation and authigenic mineralization of the organic skeleton (Briggs 1995). Physical processes affecting the organic skeleton (shells and bones) have been extensively studied, and palaeo-communities and organisms correlated with sedimentary sequences and palaeoenvironmental conditions (Brett and Baird 1986, Miller (ed) 1990).

Advances in understanding palaeoecological and evolutionary trends of communities and organisms have required an appreciation of fossilization processes. In turn this has resulted in closer examination, and in many instances a re-examination, of extraordinary and exceptionally well preserved fossil assemblages, re-defining the processes and stages of fossilization required in their formation. Two aspects have received particular attention. Firstly, taphonomic processes, those between death and burial of an organism, including chemical and mechanical alteration of organic remains as they become incorporated into the rock record. Secondly, the recognition and study of preserved soft tissues. The study of soft tissues is not a new aspect of palaeobiological investigation (Dean 1902). Neither is it confined to any one particular fossil group or time period (see Whittington and Conway Morris (eds) 1985, Briggs and Crowther (eds) 1990, chapter 3 - Taphonomy, for case-study reviews; Allison and Briggs 1993 for Phanerozoic distributions), although vertebrate specimens may provide the more photogenic elements popularly known (*e.g.* Frickhinger 1994).

Consequently, the study of extraordinarily preserved fossil biotas, and an understanding of the mode and stages of preservation is a dynamic and vibrant topic within palaeobiology, with case studies recorded in specialized volumes (*e.g.* Allison and Briggs (eds) 1991, Donovan (ed) 1991). The advances in understanding taphonomic processes have resulted from the careful study of fossil material and the simulation of fossilization procedures resulting in their preservation. These procedures have been examined in the field in an attempt to understand palaeoenvironmental conditions influencing death and decay (Allison and Pye 1994, Plotnick 1986). More quantifiable, and arguably more successful, have been the numerous laboratory experiments, and subsequent examination of soft tissue and mineralized components (Briggs and Kear 1993a, b & c, 1994, Briggs *et al.*, 1993, 1995, Kear *et al.*, 1995, Martill and Harper 1990, Martill *et al.*, 1992). Techniques and procedures used in taphonomic laboratory experimentation are detailed in this section. The procedure's centre on the killing, preserving and decay of living material under controlled conditions, and its subsequent dehydration and drying. Scanning electron microscope (SEM) examination techniques are detailed in a later section (see 34 ELECTRON MICROSCOPY TECHNIQUES).

11.2. KILLING AND PRESERVING OF RECENT SOFT TISSUE

Careful consideration must be given to the method of killing and preserving material. The primary objective of palaeobiological soft-tissue decay experiments is the detailed ultrastructural observations of tissues from the instance of death. The critical points of determining the timing of death and the initiation of fossilization are difficult to establish. Using standard zoological specimen preservation procedures may prove inappropriate, particularly amongst some invertebrate groups where muscle tissues instantly contract upon death (Lincoln and Sheals 1979).

The preferred method of killing and preserving marine specimens for taphonomic experiments is by anoxia (Baas *et al.*, 1995, Briggs and Kear 1994), avoiding the introduction of toxins that are unlikely to occur in natural conditions (Briggs *et al.*, 1995). This procedure involves maintaining live material in an aquarium filled with artificial sea water (35-37‰, normal marine salinity), regulated at a temperature of between 3-6°C. Killing was as follows:

1. Place the organism in an empty beaker in the air-lock of an anaerobic cabinet.
2. Evacuate the chamber.
3. Flush the chamber with oxygen-free nitrogen.
4. Evacuate and repeat stage 3.
5. Evacuate the chamber and fill with an anaerobic gas mixture (CO_2, N_2, H_2).
6. Leave organisms in the chamber for 40-60 minutes to ensure death. Proceed rapidly with the experiment, avoiding excess dehydration that may result in soft-tissue damage (Briggs and Kear 1994).

This method was successful with the shrimp *Crangon crangon* and the prawn *Palaemon eleganst* (Briggs and Kear 1994), but proved unsuccessful on the cephalochordate *Branchiostoma lanceolatum* (Briggs and Kear 1993c), the polychaete *Neris virens* (Briggs and Kear 1993a) and the pterobranch *Rhabdopleura compacta* (Briggs *et al.*, 1995). These animals can tolerate low oxygen levels. Consequently, it proved difficult to accurately determine the time of death and the onset of decay. Either the decay experiment or the procedure for killing must either be modified to accommodate this (*e.g.* Briggs *et al.*, 1995). The killing of *Neris virens* (Briggs and Kear 1993a) was achieved by:

- Gripping them behind the pharynx region with blunt forceps, and suspending the head in a stream of hot water (50-54°C) for 12-15 seconds.

For the purposes of the decay experiments, Briggs and Kear (1993a) considered the time of death set at the time of immersion in steam, although the posterior remained in motion for many hours. This process destroyed the worm's brain, leaving the cuticle un-ruptured, and does not introduce toxins or fixatives into the tissue (Briggs and Kear 1993a). Where sterile carcasses were required, specimens were transferred to a beaker containing artificial seawater, and exposed to a gamma irradiation dose of 24-32 kilograys (kGy) for a period of 8 hours (Briggs and Kear 1993a).

The killing of the cephalochordate *Branchiostoma lanceolatum* proved even more difficult to achieve (Briggs and Kear 1993c). Short term anoxia (2 hours) was ineffective, as was poisoning with carbon monoxide (CO) and anaesthetizing (narcotization) with a solution of magnesium chloride ($MgCl_2$ - see below). Killing was eventually achieved by placing the organisms in vessels of degassed sea water sealed in an anoxic atmosphere. Specimens examined after 22 hours were dead (and what is more important showed no signs of decay). "Death" and the onset of decay were timed from 24 hours after the experiment was sealed (Briggs and Kear 1993c).

Where material must be preserved in a relaxed condition, the pre-preservation procedures of slow anaesthetization can be considered (Lincoln and Sheals 1979). These procedures are more commonly used in zoology where immersion in either a strong alcohol or a 3% formaldehyde (HCHO) solution causes tissue to rupture. A summary of the chemicals commonly used in anaesthetizing marine and fresh water organisms is presented in table 11.1. This list is not exhaustive, and the methods are not described in detail. Many of the chemicals are toxic, and should only be used with the preparator using appropriate personal protective equipment.

By adding a few drops to cotton wool and placing over the open end of a tube enclosing the animal ether, chloroform and ethyl acetate vapour can also be used for anaesthetizing and killing terrestrial arthropods (Lincoln and Sheals 1979). Tobacco smoke is also effective on many small aquatic animals when slowly bubbled through a fine glass tube into water at the bottom of the container. Coelenterata and Echinodermata have been successfully anaesthetized by the introduction of carbonic acid gas introduced by squirting soda water from a siphon into the water containing the specimens (Lincoln and Sheals 1979).

11.3. DEHYDRATING AND DRYING RECENT SOFT TISSUES

Once organisms have been anaesthetized, killed, fixed and preserved they can be prepared for study. Details of fixing and preserving samples are discussed in section 4 COLLECTING TECHNIQUES FOR MICROFOSSIL AND LIVE FORAMINIFERA SAMPLES). Critical to both the chemical and visual (optical or scanning microscopy) analyses of material are the processes of drying and dehydrating specimens. The method of drying can have implications for certain kinds of chemical analysis (Briggs 1995). Conventional air drying of unsupported tissues results in fluid evaporation. During the final stages of this process the surface tension exerted by the meniscus on small organelles is relatively enormous, disrupting them from their life position and forcing them flat against cell walls. Temperatures employed during oven drying can alter the composition of organic materials like cuticle. This disadvantage is not associated with the methods of freeze-drying, freeze-substitution or critical point drying (Briggs 1995, Hayat 1989, Martill and Harper 1990). However, different tissues and structures may preserve better using one particular procedure (Nation 1983), and should be tested when initiating new ultrastructural research.

11.3.1. Freeze drying

This method involves the dehydration of frozen tissue through the sublimation of ice (Hayat 1989). Organic solvents are not used, differentiating it from the method of

ANAESTHETISING AGENT	CONCENTRATION, METHOD & ORGANISMS EFFECTED
Menthol	Allow animals to expand in clean water (fresh or sea water), and scatter crystals of menthol on surface. For sessile animals, e.g. Polyzoa. May require 12 hours to be effective.
Magnesium sulphate	(1) Completely immerse animals. (2) Sprinkle crystals on water containing specimen. (3) As a 20-30% solution (150g/l) gradually introduced into the water over some hours. For giant nudibranchs, chitons and madreporaria. Duration of treatment can cause tissues to rupture.
Magnesium chloride	Immerse animals in an isotonic solution, mixed in fresh or distilled water. Decant and replace with formalin. Suitable for most marine animals.
Chloral hydrate	(1) Sprinkle crystals on the surface of water containing animals. (2) Immerse animals in a fresh 2% solution. Suitable for most marine and fresh water animals.
Ethyl m-aminobenzoate	Immersion time varies from a few seconds to 10-15 minutes. Suitable for cold-blooded vertebrates, and certain invertebrates (e.g. Crustacea).
Propylene phenoxetol	Immerse animals in clean water (fresh or sea water), and add propylene phenoxetol (not exceeding 1% vol) until large viscous globules form on the bottom of the container. Treatment time varies from a few minutes to several hours. Suitable for a large number of invertebrates.
Benzamine hydrochloride & cellosolve mixture	Prepared as follows: 3 parts 2% benzamine hydrochloride; 1 part cellosolve (ethylene glycol monoethyl-ether); 6 parts distilled water. Suitable for small aquatic animals.
Eucaine	Prepared as follows: 1g Eucaine; 10ml 90% alcohol; 10ml distilled water. Add gradually to specimens. Suitable for small aquatic animals.
Stovaine (Amyl chlorohydrin)	Prepared as follows: 1g Stovaine; 10ml 90% alcohol; 10ml distilled water. Add gradually to specimens. Suitable for small aquatic animals.

Table 11.1. Commonly used anaesthetizing solutions for zoological marine and fresh water specimens. Data from Lincoln and Sheals (1979).

freeze-substitution (see below). The pressure within the apparatus determines the mean-free path (the distance travelled by a molecule in a gas between collisions) of the water vapour leaving the specimen. Providing the mean-free path is comparable or greater than the specimen-condenser spacing, water vapour leaving the specimen is immediately trapped on the surface of a nearby condenser. The technique is thus a vacuum distillation process and not a vacuum drying process (Hayat 1989), involving the specimen passing through two phase boundary reactions. First during freezing, a liquid-solid boundary rapidly passes through the specimen. Second as drying commences a solid-vapour phase boundary affects the specimen.

Freeze-drying large blocks requires special care, and the rise to the ambient temperature is gradual, ensuring complete drying to the core of the specimen. After abrupt warming, large specimens may contain undried portions that can cause rehydration. An increase in temperature may cause ice recrystallization, damaging the ultrastructure and other fine detail. Furthermore, intracellular, whole-cell and entire tissue shrinkage may be evident in specimens that are ultra-rapidly cryofixed. Such specimens are thermodynamically unstable at low temperatures, where the formation of ice crystals causes partitioning in the specimen as dehydration of cell organelles occurs. Between 7% and 20% shrinkage in cells and tissues has been reported (Hayat 1989).

Preparation must also ensure that specimens are thoroughly washed in distilled water to remove all salts in solution before they are frozen. Any salt present as freezing is initiated may result in artefacts of filament-like structures forming within the cells and tissues (Miller *et al.*, 1983). However, despite the action of washing carrying the risk of osmotic and physiological damage to the specimen, the advantages outweigh the risks (Hayat 1989).

Commercially available freeze-drying equipment is described and illustrated by Hayat (1989, p. 393-395), although many laboratories may construct their own. At its most basic, the freeze-drying apparatus consists of three glass chambers. A small chamber holds crystals of osmium tetroxide (OsO4) required for vapour fixation. A second chamber (condenser) holds the specimen, and a third, containing calcium aluminosilicate, acts as a molecular sieve, and prevents rehydration. The specimen can be transferred from one chamber to the next, or alternatively dried under vacuum within apparatus connected by a manifold (Chiovetti *et al.*, 1987). An outline of the procedure using this type of apparatus is as follows:

1. Specimens rapidly frozen in liquid nitrogen, are transferred to an aluminium specimen holder and held at a temperature of -123°C under a vacuum of 10^{-3} Torr. Pressure must be maintained to prevent specimen rehydration.

2. Retaining the vacuum for 24 hours and dry the specimen at -110°C.

3. Continue warming the specimen by rising the temperature at intervals of 10°C per hour until the temperature reaches -55°C.

4. Leave the apparatus for at least 36 hours to reach room temperature and prevent rehydration. Specimens are now suitable for resin impregnation and thin sectioning.

The success obtained in using this procedure requires practice, and an understanding of the operation and familiarisation with the apparatus. A portable unit, using solid carbon dioxide and alcohol to freeze material, has been developed for both laboratory and field use (Spicer *et al.*, 1974). Initially used to dry freeze-fractured angiosperm leaves for SEM observation, it was also successfully used to stabilize recent algal assemblages from iron rich lake sediments. This study of early element exchange between organic matter and sediment required the freeze-drying of samples immediately upon collection to preclude soluble ion movement. Use of this equipment and procedure in other ecological studies is not known. Spicer *et al.*, (1976) successfully used it in palaeobotanical studies in preparing pteridophyte spores and sporangia for SEM examination, while Hill (1987) freeze-dried Jurassic plant remains from north Yorkshire for comparative anatomical studies.

11.3.2. Freeze-substitution

This is another method successfully applied to biological specimens. It involves the replacement of ice, frozen within tissues, with an organic solvent (Hayat 1989). The critical factor is that this occurs at a higher temperature than that at which the specimen was cryogenically fixed. In turn the solvent, most commonly acetone, is replaced by a resin. Other alternatives used in biological preparation include freeze-drying and embedding and freeze-substitution and embedding, although none of these appear to have been applied to a palaeobiological study.

11.3.3. Critical point drying

A superior alternative to freeze-drying, producing little soft-tissue distortion, is critical point drying (Anderson 1951). This procedure, developed and described by Anderson (1950a & b, 1951), involves the dehydration of pre-fixed soft tissue by the process of immersion in a series of water-alcohol mixtures culminating with immersion in amyl acetate. The amyl acetate is flushed out and replaced by liquid carbon dioxide below its critical temperature of 32°C using temperature and pressure apparatus. By slow warming above 32°C the carbon dioxide is transformed to a gas. The critical point at which phase change from liquid to gas occurs, results in the elimination of surface tension within soft tissues and cell walls, and the passage of gases through cell walls can proceed with minimal amount of morphological damage.

In resolving early diagenetic soft tissue features present in phosphatized Cretaceous fish, Martill and Harper (1990) successfully adapted the method for their comparative morphological study of recently killed fish.

The relatively simple apparatus, consists of a specimen pressure chamber (known as a bomb) connected by high pressure stainless steel tubing to a liquid carbon dioxide supply, and a venting valve with pressure gauge (see supplier's list). The method, is as follows:

1 Fix specimen for 1.5 hours in 1% osmium tetroxide (OsO4), buffered in phosphate to a pH of 7.2 to prevent further post-mortem degradation. Freshly killed marine samples should be pre-treated by washing in a fresh sea water bath of normal salinity.

2 Repeat the buffered fixation stage.

3 Dehydrate samples by immersion in either ethanol or acetone in a series of progressively higher concentrations (30%, 50%, each for 10 minutes; then 70%, 85%, 90%, 95% and finally 100%, each for 15 minutes). Repeat the 100% wash to ensure the complete removal of water. Anderson (1951) replaced the alcohol with *iso*-amyl acetate ($CH_3COO(CH_2)_2CH(CH_3)_2$).

4 Transfer sample to a porous vessel, flood with alcohol (or *iso*-amyl acetate) and load into the drying bomb. Ensure both inlet and outlet valves are closed.

5 Attach high pressure tubing to the tank containing liquid carbon dioxide. Open the tank supply, and slowly fill the bomb with liquid CO_2. The pressure gauge of the bomb must register between 700-900psi. Slowly open the outlet valve to exhaust the tank of the CO_2 and acetone mixture. While the pressure in the bomb remains constant, repeat the flushing through process three or four times over a 3.5 hour period. This process will expel all the alcohol, and replace it with CO_2. Finally, close both the outlet valve and the carbon dioxide inlet valve. **Do not over tighten, or the seals may become damaged.**

6 Gently heat the bomb to a temperature of 35°C, the critical point for CO_2.

7 Gently open the outlet valve, and allow the gaseous carbon dioxide to escape. This venting should take about 2 minutes.

8 Carefully remove the specimen, and transfer to an evacuated desiccator. Because of the danger of rehydration, long term storage in this state is not recommended.

The rigid specimen can now be gold sputter coated and examined under the scanning electron microscope (SEM). Biological soft tissue prepared in this way provides palaeobiologists with the opportunity to observe material in the familiar three-dimensional way comparable to the majority of fossils viewed by electron microscopy. Although tissue shrinkage is not as pronounced as in specimens that have been freeze-dried, some shrinkage is evident (Martill and Harper 1990). A further problem observed by Martill and Harper (1990) is the recognition of osmium tetroxide crystals on the surface of samples, although these are only evident at very high magnifications.

Critically point dried material must be stored in a desiccator cabinet to prevent rehydration. If stored correctly prepared material may have a shelf life in excess of 12 months, without any noticeable deterioration (Martill and Harper 1990).

11.3.4. Hexamethyldisilazane $(CH_3)_3Si.NH.Si(CH_3)_3$

Safety note: Avoid skin contact with hexamethyldisilazane as it may react with tissue compounds. Wear gloves and eye protection, and use only in a fume cupboard.

The results of this procedure compared favourably with critical point drying (Nation 1983), and as the treatment time is only a matter of minutes for each specimen (in comparison with many hours for critical point drying), it is the favoured method of dehydration employed by Briggs and Kear (1993b, 1993c, 1994) and Kear *et al.*, (1995). Soft tissues are initially fixed in 1% glutaraldehyde ($CH_2(CH_2.CHO)_2$) solution (see section 4 COLLECTING TECHNIQUES FOR MICROFOSSIL AND LIVE FORAMINIFERA SAMPLES), and then dehydrated through a graded series of ethanol solutions before immersion for 5 minutes in hexamethyldisilazane (HMDS, $(CH_3)_3Si.NH.Si(CH_3)_3$). Finally, specimens were air dried prior to gold sputter coating and examination by scanning electron microscope (SEM).

The method outlined by Nation (1983), including the fixation stage, is as follows:

1 Dissected tissues immersed in physiological saline, are immediately transferred to 1% glutaraldehyde in 0.1 M cacodylate buffer (pH7) for 5 minutes.

2 Wash tissue in distilled water.

3 Dehydrate through a series of ethanol solutions (70%, 85%, 95%, 100%), immersing the specimen in each for at least 5 minutes.

4 Transfer the specimen to HMDS, immerse for 5 minutes.

5 Allow the specimen to air dry at room temperature.

6 Prepare material on an SEM stub for sputter coating and examination (see section 34 ELECTRON MICROSCOPY TECHNIQUES for details).

For comparison Nation (1983) treated similar material by critical point drying (treated to stage 3 above), and then transferred to a Critical Point Dryer and flooded with liquid carbon dioxide. This was slowly vented, and the process repeated until all traces of ethanol had been removed. From the final immersion in CO_2 the temperature was raised over a 10 minute period to 42°C, and then slowly vented over the next 20 minutes.

Specimens prepared by the HMDS procedure gave results as good as those from critical point drying (Figure 1 in Nation 1983), but with the advantage of equipment cost and saving of time.

11.4. EXPERIMENTAL STUDIES IN SOFT TISSUE FOSSILIZATON

Two main approaches have been adopted by palaeobiologists in the study of soft tissue fossilization. First, extensive studies of the sixty plus exceptionally preserved biotas occurring within the Phanerozoic have developed with comparative anatomical observations between fossil and recent material. These studies have seen sub-micrometer levels of resolution providing the detail (*e.g.* Martill and Harper 1990, Martill *et al.*, 1992). Secondly, in an attempt to understand the environmental setting and mechanisms of preservation of these biotas, laboratory experimental work on inhibiting decay rates and promoting soft tissue preservation of recent organisms has been undertaken (*e.g.* Allison 1990, Briggs and Kear 1993a, b, c). Both lines of research provide the means to extract the maximum amount of data in relation to taphonomic bias within an assemblage and differentiate it from ecological factors (Briggs and Kear 1993a).

Field studies indicate that there is a marked difference between the taphonomic processes operating in shallow and deep-water environments (Allison *et al.*, 1991). From a comparison of the decay processes affecting recently dead whales, a model relating gases produced during decomposition to carcass buoyancy with depth was presented. Allison *et al.*, (1991) also suggested that the model could be linked to the degree of skeleton articulation and be used as a rough index of palaeobathymetry.

The preservation potential of organisms ranges from complete decay (leaving only shelly fossils), to the diagenetic replication of muscle and other soft tissues. The degree of information lost immediately following death and before burial, is characterized into one of five stages of decay attained by the organism (Briggs and Kear 1993a). The decay stages can be used as a taphonomic threshold, reflecting the significance of diversity of a fossil assemblage. Exceptionally preserved biota have a higher threshold, recording community completeness and diversity (Briggs and Kear 1993a). Through a series of experimental studies on numerous aquatic organisms, Briggs and co-workers have observed decay rates (*e.g.* weight loss and chemical compositional changes) and correlated them to various morphological stages.

Briggs and Kear (1993a, 1994) illustrate twelve artificial sea water experiments (Table 11.2), reflecting oxygenated and deoxygenated conditions in normal and controlled (N_2/CO_2) atmospheres, transportation, and the presence of sediment. In some experiments the artificial sea water was "inoculated" with water (50ml/l) from the Tay Estuary, Scotland and yeast extract (0.1g/l) to provide a bacterial substrate to initiate decay. This "standard" was chosen because (i) the area is characterized by high rates of organic matter degradation, generating both aerobic and anaerobic sulphate reductions, and (ii) the bacteria can tolerate wide salinity ranges. Inoculated water was allowed to incubate for at least 48 hours at room temperature, and adjusted to pH 8. Sediment used in the experiments was also from this locality. Artificial sea water used in sterile experiments was autoclaved and regassed with either sterile air or oxygen free nitrogen.

Once killed, carcasses were dried by blotting with tissue, weighed, and individually transferred to wide-neck screw-top jars and immersed in 50-100ml of artificial sea water (standard or sterile). Sampling is destructive, and consequently individual specimens can only be sampled once. Sampling times varied, depending on the organisms studied, but in all cases the sampling interval increased during the experiment as decay ceased. All experiments were terminated when all the major changes were considered to have occurred. Sampled material was subsequently analysed by the scanning electron microscope, electron microprobe, x-ray diffraction and the elemental

EXPERIMENT	CHARACTERISTIC MONITORED	STATE OF ARTIFICIAL SEA WATER (Normal/sterile; oxygenated/deoxygenated)	OXYGEN AVAILABILITY	ADDITIONAL NOTES
1a	Oxygen content	normal, oxygenated	rapid diffusion	
1b	Oxygen content	normal, oxygenated	slow diffusion	
1c	Oxygen content	normal, oxygenated	absent (N_2/CO_2 atmosphere)	
1d	Oxygen content	normal, deoxygenated	absent (N_2/CO_2 atmosphere)	
1e	Oxygen content	normal, oxygenated	restricted slow diffusion (O_2/CO_2 atmosphere)	
1f	Oxygen content	normal, deoxygenated	restricted slow diffusion (O_2/CO_2 atmosphere)	
2	Transportation	normal, oxygenated	slow diffusion	shaker 110 rpm
3	Sediment	normal, oxygenated	slow diffusion	sediment added
4	Degrading agent	normal, oxygenated	slow diffusion	bacterial activity eliminated; whole jar sterilized by gamma radiation
5	Degrading agent	normal, oxygenated	slow diffusion	water column bacteria only; worm sterilized by gamma radiation
6a	Degrading agent	sterile, oxygenated	rapid diffusion	bacterial associated with carcass only
6b	Degrading agent	sterile, oxygenated	slow diffusion	bacterial associated with carcass only

Table 11.2. The experimental conditions used by Briggs and Kear (1993a, 1994) in quantifying soft tissue decay and early diagenetic mineralization. Results indicate that although the oxygen content of the sea water may not have a pronounced effect on soft tissue decay rates, the open or closed nature of the system influences pH and determines the type of early mineralization, and the subsequent nature of preservation.

analyzer. Details of the procedures and variations of the method used can be found in Briggs and Kear (1993a, c, 1994) and Kear *et al.*, (1995). Additional data on the chemical analysis of experimental decomposition products, in particular Curie-point-gas-chromatography, can be found in Baas *et al.*, (1995) and Briggs *et al.*, (1995).

REFERENCES

ALLISON, P. A. 1990. Variations in rates of decay and disarticulation of echinodermata: implications for the application of actualistic data. *Palaios*. **5**, (5), 432-440.

ALLISON, P. A, and BRIGGS, D. E. G. (Eds). 1991. *Taphonomy: releasing the data locked in the fossil record*. Plenum Press (New York). 560p.

ALLISON, P. A, and BRIGGS, D. E. G. 1993. Exceptional fossil record: distribution of soft-tissue preservation through the Phanerozoic. *Geology*. **21**, (6), 527-530.

ALLISON, P. A, and PYE, K. 1995. Early diagenetic mineralization and fossil preservation in modern carbonate concretions. *Palaios*. **9**, (6), 561-575.

ALLISON, P. A, SMITH, C. R, KUKERT, H, DEMING, J. W and BENNETT, B. A. 1991. Deep-water taphonomy of vertebrate carcasses: a whale skeleton in the bathyal Santa Catalina Basin. *Paleobiology*. **17**, (1), 78-89.

ANDERSON, T. F. 1950a. A new technique for the study of biological structures in the electron microscope (Abstract only). *Biological Bulletin*. **99**, 315-316.

ANDERSON, T. F. 1950b. The use of critical point phenomena in preparing specimens for the electron microscope. *Journal of Applied Physics*. **21**, (7), 724.

ANDERSON, T. F. 1951. Techniques for the preservation of three-dimensional structure in preparing specimens for the electron microscope. *Transactions of the New York Academy of Sciences Series II*. **13**, (4), 130-134.

BAAS, M, BRIGGS, D. E. G, van HEEMST, J. D. H, KEAR, A. J and de LEEUW, J. W. 1995. Selective preservation of chitin during the decay of shrimp. *Geochimica et Cosmochimica Acta*. **59**, (5), 945-951.

BRETT, C. E and BAIRD, G. C. 1986. Comparative taphonomy: a key to paleoenvironmental interpretation based on fossil preservation. *Palaios*. **1**, (3), 207-227.

BRIGGS, D. E. G. 1995. Experimental taphonomy. *Palaios*. **10**, (4), 539-550.

BRIGGS, D. E. G and CROWTHER, P. R. (Eds). 1990. *Palaeobiology - A synthesis*. Blackwell Scientific Publications (Oxford). xiii + 583p.

BRIGGS, D. E. G and KEAR, A. J. 1993a. Decay and preservation of polychaetes: taphonomic thresholds in soft-bodied organisms. *Paleobiology*. **19**, (1), 107-135.

BRIGGS, D. E. G and KEAR, A. J. 1993b. Fossilization of soft tissue in the laboratory. *Science*. **259**, 1439-1442.

BRIGGS, D. E. G and KEAR, A. J. 1993c. Decay of *Branchiostoma*: implications for soft-tissue preservation in conodonts and other primitive chordates. *Lethaia*. **26**, (3), 275-287.

BRIGGS, D. E. G and KEAR, A. J. 1994. Decay and mineralization of shrimps. *Palaios*. **9**, (5), 431-456.

BRIGGS, D. E. G and KEAR, A. J, BASS, M, de LEEUW, J. W and RIGBY, S. 1995. Decay and composition of the hemichordate *Rhabdopleura*: implications for the taphonomy of graptolites. *Lethaia*. **28**, (1), 15-23.

BRIGGS, D. E. G, KEAR, A. J, MARTILL, D. M and WILBY, P. R. 1993. Phospatization of soft-tissue in experiments and fossils. *Journal of the Geological Society, London*. **150**, (6), 1035-1038.

CHIOVETTO, R, McGUFFE, L. J, LITTLE, S. A, WHEELER-CLARK, E and BRASS-DALE, J. 1987. Combined quick freezing, freeze-drying, and embedding tissue at low temperature and in low viscosity resins. *Journal of Electron Microscopical Techniques*. **5**, (1), 1-15.

DEAN, B. 1902. The preservation of muscle fibres in sharks of the Cleveland Shale. *American Geologist*. **30**, 273-278.

DONOVAN, S. K. (ed). 1991. *The Processes of Fossilization*. Belhaven Press (London). ix + 303p.

FRICKHINGER, K. A. 1994. *The fossils of Solenhofen*. Goldschneck-Verlag (Korb). 336p.

HAYAT, M. A. 1989. *Principles and techniques of electron microscopy, Biological applications* (3rd edition). Macmillan Press (Basingstoke). xv + 469.

HILL, C. R. 1987. Jurassic *Angiopteris* (Marattiales) from north Yorkshire. *Review of Palaeobotany and Palynology*. **51**, (1/3), 65-93.

KEAR, A, J, BRIGGS, D. E. G and DONOVAN, D. T. 1995. Decay and fossilization of non-mineralized tissue in coleoid cephalopods. *Palaeontology*. **38**, (1), 105-131.

LINCOLN, R. J and SHEALS, J. G. 1979. *Invertebrate animals. Collection and preservation*. British Museum (Natural History), (London and Cambridge University Press). vii + 150p.

MARTILL, D. M and HARPER, E. 1990. Critical point drying, a technique for palaeontologists. *Palaeontology*. **33**, (2), 423-428.

MARTILL, D. M, WILBY, P. R and WILLIAMS, N. 1992. Elemental mapping: a technique for investigating delicate phosphatized soft tissues. *Palaeontology*. **35**, (4), 869-874.

MILLER, III, W. (ed). 1990. *Paleocommunity temporal dynamics: the long-term development of multispecies assemblies*. The Paleontological Society Special Publication. **5**. vii + 421.

MILLER, K. R, PRESCOTT, C. S, JACOBS, T. L and LASSIGNOL, N. L. 1983. Artifacts associated with quick-freezing and freeze-drying. *Journal of Ultrasturctural Research*. **82**, (2), 123-133.

NATION, J. L. 1983. A new method using hexamethyldisilazane for preparation of soft insect tissues for scanning electron microscopy. *Stain Technology*. **58**, (6), 347-351.

NORRIS, R. D. 1989. Cnidarian taphonomy and affinities of the Ediacara biota. *Lethaia*. **22**, (4), 381-393.

PLOTNICK, R. E. 1986. Taphonomy of a modern shrimp: Implications for the arthropod fossil record. *Palaios*. **1**, (3), 286-293.

SPICER, R. A, GRANT, P. R and MUIR, M. D. 1974. An inexpensive portable freeze-drying unit for SEM specimen preparation. In: O. Johari and I Corvin (eds), *Proceedings of the Seventh Annual Scanning Electron Microscopy Symposium*. IITRI Chicargo. 290 & 299-304.

SPICER, R. A, SENGUPTA, S and PHILLIPS, L. 1976. Low temperature preparation of pteridophyte spores and sporangia for SEM. *Journal of Microscopy*. **106**, (3), 351-354.

WHITTINGTON, H. B and CONWAY-MORRIS, S. 1985. Extraordinary fossil biotas: their ecological and evolutionary significance. *Philosophical Transactions of the Royal Society of London*. **B311**, 1-192.

PART III

LABORATORY TECHNIQUES

B. PHYSICAL PROCEDURES

12. Mechanical methods for preparing fossil specimens.

13. Fossil extraction techniques by thermal disintegration.

14. Disaggregation and dispersal of partially consolidated and unconsolidated sediments.

15. Microfossil sample contamination and reliability problems.

16. Washing and sieving techniques used in micropalaeontology.

17. Centrifuge techniques used in micropalaeontology.

18. Mechanical separation of microfossil residues.

19. Flotation and liquid separation techniques.

20. Thin section and slide preparation techniques of macro- and microfossil specimens.

21. Staining techniques used in micropalaeontology.

22. Preparation of carbonate stained acetate peels and thin sections.

23. Preparation of amber specimens containing fossils.

24. Preparation and conservation of vertebrate fossils.

12. MECHANICAL METHODS OF PREPARING FOSSIL SPECIMENS

12.1. INTRODUCTION

Preparation of material by mechanical methods employs a variety of tools to assist in the stock removal of matrix. Most of the techniques illustrated are applied to macrofossil material, although, with care and patience, they can equally be used in the preparation of some microfossil specimens (Berdan 1989). Furthermore, the techniques are not mutually exclusive, and can be used in association with chemical and thermal disintegration procedures. Mechanical cleaning procedures used in the preparation of individual micropalaeontological specimens (*e.g.* Martinsson 1965), are discussed in a later section (see 32 SPECIALIST TECHNIQUES USED IN MICROPALAEONTOLOGY). The mechanical extraction of fossils in the field is outlined in part II FIELD TECHNIQUES.

In addition to matrix removal, these techniques are occasionally used in the removal of external layers of fossil material, exposing hidden features beneath. The prime objective is always to reveal the maximum amount of information while inflicting the least amount of damage to the specimen. What may appear as simple easy procedures require maximum dexterity, time and patience, although experience provides the key in deciding which tools to use, and the extent of preparation that should be undertaken. A sound appreciation of the fossil's geometry, and the lithology it is embedded in are essential for the success of these procedures. The misuse of these techniques may result in an irreparable information loss.

A wide variety of mechanical procedures are outlined in this section, from the hammer, chisel, pick and knives, to the more specialized airbrassive units, pneumatic percussion and rotary drills, and ultrasonic tanks and pens. In general, procedures are outlined in order of descending stock removal and "damage" that can be inflicted upon the specimen.

12.2. PERCUSSION TOOLS

Employing procedures allied to the craftsmanship exhibited by sculptors and stonemasons, Sohl (1989) justifiably considers these to be some of the oldest methods used by palaeontologists in fossil preparation. A variety of hand-held hammers/mallets, chisels, picks and knives can be used in matrix removal (sometimes referred to as "development"). For small specimens and microfossils, specialized pneumatic and electrical equipment (described below) is employed (Berdan 1989, Palmer 1989). A number of specialized tools were developed by Lörcher and Keller (1985) for work on the Upper Liassic Posidonienschiefer of Germany. Mechanical techniques were used in combination with chemical techniques for the delicate extraction of *Ichthyosaur* fragments.

12.2.1. Hammers and mallets

The variety of tools that can be used is virtually limitless, although three basic types can be employed, according to need:

(1) Wooden mallet: gives a gentle blow.
(2) Iron mallet: gives a hard full blow.
(3) Steel hammer: delivers a powerful hard blow.

Obviously a great deal of variation can be applied from the force delivered by hand. The best type to work with are the two-faced kind, varying in weight from 150g to 1kg. Sculptor's hammers tend to have sharp corners, and should only be used by the more experienced preparator (Rixon 1976). The effective use of these basic tools in specimen preparation can only be achieved by skill and patience.

12.2.2. Chisels, knives, mounted needles and dental picks

Most commercially available chisels are too broad for the delicate work that is required, and have to be modified by the preparator (Lörcher and Keller 1985). Chisels must have steel shafts, while sculptor's knives are usually mounted with wooden handles. Both knives and chisels can be ground down on carborundum grinding wheels, and finished by hand on an oilstone to obtain the desired shape and bevel for any specialized job. The most important point is to ensure that the sloping faces are not too steep, restricting the effectiveness of the tool in cutting through the rock. Chisels should be of tempered steel, ranging in size from 5-15mm, although this depends very much on the job in hand.

Before knives and chisels are used it is important to ensure that the specimen is well supported and consolidated, and that undue shock will not result in its fragmentation. Chisels should be held at an angle of approximately 60° to the specimen surface, with the blade placed on the edge of the matrix about 3mm or less away from the exposed fossil. Hand held chisels should be resting on a block, with the wrist tensed in a way that tends to pull the chisel back. When hit with the hammer a small flake of matrix should spring off, revealing more of the fossil beneath. Move the chisel sideways and repeat the process. This action exploits pre-existing planes of weakness between the matrix and fossil.

The use of knives requires a particular dexterity that should be practised before attempting to prepare a sample. This is because the tool is guided by both hands. The knife is held between the middle finger and thumb of one hand, and rested on the other hand near the area to be worked. Thus the lower hand acts as a fulcrum for the knife, which is used to lever away matrix. Care must be taken to ensure that the tool is used only for the levering of the matrix, and not pressed into the specimen, as this may result in a pressure point damaging the fossil. This work should be carried out under a binocular microscope or a magnifying light (Whybrow 1978). In general work away from the fossil towards the edge of the sample, taking advantage of the natural planes of weakness (*e.g.* bedding, fissile, jointing and cleavage) when removing matrix (Sohl 1989).

Mounted steel needles and dental picks can be used in the preparation of small fossils where only a small amount of matrix is required to be removed by a gentle scrapping action (Palmer 1989, Sohl 1989). The needles are usually 3cm in length and can be straight, or have the lower 1cm curved. With continued use they become blunt and have to be re-sharpened.

During the preparation of a fossil, small flakes of shell may become dislodged. These can be re-stuck to the specimen later, while fragile surfaces can be strengthened by a surface application of an acrylic polymer, polyvinyl butyral (PVB) or polyvinyl acetate

(PVA) solution. The application of this may have to be repeated as new fossil material is exposed.

12.3. PNEUMATIC AND ELECTRICAL ROTARY AND RECIPROCAL TOOLS

Most well equipped laboratories now have a variety of specialized hand-held pneumatic and electrical equipment used in the stock removal of matrix from specimens. Accounts of this type of equipment used by palaeontologists in museum laboratories have been described (*e.g.* Bruton 1974, Rixon 1976, Whybrow 1978, Howie 1979a, Wilson 1987, Chaney 1989a). This type of equipment is particularly useful for working on specimens with hard matrices, or where the sample is too small for use of a hammer and chisel.

A number of tools are available, powered by either electricity or compressed air, and rotary or reciprocal (percussion) in action. Preparators may find it necessary to modify standard equipment to suit their own laboratory requirements (Palmer 1989).

Electrical dental drills can be adapted for geological preparation use. One of the most versatile on the market is the German designed *ELCO-S* rotary hand working machine. It is available in the U.K. through either *Brooks and Walker Tools* or *Claudius Ash* (see list of suppliers) This versatile chunky pen type tool is virtually maintenance free, has a variable motor speed running between 2, 500 to 32, 000 rpm, in either a clockwise or anti-clockwise motion. As power is applied using a foot control pad, both hands are free to hold the specimen and manipulate the hand-tool. A rapid exchange of working bits is possible by a quarter turn of the hand piece grip and interchanging $^1/_8$" $^3/_{32}$" and 3mm collets and fitting one of the vast range of diamond wheels, points, tungsten carbide cutters, polishing or grinding abrasive wheels. Use can be extended by the fitting of a *Diprofil* right-angle pistol hand piece. This converts the machine from a rotary type to a reciprocating action, capable of taking points and chisels.

Burgess Tools electrical hand engravers have for many years been used by hobbyist's. A number of variations of these tools are available, the most common being a two speed version. The high speed setting is 100 strokes per second and the low speed setting is 50 stokes per second. The former setting is probably the most useful in palaeontological preparation, utilizing a variety of hard points and chisel ends. Specimens should be supported on a shock absorbent base (foam-rubber or sand bag), as excessive pressure can result in damage to the surface of the fossil, or splitting of the specimen along pressure induced fissures. The successful development of fossils is done by lightly placing the point of the engraver on the matrix and holding it like a pen. Light pressure only should be used in a sideways action for surface smoothing, or back and forth for large matrix removal (Palmer 1989). An air line can be adapted to blow away dust and freed rock particles (Robison 1965, Palmer 1989).

The use of pneumatic tools in palaeontology was first documented over 30 years ago (Jones 1969, *in* Rixon 1976). They can be used for rapid stock removal and development of fossils. A compressed air supply of between 80-100 psi, and a sufficient flow rate (7-8 cfm) is required. Simultaneous use with other pneumatic tools also has to be taken into account when installing an air line system in the laboratory. A three horse power compressor will be able to deliver 12-15 cfm, while larger compressors usually require a three phase electricity supply. All compressors can be very noisy if they are in constant use. It can make the working environment better if they are housed outside the laboratory, although if they have a large enough reservoir, and can re-charge quickly enough, this is usually not a problem. Air lines should be fitted with air filters and lubricators. However, if an airbrassive unit (see below) is also used, an oil filter will be required on the line. It is important to get these fitted in the correct order and position along the line to ensure maximum efficiency from all the tools that are used.

Hard wearing tungsten tipped points or chisels are the usual bits used on these tools. The procedure is similar to that outlined for the electric engraver. Two models in particular have found extensive use in palaeontological laboratories, the *Desoutter* VP2-X, and the *Chicago Pneumatic Co.*, engraving tool. In both examples, an air port adjacent to the tip of the point acts as an exhaust that blows away debris recently removed from the fossil, although the latter tool has the advantage of having variable strike rate, controlled by a collar at the top of the handle.

All the tools described above should only be used with the operator wearing suitable eye protection, and working in a dust-box connected to a dust extracting system. Noise generated by the compressor and dust extraction motors, as well as the tools, means that ear protection is also essential when using this type of equipment.

12.4. AIR-ABRASIVE TOOLS

The *Airbrasive* system available from *S. S. White* has been in use in palaeontological laboratories for over thirty years (Stucker 1961, *in* Rixon 1976; Stucker *et al.* 1965; Hannibal 1989). Fundamentally, a compressed air stream, in the range of 40 to 140 psi, and saturated with a fine jet of powder, is directed under pressure towards the matrix surrounding the specimen. Unlike the pneumatic tools outlined above, where the air supply must be oil lubricated for the bit to work effectively, in an *Airbrasive* system, the air supply must be dry and free of moisture to allow for the free flow of the powder. The system now available (*Airbrasive 6500* system 2 plus), consists of a powder delivery unit, an illuminated work chamber, foot switch and hand piece. The work chamber is connected via an exhaust hose, to a *Torit* dust collector. A variety of powders are available, ranging in size from 10μm to 100μm, and composed of either aluminium oxide, sodium bicarbonate, silicon carbide or glass (Table 12.1). Dolomite powder (Hannibal *et al.*, 1988) can also be used. In practice, it is advisable to remain with just one type of powder thus avoiding contamination in the powder reservoir.

The hand piece can be fitted with a variety of nozzles, either straight or angled, and either round or rectangular in diameter. For effective use it is important to use a nozzle of sufficient diameter to allow for the free flow of powder. Failure to do this results in the system clogging, and increased wear on the powder delivery airline. With heavy use the nozzle, its holder, hose and pinch valves should be regularly inspected and replaced when signs of wear are evident.

The system can be used to clean specimens and cut away matrix (Table 12.1). Cutting and cleaning rates are a function of the powder composition, operating pressure, powder flow rate, nozzle size and nozzle tip distance from the surface of the specimen. If there is any danger of damaging the fossil, the pressure or powder flow rate must be lowered. The system is very easy to operate, but like other mechanical procedures, requires patience, dexterity and practice to obtain the best results. A general outline of use is as follows:

1. Ensure that the powder chamber is full.

2. Turn on the dust collecting unit (remember first on and last off).

POWDER GRAIN SIZE & COMPOSITION	APPLICATION	COMMENTS
10μm, Aluminium oxide	Cutting, polishing	Fine finish, difficult to flow
17μm, Aluminium oxide	Cutting, deburring, etching, demarking	Fine finish
27μm, Aluminium oxide	Cutting, drilling deburring, etching	Medium cutting speed; smooth finish
50μm, Aluminium oxide	Cutting, drilling deburring, etching	Coarse finish
50μm, Silicon carbide	Cutting	Fastest of all powders, may discolour workpiece
50μm, Glass bead	Cleaning, surface finishing	Mild cleaning
75μm, Crushed glass	Cleaning, light cutting	Coarser finish than glass bead
100μm, Sodium Bicarbonate	Cleaning, enamel removal, polishing	Cool, dry storage required

Table 12.1. Size range and composition of powders available from *S. S. White* Inc., for use in an *Airbrasive* unit. The effectiveness of the powder used is dependent on the shape of the nozzle used, and the distance of the specimen from the nozzle tip. (Data for table from *S. S. White*).

3 Check the nozzle to ensure it is the correct type for the work-in-hand. If necessary select an alternative and fit it to the hand piece (Checking for wear can also be done at this stage). Place the hand piece in the work chamber with the specimen. Turn on the lights within the chamber. Additional lighting can be obtained with use of a free standing magnifying light.

4 Turn the "on-off" switch of the powder delivery unit on, (the ball indicator above the switch turns red when the powder chamber is pressurized). Observe the pressure gauge and adjust to about 60 psi. Adjust the powder flow setting to coincide with the nozzle type selected, (in general small diameter circular or rectangular nozzles, ranging in size from 0.007 to 0.017mm should be used for flow rates not exceeding 3).

5 Direct the nozzle at the specimen surface and fully depress the foot switch. A fine jet of powder will flow and strike the specimen surface. Ensure that damage is limited to the matrix only. Release the foot switch after a few seconds to observe the results.

6 The powder flow rate, operating pressure and distance the nozzle is away from the specimen surface can all be adjusted to accommodate varying degrees of stock removal.

To avoid contamination problems with different powders in the same reservoir, and to provide greater versatility with this procedure, a miniature air-abrasive unit of the type described by Gunther *et al.*, (1979) can be fitted into the air line system and used in the work chamber. Its small powder chamber is ideal to contain the finest abrasive powder, and when used in tandem with the *Airbrasive* unit described above it provides the final polish and finish to a specimen.

Air-abrasive techniques are best avoided on specimens that may require scanning electron microscopy (SEM) examination. The use of even the finest powder for cleaning will leave evidence that is detectable from SEM observations as fine surface pitting. Surface cleaning of robust specimens to be examined by SEM is best undertaken using and ultrasonic tank.

12.5. ULTRASONIC TOOLS

12.5.1. Ultrasonic tanks

Before to the development of ultrasonic tanks chemical methods were employed in specimen cleaning (*e.g.* Bassler 1953, Rasetti 1947). However, these methods have a disadvantage in that they may alter the composition of the specimens outer surface. The means by which ultrasonic vibrations could be used in the preparation of specimen surfaces, without inadvertently altering them, was most welcome by Palaeontologists and Preparators alike. Nowadays, all are aware of their existence, and most will have used them (see Stevens *et al.* 1960, for a review of early applications). There are, however, some justifiable reservations in their use, particularly in micropalaeontology (Hodgkinson 1991). Ultrasonic tanks should only be used for fossil cleaning when there is no danger of damage to the specimen surface. Broken or cracked material may be damaged further by prolonged use of this action (Stevens *et al.* 1960). The method is ineffective on soft materials, such as rubber, cloth and fibres (Pojeta and Balanc 1989a)

Safety Note: Although appearing innocuous, never immerse fingers or hands in an active ultrasonic tank as serious damage to subcutaneous soft tissue can result. Always switch the machine off when moving, changing or examining the specimen being cleaned.

Tanks (sometimes referred to as baths) come in various sizes, but all work on the same principle, with transducers fixed to either the tank bottom or sides. These are activated by ultrasonic oscillators, converting the electrical energy into mechanical energy and generating waves (Fig. 12.1a). As the waves radiate through the liquid (usually water plus detergent), they cause high and low pressure areas. During the low pressure stage millions of bubbles form and grow in size (Fig. 12.1b). This action is known as cavitation (meaning "formation of cavities"). The bubbles implode or collapse, during the high pressure stage (Fig. 12.1c), releasing energy and producing shock waves that dislodge loose material from surfaces, cracks and fissures within the specimen. This action all occurs at about 55,000 cycles per second (Pojeta and Balanc 1989a).

Specimens are treated by placing them in water with a few drops of detergent or other wetting agent, which accelerates cleaning. Experimental work involving the addition of abrasives to solutions resulted in excessive pitting of specimens (Stevens *et al.*, 1960), and is not recommended. Time required is dependent on the robust nature of the fossil and its matrix. Some corals and brachiopods may require many hours (or even days) to remove all matrix, while individual microfossils (usually treated in small glass vials) require only a few seconds, depending on matrix hardness. Ultrasonic tanks can be used for disaggregating sediments (Gipson 1963), although this usually results in the destruction of any microfauna (Sohn 1960). They have also been used in the

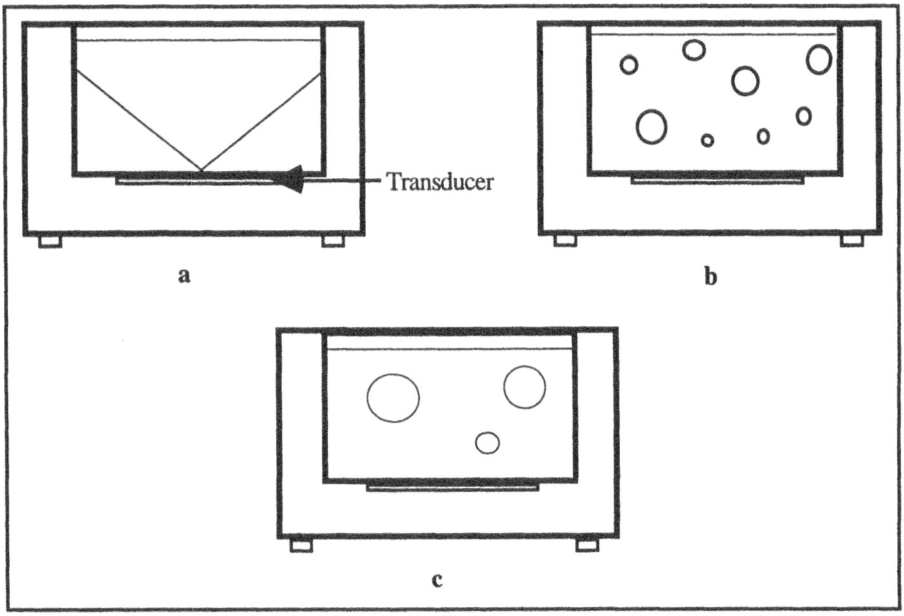

Figure 12.1. Diagrammatic illustrations of the main phases recognised during ultrasonic cleaning. (a), generation of sound waves from a transducer attached to the base of the tank. (b), low pressure stage (cavitation) - during which millions of microscopic bubbles are formed. (c), high pressure stage - bubbles implode and energy is released. (Re-drawn from Anonymous 1986).

deflocculation of pollen and spore rich residues (Gray 1965b). The procedure is excellent for removal of dust from specimens, and particularly efficient in removing surface coatings of magnesium oxide from megafossils (Pojeta and Balanc 1989a)

One of two methods can be employed in the cleaning of material, the choice of which is dependent on specimen size, number and fragility of the material being cleaned. The two methods are (i) direct, (ii) indirect. Each will be briefly outlined.

(i) Direct method

In this method the cleaning solution is poured directly into the tank, and the specimens placed on a perforated tray lowered into the tank (Fig. 12.2). In some circumstances' specimens may be placed directly on the floor of the tank, although this is not recommended (see below). The technique is best suited to cleaning larger specimens. A disadvantage is that only one cleaning solution can be used at a time, and that highly acidic or caustic solutions in direct contact with the tank may result in corrosion of the metallic surface.

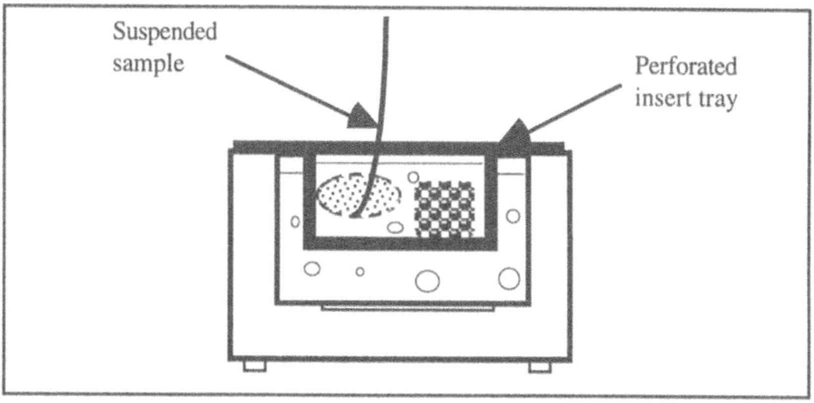

Figure 12.2. Direct method of cleaning specimens. Specimens placed in a perforated insert tray or suspended in the cleaning solution in the tank. (Re-drawn from Anonymous 1986).

(ii) Indirect method

Small or delicate specimens and specimens from different samples are not placed directly in the ultrasonic bath, but treated separately in fluid filled beakers containing a cleaning solution (Fig. 12.3). The beakers are then placed within rubber ringed positioning inserts (Fig 12.3a), or on an insert tray supported by the tank rim (Fig. 12.3b). Many types of insert are available from suppliers. The advantage of using this method is that multi-specimen preparation can be undertaken, and one or more cleaning solutions can be used simultaneously.

(iii) Precautions with ultrasonic tanks

Manufacturers recommend not to rest samples directly on the base of the tank. Excessive weight on the tank bottom dampens the energy sound, and therefore the efficiency of the machine, and may result in damage to the transducer (Pojeta and Balanc 1989a). Reduce the sample size or number being cleaned and if possible use a perforated insert tray. It is faster and more efficient to clean several small loads. Effective ultrasonic cleaning depends on vigorous cavitation in a large fluid volume in relation to the surface area of the object being cleaned.

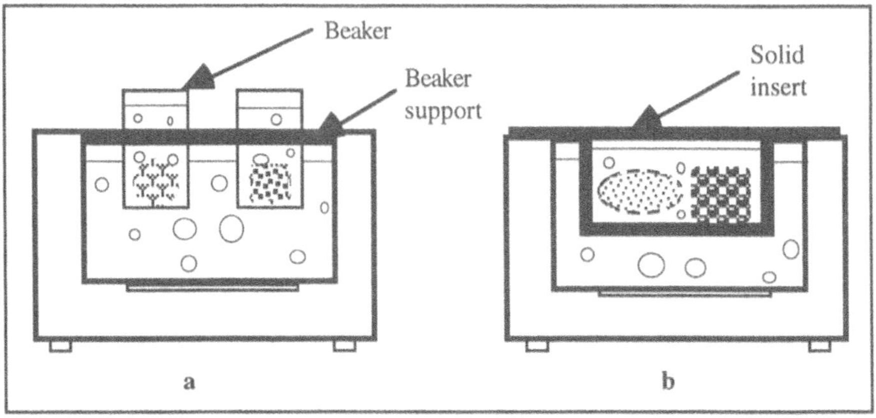

Figure 12.3. Indirect methods of cleaning specimens. (a), specimens in beakers containing the cleaning solution, and supported by a rubber ringed beaker support. (b), specimens in a solid insert tray. (Re-drawn from Anonymous 1986).

Soft tissues can be permanently destroyed by this action, so never place hands or fingers in the tank while it is switched on (see safety note above). Always ensure that sufficient water is in the tank, and never operate it dry. Monitor operation times so that it is not allowed to run for excessive (4 hours +) periods of time, as water will begin to boil. Do not allow long periods of running time unattended (for the safety of the machine and the specimens). Never use solvents either directly in the tank, or as the medium in which to place specimens (*i.e.* recent material stored in alcohol should be thoroughly washed first).

(iv) Ultrasonic descalers

Ultrasonic descalers, as used by dental practitioners, are becoming particularly popular amongst preparators for very fine cleaning of delicate areas (*e.g.* hinge lines and ribs on molluscs, veins on palaeobotanical specimens). One such suitable machine is the *Bondent* periosonic ultrasonic descaler. Operating on a similar principal to the ultrasonic tank, except the transducer is set in a pen-type handle, and the ultrasonic pulse is emitted in a fine jet of water. This compact, user friendly tool consists of a small control box in which both the water supply and power supply can be varied. Once switched on, the tool, held like a pen, is foot operated. By carefully drawing the tool over the surface of the fossil it is possible to remove the finest of matrix without causing any damage to the fossil, although as with ultrasonic tanks it is advisable to test on a small sample. This process offers the least stock removal of matrix from fossil material.

European suppliers of this and other mechanical equipment discussed in this section can be found in the supplier's section of Appendix 1. Equipment suppliers in North America are listed in Chaney (1989) and Feldman *et al.*, (1989).

REFERENCES

ANONYMOUS. 1986. *Bransonic ultrasonic cleaner, operator's manual.* Branson Europa, Netherlands. 80p.

BASSLER, R. S. 1953. Techniques, pp G16-G21: In; R. C. Moore (ed),.*Bryozoa, Part G of Treatise on Invertebrate Paleontology*. Geological Society of America and the University of Kansas.
BERDAN, J. M. 1989. Mechanical extraction of microfossils, pp 99-100; In: R. M. Feldman, R. E. Chapman and J. T. Hannibal (eds), *Paleotechniques*. The Paleontological Society Special Publication. **4**. iv + 358.
BRUTON, T. 1974. Pneumatic hand tools. *Newsletter of the Geological Curators Group*. **1**, (2), 70-72.
CHANEY, D. S. 1989a. Hand-held mechanical preparation tools, pp 186-203; In: R. M. Feldman, R. E. Chapman and J. T. Hannibal (eds), *Paleotechniques*. The Paleontological Society Special Publication. **4**. iv + 358.
FELDMAN, R. M, CHAPMAN, R. E and HANNIBAL, J. T. (Eds). 1989. *Paleotechniques*. The Paleontological Society Special Publication. **4**. iv + 358.
GIPSON, Jr, M. 1963. Ultrasonic disaggregation of shale. *Journal of Sedimentary Petrology*. **33**, (4), 955-958.
GRAY, J. 1965b. Extraction techniques, pp 530-587: In; B. Kummel and D. Raup (eds), *Handbook of Paleontological Techniques*. W. H. Freeman and Co , (San Francisco). xiii + 852p.
GUNTHER, L. F, GUNTHER, V. G and RIGBY, J. K. 1979. An economic miniature sandblaster for preparation of fossils. *Journal of Paleontology*. **53**, (2), 506-507.
HANNIBAL, J. T. 1989. The air-abrasive technique, pp 227-236; In: R. M. Feldman, R. E. Chapman and J. T. Hannibal (eds), *Paleotechniques*. The Paleontological Society Special Publication. **4**. iv + 358.
HANNIBAL, J. T, WILLIAMS, M. E and JACKSON, G. L. 1988. An inexpensive source of dolomite powder for use with airbrassive units. *Journal of Paleontology*. **62**, (2), 316.
HODGKINSON, R. L. 1991. Microfossil processing: a damage report. *Micropaleontology*. **37**, (3), 320-326.
HOWIE, F. M. P. 1979a. Physical conservation of fossils in existing collections. *Newsletter of the Geological Curators Group*. **2**, (5), 269-280.
JONES, M. D. 1969. A pneumatic power-tool for the palaeontologist. *Museum Assistance Group Newsletter, March*.
LÖRCHER, F and KELLER, T. 1985 (for 1984). Preparation techniques for material from the Posidonienschiefer (Lias Epsilon, Upper Liassic) of Germany. *The Geological Curator*. **4**, (3), 164-168.
MARTINSSON, A. 1965. Preparation of individual microfossils on cellulose acetate films, pp 247-251: In; B. Kummel and D. Raup (eds), *Handbook of Paleontological Techniques*. W. H. Freeman and Co., (San Francisco). xiii + 852p.
PALMER, A. R. 1989. Techniques for mechanical extraction of smaller fossils, pp 208-212; In: R. M. Feldman, R. E. Chapman and J. T. Hannibal (eds), *Paleotechniques*. The Paleontological Society Special Publication. **4**. iv + 358.
POJETA, Jr., J and BALANC, M. 1989a. Uses of ultrasonic cleaners in Paleontological laboratories, pp 213-217; In: R. M. Feldman, R. E. Chapman and J. T. Hannibal (eds), *Paleotechniques*. The Paleontological Society Special Publication. **4**. iv + 358.
RASETTI, F. 1947. Notes on techniques in invertebrate paleontology. *Journal of Paleontology*. **21**, (4), 397-399.
RIXON, A. E. 1976. Chapter 4, Mechanical development, pp 59-83; In: *Fossil Animal Remains*, their preparation and conservation. Athlone Press, (London). viii + 304p.
ROBISON, R. A. 1965. Use of the Vibro-tool for mechanical preparation of fossils, pp 267-270: In; B. Kummel and D. Raup (eds), *Handbook of Paleontological Techniques*. W. H. Freeman and Co., (San Francisco). xiii + 852p.
SOHL, N. F. 1989. Mechanical extraction of fossils extraction by hammers, chisels, and picks, pp 204-207; In: R. M. Feldman, R. E. Chapman and J. T. Hannibal (eds), *Paleotechniques*. The Paleontological Society Special Publication No. **4**. iv + 358.
SOHN, I. G. 1960. Cleaning ostracode valves with ultrasonic vibrations. (Abstract only). *Bulletin of the Geological Society of America*. **71**, 1982.
STEVENS, C. H, JONES, D. H and TODD, R. G. 1960. Ultrasonic vibrations as a cleaning agent for fossils. *Journal of Paleontology*. **34**, (4), 727-730.
STUCKER, G. F. 1961. Salvaging fossils by jet. *Curator*. **4**, (4), 332-340.
STUCKER, G. F, GALUSHA, M. J and MCKENNA, M. C. 1965. Removing matrix from fossils by miniature sandblasting, pp 273-274: In; B. Kummel and D. Raup (eds), *Handbook of Paleontological Techniques*. W. H. Freeman and Co., (San Francisco). xiii + 852p.
WHYBROW, P. J. 1978. Fibre optics and foot operated microscope control equipment to assist in the preparation of fossils. *Newsletter of the Geological Curators Group*. **2**, (3), 141-143.
WILSON, J. 1987. Approaches to the preparation and development of calcareous fossils from calcareous matrices, pp 447-450; In: P. R. Crowther and C. J. Collins (eds), The Conservation of Geological Material. *The Geological Curator*. **4**, (7), 375-474.

13. FOSSIL EXTRACTION TECHNIQUES BY THERMAL DISINTEGRATION

13.1. INTRODUCTION

An alternative method of fossil preparation to acid and mechanical tool extraction involves the use of rapid temperature changes. This may be in the form of either heating or rapidly cooling rocks to free matrix enclosed fossils. The principle behind these procedures is similar, exploiting the differential thermal expansion rate between the fossil and surrounding matrix. The fossil-matrix interface represents a natural line of weakness along which a fracture is initiated and extended. The procedures are of particular use where the fossil and matrix composition are similar, inhibiting the effectiveness of partial matrix dissolution by acids and an ill-defined fossil-matrix boundary exposed during mechanical tool preparation. Despite compositional similarities between the fossil and matrix, they frequently have different micro-crystalline orientations, and by lowering and raising the temperature above a mineral inversion or transformation point, a parting of the two may result. This feature can be exploited following methods detailed below. Variations discussed are:

(1) high temperature fracturing,
(2) high temperature heating and quenching,
(3) freeze-thaw fracturing.

The procedures have been used in the extraction and preparation of both macro and microfossil material.

Safety Note: All procedures must be undertaken in a well-ventilated fume cupboard or fume hood, with the operator wearing appropriate PPE (heat resistant gloves, apron, goggles).

13.2. HIGH TEMPERATURE FRACTURING METHOD

A successful method of extracting acid soluble macrofossils from limestones, dolomites and sandstones was developed and described by MacVicar (1951a; 1951b; 1952), and used by Adamczak (1961) for the extraction of ostracods from limestones. This little known technique was originally devised for the extraction of partially altered trilobite carapaces, of phosphatic and carbonate composition, from limestones. It was also successfully used to release fossils of phosphatic and siliceous composition held by a calcareous cement from dolomitic limestones. A successful disaggregation requires that only partial replacement of primary biogenic phosphate (*e.g.* trilobites) and silica (*e.g.* porifera) by carbonate has occurred. MacVicar (1951b) favoured this technique for fossils of siliceous composition, considered too delicate for standard acid extraction techniques.

MacVicar (1951b) describes the process as selective chemical decomposition of granular rocks. Temperatures are maintained long enough for the calcite matrix (or cement) to convert into a soft powdery calcium oxide, in the process releasing the harder phosphatic and siliceous fossils (MacVicar 1952). The only adverse effect to trilobites observed was the partial thermal reorganisation of the tri-calcium phosphate carapace, resulting in a glossy appearance, and limiting use of such material in shell ultrastructure studies.

For thermal fracturing to be successful a number of criteria must be observed.

- Slight compositional difference between the fossil and matrix must exist, and reflect different thermal expansion rates.
- The composition of the fossil must contain significant traces of either phosphate or silica.
- The matrix binding cement must be calcareous in composition.

13.2.1. Heating procedure

1 Heat the sample to 100°C in an electric furnace, and maintain the temperature for at least an hour to ensure that all interstitial water is removed.

2 Steadily raise the temperature to 1000°C, at a rate of 4°C per minute. Maintain the temperature for 1-3hours, depending upon specimen. This ensures total conversion of the calcite.

3 Switch the furnace off, and with the door shut allow to cool to room temperature. **N.B. Do not attempt to speed up this process by quenching the sample in water. Internal stresses within the sample will cause it to shatter, and might damage the fossil.**

4 Once cool enough to handle, remove the sample from the furnace and place it in a desiccator. This procedure will prevent sample rehydration, and potential fossil damage. Robust microfossils can be retrieved from the crumbled matrix within 48 hours by allowing the specimen to cool under normal atmospheric conditions.

13.2.2. Removal of fossil material

A great deal of care and patience is required in removing rock matrix from fossil material.

1 Removal of the matrix can be aided by using a mounted needle, scraping at the surface, and then gently brushing it away with a sable hair paint brush, or by use of a mechanical air duster.

2 For particularly resistant areas, a drop of water administered from a pipette, hydrates the matrix, causing sudden local expansion. Too much water may result in damage and loss of the fossil.

3 Once cleaned of matrix, microfossil material can be removed from the desiccator. Macrofossils can be coated with a PVA preservative as they are being extracted, this will prevent disruption along the fractures.

Micropalaeontological applications are limited to radiolaria, ostracods and foraminifera, particularly those which are silicified or phosphatized. MacVicar (1951b) also speculates that the technique might be applicable to Precambrian material. Heating of partially consolidated sediments was successfully used by Patrick and Reimer (1966) to induce oxidation of organic matter and release slightly charred fossil diatoms and other siliceous skeletons.

A variation on the procedure to assist in micropalaeontological preparations, using a domestic pressure cooker was outlined by Driver (1928). This method allows for multiple samples to be processed simultaneously, and proved successful in breaking down all but the most indurated rocks tested. However, care must be employed when slowly releasing the pressure valve to avoid excess boiling increasing the possibility of cross sample contamination. Treated samples are then wet sieved or processed further by standard micropalaeontological procedures. Driver (1928) further speculated on the possible use of a modified autoclave in the procedure, offering greater pressure and temperature variations, increasing the effectiveness of the technique.

13.3. HIGH TEMPERATURE HEATING AND QUENCHING METHOD

Despite the reservations of quenching heated specimens in cold water expressed by MacVicar (1952), the method has been successfully used by Pojeta and Balanc (1989b). They have extracted Upper Cambrian mollusc specimens predominantly in the size range 5-7mm, although specimens up to 5cm in length have also been released. The method will be familiar to many European micropalaeontologists who have prepared and examined split sections of Tertiary *Nummulites* sp., from the Anglo-Paris Basin (see section 20 THIN SECTION AND SLIDE PREPARATION TECHNIQUES OF MACRO AND MICROFOSSIL SPECIMENS for details).

13.3.1. Heating procedure

1 A wire gauze, placed on a standard laboratory tripod, supports the specimen above a bunsen burner. Pojeta and Balanc (1989b) suggest a *Meker*-type high temperature (butane gas) burner, which produces a wide flame of uniform intensity.

2 Cover the specimen with a metal cylinder perforated at the base. Pojeta and Balanc (1989b) used a 15cm, 1kg coffee tin with holes punched in the base. This ensured the even distribution of heat, and offered some protection should the rock explode.

3 After approximately 1 hour of direct heating, remove the specimen using tongs or forceps, and immediately place in a stainless steel beaker or saucepan.

4 Preferably working in a deep sink, flood the container with a stream of cold running water. When the temperature of the specimen has reached that of the water, the procedure can be repeated.

13.3.2. Removal of fossil material

Repeating the procedure several times may result in the complete removal of matrix by simply crumbling between the fingers. For larger specimens mounted needles or a small chisel and hammer can be used.

13.4. PREPARATION INDUCED ALTERATION

13.4.1. Fracturing

Although Pojeta and Balanc (1989b) found no direct evidence to confirm MacVicar's (1952) findings of fractures within material resulting from the heating and quenching procedure, some specimen modifications were observed. Mollusc and articulated brachiopods separated from the internal moulds of the fossils, and the shell material was often fragmented. In some cases this was considered advantageous, as the internal mould provided additional information obscured in shelled examples.

Fractures observed in larger fossils that might have been formed by heating and quenching could not be differentiated from those evident in mechanically prepared specimens, or even unprepared specimens collected at outcrop (Pojeta and Balanc 1989b, p. 220). Diagenesis, weathering and taphonomic factors are all considered to influence the preservation potential of specimens from Antarctica (Pojeta and Balanc 1989).

The heating and fracturing procedure worked best on dense argillaceous coquina limestones with sparite filled pore spaces. Some success was also seen in limestones and dolomites in which the fossils are phosphatic or siliceous in composition.

13.4.2. Chemical transformations

Success of fossil extraction in using one of these methods requires knowledge and understanding of how material has been preserved, and subsequent modifications induced by weathering. Furthermore, establishing the chemical composition of the fossil material, and determining physical and chemical changes that might be introduced when either heating or heating and quenching is undertaken. Quantifiable analytical determinations of chemical and physical changes of the fossil or its matrix have yet to be undertaken. Both MacVicar (1952) and Pojeta and Balanc (1989b) note the production of calcium oxide (CaO) on carbonate fossils from which carbon dioxide (CO_2) has been liberated.

13.5. FREEZE-THAW FRACTURING

The action of freeze-thawing rocks to assist in their breakdown has been used by geologists for nearly 70 years, and is recognised as an accelerated form of mechanical weathering (Hanna and Church 1928). Its adaptation to palaeontology was soon described (Camp and Hanna 1937), although more recent successes have been in disaggregating clay-rich sediments containing microfossils (*e.g.* Sohn 1961, Sohn *et al.*, 1965, Abelmann 1988, Hinchey and Green 1994).

13.5.1. Methods

A variety of solutions have been successfully used in this procedure, from tap water (Hanna and Church 1928), to water plus a dissolved sodium salt (Sohn 1961, Sohn *et al.*, 1965), and liquid nitrogen (Hinchey and Green 1994). Regardless of the rock type or solution used, the principle of the technique is similar: (i) specimens are soaked in a solution until all pore spaces are fluid filled; (ii) the fluid is cooled until it solidifies, a process that results in the formation of crystals within the pores, and with continued growth disaggregate the rock (Pojeta and Balanc 1989c). However, for obvious reasons the procedure will not work well on impervious rocks (Pojeta and Balanc 1989c).

13.5.1a. Tap water (H_2O) (Hanna and Church 1928)

1 Place specimens in a suitable container in which when filled with fluid, specimens are covered.

Pojeta and Balanc (1989c) suggest some thought should be given in choosing the container. They favour inexpensive domestic saucepans (aluminium metal containers with long handles, preferably with a pouring lip), as these are inexpensive and allow for solutions to be easily decanted. Some plastic containers may be suitable, but glass or porcelain are susceptible to fracturing when the solution freezes, while paper containers have a shortened life expectancy as they tend to absorb the solution (Pojeta and Balanc 1989c, p 223). Ensure a variety of container sizes are available to accommodate all specimen sizes.

2 Soak specimens in a solution until thoroughly saturated.

3 Place in a deep freeze, and periodically observed until frozen. (Plan the procedure so that specimens can be left overnight).

4 Remove the sample from the freezer, and bring the sample back to room temperature. Decant the excess solution, remove and dry the sample fragments and examine.

Before repeating the procedure, wet sieve and remove disaggregated fragments. Porous fossil material may be damaged if subjected to repeated freeze-thaw action (Sohn 1961). This procedure was favoured by Baird *et al.*, (1985) for splitting a large number of the 285,000 Pennsylvanian concretions containing Mazon Creek fossils. They considered (p. 258) "this process was found to be superior to hammer pounding, which damages specimens and results in poorer fossil yields".

13.5.1b. Water with sodium acetate *3-hydrate* ($CH_3.COONa.3H_2O$) (Sohn 1961, Sohn *et al.*, 1965)

This method is considerably quicker than freeze-thaw with water (Sohn 1961).

1 Oven dry samples at 120°C for 1 hour.

2 Break sample into $2cm^3$ fragments and place in a heat resistant container.

3 Cover the sample with sodium acetate *3-hydrate* (sodium acetate trihydrate).

4 Add a few drops of water. Cover the container and place on a hotplate set at a low temperature (sodium acetate melts at 120°C). Maintain the temperature until the sodium acetate has melted, and soaked into and saturated the sediment.

5 Remove from heat and add a few fresh crystals of sodium acetate to assist in crystal nucleation. Cool the container by immersing in cool running water. The sodium acetate will crystallise, and disaggregate the rock.

6 Slowly re-heat the specimen, and carefully decant excess sodium acetate, which can be reused in repeating the procedure together with fragments requiring further disaggregation.

7 It is essential to remove disaggregated fragments before repeating the procedure to avoid the destruction of fossils. Treated fragments should be boiled in water and rinsed through a nest of sieves to remove excess sodium acetate. The repeated process is aided by oven drying fragments before sodium acetate immersion and re crystallisation.

13.5.1c. Water with sodium sulphate *10-hydrate* ($Na_2SO_4.10H_2O$) (Surlyk 1972)

1 Oven dry samples at 30°C, temperatures in excess of this may cause fine grained calcareous sediments (particularly chalk) to harden.

2 Cover samples with a 40°C (maximum solubility at this temperature) supersaturated solution of hydrated sodium sulphate (sodium sulphate decahydrate, Glauber's Salt).

3 Allow the solution to cool (may take several hours), and decant excess solution.

4 Place sample in a freezer.

5 Re-heat the sample to 40°C and repeat the process 16 to 18 times.

6 Following the final process, wet sieve the residue through a nest of sieves. Dry the residue, and dry sieve if required before examining.

Surlyk (1972) used this procedure to extract small brachiopods from 10kg samples of soft friable Cretaceous chalk. However, this is a particularly long and time consuming procedure, which Sohl reports (in Pojeta and Balanc 1989c, p. 225) can be accelerated by performing the heating and cooling stages (2 and 3) under vacuum.

13.5.1d. Water with *tri*-sodium citrate ($Na_3C_6H_5O_7.2H_2O$) (Pojeta and Balanc 1989c)

Plaster of Paris has many diverse applications in fossil preparation, used in constructing plaster jackets in the field before transportation to the laboratory, providing supporting bases for material in laboratory preparation and on display, as well as use as a filler (Rixon 1976, Sohl 1989, Pojeta and Balanc 1989c). Frequently the removal of the plaster is required to enable on going work to continue. Hydrated *tri*-sodium citrate provides a method of achieving this, although it is not strictly a "freeze-thaw" method.

1. Make up a saturated solution by dissolving *tri*-sodium citrate in a beaker of warm water.

2. Cover the specimen, and let it stand until the plaster is reduced to a thick watery paste.

3. Remove the specimen, and wash under warm running water to remove any residual paste adhering to undercut areas.

4. Repeat the process if required, and after the final wash, dry the specimen.

REFERENCES

ABELMANN, A. 1988. Freeze-drying simplifies the preparation of microfossils. *Micropaleontology*. **34**, (4), 361.

ADAMCZAK, F. 1961. Eridostraca - a new suborder of ostracods and its phylogenetic significance. *Acta Palaeontologica Polonica*. **6**, (1), 29-104.

BAIRD, G. C, SHABICA, C. W, ANDERSON, J. L and RICHARDSON, Jr., E. S. 1985. Biota of a Pennsylvanian muddy coast: habitats within the Mazonian delta complex, Northeast Illinois. *Journal of Paleontology*. **59**, (2), 253-281.

CAMP, C. L and HANNA, G. D. 1937. *Methods in Paleontology*. University of California Press, (Berkeley, California). xxiii + 153p.

DRIVER, H. L. 1928. Aid in disintegrating samples for micro-organic study. *Journal of Paleontology*. **1**, (4), 253-254.

HANNA, G. D and CHURCH, C. C. 1928. Freezing and thawing to disintegrate shales. *Journal of Paleontology*. **2**, 131.

HINCHEY, J. V and GREEN, O. R. 1994. A guide to the extraction of fossil diatoms from lithified or partially consolidated sediments. *Micropaleontology*. **40**, (4), 368-372.

MACVICAR, D. G. 1951a. Heat treatment frees Pre-Cambrian fossils. *Science News Letter*. **59**, (17), 261.

MACVICAR, D. G. 1951b. Extraction of fossils by heat. *Micropaleontologist*. **5**, (3), 15.

MACVICAR, D. G. 1952. Thermal disintegration of sedimentary rocks. *American Journal of Science*. **250**, 271-274.

PATRICK, R and REIMER, C. W. 1966. The diatoms of the United States. Volume 1. *Monographs of the Academy of Natural Sciences of Philadelphia*. 688p.

POJETA, Jr., J and BALANC, M. 1989b. Heating and quenching fossils, pp 218-222; In: R. M. Feldman, R. E. Chapman and J. T. Hannibal (eds), *Paleotechniques*. The Paleontological Society Special Publication. **4**. iv + 358.

POJETA, Jr., J and BALANC, M. 1989c. Freezing and thawing of fossils, pp 223-226; In: R. M. Feldman, R. E. Chapman and J. T. Hannibal (eds), *Paleotechniques*. The Paleontological Society Special Publication. **4**. iv + 358.

RIXON, A. E. 1976. *Fossil animal remains: their preparation and conservation*. Athlone Press, (London). viii + 304p.

SOHL, N. F. 1989. Mechanical extraction of fossils extraction by hammers, chisels, and picks, pp 204-207; In: R. M. Feldman, R. E. Chapman and J. T. Hannibal (eds), *Paleotechniques*. The Paleontological Society Special Publication. **4**. iv + 358.

SOHN, I. G. 1961. Techniques for preparation and study of fossil ostracodes, pp Q64-Q70: In; R. C. Moore (ed), *Treatise on Invertebrate Paleontology. Part Q; Arthropods 3*. Geological Society of America and the University of Kansas (Lawrence, Kansas). xxiii + 442p.

SOHN, I. G, BERDAN, J. M and PECK, R. E. 1965. Ostracods, pp 75-89: In; B. Kummel and D. Raup (eds), *Handbook of Paleontological Techniques*. W. H. Freeman and Co., (San Francisco). xiii + 852p.

SURLYK, F. 1972. Morphological adaptations and population structures of Danish chalk brachiopods. *Kogelike Dansk Videnskabernes Selskab, Biologiske Skrifter*. **19**, (2), 1-57.

14. DISAGGREGATION AND DISPERSAL OF PARTIALLY CONSOLIDATED AND UNCONSOLIDATED SEDIMENTS

14.1. INTRODUCTION

Outline disaggregation and dispersal procedures for partially consolidated sediments have been provided in the excellent works of Gray (1965b), Allman and Lawrence (1972) and Harris and Sweet (1989). The following section updates and modifies the five main stages of treatment that can be applied to unlithified or partially lithified sediments to break them down to their component parts and free macro- and microfossils. Not all stages need necessarily be used on every sample or rock type. The procedures used on a sample are very much dependent on the mineral composition of the fossilized material and its state of preservation. The following five stages will be considered:

(1) Mechanical treatment
(2) Chemical treatment : with acids to remove carbonates, sulphates and sulphides in solution.
(3) Chemical treatment : with oxidizing agents to destroy organic compounds.
(4) The removal of electrolytes by washing.
(5) The addition of peptizers to produce a deflocculated residue.

A survey on the effects of mechanical and chemical processing techniques used in micropalaeontology has been presented by Hodgkinson (1991). It becomes clear from this work that the majority of processes have some lasting, often damaging effect on the microfauna, and that the damage is related to the mode and degree of preservation, besides the time and intensity a sample is subjected to a particular process. In order to minimise the effects it is desirable to use as few processing stages as possible, and to constantly monitor the state of preservation. A full processing history of the sample should be available to the preparator and person working on the fauna so that all possible effects can be considered.

14.2. MECHANICAL TREATMENT

In all the methods outlined below the sample should be thoroughly dried before any procedure described is adopted. The treatments are described in order of increasing severity. Ensure that the surfaces of all consolidated specimens are clean (wash if necessary) and free of loose debris and recent vegetation.

(i) Crumble the dried sample between fingers and sieve through a nest of sieves into various size fractions.

(ii) Immerse the sample in warm water and gently crumble between fingers, then either wet sieve under a directed stream of water, or dry and sieve. Partially consolidated rocks can be gently scrubbed with a brush. This action will generate a suspension that can be wet sieved to obtain a residue.

(iii) Immerse the sample in water and gently bring to the boil. The addition of a concentrated detergent will aid in the breakdown. When disaggregation is complete wet sieve the residue to remove the detergent.

(iv) Soak the sample in petroleum spirit or white spirit for 1-24 hours. Decant off, and filter (for reuse) the excess spirit, then add boiling water to the sample. Once a mud has been produced, wet sieve the residue.

(v) Soak the sample in a 15% (w/v) solution of hydrogen peroxide (H_2O_2) for 24 hours. Top up with water and gently bring to the boil until a mud is produced. Wet sieve the residue. Avoid using this procedure on delicate calcareous material (*e.g* nannofossils), or calcitic material that may subsequently be used in geochemical analyses as some dissolution or etching of material may result (Pingitore *et al.*, 1993).

(vi) Soak the sample in sodium hypochlorite (NaClO) for up to 4 weeks, until a mud is formed. Boil the sample in water with a little detergent then wet sieve the cleaned residue.

In all methods except (iv, which utilizes a flammable chemical), an ultrasonic tank or mechanical shaker can be used at any stage to assist in the breakdown of the sediment. The sediment, together with water, detergent, hydrogen peroxide and sodium hypochlorite solution is placed in a beaker within the ultrasonic tank. Between 5-30 minute's treatment is usually sufficient, this being dependent on the preservation and nature of the fossiliferous material which in some cases may be destroyed during excessive treatment. Ensure that all equipment is thoroughly cleaned between samples.

A pestle and mortar can be used to break down partially consolidated sediments before any of the above treatments being applied. Place the fragments in a mortar with a little distilled water, and crush (do not grind) with a pestle until they are in the range size of 2-5mm in diameter. Some laboratories may, however, have developed customised mechanical preparation equipment for the processing of large numbers of samples (Hussey and Campbell 1951).

Lund (1970) suggested boiling as an alternative procedure. It was successfully used in the extraction and retrieval of small fossil vertebrate and invertebrate remains from shales, claystones and limestones. Boiling with sodium carbonate was considered too violent, resulting in faunal fragmentation, while the petroleum spirit method left an excessive amount of residue. The method described by Lund (1970), undertaken at room temperature, separates "bone from silt by foam", reducing the matrix volume by 60-99% in one treatment. The principle chemical (*Amine 220,* a stearic acid based imidazoline, chemically known as 2-(heptadecenyl)-1-(hydroxyethyl)-2-imidazoline) was manufactured by Union Carbide, but is no longer available (A. J. Garlick, Union Carbide, *pers. com.*). However, similar surfactants can be obtained from other suppliers, although careful testing is required on palaeontological specimens to assess chemically induced damage. Lund (1970) found the best results were obtained after sediments had been partially broken down using either the sodium carbonate or petroleum spirit methods. No pre-treatment was required on claystone lithologies. A brief outline of the procedure, which should be undertaken in a fume cupboard, is given below:

1. One part Methyl Amyl Alcohol is mixed with 4 parts *Amine 220* in a large beaker.

2. To this, add 100 parts warm water via a jet nozzle, making a frothy milky white liquid.

3. Add approximately 30% by volume partly disaggregated sediment to the solution.

4. Occasionally briskly stir the residue in the beaker.

5. Disaggregation ceases when the liquid in the flask forms a thick slurry. This usually takes 1 day.

6. Wash the residue over a 63µm sieve, until thoroughly clean.

7. Allow residue to dry before examining under a stereozoom binocular microscope.

A modified method of acetylation is extremely successful in breaking down indurated limestones and releasing shell debris. The method is particularly effective on marly and sparry calcitic limestones, and less so on micritic limestones (Lethiers and Crasquin-Soleau 1988). The method (see section 30 EXTRACTION TECHNIQUES FOR CALCAREOUS MICROFOSSILS FROM CARBONATE SEDIMENTS for details) uses concentrated acetic acid ($CH_3.COOH$), and can take several weeks to complete. Strong fumes are generated during the process, **which must be undertaken in a fume cupboard with an adequate extraction system removing any vapours.**

14.3. CHEMICAL TREATMENT : ACID REMOVAL OF CARBONATES, SULPHATES, SULPHIDES AND SILICATES

14.3.1. Carbonates

Calcite and dolomite are the two most common carbonate minerals. The complete removal of these minerals is essential in acid-resistant residue preparations that require secondary chemical stages. Palynological preparations can be adversely effected if the carbonate fraction is not removed before the hydrofluoric (HF) acid stage, with the formation of insoluble calcium fluoride (see section 25 EXTRACTION TECHNIQUES FOR PALAEOBOTANICAL AND PALYNOLOGICAL MATERIAL). Conodont preparations may be affected by the inadequate removal of all carbonate from residues before liquid separations using sodium polytungstate. Any residual carbonate, particularly more resistant dolomite, within the residue must be removed to avoid the formation of insoluble calcium polytungstate (Jeppsson and Anehus 1995, see section 27 EXTRACTION TECHNIQUES FOR PHOSPHATIC FOSSILS).

A 10% hydrochloric acid (HCl) will remove carbonate from a residue, while gentle heating of the solution will aid in the removal of dolomite. If the reaction is very vigorous, a few drops of acetone will reduce the surface tension of the liquid, preventing "boiling". Other acids have been used in the removal of carbonates, and should be considered particularly if fossil material vigorously reacts with the acid. These include 10% nitric acid (HNO_3), 10% acetic acid ($CH_3.COOH$) and 6% formic acid (H.COOH). Both acetic and formic acids produce a gentler reaction than either hydrochloric or nitric. Bulk processing for macrofossils using these acids is described by Grant (1989).

Although successfully used in processing conodont rich samples, acetic acid etching of phosphatic macrofossils (brachiopods) was fortuitously observed by Bell in 1942. However, the onset of the Second World War delayed the formal reporting of the procedure (Bell 1948). Little modified, the procedure continues in use today (*e.g.* Lindsay 1987, Grant 1989).

1 Fragmented sediments (slightly larger than the fossils) are placed in a large (1000ml) beaker, and covered with a 20% solution of acetic acid.

2 The solution is left at room temperature for between 48-96 hours, until the reaction has ceased. (The maximum time is now considered excessive, as acetic is known to effect phosphatic material when contact is maintained for a long time - see below).

3 Top up with distilled water. Allow sediment to settle before decanting.

4 If matrix is still visible, repeat the procedure. No advantage is gained by adding fresh acid to the exhausted solution. The reaction will be very slow.

5 Repeat stage 3, washing material with a gentle stream of water over a 63μm sieve. Etched shells are often very fragile.

6 Oven dry at 60°C.

7 Examine under a binocular microscope.

Experimental processing techniques undertaken by Ziegler *et al.*, (1971) have alerted preparators to the dangers of leaving conodont rich residues in contact with acidic solutions for extended periods of time. Selective corrosion was evident on many of the conodonts, particularly compound (ramiform and prioniodiform) elements, which were processed in formic and monochloracetic acids. Ziegler *et al.*, (1971, p 585) concluded that the acid concentration was not solely responsible for the damage, but that the immersion time between residue and acid was also significant. Subsequent work (Jeppsson *et al.*, 1985, and Jeppsson and Anehus 1995) has also alerted to the dangers of acetic and formic acids in conodont processing. However, buffered and double buffered acidic solutions respectively reduce the corrosive action on phosphatic microfossils (see section 27 EXTRACTION TECHNIQUES FOR PHOSPHATIC FOSSILS).

Argillaceous sediments can be treated with a saturated solution of oxalic acid ($(COOH)_2.2H_2O$), while the procedure outlined by Rasetti (1947), proved successful on sediments containing partly weathered fossils (brachiopods, trilobites, corals, echinoderms, bryozoans), requiring additional surface cleaning of excess attached matrix. The chemical dissolution of carbonate minerals is linked to its preservation, so careful examination of the material must be undertaken before proceeding.

1 Immerse specimens in a beaker containing a 20% oxalic acid solution. At room temperature pure crystalline calcium carbonate is hardly attacked, but increased matrix porosity aids in the dissolution.

2 Continually observe the process if the reaction is rapid. The process may take from 2-12 hours, although if specimens remain too long in the acidic solution a white calcium oxalate precipitation may be deposited on the surface. A small soft brush can be used to help with the cleaning of both the matrix and precipitate.

3 Thoroughly wash and dry material.

14.3.2. Sulphates

The removal of gypsum and anhydrite from sediments can be achieved by gently warming the residue in 10% hydrochloric or nitric acid. When the reaction is performed in the ultrasonic tank disintegration times are shorter. Both gypsum and anhydrite are soluble in anhydrous sodium thiosulphate ($Na_2S_2O_3$), one part dissolved in four parts water, and gently heated.

14.3.3. Sulphides

Pyrite and marcasite are responsible for the blue or blue-grey colour of many clays. They are either finely disseminated or in small granular particles. They can be removed by soaking the sample in cold concentrated nitric acid for 10 minutes to 24 hours. It may be necessary to boil the acid if marcasite is present. Sulphur is released during this reaction, which can be removed by dissolving in xylene (C_6H_4 $(CH_3)_2$).

14.3.4. Silicates

Silicates can be removed by using hydrofluoric acid (HF). Before the addition of HF, carbonates must be removed using HCl acid. This is because finely disseminated calcium fluoride is produced following the interaction of HF with calcite. A 50% aqueous solution of HF is sufficient to remove silicates. A 1% solution of HF in 30 to 50% ethyl alcohol is also effective in some cases. The reaction may take a few minutes or many hours (8 to 12 hours is usual). Because HF is very corrosive it is necessary that all work be undertaken in a fume cupboard using polythene containers (see section 25 EXTRACTION TECHNIQUES FOR PALAEOBOTANICAL AND PALYNOLOGICAL MATERIAL for specific use and safety precautions). A brief outline of the procedure is given below:

1 The sample together with a little water is placed in a polythene beaker in a fume cupboard.

2 Add HF a few drops at a time, and stir after each addition. Watch the reaction carefully, and at the least sign of becoming over vigorous or boiling, cease the addition of acid for at least 10 minutes.

3 Loosely cover the beaker. Add more acid periodically until the reaction is complete.

4 Decant off the spent acid into a beaker and neutralize with sodium carbonate (washing soda, Na_2CO_3).

5 Neutralize the residue and rinse with warm 10% HCl then wash with distilled water half a dozen times.

When some sediments are digested in HF the formation of a dense gel may occur. The gel can be precipitated using ammonium hydroxide (NH_4OH), and will dissolve in presence of nitric acid (HNO_3).

14.4. CHEMICAL TREATMENT : OXIDATION OF ORGANIC MATTER

14.4.1. Bituminous material

Although fats, waxes and resins are not common in plants they are resistant to decomposition. Thus they may form up to a third of the organic material in sediments, with the proportion frequently increasing with sediment age. Most bituminous substances are soluble in acetone ($(CH_3)_2CO$), benzene (C_6H_6), petroleum spirit or chloroform ($CHCl_3$). A combination of methanol (CH_3OH), acetone and benzene in the proportion 15:15:70 are suitable for the removal of most bituminous material. This mixture is also used to soften coals and will remove asphalt and crude oil from sediments. One major problem is that it is difficult to filter even dilute solutions of crude oil. Two applications of the solvent separated by treatment in hot dilute HCl (2-4%) may improve the effectiveness of the process because acidic treatment will help to remove even the combined fats and fatty acids.

14.4.2. Proteinaceous material

Many of the proteins of decomposing plants are soluble in water. Others can be removed after 10-30 minute's treatment in dimethylformamide ($H.CO.N(CH_3)_2$).

14.4.3. Cellulose

Forms a major component of the cell wall of plants. Two types of treatment can be used for its removal.

i **Acid-hydrolysis :** A slow process, little used now. Crush the sample and allow it to stand for three hours or more in a small quantity of cold 80% sulphuric acid (H_2SO_4). This process can take 24 hours. **N.B. water must be present for hydrolysis to occur, do not use concentrated (90-95%) H_2SO_4.**

ii **Acetylation :** This is the prime method used for the removal of cellulose from samples. The reagent should be made immediately before use by adding one part concentrated H_2SO_4 to nine parts acetic anhydride $(CH_3.CO)_2O$. The reaction is exothermic and the solution must be handled with care. Water should not be allowed to come into contact with the solution that is potentially explosive (see section 25 EXTRACTION TECHNIQUES FOR PALAEOBOTANICAL AND PALYNOLOGICAL MATERIAL for specific use and safety precautions). A brief outline of the procedure is given below:

1 The washed residue is placed in a 15ml centrifuge tube.

2 Add acetic acid ($CH_3.COOH$), or an organic solvent such as acetone, and mix in with the sample. Allow it to settle and centrifuge, then decant off the excess liquid. Repeat if there is a large amount of sample to dehydrate.

3 Add the freshly mixed acetylation reagent (about 10ml) to the centrifuge tube, and place in a water bath set at 90°C for between 5-30 minutes.

4 Remove the tube and centrifuge. Decant off the excess liquid.

5 Wash the residue in acetic acid to prevent precipitation of cellulose acetate, stir, centrifuge and decant off excess liquid.

6 Wash three or four times with water.

14.4.4. Lignin, Humus and Amorphous compounds

Lignin lends mechanical support to the secondary thickening of the walls of cells, but is chiefly present in the cementing material between cells in plants. Organic material becomes increasingly soluble in alkaline solutions as it decomposes. Un-decomposed lignin must be oxidized in order to remove it.

i **Alkalines :** Dilute solutions (<10%) of KOH, NaOH and NH_4OH have been used. The sample, together with a quantity of the alkali is placed in a porcelain evaporating dish. This is then heated to 100°C in a constant temperature water bath for 5-10 minutes.

ii **Oxidation :** Various oxidizing agents have been used to break down coals, lignites and carbonaceous shales.

14.4.1a. Nitric acid

1 Place 10-20 grams of fragmented sample into a one litre flask.

2 Cover with a solution of 50ml of distilled water and 50-60ml of 50% nitric acid (HNO_3).

3 Leave for 24 hours, or until the lignite is easily broken down with a stirring rod.

4 Fill the flask with water and leave for 24 hours.

5 Decant the excess HNO_3 solution and add fresh water.

6 Wet sieve and wash the residue thoroughly, complete the maceration with an alkali solution as outlined above.

14.4.1b. Schultze's solution

One part aqueous saturated solution of $KClO_3$ or $NaClO_3$ combined with 2 to 3 parts of cold concentrated HNO_3. This solution is used for the maceration of high rank coals. Low rank coals or the organic content of shales or carbonates may be broken down in about 5 minutes. Dilute Schultze's solution can be used, made by using 40% HNO_3, increasing the reaction time, but reducing the vigorous nature of the reaction.

1 Place 10 to 20 grams of fragmented sample into a beaker.

2 Pour in the potassium or sodium chlorate.

3 Slowly add concentrated HNO_3. The reaction is extremely exothermic, it may be necessary to cool the beaker under running water. The more rapidly the reaction occurs the sooner it will be completed.

4 Add water to the beaker, allow the residue to settle then decant.

14.4.1c. Sodium hypochlorite

1 Add a 5% solution of sodium hypochlorite ($NaClO$) to a beaker containing the fragmented sample.

2 Add a few drops of concentrated HCl until the chlorine is released.

3 Oxidation may take from a few hours to a few days.

4 When maceration is complete decant off the excess liquid, and add water.

5 Wet sieve residue through a nest of sieves.

14.4.1d. Hydrogen peroxide

Generally 10 to 30% concentration of hydrogen peroxide (H_2O_2) is used. If the sample is being prepared for palynomorphs care must be taken as hydrogen peroxide is a powerful oxidant and may destroy pollen and spore grains. The continued use of hydrogen peroxide in this role for palaeontological preparation is now in doubt. Recent evidence (Gaffey and Bronnimann 1993) reveals that the breakdown of organic matter to simple gaseous products or easily removed soluble phases does not occur, but instead, a variety of intermediate products are formed.

1 Add 50ml of 30% H_2O_2 to 0.5g of sample in a beaker.

2 Boil gently for several hours, maintaining a constant volume of liquid by the addition of further H_2O_2.

3 When the reaction is complete the liquid will become colourless.

4 Cloudiness in the final solution can be removed by the addition of a few drops of concentrated HCl.

Oxidized humic matter must be removed after oxidation is complete in all, except the hydrogen peroxide method, by using KOH or NaOH in 5 to 10% aqueous solutions. When oxidation is complete the alkaline solutions will turn dark brown, evident if a small sample is tested on a microscope slide. Following alkaline treatment, the sample must be thoroughly washed, by wet sieving, to remove all traces of alkali.

Most oxidation techniques can result in partial dissolution of calcareous material, and consequently must be used with great care if this type of material is to be studied (Hodgkinson 1991, Gaffey and Bronnimann 1993, Pingitore *et al.*, 1993).

14.5. REMOVAL OF ELECTROLYTES BY WASHING

Washing in distilled water removes excess electrolytes from the suspension and thus de-coagulates the residue. In the absence of carbonates, suspensions treated in this way are mostly stable and give good dispersions even without the addition of a peptizer. A peptizer is necessary however, if maximum dispersion is to be obtained. When washing is carried out with the beaker in an ultrasonic tank, dispersion may be enhanced, but destruction of microfauna will be inevitable as the process is largely uncontrollable (Hodgkinson 1991).

14.6. DISPERSION WITH THE ADDITION OF PEPTIZERS

With fine grained or colloidal materials, chemical dispersers are often necessary to get particles into suspension. A residue may be dispersed only with an exact quantity of a disperser. It will flocculate if either too much or too little disperser is used. The best dispersers are those that are effective over a wide range of concentrations. Not all dispersers act with all matrices. It may be necessary to try several dispersers before a satisfactory one for a specific matrix can be found. The following dispersers have been used for clays:

i 5 to 10% aqueous solution of sodium hydroxide (NaOH), potassium hydroxide (KOH), or ammonium hydroxide (NH_4OH) for up to 24 hours.

ii 10% aqueous solution of sodium carbonate (Na_2CO_3), and boil (time unspecified).

iii 1 to 5% aqueous solution of *tetra*-sodium pyrophosphate ($Na_4P_2O_7.10H_2O$) for 3 to 5 hours.

iv 5 to 10% aqueous solution of hydrogen peroxide (H_2O_2) for 10 to 20 minutes at room temperature. About 20ml for every gram of sediment recommended.

Dilute solutions of sodium oxalate ($Na_2C_2O_4$), lithium chloride (LiCl), lithium hydroxide (LiOH) (both very expensive), *penta*-sodium triphosphate ($Na_5PO_3O_{10}$), and sodium hexametaphosphate ($Na(PO_3)_6$), have also been used as peptizers. Work by Kontrovitz (1991) has shown the last to cause partial dissolution of ostracod valves, even at low concentrations (2.5%), if prolonged contact is maintained (6+ hours). Misleading taphonomic conclusions may be made if the effects are not realized.

It should be apparent from these short notes on the dispersal, disaggregation and maceration of sediments that many of the chemicals used serve more than one purpose. Care must therefore be taken to ensure that a chemical chosen for one purpose does not, at the same time perform a secondary unwanted task, and result in the loss of valuable fossilized material. Sub-sample sediments before processing, thus allowing for different techniques to be attempted, and if time permits experiment with chemicals of varying concentrations and contact times. Observations reveal the effects of many chemicals used in various preparation stages can result in irreversible damage commencing after specimens have been transferred to micropalaeontological slides (Green 1995b). Calcareous test destruction within micropalaeontological slides can be attributed to the partial or ineffective stabilization of the celluloid used as a cover slip (Serandrei Barbero and Toffoletto 1996), resulting in the transfer of nitric acid from the celluloid to the fossil. This, however, cannot be the cause of destruction seen in glass covered micropalaeontological slides. In these examples a possible cause is the inadequate buffering of formalin based preservatives used when collecting organic carbon-rich sediments (Corliss and Emerson 1990, Jonasson and Patterson 1992), or the inadequate mixing of formalin with gum tragacanth used in attaching the tests to the slide (Green 1995b). This probably accounts for why calcareous tests from the same sample mounted on the same glass covered micropalaeontological slide exhibit various stages of destruction (Fig. 7, p 163; Green 1995b).

Techniques for the extraction of individual fossil groups of differing biomineralization, and from varying rock compositions are detailed in subsequent sections.

REFERENCES

ALLMAN, M and LAWRENCE, D. F. 1972. *Geological Laboratory Techniques*. Blandford, (London). 1-335p.

BELL, W. C. 1948. Acetic acid etching technique applied to Cambrian brachiopods. *Journal of Paleontology*. **22**, (1), 101-105.

CORLISS, B. H and EMERSON, S. 1990. Distribution of rose bengal stained deep sea benthic foraminifera from the Nova Scotia continental margin and Gulf of Maine. *Deep Sea Research*. **37**, (3), 381-400.

GAFFEY, S. J and BRONNIMANN, C. E. 1991. Effects of bleaching on organic and mineral phases in biogenic carbonates. *Journal of Sedimentary Research*. **63**, (4), 752-754.

GRANT, R. E. 1989. Extraction of fossils from carbonate by acid, pp 237-243; In: R. M. Feldman, R. E. Chapman and J. T. Hannibal (eds), *Paleotechniques*. The Paleontological Society Special Publication. **4**. iv + 358.

GRAY, J. 1965b. Extraction techniques, pp 530-587: In; B. Kummel and D. Raup (eds), *Handbook of Paleontological Techniques*. W. H. Freeman and Co., (San Francisco). xiii + 852p.

GREEN, O. R. 1995b. Pitfalls, problems and procedures in micropalaeontological preparation and conservation. *The Geological Curator*. **6**, (4), 157-166.

HARRIS, A. G and SWEET, W. C. 1989. Mechanical and chemical techniques for separating microfossils from rock, sediment and residue matrix, pp 70-86; In: R. M. Feldman, R. E. Chapman and J. T. Hannibal (eds), *Paleotechniques*. The Paleontological Society Special Publication. **4**. iv + 358.

HODGKINSON, R. L. 1991. Microfossil processing: a damage report. *Micropaleontology*. **37**, (3), 320-326.

HUSSEY, K. M and CAMPBELL, C. B. 1951. A new method of sample preparation. *Journal of Paleontology*. **25**, (2), 224-226.

JEPPSSON, L, FREDHOLM, D and MATTIASSON, B. 1985. Acetic acid and phosphatic fossils - a warning. *Journal of Paleontology*. **59**, (4), 952-956.

JEPPSSON, L and ANEHUS, R. 1995. A buffered formic acid technique for conodont extraction. *Journal of Paleontology*. **64**, (4), 790-794.

JONASSON, K. E and PATTERSON, R. T. 1992. Preservation potential of salt marsh foraminifera from the Fraser Delta, British Columbia. *Micropaleontology*. **38**, (3), 289-301.

KONTROVITZ, M, SLACK, J. M and YUHONG, Z. 1991. On the use of some phosphates in the preparation of ostracod shells. *Journal of Micropalaeontology*. **10**, (2), 121-126.

LETHIERS, F and CRASQUIN-SOLEAU, S. 1988. Comment extraire les microfossiles a tests calcitiques des roches calcaires dures. *Revue de Micropaléontologie*. **31**, (1), 56-61.
LINDSAY, W. 1987. The acid technique in vertebrate palaeontology: a review, pp 455-454; In: P. R. Crowther and C. J. Collins (eds), The Conservation of Geological Material. *The Geological Curator*. **4**, (7), 375-474.
LUND, R. 1970. A new technique for chemical preparation of fossils. *Journal of Paleontology*. **44**, (3), 578.
PINGITORE, Jr, N. E, FRETZDORFF, S. B, SEITZ, B. P, ESTRADA, L. Y, BORREGO, P. M, CRAWFORD, G. M and LOVE, K. M. 1993. Dissolution kinetics of $CaCO_3$ in common laboratory solvents. *Journal of Sedimentary Research*. **63**, (4), 641-645.
RASETTI, F. 1947. Notes on techniques in invertebrate paleontology. *Journal of Paleontology*. **21**, (4), 397-399.
SERANDERI BARERO, R and TOFFOLETTO, D. 1996. Dissolution of calcareous foraminifera in microfossil slides. *Micropaleontology*. **42**, (4), 412-414.
ZIEGLER, W, LINDSTRÖM and McTAVISH, R. 1971. Monochloracetic acids and conodonts - a warning. *Nature*. **230**, 584-585.

15. MICROFOSSIL SAMPLE CONTAMINATION AND RELIABILITY PROBLEMS

15.1. INTRODUCTION

There are two aspects of processing micropalaeontological samples that have to be considered when interpreting and evaluating the fauna within the residue. Firstly, the effects of mechanical and chemical techniques employed during collection, disaggregation, maceration and residue dispersal must be eliminated. Secondly, sample reliability in reflecting a fauna contemporaneous with the time of deposition must be assessed. Discussions of the effects of sample processing can be found in other sections (see sections 14 DISAGGREGATION AND DISPERSAL OF PARTIALLY CONSOLIDATED AND UNCONSOLIDATED SEDIMENTS, 16 WASHING AND SIEVING TECHNIQUES USED IN MICROPALAEONTOLOGY and 17 CENTRIFUGE TECHNIQUES USED IN MICROPALAEONTOLOGY), and will only briefly be outlined here.

15.2. SOURCES OF CONTAMINATION

Sample reliability in relation to field collecting (induced) and sedimentary contamination (natural) is discussed in detail in this section. Broadly speaking induced unreliability may result from poor field and laboratory practices. These can occur while collecting, packing and transporting, and in the laboratory during unpacking, storing, handling and sample processing. This problem has already been mention in previous sections (see sections 1 PALAEONTOLOGICAL TECHNIQUES - AN INTRODUCTION TO PRACTICAL PROCEDURES and 4 COLLECTING TECHNIQUES FOR MICROFOSSIL AND LIVE FORAMINIFERA SAMPLES). Naturally occurring displacement of microfossils, through recycling and stratigraphic leakage, is a problem that micropalaeontologists' must use all their experience to recognize and correctly interpret. Knowledge of a sample's history, from collecting to processing, can assist in identifying contamination. The basis for recognizing the effects lie in identifying mixed microfossil assemblages, and differentiating between authochthonous and allochthonous faunal elements (Austin 1994).

A general overview on all aspects of sample unreliability is provided by Riley and King (1973), while sedimentological contamination has been described in detail by Jones (1958). A review of sample un-reliability through field collecting, laboratory procedures and sedimentary displacement follow.

15.3. FIELD CONTAMINATION FROM SAMPLE COLLECTING

Collecting techniques are particularly important to the micropalaeontologist. Methods for both hand and mechanically collected samples are outlined in the section 4 COLLECTING TECHNIQUES FOR MICROFOSSIL AND LIVE FORAMINIFERA SAMPLES, and are not repeated. However, the problems that can lead to erroneous conclusions will be reiterated. A lack of knowledge of the collecting area may result in samples collected from a horizon different from which they are assigned (Riley and King 1973).

Natural weathering, diagenesis or metamorphism may all present problems, resulting in the partial dissolution, complete destruction or leaching of microfossils. These processes cause changes in the oxidation-reduction potential (Eh), and hydrogen ion concentration (pH) within sediments. To some extent the degree of weathering is controlled by sediment lithology, which in turn can affect the organism's preservation potential. Silicified rocks are invariably more resistant than calcareous rocks or unconsolidated sediments.

Removal of weathered surfaces is an important first step in reducing sample unreliability. It will also avoid contamination from the introduction of recent micro-organism *e.g.* modern spores and pollen, as well as material washed down slope. Hay (1977), collecting for nannofossils, reports an example of contamination related to a locality's site and situation. From a stream section, samples were collected high above the water level, and to avoid contamination, fresh surfaces were cut. However, on processing, the samples were found to contain specimens that occurred only near the top of the section. Careful preparation of samples from both the surface and interior of the hand specimens revealed the anomalous assemblage was confined to surface fragments. It was shown that the nannofossils had become impregnated into the surface rocks during periods of high water, having been transported from stratigraphically higher parts of the section down river, and splashed above the high water level. Capillary action had then drawn the nannofossils several centimetres down cracks into the rock. Further examples of palynomorph contamination are cited by Wilson (1964) from North America. In the first example, a previously barren Ordovician chert was shown to contain abundant recent algal spores. The processed samples had been collected from the surface by an inexperienced palynologist. In the second example, a coal was processed, revealing abundant Carboniferous spores and recent pollen from the surrounding forest.

Great care must be taken when collecting unconsolidated samples, ensuring that all collecting tools (spades, hammers, augers, piston corers) are thoroughly cleaned between sampling. Potential problems from mechanically collected samples (drill chippings, cores), are detailed in the section 4 COLLECTING TECHNIQUES FOR MICROFOSSIL AND LIVE FORAMINIFERA SAMPLES. The recirculating mud is one of the main sources of sample contamination, although caving within the core can result in re-working and stratigraphical leakage of microfossils (Wilson 1964). Some coring devices produce a disturbed and contaminated outer layer adjacent to the core barrel, around which sediments are "smeared" downwards. Although this contaminated zone is only rarely greater than 0.5cm thick, if unrecognized sharp faunal boundaries may be interpreted as gradational (Austin 1994).

15.4. PACKING AND TRANSPORTATION CONTAMINATION

Contamination problems arising from the labelling and transportation of samples back to the laboratory, and packaging materials have been adequately covered in section 1 PALAEONTOLOGICAL TECHNIQUES - AN INTRODUCTION TO PRACTICAL PROCEDURES.

15.5. LABORATORY CONTAMINATION

For the palynologist, acid insoluble microfosssil and calcareous nannofossil worker - in fact any micropalaeontologist working on material of the 1-20μm size range - airborne contaminants such as spore and pollen may cause some concern, particularly when examining recent residues. The importance of recognizing modern contaminants within a residue depend to a large extent, upon its age. Recent spores and pollen can be readily differentiated from older types by colour, sheen, staining reaction, and shape, in particular the lack of compression. However, when working on contemporary assemblages use a greased slide placed near the sample processing area, and periodically examine it to see what contaminants have collected. It should also be remembered that different contaminants may be encountered seasonally, and with the wind in different directions.

Airborne dust from other samples may also be a source of contamination, particularly if mechanical rock crushing has formed any part of the processing. Recognition and elimination of this form of contamination is extremely difficult. When undertaking processing be aware of other techniques within the laboratory that might generate airborne contaminants, and take care to ensure all samples are covered during long chemical maceration or settling stages.

Good Laboratory Practice (GLP) and processing "hygiene" are essential if cross sample contamination is to be reduced. Hands, nails, small cuts and even natural skin crevices provide areas in which samples can be inadvertently mixed. Hands should be washed between the handling of different samples, particularly if sediments are partly consolidated. Disposable gloves should be worn, and discarded between samples or when they become damaged or show signs of ware.

Another area of contamination within the laboratory is the water supply, which may contain modern pollen, spores, dinoflagellates, desmids and diatoms. This can be eliminated by using good quality filtered distilled water, but care must be taken to ensure that the stocks are stored in closed reservoir containers. All glassware and equipment used in processing should be cleaned in acid (a strong oxidizing agent will destroy all palynomorphs), washed and rinsed in distilled water, as palynomorphs adhere to the glass surfaces. Ideally all glassware should be kept immersed in a bath of acid until it is required. Equipment, such as plastic pipettes, should only be used once. Sinter funnels can be cleaned by prolonged immersion in an ultrasonic tank. Sieve contamination is covered in section 16 WASHING AND SIEVING TECHNIQUES USED IN MICROPALAEONTOLOGY.

Pestle and mortars must be scrubbed and ultrasonically cleaned between sample processing. Care should be taken when using these implements to avoid pitting the surfaces. If pitting is evident the instruments must be replaced immediately. Pitting can however, be reduced by lining the pestle with a heavy duty polythene sheet covered with several layers of aluminium foil before crushing commences (Riley and King 1973). Discard and replace the lining between each sample.

Laboratory and general purpose reagent (GPR) grades of acid and solutions used in chemical preparations, are a further source of contamination. A fine sediment at the bottom of containers may contain spore and pollen material derived from the water supply during preparation. This source of contamination can be eliminated by use of high purity analytical grades of acid or solution. Alternatively, carefully decant and filter a laboratory grade reagent, ensuring not to disturb any sediment at the bottom of the container (Riley and King 1973).

Confidence in the results of micropalaeontological analyses has, in part, long been known to depend upon laboratory cleanliness (Williams-Mitchell 1948). During all stages of micropalaeontological processing strict attention must be given to the cleaning and maintenance of all equipment and solutions used.

15.6. STRATIGRAPHIC CONTAMINATION

No matter what amount of care is taken during the field and laboratory preparation of samples, natural recycling of microfossils through the erosion, transportation and re-deposition of sediments to form younger rock is unavoidable. Particularly in palynology, displaced microfossils can provide a major part of the faunal assemblage. The recognition of any displaced microfossils is vital before an investigation is undertaken, and requires a great deal of geological skill and knowledge of the sample's collection history.

Palaeoenvironmental studies reveal assemblages can generally be assigned to one of six different groups (Wilson 1964, and Table 15.1), which may indicate varying degrees of vertical or horizontal displacement (Jones 1958). These are defined below, followed by a brief discussion of examples from the literature where microfossil reworking has been recognized.

Fossil assemblage contains:		Characteristics & mode of formation
1	Microfossils of more than one age.	Sediments from tectonically active areas. Contaminants may be difficult to recognize, and source of origin difficult to define.
2	Microfossils showing differential preservation.	Optically observing the amount of damage taxa exhibit.
3	Fossil assemblages differentiated by biological stain reactions.	Revealed when a processed sample is stained, results detailed in table 15.2. May reflect diagenesis.
4	Microfossils recognized to be older than the strata they now occur in.	Rapidly deposited sediments, containing clasts of older material (and therefore older organisms).
5	Microfossils recognized to be younger than the strata they now occur in.	Stratigraphic leakage (see below: vertical displacement).
6	Microfossils recognized as marine now occurring in non-marine strata.	Fossils (of any age) known to be marine in origin, now recycled into non-marine sediments.

Table 15.1. Characteristics of the palynomorph assemblages noted by Wilson (1964) for the recognition of reworked or suspect material.

15.6.1. Vertical displacement

Two assemblage types can be recognized as having undergone vertical displacement.

1 **Reworked assemblages**: contain fossils of more than one geological age (assemblage 1, table 15.1), or fossils known to be older than the sediments in which they now occur (assemblage 4,

table 15.1). Isolated individuals can be recognized by partial corrosion and are less well preserved than contemporaneous fossils (assemblages 2 or 3, table 15.1). However, in some cases there may be an abundance of well-preserved reworked material that may dominate over the contemporaneous assemblage.
Rapid deposition of sediments, particularly coarse clastics, may result in reworked clasts containing older fossils plus matrix, a relationship best seen in thin section (assemblage 4, table 15.1). Bioturbation may also be evident.

Environments that can generate this type of assemblage have been discussed in detail by Jones (1958), and include coastal wave and current action (in both marine and non-marine environments), turbidity currents, aeolian and glacial action, although each of these processes is not exclusive to vertical displacement. Sedimentological evidence must be carefully examined when assessing an assemblage that may have been reworked. Reworked species may not readily be identifiable as such, particularly if they are microfossils resistant to abrasion. Generally, reworking will cause an apparent extension of the range of a species (Baumgartner *et al.*, 1981).

2 **Stratigraphic leakage**: involves the deposition of younger fossils in older sediments (assemblage 5, table 15.1). It occurs when fossils are deposited in cracks, fissures and joints of older sediments, usually in karst environments, along fault planes, along rivers, or by organism displacement. Evidence from boreholes is seen in duplicated assemblages, with many of the specimens either destroyed or, in the case of palynomorphs, highly carbonized.

This type of contamination is usually readily identified, and is less common than reworking. It can usually be avoided through careful field mapping when collecting samples.

15.6.2. Horizontal displacement

Horizontal or geographic displacement results in the mixing of assemblages from laterally equivalent environments. Three sub-divisions have been recognized by Jones (1958):

(i) Intracontinental displacement (assemblages 1 and 4, table 15.1).
(ii) Intramarine displacement (assemblages 1 and 4, table 15.1).
(iii) Displacement between marine and continental environments (assemblage 6, table 15.1).

Clearly environments that can cause vertically displaced assemblages, can also induce contemporaneous horizontal displacement, resulting in reworked assemblages.

The displacement of marine organisms such as tasmanites, chitinozoa, dinoflagellates, marine diatoms, foraminifera, silicoflagellates and discoasterids into a non-marine sequence may indicate a tectonically active area, in which case the assemblage may show signs of differential preservation. Wilson (1964) cites examples where less well preserved older and younger fossils than the contemporary assemblage occur at a particular horizon. In general, the closer two assemblages are phylogenetically, in geological history and age, and palaeoecologically, the more alike their preservation. Further evidence is indicated by the ability of palynomorphs response to staining.

Particularly useful in studying fossil pollen and spores, staining provides an objective means of determining reworked palynomorphs (Table 15.2). Wilson (1964) described a range stain colour intensities seen in palynomorphs of different preservational states and age. Deeply buried pollen and spore grains, and material subjected to the moderate effects of induration, were recognized by their inability to absorb the stain Safranine O. Wall structures remained dark brown or black. Less deeply buried palynomorphs readily absorbed the stain and were coloured red. Modern and recent pollen stained pink.

These reactions were examined in more detail by Stanley (1966), who through extensive optical studies of the outer exine walls of recent and fossil palynomorphs recognized four distinct categories (Table 15.2). These observations illustrated morphological differences between the inner (endexine) and outer (ektexine) layers, and confirmed that age alone is not responsible for differential staining. Un-recycled older material can stain in a similar manner to Recent (group 4). Stain acceptance by pollen and spore grains appear to be a function of diagenetic and metamorphic history, possibly coupled with exine chemistry, reflecting changes that have occurred during reworking. Stanley (1966) also suggested that some of the criteria listed below could be applied in the study of recycled acritarchs and dinoflagellates.

Group	Optical Characteristic	Possible Age
1	Exine brown or light yellow after staining	Palaeozoic - Mesozoic
2	Endexine and ektexine readily take up stain	Late Mesozoic - Early Tertiary
3	Endexine stains well, ektexine does not accept stain	Mesozoic
4	Ektexine stains well, endexine does not accept stain	Recent i.e. rarely seen in recycled grains

Table 15.2. Four categories of reworked pollen and spores recognized by Stanley (1966) from optical evidence revealed by differential staining.

15.7. RECOGNITION AND APPLICATIONS OF REWORKED MICROFOSSIL ASSEMBLAGES

Microfossil studies using reworked foraminifera have been undertaken in modern estuarine environments (Brasier 1981), shallow marine environments (Nigam and Setty 1980) and deep-water marine environments (Martin 1988, Fourcade and Butterlin 1988). They have also been used as Quaternary environmental indicators (Culver and Banner 1978). River transportation and tidal regimes influenced by off-shore currents have been examined (Brasier 1981, Nigam and Setty 1980), and the source area of reworked foraminifera determined. Brasier (1981) cautions against palaeosalinity and palaeoclimatic interpretations as the proportion of exotic forms may vary along the transect line studied, distance from the source area, and quality of preservation can result in erroneous conclusions. However, careful observations can provide evidence of downslope transportation from shallow to deeper environments, and at greater depths mixing of shallow and deep-water species by sediment gravity flows (Martin 1988).

The recognition of derived benthic foraminiferal assemblages are thus critical if they are to be used as palaeo-environmental indicators. Reliability is compromised if post-mortem transportation is evident. To understand this aspect of assemblage reliability, Kontrovitz, Snyder and Brown (1978) performed a series of controlled flume experiments measuring benthic foraminiferal test traction velocities over a moveable sediment substrate. Comparisons of traction velocities of different species, exhibiting various shell characteristics were made. Results indicated that foraminifera tests average sphericity and weight correlate with the water current velocity required to move it. The tests volume, nominal diameter, effective density and maximum projection of sphericity were not significantly related to current speed. When mathematically applied to a fossil assemblage, it is possible to determine components that have been transported simultaneously. Furthermore, Kontrovitz *et al.*, (1978) through use of a multiple regression equation were able to predict that for a particular species and a combination of variables a correlation with traction velocity existed. Thus if sufficient data on the species within an assemblage is known, a predictive model can be applied instead of having to undertake the more cumbersome flume runs. However, these methods provide only an indication of differential transportation (redeposition), and not necessarily actual foraminiferal reworking.

The recognition of redeposited foraminifera within sediments can be compounded by the presence of reworked specimens. Fourcade and Butterlin (1988) examined assemblages of reworked and redeposited late Campanian to Pleistocene larger foraminifera from Bahamian basins and slopes. It was concluded that the occurrence of redeposited larger foraminifers could be linked to sea-level changes, while the reworking of taxa in two horizons was related to tectonic events. In biostratigraphic studies the recognition of reworked assemblages may be made easier by a well-constrained chronostratigraphic framework, coupled with control from lithological and faunal transitions (Austin 1994).

Reworked Tremadoc and Arenig/Llanvirn Ordovician acritarchs have been reported from the Caradoc Series of Shropshire, England (Turner 1982). The excellent preservation of these forms provided extensive palaeoenvironmental data, enabling the source area to be determined, and evidence of the lithological nature of the parent sediments. More extensive stratigraphic reworking of Ordovician acritarchs into Silurian rocks of southern Ireland (Colthurst and Smith 1977, Emo and Smith 1978), the Southern Uplands of Scotland (White *et al.*, 1991, McCaffrey *et al.*, 1992), and the Lake District of northwest England (McCaffrey *et al.*, 1992) has also been reported. Although preservation from these areas is poor, generic level identifications have been possible, prompting speculation at using reworked acritarchs as provenance indicators around the Iapetus Suture (McCaffrey *et al.*, 1992). The prerequisites for using reworked acritarchs as source indicators outlined by McCaffrey *et al.*, (1992), are equally applicable to other microfossil groups that might be considered for this purpose:

- Original distributions must show an adequately constrained provinciality.
- Preservation must permit confident identification (at least to generic level).

- It must be demonstrated that the recycled organisms were transported with their host sediments (*i.e.* evidence of ocean current up-welling and/or aeolian transportation must be eliminated) before becoming incorporated into a redeposited sediment.

Eaton *et al.*, (1989) and Lofgren *et al.*, (1990) have examined reworked microvertebrate assemblages across the Mesozoic-Tertiary boundary from eastern Montana, USA. Fiorillo (1998) has quantified this and additional data from museum collections, and demonstrated that a minimum of 10-15% of the specimens represents reworked material. A feature that could have direct implications for conclusions arrived at in palaeoecological studies comparing the relative frequencies of various taxa. Furthermore, the data suggested that a river eroding its own channel banks can accurately sample the paleofauna when incorporating the fossil remains into the bedload.

REFERENCES

AUSTIN, W. E. N. 1994. Disturbed foraminiferal stratigraphies - a cautionary tail. *Cushman Foundation Special Publication.* **32**. 155-159.

BAUMGARTNER, P. O, BJØRKLUND, K. R, CAULET, J-P, DE WEVER, P, KELLOGG, D, LABRACHERIE, M, NAKASEKO, K, NISHIMURA, A, SCHAAF, A, SCHMIDT-EFFING, R and YAO, A. 1981. EURORAD II, 1980 - Second European meeting of radiolarian paleontologists: current research on Cenozoic and Mesozoic radiolarians. *Eclogae Geologicae Helvetiae.* **74**, (3), 1027-1061.

BRASIER, M. D. 1981. Microfossil transport in the tidal Humber basin, pp 314-322: In; J. W. Neale and M. D. Brasier (eds), *Microfossils from Recent and Fossil Shelf Seas.* Ellis Horwood (Chichester). 380p.

COLTHURST, J. R. J and SMITH, D. G. 1977. Palaeontological evidence for the age of the Lower Palaeozoic rocks of the Slievenamon Inlier, County Tipperary. *Proceedings of the Royal Irish Academy.* **77B**, 143-148.

CULVER, S. J and BANNER, F. T. 1978. Foraminiferal assemblages as Flandrian environmental indicators. *Palaeogeography, Palaeoclimatology, Palaeoecology.* **24**, 53-72.

EATON, J. G, KIRKLAND, J. I and DOI, K. 1989. Evidence of reworked Cretaceous fossils and their bearing on the existence of Teriary dinosaurs. *Palaios.* **4**, 281-286.

EMO, G. T and SMITH, D. G. 1978. Palynological evidence for the age of the Lower Palaeozoic rocks of Slieve Aughty, Counties Clare and Galway. *Proceedings of the Royal Irish Academy.* **78B**, 281-292.

FIORILLO, A. R. 1998. Measuring fossil reworking within a fluvial system: an example from the Hell Creek Formation (Upper Cretaceous) of eastern Montana, pp 243-251: In; Y. Tomida, L. J. Flynn and L. L. Jacobs (eds), *Advances in Vertebrate Paleontology and Geochronology.* National Science Museum Monographs (Tokyo). xii + 292.

FOURCADE, E and BUTTERLIN, J. 1988. Reworked and redeposited larger foraminifers on slopes and in basins of the Bahamas, Leg 101. *Proceedings of the Ocean Drilling Program, Scientific Results.* **101**, 47-61.

HAY, W. W. 1977. Calcareous nannofossils, pp 1055-1200 (techniques 1056-1069): In; A. T. S. Ramsay (ed), *Oceanic Micropalaeontology* (2 Volumes). Academic Press (London). xii + 1453.

JONES, D. J. 1958. Displacement of microfossils. *Journal of Sedimentary Petrology.* **28**, (4), 453-467.

KONTROVITZ, M, SNYDER, S. W and BROWN, R. J. 1978. A flume study of the movement of foraminifera tests. *Palaeogeography, Palaeoclimatology, Palaeoecology.* **23**, 141-150.

LOFGREN, D. L, HOTTON, C. L and RUNKEL, A. C. 1990. Reworking of Cretaceous dinosaurs into Paleocene channel deposits, upper Hell Creek Formation, Montana. *Geology.* **18**, 874-877.

MARTIN, R. E. 1988. Benthic foraminiferal zonation in deep-water carbonate platform margin environments, Northern Little Bahama Bank. *Journal of Paleontology.* **62**, (1), 1-8.

McCAFFREY, W. D, BARRON, H. F, MOLYNEUX, S. G, KNELLER, B. C. 1992. Recycled acritarchs as provenance indicators: implications for Caledonian terrane reconstruction. *Geological Magazine.* **129**, (4), 457-464.

NIGAM, R and SETTY, M. G. A. P. 1980. Paleogene reworked foraminifera in Recent sediments off Daman, Western India. *Proceedings of the 3rd Indian Geological Congress, Poona.* 273-280.

RILEY, L. A and KING, A. D. 1973. Causes of unreliability in microfossil samples. *The Mercian Geologist.* **4**, (3), 197-203.

STANLEY, E. A. 1966. The problems of reworked pollen and spores in marine sediments. *Marine Geology.* **4**, 397-408.

TURNER, R. E. 1982. Reworked acritarchs from the type section of the Ordovician Caradoc Series, Shropshire. *Palaeontology.* **25**, (1), 119-143.

WHITE, D. E, BARRON, H. F, BARNES, R. P and LINTERN, B. C. 1991. Biostratigraphy of late Llandovery (Telychian) and Wenlock turbidite sequences in the SW Southern Uplands, Scotland. *Transactions of the Royal Society of Edinburgh: Earth Sciences.* (at press).

WILLIAMS-MITCHELL, E. 1948. The zonal value of foraminifera in the Chalk of England. *Proceedings of the Geologists' Association.* **59**, (2), 91-112.

WILSON, L. R. 1964. Recycling, stratigraphic leakage, and faulty techniques in palynology. *Grana Palynologica.* **5**, (3), 425-436.

16. WASHING AND SIEVING TECHNIQUES USED IN MICROPALAEONTOLOGY

16.1. INTRODUCTION

A wide variety of wet and dry sieving techniques are used in micropalaeontology for the separation of calcareous, siliceous and phosphatic microfossils, and palynomorphs in palynology. Sieving sorts and concentrates microfossils from processed residues, as the material is size classed, assisting in the examination of prepared material under the microscope. For the operator less eye strain occurs when material of the same size grade is examined, as there is less need re-focus and adjust the microscope. Residue sub-samples derived from either wet or dry sieving can be further treated by concentrating techniques (*e.g.* decanting, heavy liquid separation, centrifuging, electromagnetic separation), to obtain even richer residues through removal of the non-fossiliferous fraction. These specialist techniques are outlined in subsequent procedures (see sections 18 MECHANICAL SEPARATION OF MICROFOSSIL RESIDUES and 19 FLOTATION AND LIQUID SEPARATION TECHNIQUES).

16.2. SIEVE SIZES AND TYPES

Well-equipped laboratories will have two sets of Test Sieves, one dedicated for dry sieving (without the aid of a liquid) and one for wet sieving (liquid assisted). The most commonly used sieves are constructed of brass or stainless steel and 200mm (8in) diameter. For smaller amounts of residue (<50g), as might be generated from micropalaeontological processing techniques, 100mm (4in) diameter sieves are more suitable. Brass sieves are preferable for dry sieving and stainless steel sieves for wet sieving, particularly if solutions are likely to corrode the delicate brass sieve mesh.

It is essential that all sieves are constructed to fit into one another. When assembling a series (or nest) of sieves ensure that they are all the same make, diameter (metric diameter (mm) sieves will fit into imperial diameter (in) sieves, but with approximately 5mm of lateral movement), and constructed of similar material. For dry sieving a lid and receiver (pan) are required to fit onto the top and bottom of the nest respectively. The mesh of each sieve is normally constructed of phosphor bronze for sieves less than 250µm aperture size, brass for sieves of aperture size 250µm to 1mm, and mild steel for sieves with an aperture size greater than 1mm. The mesh or "cloth" is wire of circular cross section, and woven in such a way that the aperture is square. The mesh is attached to the frame by solder. For analytical work both the sieve frame and mesh must conform to British Standard 410:1986, with the words 'Test Sieve to B.S. 410:1986', appearing on the manufacturers label attached to the sieve frame.

Sieves range in aperture size from 16mm to 39µm (Fig. 16.1). Finer mesh sieves (20µm and 10µm) used in sieving palynological samples are usually constructed of a monofilament nylon mesh. The sizes within the standard range are all related in the aperture width to a fourth root of two (Rittinger) series. This scale has been converted to logarithmic moments by Krumbein (1936), and can be read as the phi (Ø) scale (Fig. 16.1). The phi scale is simpler to visualize within a stack of sieves, as coarse mesh sizes are given negative values, and fractions less than 1mm (0.0Ø) have positive values, increasing as the mesh becomes finer. However, the phi scale convention is little used outside the field of geology, and in particular sedimentology, where values are simpler to manipulate in statistical calculations than metric fractions. There is unquestionably strong reasoning for providing all units as metric values (Lindholm 1987, Pierce and Graus 1981). Fortunately, this problem does not effect palaeobiologists using sieving as a method of size grading sediments or residues to facilitate optical examination. However, sieves marked with phi values are easier to stack in the correct order, as quarter, half or whole phi intervals are used in a nest (Fig. 16.1) to ensue equal size divisions, and avoid potential loss of residue from incorrectly stacked sieves.

Specialized sieves and screens used in grain size analysis and sediment grading may also be encountered in some laboratories. These are often used in the field for sieving coarse gravels and concentrating macrofossil faunal assemblages (Miller 1989), or in the agricultural industry for sieving cereals. The sieve mesh may consist of a plate with punched square, oblong or round holes. This type of sieve should not be used for micropalaeontological residues.

16.3. SIEVE SHAKERS

Sieve shakers can hold up to 10 standard type sieves (200mm or 100mm diameter) plus the receiver and lid. The mechanical action results in a gyratory motion of grains on the sieve mesh, while at the same time a reciprocating vertical movement is performed on the sieves. This ensures the movement of grains at all possible angles to the mesh, and only those grains larger than the mesh size is retained on it, while all smaller particles pass through.

Wet sieving attachments can be fitted to a nest of sieves, replacing the lid and collecting pan. The lid attachment connects to a water supply via a rubber hose, and three angled directing nozzles ensure a fine jet of water washes the sample from the coarsest sieve through the nest. Very fine wet sieving can be aided by inserting a spacer within the nest of sieves to provide a second supply of water (again directed from three fine jets). This is also suitable for large samples, where there is a danger of the sieve mesh becoming clogged, resulting in overflowing and sample loss. These attachments are more commonly used in sedimentological grain size analysis procedures.

16.4. SIEVE CLEANING AND MAINTENANCE

One of the most important (and often most neglected) steps in sieving is the correct care and maintenance of the sieves. Test Sieves are precision measuring instruments, and must be treated with care, regularly cleaned, and stored in a dry place. Particles should not be forced through the mesh, although occasionally gentle brushing of fine material may be required. This is generally undesirable, but unavoidable for material difficult to sieve. Care should be taken to avoid damage to the sieve mesh or particle breakage. Sieves frequently used should be inspected regularly to detect any defects in the mesh, solder or rim. Before use all sieves should be visually checked with particular care given to the underside and the solder around the edge securing the mesh to the frame. This is often the first area that becomes damaged on fine mesh sieves used for wet sieving. The mesh should also be checked for any distortions, and cleaned if required. The procedure is as follows:

Phi (Ø)	Millimetres (mm)	Nominal aperture size (µm)		Wentworth size class
-12	4096			
-10	1024			Boulder
-8	256	256		
	128			Large, Small Cobble
-6	64	64		
	32			Large
-4	16			Medium Pebble
-2	4	4	4000	Small
-1.75	3.36		3350	
-1.5	2.83		2800	Granule
-1.25	2.38		2360	
-1.0	2.00	2	2000	
-0.75	1.68		1700	Very
-0.5	1.41		1400	coarse
-0.25	1.19		1180	sand
0.0	1.00	1	1000	
0.25	0.84		850	
0.5	0.71		710	Coarse
0.75	0.59		600	sand
1.0	0.50	1/2	500	
1.25	0.42		425	Medium
1.5	0.35		355	sand
1.75	0.30		300	
2.0	0.25	1/4	250	
2.25	0.210		212	
2.5	0.177		180	Fine
2.75	0.149		149	sand
3.0	0.125	1/8	125	
3.25	0.105		106	Very
3.5	0.088		90	fine
3.75	0.074		75	sand
4.0	0.0625	1/16	63	
4.25	0.053		53	Coarse
4.5	0.044		45	silt
4.75	0.037		38	
5.0	0.031	1/32		
				Medium
6.0	0.0156			Silt
				Fine
7.0	0.0078			
				Very fine
8.0	0.0039	1/256		
9.0	0.0020			
10.0	0.00098			Clay
11.0	0.00049			

Figure 16.1. Phi-mm-mesh size conversion chart with class intervals for grade scales. (Re-drawn with modifications from Pettijohn, Potter and Siever 1972).

16.4.1. Cleaning of dry sieves

1 Start at the top (coarsest mesh fraction) of the nest of sieves. Remove the sieve and invert on a clean piece of paper, larger than the diameter of the sieve (ensures all material is retained).

2 Large mesh sieves can be gently tapped on the bench, but sieves with a mesh size finer than 1mm should be gently brushed. Brushing should be done using a fine brass or nylon bristle brush, in an action diagonally across the mesh, and circular around the very edge. The diagonal action ensures that the mesh is not distorted from its true square shape.

3 Final cleaning of fine mesh sieves can be done using a compressed air supply, by gently directing the jet stream over the surface. Sieves of fine mesh can be cleaned by immersing in an ultrasonic tank.

4 For long term maintenance the sieves can occasionally be washed in warm soapy water, and air dried. Always ensure that the mesh is thoroughly dried before storage or re-use.

16.4.2. Cleaning of wet sieves

1 Remove the top sieve from the nest. With a distilled water bottle wash the residue to the edge of the sieve.

2 Carefully wash the residue into an evaporating dish.

3 Place the sieve in a sink, and invert under a tap. By using a directed jet of water via a rubber hose and brushing with a nylon brush as detailed above (diagonally across the mesh), clean the sieve. Repeat this procedure on the inside of the sieve. Ensure that no grains or mud adhere to the sieve frame, as these will result in cross-sample contamination, and invalidate results. Stubborn grains stuck in a mesh can be removed by ultrasonic treatment.

4 Periodically wash sieves in warm soapy water. Dry as above. Avoid oven drying sieves as this will distort their shape and crack the rolled rim and solder bond. Temperatures in excess of 80°C will inevitably result in mesh distortion.

The cleanliness of each sieve can be monitored by using methylene blue dye. Between each sample immerse the sieve in a bath of methylene blue. Trapped contaminants within the mesh become stained, and should they subsequently become dislodged and incorporated in a residue, will be readily visible.

After cleaning carry out a visual inspection of the sieve, checking on the frame, which may show signs of metal fatigue with age, particularly evident along the external beading round the upper edge. Damage to the sieve mesh may result in enlargement of individual apertures, and as already mentioned, the solder bond between the mesh and the frame may fail.

Once clean and dry, sieves should be stored either in their boxes, or stacked in descending size in a dry cupboard or a sieve rack. Never leave them on shelves or benches where they will become dusty, or sharp objects may damage the mesh.

6.5. DRY SIEVING

This is performed on dry sediment after mechanical or chemical disaggregation of the sample has been completed, and it has been reduced to its constituent components. It can be undertaken with the aid of a sieve shaker, or if only a few sieves are to be used, by the method of hand sieving described in B.S. 1769:1952. In micropalaeontology it is rare to have more than three or four grade sizes from each residue to examine, consequently most sieving is done by hand.

1. Visually determine the grain size. Select and assemble a nest of sieves that will retain all but the finest clay fraction. Ensure the coarsest is uppermost (negative or low phi number), and the finest sieve is at the base (highest positive number) resting on the receiver.

2. Pour the sample into the uppermost sieve and put the lid on.

3. Hold the nest so that they inclined at an angle of 30° to the left or right (operator preference).

4. Shake the tier to and fro, at the same time rotating the sieves 90°.

5. After a few minutes the sample will be separated into its component grade sizes. The time taken in shaking the sieves is not as critical as in analytical sieving, however, by ensuring times between samples remain constant the amount of mechanical abrasion inflicted on microfossil shells will be comparable.

6. Clean each sieve separately by the method outlined above and pour residues into labelled glass or plastic phials. A general guide to the type of microfossils that can be retained on the mesh of a sieve within particular grade sizes is outlined in table 16.1.

A quantitative study of some microfossil group's size ranges has been reported by Jones (1958), and the results are graphically illustrated in figure 16.2. Over 1500 specimens were used in this study, a number considered by Jones to be small, but capable of giving an approximation of the size ranges in which the microfossils occur. The wide variation in the size of miscellaneous microfossil fragments is in part a reflection of the diversity of the organisms represented (bryozoan, mollusc, echinoderm and algal remains).

Spechmann's (1981) palaeoecological study of Maastrichtian sublittoral benthonic foraminifera from western Europe illustrates how separated fractions can be used in contrasting analytical procedures. The procedure is summarized below and in table 16.2:

Microns (µm)	Phi (Ø)	COMMENTS
2000 - 1190	-1.0 - 0.25	Will retain fragments from just over 1mm to 2mm, most large benthic foraminifera, large ostracods and conodonts
1000 - 250	0.0 - 2.0	Will retain ostracods and smaller benthic foraminifera
74 - 63	3.75 - 4.0	Will retain smaller radiolarians, planktonic foraminifera and conodonts
39	4.75	Will retain many small diatoms, chitinozoa, scolecodonts and larger organic walled fossils
20 - 10	6.0 - 7.0	Will retain pollen and spore grains

Table 16.1. Generalized summary of the microfossil sizes, with the mesh size and phi grade equivalents on which they might be expected to be retained (data from Higgins and Spinner 1969).

1 Divide the sample into two parts. Process one part using conventional micropalaeontological procedures, and evaluate the thanatocoenosis >63µm. Process and analyse the second part as follows:

2 Sieve and separate the residue into seven fractions (2000µm - 630µm - 315µm - 200µm - 125µm - 63µm - receiver). Record the dry weight for comparison with sedimentological parameters.

3 Examine and pick (300-500 specimens) the three coarse fractions. Use of a micro-sample splitter is recommended for large residue volumes (Sprechmann 1981).

4 Examine the finer fractions for significant changes in the species composition.

5 Coarse fractions (>2000µm - 315µm) can be used to evaluate the foraminiferal biovolume coefficient, construct PHA (porcelaneous - hyaline - agglutinated wall compositions) triangular diagrams, evaluate the distribution of morphotypes, to test the "dominant species community concept".

Sprechmann (1981) concluded that following this procedure Maastrichtian benthonic foraminifera from sublittoral environments provided a greater resolution than other marine organisms in defining palaeobathymetric subdivisions.

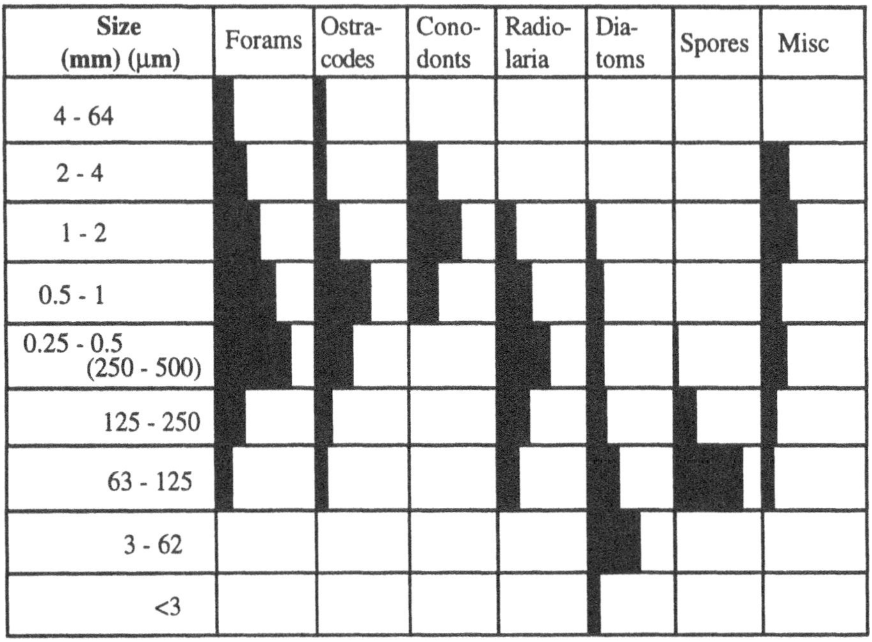

Figure 16.2. Grain size distribution and associated grade scales of some common groups of microfossils (after Jones 1958).

Microns (μm)	ANALYTICAL APPLICATION
2000 - 315	Defining associations of foraminifera using the dominant community concept
2000 - 315	Morphotype distribution
2000 - 315	Foraminiferal biovolume coefficient
630 - 315	Triangular plots of readily identifiable taxa
315 - 63	Check for significant forms absent or untypically represented in coarse fraction (2000μm - 630μm)

Table 16.2. Suggested mesh size ranges for foraminiferal palaeoecological analytical applications (data from Sprechmann 1981).

16.6. WET SIEVING

A disaggregated sample may consist of fractions ranging from clay (silty or sandy muds) to fine grained sands (<37μm - 0.25mm). The clay fraction may well obscure morphological details of the microfossils during initial optical examination. It can be removed by wet sieving the residue. However, this is not without risk, as wet sieving can cause considerably more damage than dry sieving. To off-set potential damage, process a larger volume of sample.

1 Select a series of sieves with a range of apertures that will retain all but the finest clay fraction (1mm, 250μm, 125μm, and 63μm) and arrange them to form a nest, with the coarsest (1mm) at the top and the finest (63μm) at the base.

2 Place the residue in the top sieve and using a directional jet of water, via a rubber hose attached to the tap, gently and thoroughly wash it. Care must be taken in washing the residue to the sieve centre. Excessive water pressure towards the side may splash sediment over the sieve rim. Continue washing for approximately 5 minutes or until clear water passes through the sieve.

3 This process will effectively split the residue into its component size classes, and remove the <63μm fraction.

4 Each fraction is then washed into a labelled evaporating dish, which is allowed to stand until all the grains have settled.

5 Decant off excess liquid, and allow to dry in an oven pre-set to 60°C.

6 Once dried, each fraction can be poured into a glass or plastic labelled phial, indicating the sample number and retained size fraction.

7 Ensure that the sieves are thoroughly cleaned prior to washing the next sample, by following the procedures detailed above.

A modified method of breaking down and washing a large number of samples has been described by Hulme (1961). If employing such a process, ensure that cross sample contamination is eliminated during processing and oven drying.

REFERENCES

ALLMAN, M and LAWRENCE, D. F. 1972. *Geological Laboratory Techniques*. Blandford, (London). 1-335p.

BRASIER, M. D. 1980. *Microfossils*. Allen and Unwin (London). xii + 193p.

BRITISH STANDARD 410:1986. Metric units specification for test sieves. *British Standard Institute*.

BRITISH STANDARD 1769:1952. Methods for the use of B.S. fine mesh test sieves. *British Standard Institute*.

HIGGINS, A. C and SPINNER, E. G. 1969. Techniques for the extraction of selected microfossils. *Geology - Journal of the Association of Teachers in Geology*. **1**, 12-28.

HULME, S. G. 1961. A mechanized method for breaking down and washing foraminiferal rock samples. *Micropaleontology*. **7**, (1), 107-113.

JONES, D. J. 1958. Displacement of microfossils. *Journal of Sedimentary Petrology*. **28**, (4), 453-467.

KRUMBEIN, W. C. 1936. Application of logarithmic moments to size frequency distributions of sediments. *Journal of Sedimentary Petrology*. **6**, (1), 35-47.

LINDHOLM, R. C. 1987. *A Practical Approach to Sedimentology*. Allen and Unwin (London). xviii + 276p.

MILLER, B. B. 1989. Screen-washing unconsolidated sediments for small macrofossils, pp 260-263; In: R. M. Feldman, R. E. Chapman and J. T. Hannibal (eds), *Paleotechniques*. The Paleontological Society Special Publication. **4**. iv + 358p.

PETTIJOHN, F. J, POTTER, P. E and SIEVER, R. 1972. *Sand and Sandstone*. Springer-Verlag (New York). xvi - 618p.

PIERCE, J. W and GRAUS, R. R. 1981. Use and misuses of the phi-scale: discussion. *Journal of Sedimentary Petrology*. **51**, (4), 1348-1350.

SPRECHMANN, P. 1981. Paleocommunities and paleobathymetry of Maastrichtian sublittoral benthonic foraminifera from western Europe. *Neues Jahrbuch für Geologie und Paläontologie, Abhandlungen*. Band **162**, Heft 2, 188-230.

17. CENTRIFUGE TECHNIQUES USED IN MICROPALAEONTOLOGY

17.1. INTRODUCTION

Centrifuge techniques play an important role in microfossil separation and concentration procedures. A wide variety of microfossil groups, particularly the smaller specimens, utilize procedures involving this specialized equipment (see sections 25 EXTRACTION TECHNIQUES FOR PALAEOBOTANICAL AND PALYNOLOGICAL MATERIAL, 26 EXTRACTION TECHNIQUES FOR ACID INSOLUBLE MICROFOSSILS and 33 EXTRACTION TECHNIQUES FOR CALCAREOUS NANNOFOSSILS). An application is also described for the concentration of live foraminiferal cultures, when it is essential to retain all specimen tests (Arnold 1965b). The purpose of this section is twofold; firstly to briefly outline the many uses of the centrifuge within micropalaeontology, and secondly to provide a practical data base of information relating particle size to centrifuge times at varying speeds using heavy liquids (*e.g.* bromoform), and centrifuge washing times in either water or acetone for use with the K-factor (centrifuge constant).

17.2. PARTICLE SIZE SEPARATION

The primary use of the centrifuge is in particle size separation, usually in mineral residue studies (Allman and Lawrence 1972, Hutchison 1974). The principle is the same for techniques and adaptations in micropalaeontology, with the additional advantage that a specified grade size contains a concentration of microfossils (Edwards 1963, Katz 1978).

Procedures for the separation of two main grade sizes of particles are outlined; firstly, particles larger than 50μm in diameter, and secondly particles with a mean diameter of less than 50μm.

17.2.1. Particles >50μm

The usual means of separating particles greater than 50μm involves gravity (liquid or mechanical) procedures. The centrifuge may be used if either the residue size is very small, or the specific gravity of the residue is similar to that of the separating liquid. Both these scenarios can occur during micropalaeontological processing, and in particular during the preparation of palynological and nannofossil residues. A centrifuge with a swing-out head, capable of holding a minimum of four tubes (20ml capacity) and attaining speeds of 2000 rpm is adequate for this type of separation.

The standard types of centrifuge tubes used in particle size separation and residue washing, are either the round bottom or taper bottom tubes (Fig. 17.1). However, for low rotational force separation, adapting tubes to assist the decanting stage is just as effective. If suitable glass making facilities are available construction of either the Hutton or Taylor tubes (Figs. 17.2a, b) should be considered. The adaptation of a normal round tube to a pipette tube (Fig. 17.2c) is adequate and probably the simplest solution. Modified tubes described by Yakovley and Shterenberg (1967) are similar to the pipette tubes (Fig. 17.2c) described above, and have been successfully used in the separation of conodonts using a cadmium iodide and barium iodide solution. The solution's viscosity means centrifuge times of 5 minutes at 2000rpm are required for separations. However, both light and heavy fractions can be washed using warm water (Yakovley and Shterenberg 1967).

Using a water suspension for their samples, Funkhouser and Evitt (1959) employed a 45 second centrifuge time at 1500 rpm, followed by decanting of the supernate containing fine debris that did not settle as fast as the fossils. This method was particularly effective after employing ultrasonic treatment. Pre-treatment of samples before centrifuging is necessary as centrifuge times of around 30 seconds are usually ineffectual, particularly if a heavy liquid is used. In the majority of procedures to concentrate a residue at the bottom of a tube using a heavy liquid, and then decant off the fines, usually takes about 3 minutes at 2000 rpm.

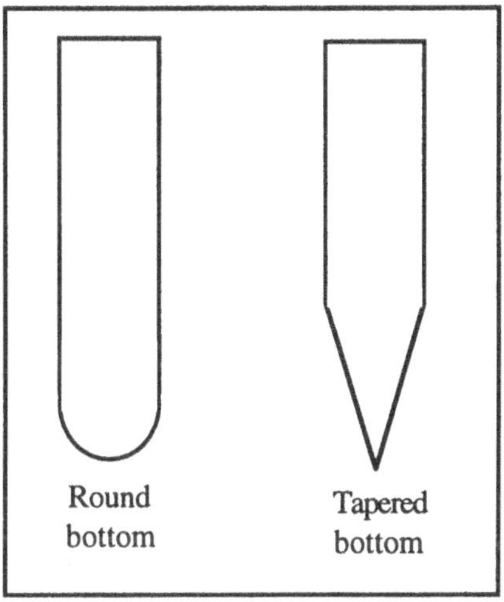

Figure 17.1. The two types of centrifuge tube normally used for the sizing or washing of residues. The round bottom tube is ideal for modifying into a pipette tube (see Fig. 17.2c), (after Allman and Lawrence 1972).

The most important factor that has to be taken into consideration when centrifuging large diameter particles is the ratio of residue weight to the volume of heavy liquid. If the volume of residue is large then there is a danger that some of the denser grains will become trapped within the concentration of less dense grains, while clumps of the heavier grains entrap lighter grains forcing them to the base of the tube. A residue to liquid ratio of 1:10 significantly reduces this problem.

The efficiency of the separation can be further effected by the surface tension of the density liquid used. This can be partly eliminated by ensuring that the sample is completely wet before centrifuging, achieved by gently stirring or agitating the tube. It is also important to ensure that the sample is free of moisture before coming into contact with the heavy liquid. This can be achieved by washing and centrifuging residues in acetone, or thoroughly oven drying for at least 30 minutes. However, continue with caution when working with small volumes of residue, avoiding loss in fan assisted ovens. When washing with acetone, ensure the sample is saturated (*i.e.* agitate the acetone soaked residue in the tube), then centrifuge for 1 minute at 2000 rpm, decant the excess liquid and allow the residue to air dry.

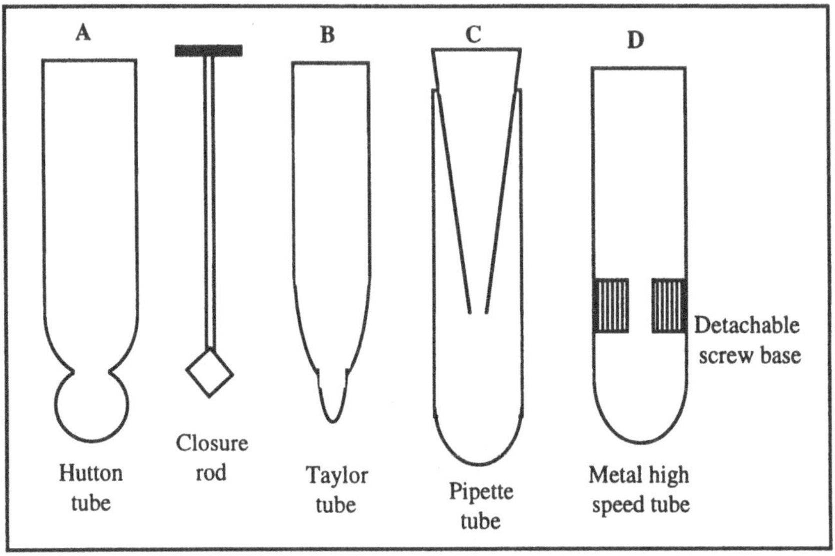

Figure 17.2. Specialized separating tubes used for centrifuging. Steel supporting buckets should be used with examples a-c, (after Allman and Lawrence 1972).

Despite these precautions it is unlikely that any separation will be 100% successful with one attempt. The centrifuge procedure may have to be repeated a number of times to obtain a separation suitable for further work.

17.2.2. Particles <50µm

Much more complicated procedures are required in the centrifugation of particles of less than 50µm. Gravity settling techniques are ineffective, as the grains form a layer on the surface of the liquid. Heavy liquids have long been used in the centrifuge to aid in the concentration of microfossil assemblages within residues (*e.g.* Malyskek 1940, Knox 1942). Many of the heavy liquids used in density separations are **toxic**, while the solvents used for cleaning the tubes is flammable and fumes hazardous to the health of the preparator. If such liquids are used, **ensure the centrifuge is situated within a fume cupboard and the appropriate safety equipment (gloves, vapour mask and eye protection) is available for use.**

The rule for centrifuging particles less than 50µm in diameter is that the higher the rotating speed, the shorter the centrifuge time, and the finer the particles that can be separated (see appendix 3). For micropalaeontological work angle head centrifuges, capable of higher speeds can be used, but these are not as versatile as the swing arm centrifuges, required for heavy mineral separation work.

Centrifuge tubes illustrated in figures' 17.1 and 17.2 are suitable, and have been discussed in detail by Allman and Lawrence (1972). A general point worth reiterating concerns the composition of the centrifuge tubes. Glass tubes should only be used at low speeds, and are therefore restricted in micropalaeontological application. Traverse (1988), discussing their use in palynology, describes how to recover samples from broken centrifuge tubes. The use of metallic (Fig. 17.2d) or polypropylene tubes (Figs. 17.1 and 17.2c) can overcome these problems.

Knox (1942) describes the multiple use of centrifuging residues in bromoform-acetone mixtures to separate foraminifera, ostracods, radiolaria, silicoflagellates, siliceous sponges, diatoms besides pollen and spores. Considerable success was claimed on samples considered 'barren' by standard micropalaeontological processing, and separations were particularly good from lake and deep ocean sediments.

The success of centrifuge separations is dependent on the residue size range. For general work four size fractions may be sufficient (Allman and Lawrence 1972):

(i) > 32μm,
(ii) 8 - 32μm,
(iii) 2 - 8μm,
(iv) < 2μm.

In practice any size range can be used, dependent upon the finest brass sieve available and the grades of monofilament nylon mesh used during wet sieving. The size grades can be obtained by washing, decanting and centrifuging residues in either water or acetone (see appendices' 3.2 and 3.3 for times and speeds). However, the times given in these tables must be multiplied by factors **K** and **A** to determine the *actual* centrifuge time. The calculation of these factors is briefly outlined below, but a more detailed account and worked examples are given in Allman and Lawrence (1972).

Factor **K**, the centrifuge constant, is expressed in the following equation:

$$K = \log_{10} \frac{x + a}{a}$$

where 'x' = the distance from the particles before fall to the bottom of the tube, and 'a' = the distance from the centre of rotation to particles before fall.

Figure 17.3 below, diagramatically illustrate how 'a' and 'x' of the equation are calculated.

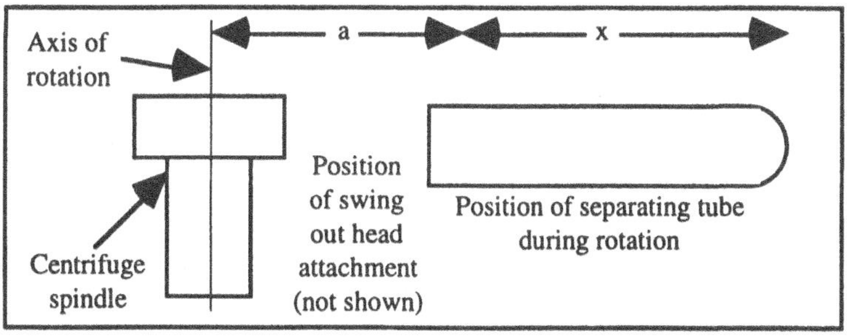

Figure 17.3. Diagram illustrating how the formula for factor **K** is determined for swing out head centrifuges, (from Allman and Lawrence 1972).

Other factors that must be taken into consideration are the grain shape factor (**A**), the time taken to reach the required speed (**R**), which for some machines may be many minutes, and finally a correction for methods where the separating liquid is not bromoform. This is because other liquids have different viscosity's, which in turn effect the settling velocity of the particles.

All these factors can be represented in the following equation:

$$T = \frac{t \times K \times A}{L \times D} + \frac{R}{2}$$

where T' = correct centrifuge time,
and 't' = time obtained from appendix 1,
and 'K' = centrifuge constant,
and 'A' = grain shape factor (usually 1.57),
and 'L' = heavy liquid factor, if other than bromoform, calculated as follows:

$$L = \frac{\text{viscosity of bromoform}}{\text{viscosity of other liquid}}$$

and 'D' = separation difference, calculated as follows:

$$D = (\text{SG of particle - SG of separating liquid}) \times 100$$

and 'R' = the centrifuge run time, usually taken as half the run up time. For long centrifuge times it can be ignored.

The procedure employed to centrifuge a sample is as follows. The method described uses the pipette tubes (Fig. 17.2c), with the sequence of steps diagramatically illustrated in figure 17.4.

1 Determine the weight of the sample and the empty tubes. Divide the sample between the tubes so that with the addition of heavy liquid a ratio of 1:10 is obtained.

2 To the outer tube add a small amount of heavy liquid (Fig. 17.4a).

3 Insert the inner pipette sealed with the closure rod (Fig. 17.4b).

4 Add the remaining amount of pre determined heavy liquid plus the thoroughly wetted sample. The level of the liquid in the inner tube should be marginally lower than that of the outer tube (Fig. 17.4c).

5 Carefully remove the closure rod. This action will allow the liquid levels within both tubes to equalize (Fig. 17.4d).

6 Ensure that all the tubes are of equal weight by the addition of a few drops of heavy liquid. This is essential if speeds in excess of 2000 rpm are to be used.

7 Determine the correct centrifuge time by using the above calculations and with reference to the appendices, and run the separation.

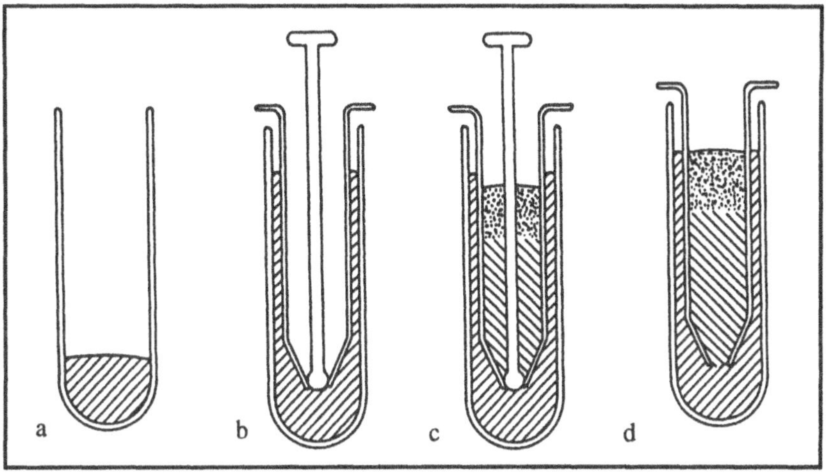

Figure 17.4. Procedure for using the pipette type centrifuge tubes, see text for explanation (from Allman and Lawrence 1972).

8 When the centrifuge has stopped, insert the closure rod and seal the inner tube.

9 Separate the two tubes and decant each fraction into separate beakers.

10 It may be necessary to wash and centrifuge the tubes with acetone to dislodge some of the heavy fraction, refer to appendix 3 for times and speeds.

11 Thoroughly wash all apparatus with acetone, repeat the procedure for each of the tubes.

The insertion of flexible rubber tubing into 100ml centrifuge tubes provides a suitable alternative if 20ml tubes are unavailable (see section 25 EXTRACTION TECHNIQUES FOR PALAEOBOTANICAL AND PALYNOLOGICAL MATERIAL for details).

REFERENCES

ALLMAN, M and LAWRENCE, D. F. 1972. *Geological Laboratory Techniques*. Blandford, (London). 1-335p.
ARNOLD, Z. M. 1965b. The centrifuge as a tool in foraminiferal preparation, pp 258-261: In; B. Kummel and D. Raup (eds), *Handbook of Paleontological Techniques*. W. H. Freeman and Co., (San Francisco). xiii + 852p.
EDWARDS, A. R. 1963. A preparation technique for calcareous nannoplankton. *Micropaleontology*. **9**, (1), 103-104.
FUNKHOUSER, J. W and EVITT, W. R. 1959. Preparation techniques for acid-insoluble microfossils. *Micropaleontology*. **5**, (3), 369-375.
HUTCHISON, C. S. 1974. *Laboratory handbook of Petrographic Techniques*. Wiley-Interscience (New York). xiii + 527p.
KATZ, B. J. 1978. Preparation of calcareous nannofossil assemblages for chemical extraction. *Journal of Paleontology*. **52**, (2), 497-500.
KNOX, A. S. 1942. The use of bromoform in the separation of noncalcareous microfossils. *Science*. **95**, 307-308.
MALYSKEK, V. T. 1940. Method of separation of microfauna from the sediments by centrifugal method on heavy liquids. *Azerbeidzhan Petrol. Industru*. **9**, 4-5.
TRAVERSE, A. 1988. *Paleopalynology*. Unwin Hyman (London). xxiii + 600p.
YAKOVLEY, N. V and SHTERENBERG, L. YE. 1967. Use of detachable centrifuge test tube for conodont separation. *Paleontological Journal*. **1**, (3), 93-94.

18. MECHANICAL SEPARATION OF MICROFOSSIL RESIDUES

18.1. INTRODUCTION

Separating microfossils from a non-fossiliferous component within residues can be a tedious, although necessary process. To aid in this the micropalaeontologist can employ a number of methods. Sieving (wet and dry), and centrifuging techniques are detailed in other sections (see sections 16 WASHING AND SIEVING TECHNIQUES USED IN MICROPALAEONTOLOGY and 17 CENTRIFUGE TECHNIQUES USED IN MICROPALAEONTOLOGY). Liquid separation and flotation procedures can also be used, and have been detailed in section 19 FLOTATION AND LIQUID SEPARATION TECHNIQUES. Liquid separation procedures are extremely successful with Recent foraminifera and ostracods with hollow or air filled shell cavities (i.e. the light fraction), or denser microfossils such as conodonts and infilled foraminifera which sink (i.e. the heavy fraction). Separating calcareous microfossils from other calcareous elements of the matrix, or within quartz rich sediments, without having to hand pick, can be difficult. The Franz Isodynamic magnetic separator provides one means of tackling this problem.

18.2. MAGNETIC SEPARATION

The use of the magnetic separator in microfossil separation was first mentioned in a short note by Schmidt (1948), in concentrating ostracods. Subsequent adaptations in general micropalaeontological work in separating and concentrating foraminifera and ostracoda, have been described by Eckert (1960) and Eckert *et al.*, (1961), and for conodonts by Dow (1960, 1965). Methods based on these descriptions are detailed below.

However, before the magnetic separation procedures are described, some indicators towards obtaining successful separations are outlined.

18.2.1. Residue preparation

Success in concentrating microfossils can be considerably enhanced by size grading the residue, and ensuring fine clays and dust are removed.

1 Wash the residue through a 0.25mm (250μm) sieve onto a 0.125mm (125μm) sieve (i.e. fine sand fraction).

2 Thoroughly oven dry sediment at 60°C for at least 12 hours.

Although this particular grade size has been found to be the most suitable to run through a magnetic separator, finer fractions can be used. However, it is essential to ensure the residue is thoroughly disaggregated, as difficulties may be encountered from electro-static charges exerted be neighbouring particles, and the resulting aggregation or "balling up" which occurs along the separator chute. Providing these problems can be overcome, good separations are possible for material as fine as 39μm.

3 Examine the residue under a stereozoom binocular microscope and determine the dominant mineralogy.

18.2.2. Instrument settings

A standard convention is applied in describing the instrument settings for the magnetic separator (Fig. 18.1). The direction at right angles to the length of the chute is called the side slope, and parallel to the length, the forward slope. Settings of most use for micropalaeontological work have been side slopes of between 15°-25°, and a forward slope of 25°. Although not critical these parameters should be tested with each new sample, as they are effected by both the current and sediment flow rate applied. A steeper forward slope may cause grains to bounce out of the chute, whereas a low slope angle may result in finer grains impeding flow along the chute. Minerals with small magnetic susceptibilities can be separated using small side slopes (2-10°).

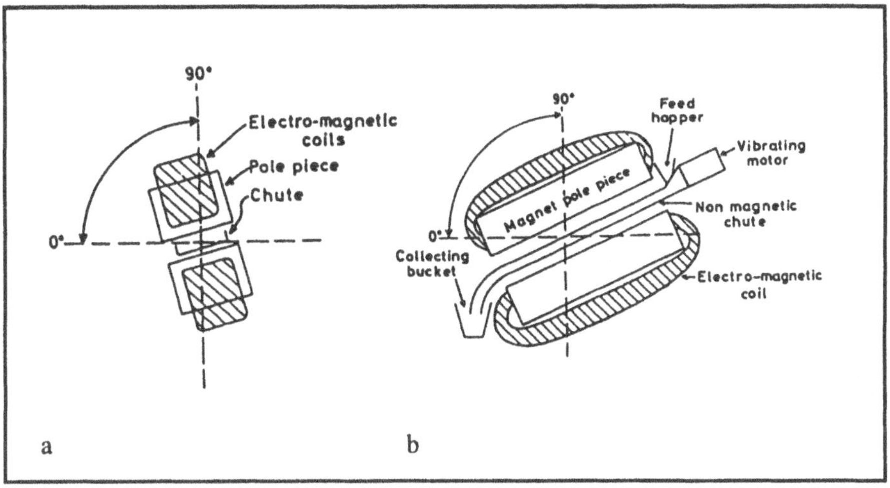

Figure 18.1. Diagrammatic cross and longitudinal sections illustrating typical operating positions of the isodynamic magnetic separator. (a) cross section showing the positive side slope; (b) longitudinal section showing the forward slope (after Allman and Lawrence 1972).

18.2.3. Separating calcareous microfossils in glauconite rich residues

The removal of glauconite and marl fragments from a residue can be achieved as follows.

1. Set the magnetic separator with a side slope of 2.5-3° (Fig. 18.1a), and a forward slope of 25° (Fig. 18.1b).

2. Switch the magnetic separator on, and set the current to 1.5 amperes.

3. Gently pour the sediment into funnel, adjusting the flow so that a single line of particles passes down the chute between the magnets.

4 Diamagnetic marl fragments and glauconite become concentrated in the magnetic bucket. Quartz, calcite and shell fragments concentrate in the non-magnetic bucket.

One run may be sufficient to remove most of the impurities. Glauconitic rich residues can be separated at the lower side slope value (2.5°) (Fig. 18.1a), and a current of 1.2 amperes (Schmidt 1948). For general applications side slope settings of between 15°-25° and a forward slope of 25°, with a current of 1.5 amperes are effective (Eckert *et al.* 1961). Smaller side slope values can be used to remove dirty specimens.

18.2.4. Separating calcareous microfossils from quartz

Separating the quartz component from calcareous microfossils is more difficult. Eckert *et al.*, (1961) successfully introduced iron into the calcareous microfossil shells so that they became magnetic. This is achieved by treating samples in the following way:

1 Add a weak (2-5%) acidic solution of ferric chloride ($FeCl_3$) to the residue. The calcareous shells react violently if a stronger solution is used, while effervescence produces carbon dioxide. The reaction should only take a few seconds, by which time the calcareous component will have a yellow-orange colour.

2 When the reaction has ceased, wash the sample with distilled water and allow the residue to slowly dry in an oven pre-set at 60°C.

If very delicate forms are present in the assemblage, neutralize the solution with the addition of a suitable buffer (*e.g.* sodium acetate ($CH_3.COONa$)).

3 The thin iron coating should be enough to allow magnetic separation. Set a side slope of 3° or less, and a current of 1.6 amperes. The microfossils are drawn along the magnetic chute and into the 'magnetic' bucket fraction. The quartz grains are concentrated in the non-magnetic bucket.

18.2.5. Separating and concentrating conodonts

The method of separating and concentrating conodonts using the isodynamic magnetic separator has been described by Dow (1960, 1965). The procedure is as follows.

1 Pre-treat the residue (washing, grading, binocular examination) as for calcareous microfossils described above. Determination of the dominant mineralogy will decide the various settings required for the first run through the separator (Table 18.1).

2 Following the first run, the magnetic and non-magnetic fractions must be examined to assess the effectiveness of the separation. The magnetic fraction collects in the outer (left) bucket.

The concentration of conodonts in the non-magnetic fraction determines whether further separations are required (Table 18.1). An initial run removes most of the magnetic material, however, if after optical examination conodonts are present within the magnetic fraction of the first run, reduce the speed of the residue feed, and the angle of the forward slope (Fig. 18.1b). Following some runs magnetic aggregates may be incorporated within the non-magnetic fraction, a result of the amperage set too high or a residue feed that is too rapid. Concentrations can be enhanced by re-setting the machine and applying second or third runs to the sample. During successive runs, a fine powder may collect within the fractions. This can be removed by dry sieving (see section 16 WASHING AND SIEVING TECHNIQUES USED IN MICROPALAEONTOLOGY). Fortunately, most samples require only one pass through the separator, with the degree of microfossil separation and concentration at the operator's discretion or determined by the lithology of the residue.

Details of conodont separations from quartz and dolomite, the two most common minerals of the residues, are outlined by Dow (1960, 1965). Quartz can be separated using similar settings to dolomite (Table 18.1, section 4). Conodont elements start to move to the non-magnetic side at 1 ampere. However, the movement of quartz to the non-magnetic side begins at 0.75 ampere, and in volume when 0.9 ampere is reached. Above this setting conodonts move with the quartz grains. Thus, if the magnetic fraction (conodont bearing) is run more than once, quartz can be removed with each pass. Only a few lithologies were tested using the instrument settings summarized in table 18.1 (Dow 1960), and further modifications, particularly during an initial run, may be required.

18.3. OTHER METHODS

18.3.1. Static electricity

In cold weather, static electricity builds up on the sheet of paper used to retain residues poured from sieves. Air filled foraminifera and ostracods, together with platy micaceous fragments adhere to the paper, and can thus be removed from the bulk of the residue. The technique described by Brooks (1954) is as follows:

1. Dry sieve the residue into a series of fractions (500µm-90µm at 0.5ø intervals).
2. Pour the fraction onto a piece of paper.
3. Alternately gently raise and lower the edges of the paper, causing the sample to slide slowly upon the paper. Try to avoid jerky movements and spilling the sample.
4. Slowly pour the sample onto a second sheet of paper.
5. Examination of the first sheet of paper surface will show it contains many of the platy minerals (mica is often a problem to separate from microfossil residues) plus fragments of the microfossils. Complete fossils often remain with the bulk of the residue.

	Fraction containing conodonts to be extracted	Forward slope	Side slope	Current in amperes	Vibrator setting	Feed	Retaining bucket conodonts are concentrated in:
1	Contains abundant limonite and magnetic particles	20 - 25	+10 - +15	1.0 - 1.3	4.0	Medium fast	Non-magnetic fraction. Re-run non-magnetic fraction at 1.3 amps
2	Contains abundant clay shale framents	10 - 15	+10 - +15	Max*	4.0	Medium fast	Non-magnetic fraction, re-run as (3)
3	Non-magnetic residue of (2) above	10 - 15	+2 - +5	Max*	4.0	Medium fast	Non-magnetic fraction, re-run as (4), (5) or (6)
4	Non-magnetic residue of (3), mainly dolomite, or residue with abundant dolomite	10	-2	0.9	2.5 - 3.0	Slow	Magnetic fraction
5	Non-magnetic residue of (3), mainly iron stained quartz, or residue with abundant iron stained quartz	25	+10	Max*	4.0	Fast	Non-magnetic fraction. Re-run magnetic fraction
6	Non-magnetic residue of (3), mainly dolomite, or residue with abundant dolomite	10	-2	Max*	2.5 - 3.0	Very slow	Non-magnetic fraction, re-run both fractions as (6)

Table 18.1. Settings for the Franz Isodynamic Magnetic Separator for use in the magnetic separation of conodonts and other phosphatic micro-fragments (data from Dow 1960, 1965).
* Prolonged use at maximum power can result in over heating of the coils, and a reduction in the magnetic field produced. Can be prevented by cooling the coils of the electro-magnet.

6 Sharply tap the paper to remove this material, to either retain for later observation or discard. For ease and speed the residue can often be examined under a binocular microscope while it is still on the paper.

7 Repeat the procedure several times if there is a high percentage of mica in the residue.

Ensure a fresh piece of paper is used for each fraction and sample to avoid the problems of contamination.

The separation of other calcareous particles provides a greater problem. The method preferred by Eckert *et al.*, (1961) was to use an inclined sheet of paper held at an angle 10°-20°. Separation of rounded foraminifera and ostracod shells from more angular calcite fragments was achieved by gently tapping the residue down the inclined slope. The success of this technique does, however, require the initial removal all quartz grains.

The methods outlined above are particularly useful in concentrating microfossils in near barren samples, where hand picking of large quantities of residue, or light microscope scanning is required to obtain a representative fauna.

18.3.2. Electro-mechanical methods

Two specialized pieces of equipment adapted to grade sizings have been described in methods for separating disaggregated fossil material from matrix. One allows for the dry separation while the other can separate wet residues.

18.3.2a. Photoelectric separation

The procedure outlined by Kuehne (1971), using a *Sortex* photoelectric separator, was successful in separating material within the grade size 1.5-2.5mm (very coarse sand to granule). Obviously, a disadvantage of this technique is that it cannot deal with sand grade size material. However, for bulk sediment residues (many kg's) which has been chemically treated (e.g. acetic acid procedure), and contain large quantities of phosphatized material (teeth, bones small shelly fossils) an adaptation can be readily seen. One run through the machine, set to separate light from dark particles, concentrated approximately 50% of the phosphatic material (Kuehne 1971). A considerable capital outlay is, however, required to purchase one of these machines, which are primarily used in the food industry to optically sort beans and nuts by biochromatic measurements of colour and shade.

18.3.2b. Vibratory separation

By using a modified *Syntron* machine, Anderson and Hoffman (1963) illustrated the mechanical separation and concentration of recent calcareous foraminifera and ostracods from clastic dominated sediments. A cast aluminium bowl with a 5mm wide spiral track vibrates 3-4g of residue. A continuous supply of water fed to the base of the bowl passes along the spirals when the machine is activated, and ensures the residue remains as a slurry. When the machine is switched on the residue vibrates up a spiral and out of the bowl via a trough. Rounded quartz grains are separated from the microfossil concentrate, and exit the trough first. Material can be collected on a suitable sieve mesh, but as a definite break occurs between the clastic and non-clastic fractions, it is possible to separate each component into an evaporating dish. Anderson and Hoffman (1963) achieved concentrations of about 80% on calcareous and agglutinated foraminifera and ostracod assemblages using this equipment, significantly reducing the time spent picking the microfauna.

18.3.3. Ignition method

The ignition method of concentrating planktonic foraminifera from aqueous samples was initially outlined by Sachs, Cifelli and Bowen (1964) and Sachs (1965b). It is a partially destructive method in that it reduces the sample bulk by burning away the combustible component, and concentrates the calcareous and siliceous microfauna. Following a comparison of methods for sorting foraminifera, Boltovskoy (1966) favoured filtration (see section 19 FLOTATION AND LIQUID SEPARATION TECHNIQUES) considering the ignition method to be time consuming, requiring expensive equipment that might not be available in every laboratory (*e.g.* muffle furnace), and in the process destroying test protoplasm, making it inappropriate for determining live-to-dead ratios. The modified ignition procedure of Smith (1967) is as follows:

1 Sub-sampling is essential. This ensures that unshelled specimens can, if required, be studied later as non shelled organisms, and a few shelled forms such as planktonic ostracoda, will not survive the method.

2 Removal of any preservative (*e.g.* formaldehyde) can be undertaken by filtration through a fast speed filter paper (Whatman No. 41 or equivalent), or if the sample is very rich, by centrifugation. Large discrete organisms (*e.g.* tunicates, jelly fish) must be removed as their presence can result in a gummy residue following ignition (Smith 1967).

3 The digestion of soft organic matter can be achieved by the addition of a 30-35% solution of hydrogen peroxide or sodium hypochlorite. In both cases it is essential to ensure that all the formaldehyde has been removed. Formaldehyde reacts in the presence of hydrogen peroxide to produce formic acid that will dissolve calcareous shells (Boltovskoy 1966), while in the presence of sodium hypochlorite poisonous chlorine gas is generated. **Ensure the procedure is undertaken in a well-ventilated fume cupboard.** Let the sample stand for one hour, or until the organic matter has turned white. Top up with additional solution if required.

4 Re-filter the sample through Whatman grade 41 filter paper, and wash with distilled water.

5 Carefully remove the filter paper, oven dry at 60°C or leave to air dry at room temperature or under an infra-red lamp.

6 Place the dried filter paper in a labelled porcelain crucible, and transfer to a pre-warmed muffle furnace. Slowly increase the temperature to 500°C, and ignite the samples maintaining the temperature for two

hours until the ash is light grey or white. The temperature of the furnace should not exceed 550°C as damage to specimens may result. If the furnace is pre-heated to 500°C, and the sample introduced, initial flashing ignition can result in the specimens being subject to temperatures in excess of 600°C. Although this may last for only a few seconds damage to the specimens can result.

7 Allow the furnace to slowly cool to room temperature. Extreme care must now be employed when handling the sample as the ash is very light. For this reason Smith (1967) suggests that ignition can be undertaken using a Pyrex petri dish with a lid.

8 Saturate the residue with distilled water and a couple of drops of detergent, and immerse for a few seconds in an ultrasonic bath to disaggregate. Wash the residue over a fine (39μm) sieve with a continuous flow of water. Finally wash the residue into a petri dish, and do the last wash with alcohol and allow the residue to dry at room temperature away from any external drafts.

9 Examine the dried residue on a picking tray under a stereozoom binocular microscope.

The final residue produced following this procedure consists of calcareous micro-molluscs, foraminifera and ostracoda, and siliceous radiolaria and diatoms. These components are very light, and can easily be lost during the final stages of preparation, particularly when the residue is dry. Great care must be employed during the transportation of the residue around the laboratory to avoid cross sample contamination, particularly in a multi-user working environment.

REFERENCES

ALLMAN, M and LAWRENCE, D. F. 1972. *Geological Laboratory Techniques*. Blandford, (London). 1-335p.

ANDERSON, L and HOFFMAN, E. 1963. Concentration of microfaunal samples by use of a *Syntron*. *Micropaleontology*. **9**, (1), 105-106.

BOLTOVSKOY, E. 1966. Methods for sorting of foraminifera from plankton samples. *Journal of Paleontology*. **40**, (5), 1244-1246.

BROOKS, H. K. 1954. A method for removing platy fragments of fossils and minerals from microfossil concentrates. *Micropaleontologist*. **8**, (4), 39-40.

DOW, V. E. 1960. Magnetic separation of conodonts. *Journal of Paleontology*. **34**, (4), 738-743.

DOW, V. E. 1965. Magnetic separation of conodonts, pp 263-267: In; B. Kummel and D. Raup (eds), *Handbook of Paleontological Techniques*. W. H. Freeman and Co., (San Francisco). xiii + 852p.

ECKERT, R. 1960. Reinigungs- und Anreicherungsversuche an Kleinforaminiferen. *Eclogae Geologicae Helvetiae*. **53**, (2), 645-648.

ECKERT, R, HAY, W. W, LORENZ, G and VOGT, P. 1961. The magnetic separator as a tool in micropalaeontology. *Journal of Paleontology*. **35**, (5), 876-877.

KUEHNE, W. G. 1971. Photoelectric separation of microfossils from gangue. *Proceedings of the Geological Society (London)*. 1664, 221-222.

SACHS, Jr., K. N. 1965b. Removal of ash from plankton samples concentrated by ignition. *Deep Sea Research*. **12**, 697.

SACHS, Jr., K. N, CIFELLI, R and BOWEN, V. T. 1964. Ignition to concentrate shelled organisms in plankton samples. *Deep Sea Research*. **11**, 621-622.

SCHMIDT, R. A. M. 1948. Magnetic separation of microfossils. *Journal of Paleontology*. **22**, (4), 530-531.

SMITH, R. K. 1967. Ignition and filter methods of concentrating shelled organisms. *Journal of Paleontology*. **41**, (5), 1288-1291.

19. FLOTATION AND LIQUID SEPARATION TECHNIQUES

19.1. INTRODUCTION

Liquid separation procedures can generally be assigned to one of two groups. (i) Gravity separation techniques: use a gravity difference between particles within a residue. The procedure's success is considerably improved by using a liquid with a specific gravity (S.G.) value between the two main mineralogical components to be separated. These techniques are most successful with sand grade or fine sand grade material. (ii) Centrifuge liquid separation techniques: undertaken with the aid of a rotary machine creating a centrifugal force to separate particles. Centrifugation can be further enhanced by using density liquids. The technique is suitable for material <50µm in diameter (coarse silt to clay grade material), and is detailed in section 17 CENTRIFUGE TECHNIQUES USED IN MICROPALAEONTOLOGY.

Separating and concentrating microfossils from prepared residues can be a relatively simple procedure, employing little more than a modified panning technique, ideal for the coarse separation of a bulk sample. More sophisticated flotation techniques, although still simple to perform, are used for smaller sample volumes of only a few grams. More complex techniques use density liquids, and when performed correctly can concentrate the smallest percentage of microfossils from near barren samples. Essentially all procedures have a similar objective - that of reducing the time required in scanning a processed residue by concentrating the microfossils. General reviews and examples of use of liquid separation procedures are provided by Allman and Lawrence (1972), Anon (1935), Brasier (1980) and Higgins and Spinner (1969).

Many of the procedures use toxic or flammable liquids. **All procedures should be performed in a fume cupboard, with the operator using adequate personnel protective equipment.**

19.2. PANNING

This technique is best applied to bulk samples of sand grade size material, and provides a reasonably good separation or bulk volume reduction (Allman and Lawrence 1972). Use a wide shallow tray or dish that will fit into a large bowl or sink (preferably one in which a separated residue can be recovered). A clean supply of water is required.

1 Fill the dish with sediment.

2 Thoroughly wet the sample, and remove large fragments or pebbles.

3 Fill a bowl with enough water to allow the dish to be immersed to a depth of at least 10cms.

4 Gently swirl the dish around whilst also applying a to-and-fro motion.

5 When performed correctly, the motions cause the light fraction to

separate from the heavy fraction, washing over the dish edge into the bowl.

6 Recover both fractions and oven dry.

This procedure is not recommended for quantitative work, as it may introduce appreciable errors in the frequency proportions of components analysed at a later stage (Allman and Lawrence 1972).

19.3. FLOTATION

Care must be taken with these separation techniques, as they can result in the selective concentration of certain types of microfossils, distorting populations present within a sample. Assemblages and populations of agglutinated foraminifera present in residues are particularly susceptible to removal by the process of liquid separation and decanting (Brooks 1954). It is therefore important to retain and examine both 'light' and 'heavy' residue fractions, knowing one will contain a richer assemblage than the other.

19.3.1. Water and detergent

This simple technique first described by Howe (1941), is usually incorporated within the general washing and sieving of a residue, and is best applied to the separation of specimens from a small amount of clean residue, typically Holocene and Tertiary clastic-carbonate rich residues.

1 Place the residue in a small (100ml) beaker, and add approximately 60ml of distilled water, with a couple of drops of detergent.

2 Seal the top and shake the mixture.

3 The light microfossils can be decanted with the detergent froth.

4 Alternatively the residue can be gently heated, and then immersed in cold water. The warm air filled microfossils will float.

5 Oven dry the fractions at 60°C.

Foraminiferal recovery from processed Holocene samples can be impaired by the presence of rootlet material and plant debris. In dried residue's plant remains may adhere to one another forming an interlocking mat enclosing microfossil tests (Lehmann and Röttger 1997). Much of the plant debris can be removed by careful decanting. Examining wet samples provides a means of recovering a wider variety of organisms, including allogromiid and monothalamous forms (those with soft tests), juvenile and brood cysts not usually seen when conventional processing is undertaken. Test recognition within the supernate is considerably enhanced when a protoplasmic stain is used (see section 21 STAINING TECHNIQUES USED IN MICROPALAEONTOLOGY), rapidly identifying isolated shells. Although, a laborious procedure, examining samples wet ensures that a greater proportion of the non-fossilized forms are preserved (Lehmann and Röttger 1997).

19.3.2. Saturated sodium chloride (NaCl)

This relatively simple procedure exploits specific gravity differences between shelled and non-shelled organisms present in marine plankton samples. Bé (1959) successfully used a saturated solution of sodium chloride (S.G. 1.20) to separate recent planktonic foraminifera and other shell organisms (S.G. 1.40) from non-shelled planktonic organisms (S.G. 1.09). The procedure is as follows:

1. Prepare a saturated solution of sodium chloride (310g NaCl per litre of distilled H_2O), to a specific gravity of 1.20 (check using a hydrometer).

2. Attach a large volume (1-2000ml) cylindrical funnel to a retort stand, ensure the stopcock is closed (and that it moves with ease, grease if required). Fill the cylinder with the sodium chloride solution.

3. Decant off any preservative (formalin or alcohol) used on the sample, and then pour the sample into the cylinder (a wash bottle filled with saturated sodium chloride is useful in washing the sample from the storage container to the cylinder). Gently stir the sample with a glass stirring rod to ensure it is completely saturated (rinse the rod of any adhering sample on removal).

4. Separation of the heavy fraction begins immediately, and will only take 2-4 minutes to settle, even following gentle agitation.

5. Remove the heavy fraction by draining through the stopcock.

6. Stages 4 and 5 may require repeating, particularly if the sample is large, or a high concentration of foraminifera is required.

7. Oven dry (60°C) or air dry the fractions. Alternatively they can be stored in preservative for later examination.

The effectiveness of the separation depends on the number of times a sample is run. Experimental work by Bé (1959) demonstrated that from some samples up to 95% (by weight and number of specimens) of planktonic foraminifera could be separated by this procedure.

19.3.3. Carbon tetrachloride (CCl_4)

Safety Note: Carbon tetrachloride is a toxic solution. Do not allow it to come into contact with skin, wear gloves, and eye protection when using the solution and carry out the process in a fume cupboard.

This procedure works best on recent faunal assemblages, where test infilling has not occurred, and fossils are significantly lighter than other associated clastic or biogenic particles and collect as a 'scum' on the surface of the liquid (Murray 1979). Foraminiferal recovery rates of up to 94% can be achieved from suitable sediments (Lutze 1968). The procedure proved unsatisfactory for certain Tertiary residues (Buzas 1965), while separations within some Holocene sediments have required modification (Lehmann and Röttger 1997). Both the "float" and "heavy" fractions should be examined as species separation and concentration prove unreliable (see discussion below).

1 Place the dried, washed residue in a beaker.

2 Add between two to three times the amount of carbon tetrachloride (S.G. 1.58).

3 Vigorously stir the mixture with a glass rod. The buoyant air filled foraminifera and paired ostracod carapaces float to the surface.

4 Decant the 'float' into a filter lined funnel (filter paper grade 1) placed in the top of a glass flask. Clean the stirring rod and the inside of the beaker of excess residue by flushing with a carbon tetrachloride squeeze bottle.

5 Repeat stages 2-4 until no further shells rise to the surface.

6 Allow the filter paper and residue to dry out until they are odourless.

7 Place a second filter paper in the funnel and flush in the remaining sediment. Allow this to drain, and dry until odourless. Reclaim as much of the heavy liquid as possible and return it to the bottle.

8 Pour the 'float' and remaining sediment into separately labelled vials.

Brasier (1980) further suggests that this technique is suitable for separating siliceous microfossils, but not thick-shelled, infilled or fragmentary foraminifera or ostracods.

Experimental studies undertaken by Gibson and Walker (1967) and Wright and Hay (1971), respectively working on Tertiary and Recent foraminifera from the Atlantic coast and Recent benthic foraminifera assemblages from South Florida, have questioned the validity of this technique. Both studies revealed selective test concentrations from sediment residues using this method, resulting in disproportionate species values significantly differing from the actual. Many species did not float, and consequently were absent from the concentrate. In both studies the method was abandoned.

Problems encountered by Lehmann and Röttger (1997) analysing organic rich salt marsh muds with low clay and sand components, were overcome by modifying the drying procedure. Washed residues must be thinly spread onto a large flat drying tray. The separated detrital particles are then gently brushed off (pushing the brush rather than pulling and crushing foraminiferal tests), and separated by flotation. A success rate of 96% recovery from a single flotation run has been claimed (Lehmann and Röttger 1997).

19.3.4. Isopropyl alcohol

The use of carbon tetrachloride and other heavy liquids is suitable when the sediment consists of dense quartz and calcite fragments. However, if the sediment is rich in organic carbonate particles (calcareous algae, foraminifera, molluscs, ostracoda), these all become concentrated together. Kornicker (1957) encountered this problem when processing a recent Caribbean sediment. Peneroplids and *Halimeda* masked smaller organisms of the fauna. The use of low specific gravity alcohol in substitution for a heavy liquid resulted in a float dominated by articulated ostracods and smaller foraminifera, with the heavier and larger fragments (Peneroplids and *Halimeda*) sinking. Results demonstrated that between 40-50% of the articulated ostracods, and practically none of the disarticulated valves, were separated. Concentrations of articulated ostracod valves were 5-16 times more than that produced by the carbon tetrachloride method.

19.4. FILTRATION

In separating recent planktonic foraminifera from marine water for quantitative studies, Boltovskoy (1966) compared four methods:

(i) "common" or "classical" method - the separation and picking of dried or wet residues under a binocular microscope using a fine brush or pipette (see section 20 THIN SECTION AND SLIDE PREPARATION TECHNIQUES OF MACRO AND MICROFOSSIL SPECIMENS);

(ii) saturated sodium chloride method of Bé (1959) - detailed above;

(iii) ignition method of Sachs *et al.*, (1964) and Smith (1967) - the burning and removal of combustible material, concentrating calcareous and siliceous tests (see section 18 MECHANICAL SEPARATION OF MICROFOSSIL RESIDUES);

(iv) filter method - considered to be the most rapid and convenient.

Foraminifera are concentrated as follows:

1. Split the sample, retaining a portion for the study of non-shelled plankton.

2. Filter the fraction using a Buchner's funnel (filter paper grade unspecified).

3. Oven dry the filter paper and residue at a temperature not exceeding 80°C. Take care if a fan assisted oven is used, as the dried residue may be dislodged from the filter paper surface, and multiple preparations may cross contaminate.

Although the method offers speed and simplicity in processing a residue, Boltovskoy (1966) recognised disadvantages in the procedure. An abundance of soft bodied organisms may cluster during residue drying, and in doing so entangle and obscure foraminiferal tests. These clusters are easily disaggregated, with no damage to the foraminifera, using a mounted needle. Delicate elements of the plankton (*e.g.* crustacea) may be destroyed during desiccation. Tunicates (marine Urochordata, *e.g.* sea-squirts) must be removed before drying the residue drying, as their gelatinous bodies may form clusters that are difficult to disaggregate. However, the relatively large size of these organisms means their separation is straight forward. The filtration method was discussed by Smith (1967), who preferred the ignition method, described in detail in section 18 MECHANICAL SEPARATION OF MICROFOSSIL RESIDUES.

19.5. DENSITY LIQUIDS

Safety Note: Bromoform and tetrabromoethane are extremely toxic solutions, and must be handled with care. Do not allow skin contact. Wear gloves, and eye protection when using these solutions and carry out the process in a fume cupboard. Consult Hauff and Airey (1980) for more detailed information on the hazards, laboratory handling and storage of these chemicals.

Density or heavy liquid separation techniques of fine and medium sand grade size material are extensively used in sedimentology (*e.g.* Allman and Lawrence 1972, Tucker 1990). The techniques have been applied and are commonly used in separating and concentrating microfossil assemblages, and in particular conodont rich residues. The two principle liquids used, Bromoform ($CHBr_3$) and Tetrabromoethane ($CHBr_2.CHBr_2$) have been frequently cited during the past sixty years (*e.g.* Hodson and Hodson 1926, Carson 1933, Knox 1942, Carson 1953, Frey 1955, Collinson 1963, Gibson and Walker 1967, Murry and Lezak 1977). The liquids individual properties can be altered by the addition of a solvent (*e.g.* acetone, Knox 1942, Carson 1953; ethyl alcohol, Von Bitter and Millar-Campbell 1984), to give a specific gravity (S.G.) within the range 2.7 to 3.1, depending on the density of the component separated. If the heavy liquid is modified with the addition of a solvent, it is necessary to periodically monitor the specific gravity during use to ensure it remains constant, as it is liable to alter with evaporation (Von Bitter and Millar-Campbell 1984). Successful separations of micro vertebrate material from sediment have been achieved using a combination of dibromoethane and tetrabromoethane (Murry and Lezak 1977, Von Hofe Grady 1979).

The high purchasing costs of bromoform and tetrabromoethane have always meant that considerable effort has been placed in attempting to reclaim used liquids (Benjamin 1971, Cohee 1937, Mileson and Jeppsson 1983, Turner 1966, Von Hofe Grady 1979), particularly if they have been contaminated by residue or acetone wash. More worrying has been the considerable concern expressed over recent years regarding the use of these carcinogenic liquids. Numerous safer non-toxic alternatives have been suggested (*e.g.* sodium polytungstate, Belka *et al.*, 1989, Callahan 1987, Krukowski 1988, Savage 1988; zinc bromide and zinc chloride, Traverse 1988; cadmium and barium iodide (used in conjunction with a centrifuge) Yakovley and Shterenberg 1967). Another alternative is the use of an interfacial method (Freeman 1982), modified and adapted for conodont separations (Belka *et al.*, 1989, Merrill 1985, 1987, see below).

1 The heavy liquid used should be reduced to a specific gravity no lower than 2.75 (this allows calcite to float), by the addition of acetone or benzene. If non-toxic sodium polytungstate is used, the specific gravity and viscosity are reduced by the addition of water.

2 Assemble two funnels (the lower lined with filter paper) above one another (Fig. 19.1). A clamp and piece of rubber tubing should be attached to the upper funnel, and a beaker placed under the lower funnel.

3 Close clamp and fill the top funnel with heavy liquid, sprinkle residue over the surface.

4 Gently stir. This allows for conodont separation from the light fraction, and movement of conodonts down the funnel sides. Savage (1988) recommends stirring every half hour for 4 hours when using sodium polytungstate.

5 Allow liquid to stand for 30 minutes. Conodonts should sink.

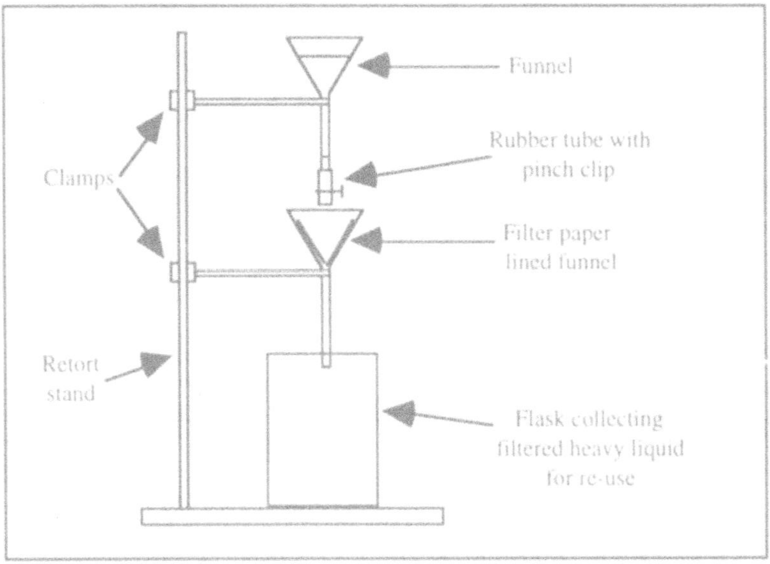

Figure 19.1. Diagrammatic sketch showing the position of equipment used in a heavy liquid separation.

6 Gently release clamp and allow heavy conodont rich fraction to drain into the lower filter paper-lined funnel.

7 Replace lower flask and thoroughly flush residue with acetone. Allow residue to dry at room temperature.

8 With a second piece of filter paper in the lower funnel, drain remaining heavy liquid and light fraction into first flask containing used heavy liquid only. Heavy liquid can be reclaimed.

9 Wash light fraction as before, with acetone, and allow it to drain into second flask. The light fraction should be retained for examination - it may contain conodonts that have not separated.

If a large number of liquid separations are to be undertaken, then Charlton's (1969) modification to the tedious and time consuming stirring stages can be considered. A series of weighted plastic paddles was made, and immersed in the top funnel. Motion was achieved with the aid of a compressed air jet, which when activated ensured the residue was kept in constant suspension within the liquid. The procedure is as follows:

1. Fill the top funnel with heavy liquid.
2. Drop in the stirring device.
3. Gently sprinkle on the sediment, and manually stir to ensure that it is completely saturated.
4. Turn on the compressed air supply, adjusting flow so that stirring device is in gentle motion.
5. Add additional heavy liquid if required.
6. Run the apparatus for 1 hour.
7. Turn off the air flow, and allow the apparatus to remain idle for 2 minutes.
8. Draw off the heavy fraction, as in stages 6 and 7 of main method.
9. Remove the stirring device and clean with a solvent.
10. Draw off the light fraction following stages 8 and 9 of main method.

Controlled experimental work by Charlton (1969), comparing manual and automatic stirring methods, demonstrated the efficiency of this modification. After 30 minutes the automatic method was giving results comparable to a manual method (stirring once every 5 minutes) over a 2 hour time period (Table 19.1). Furthermore, it was possible to demonstrate that sample size and the time period of liquid separation contributed to efficient conodont recovery. Samples of 25 grams were preferable to those of 70 grams, while smaller conodonts were more difficult to remove through liquid separation than large conodonts (Table 19.1).

A modified stirring procedure was subsequently been employed by Davis and Webster (1985). By mounting an angled piece of glass tubing below the rim of the separating funnel, a directed air current from a small pump was applied to the surface of the heavy liquid. This forms a circulatory current on the liquid surface, removing the need to use paddles in stirring and a possible source of residue contamination if they are not thoroughly cleaned between samples. Successful separations were achieved within 15-20 minutes of circulation, although care has to be taken in setting the initial volume of the air flow, as a high volume results in residue splattering around the sides of the funnel. Conodonts trapped in this portion of the residue would remain in the light fraction. A further problem results if a large amount of residue is added to the heavy liquid. Particles form an impermeable unsaturated layer on the surface of the liquid, and fail to mix and thus separate. This can be eliminated by separating smaller quantities of residue. However, extreme care must be employed by the preparator when using a toxic heavy liquid with this apparatus, as excess fumes are generated by vigorous stirring. Ensure the equipment is assembled and used in a fume cupboard with suitable ventilation.

CONODONT SIZE		SEDIMENT WEIGHT (GRAMS)							
		25		70		70			
	TIME (mins)	30	60	30	60	30	60	90	120
SMALL SPECIMENS (x20)	% RECOVERED	85	97.5	80	87.5	52.5	65	70	75
LARGE SPECIMENS (x20)	% RECOVERED	95	97.5	95	97.5	70	87.5	90	97.5
TOTAL % OF CONODONTS RECOVERED		90	97.5	87.5	92.5	61.25	76.25	80	86.25
STIRRING METHOD		AUTOMATIC				MANUAL			

Table 19.1. Comparison of the percentage of conodont material recovered using automatic and manual methods. Size divisions are based on the arithmetic mean of the two largest components. The same 40 specimens were used in all 8 runs. (Data from Charlton 1969).

19.6. INTERFACIAL TECHNIQUES

Interfacial techniques were first used by Freeman (1982) to aid in the concentration of coarse sand grade size (0.5-1.0mm) micro-vertebrate (bone) fragments. This procedure exploits the lipophilic properties of calcium phosphate grains. This is the grain's ability to selectively attract oils to their surfaces. The oil-coated grains can then be removed from a residue by adhering to the surface of a substrate to which the oil is a solvent. It is the reaction at the interface of the two mutually immiscible liquids which Freeman (1982) used to name the technique. Three variations of the technique were described by Freeman (1982). Two of the procedures involved the adhesion of bone particles to a solid or gelatinous substrate (polystyrene or petroleum jelly/wax), while the third method relies on two-phase liquid mixtures.

The three procedures differ in the way in which the interfacial properties are exploited. In Procedures A and B the bone fragments are gently agitated in a suspension of water-insoluble organic liquid (the "solvent", *e.g.* tetrachloroethylene, $CCl_2{:}CCl_2$; kerosene) plus a dilute aqueous detergent. The bone particles become coated with the organic liquid, and are selectively removed from the residue when they encounter the solid substrate and attach by means of their organic coatings. In Procedure A the organic liquid used is tetrachoroethylene and the substrate is polystyrene. The bone fragments are removed from the substrate by gentle brushing or dissolving the substrate with a suitable solvent. In Procedure B the organic liquid is kerosene (paraffin) and the substrate is an inert support coated with petroleum jelly. Fragments are recovered by melting the substrate in hot water containing a small amount of detergent.

In the third procedure the residue is agitated in a solution of water (without detergent) and an immiscible organic liquid (*e.g.* kerosene). This is allowed to settle, from which the interface between the two liquids is readily visible, and that the bone fragments are held in suspension within the kerosene. The kerosene can then simply be decanted. Although recovery is low following this procedure, the speed and simplicity of the method makes it ideal for reconnaissance studies.

In detail, the procedures are as follows:

(i) Procedure A

1 Disaggregate the sample to obtain a minimum 5g of residue in the 0.5-1.0mm size grade. Place this in a 250ml glass beaker.

2 Add a saturated solution of tetrachloroethylene and 50ml of 1% detergent.

3 Briefly agitate the mixture, and then rapidly pour it into a 160ml container made of unfoamed polystyrene (*e.g.* yoghurt carton). Leave for approximately 15 seconds, and then transfer the mixture to a second similar container. The surface of the polystyrene in contact with the mixture will be softened and pitted by the tetrachloroethylene, embedded in which will be a concentrate consisting predominantly of bone particles.

4 Repeat this process until the yield of the concentrate attached to the polystyrene is negligible (Freeman 1982, suggests at least five times, giving a total of six concentrates).

Quantitative experimental work by Freeman (1982, p 475) revealed recoveries of bone material ranging from 18-97%, depending on the percentage of bone present in the original residue and the main component of the matrix.

(ii) Procedure B

1 Prepare the sample as in Procedure A. For experimental work Freeman (1982) tested 25g of sample from the 0.5-1.0mm size grade.

2 Saturate the residue with kerosene, and wash into a 1 litre screw top bottle with 250ml of a 1% detergent solution. Use a plastic bottle with a push in seal, modified by having its base cut off. Heavily smear both inside and outside a plastic cup, with petroleum jelly. Place the insert in the neck of the bottle, and seal the base. Agitate the mixture and residue for 1 minute ensuring that it comes into contact with the petroleum jelly coated insert. The bone rich residue adheres to the surface of the insert.

3 Remove the insert from the bottle and flush it under cold water. This removes any non-adherent particles of matrix. Freeman (1982) states this stage is essential in maintaining the richness of the concentrate.

4 Removal of the bone rich residue is completed by individually washing the inserts in a beaker of hot water containing a small amount of detergent. While the bone particles sink to the bottom of the beaker, the petroleum jelly will melt and float on the surface where it can be recovered and reused.

(iii) Procedure C

1 Prepare a 5g sample from the 0.5-1.0mm size grade as in Procedure A.

2 Place the sample in a 250ml glass beaker, and cover with 25ml of kerosene and 25ml of cold water. The action of adding the two liquids is sufficient to gently agitate the residue. Allow the mixture to stand for a few seconds. The interface between the liquids will be readily apparent, with the bone rich residue concentrated in the upper layer of kerosene.

3 Immediately decant the upper kerosene rich portion of the mixture through a 63μm mesh sieve, ensuring that the bone-poor residue at the bottom of the beaker remains.

4 Repeat the procedure to concentrate the bone particles. Freeman (1982) repeated the procedure 100 times in the space of 15 minutes.

5 Wash the residue in running water with a few drops of detergent to remove all traces of kerosene.

Freeman (1982, pp 480-482) went on to describe "scaled-up" versions of the procedures that could be adapted for mass-production of palaeontological samples. Procedure's B and C have been adapted for conodont separations (Merrill 1985, 1987 and Belka *et al.*, 1989 respectively). The modified procedure B for separating conodonts is as follows:

1 The procedure is undertaken on a disaggregated sample of coarse or medium sand grade size (0.25-1.0mm), which is evenly spread over the base of a large flat enamelled pan (25x40x5cm).

2 Cover the sediment with a 1% aqueous solution of detergent, and 5ml of kerosene.

3 Gently agitate the pan to disperse the kerosene into droplets.

4 Add flakes of wax to the pan (prepared earlier by pouring melted paraffin wax onto a flexible surface, allowing it to harden and breaking it into thin 1-2cm pieces).

5 Agitate the pan for several minutes to ensure all the flakes mix with the residue. The oil coated phosphatic grains preferentially adhere to the wax flakes.

6 Skim off the wax flakes into a beaker, and repeat the process up to six times, adding additional wax and kerosene as required, to ensure all particles have a chance to come into contact with the wax.

7 When all the wax embedded flakes have been removed, top up the beaker with additional detergent solution and gently heat the solution until all the wax melts. Stirring the solution will aid in the dispersion and emulsifying of any remaining oil droplets.

8 Wash the final concentrate with heavy (petroleum) distillate. Dry the residue and wash with acetone. Re-dry and finally wash with water. The clean and dried residue can be examined under a binocular microscope in the conventional manner.

Interfacial methods are considerably safer and cheaper than using toxic heavy liquids. Continued experimental work will eventually provide safer and more efficient methods of concentrating microfossils. Freeman (1982) noted that fossil recovery was considerably higher from quartz-rich residues than carbonate-rich residues. Belka *et al.*, (1989) modified Freeman's (1982) Procedure C by using a kerosene and 30% hydrogen peroxide (H_2O_2) combination instead of a kerosene and water mixture. The use of this mixture considerably increased the surface tension at the interface between the two liquids, enhancing the separation of conodonts from dolomite-rich residues. The modified procedure is as follows:

1 Place the thoroughly dried disaggregated sediment (approximately 50g) in a 500ml graduated measuring cylinder.

2 Cover the sediment, to a depth of 5mm, with kerosene. Agitate, ensuring complete sediment dispersion. If a glass stirring rod is used take care not to grind the particles as this may destroy fossils.

3 Add 100-150ml of 30% hydrogen peroxide solution.

4 Insert a bung in the cylinder and gently agitate for about 20 seconds.

5 Allow the cylinder to stand for 1 minute. During this time the separation between the two liquids becomes apparent, with the kerosene floating on the hydrogen peroxide. The interface between the liquids, forming 3-4cm above the sediment surface holds a concentrate rich in conodonts and other phosphatic particles.

6 When the separation is complete, decant off the kerosene rich layer containing the microfossils into a filter paper lined funnel (Whatman grade No., 41). Belka *et al.*, (1989) successfully used a 75μm nylon mesh for filtration. Filtered kerosene and hydrogen peroxide can be re-cycled for later use.

7 Wash the filtered residue with distilled water and a few drops of detergent into an evaporating dish. Complete the final washing with acetone, and oven dry the cleaned residue at 60°C.

8 The procedure should be repeated with "heavy" fraction (tailings) concentrated in the hydrogen peroxide. However, for the most efficient results the residue is best filtered, washed and thoroughly oven dried.

Conodont recovery ranged from 43-93%, comparing favourably with toxic heavy liquid separations, with the advantage's of safety, cost effectiveness, and rapidly generating a clean residue (Belka *et al.*, 1989). However, the disadvantage of this procedure, is that organic rich samples and those with a high percentage of pyrite, react violently when in contact with hydrogen peroxide. The resulting endothermic reaction reduces the liquid's surface tension, inhibiting the separation of phosphatic material, although there is no evidence of damage to individual conodonts (Belka *et al.*, 1989).

REFERENCES

ALLMAN, M and LAWRENCE, D. F. 1972. *Geological Laboratory Techniques*. Blandford, (London). 1-335p.

ANONYMOUS. 1935. Recovery of foraminifera by means of flotation, Laboratory of N. V. Bataafsche Petroleum Maatschappij. *Journal of Paleontology*. **9**, (8), 745-746.

BÉ, A. W. H. 1959. A method for rapid sorting of foraminifera from marine plankton samples. *Journal of Paleontology*. **33**, (5), 846-848.

BELKA, Z, SKOMPSKI, S and WAKSMUNDZKI, B. 1989. A new interfacial method for conodont separation. *Journal of Paleontology*. **63**, (2), 253-255.

BENJAMIN, R. E. K. 1971. Recovery of heavy liquids from dilute solutions. *American Mineralogist*. **56**, (3/4), 613-619.

BOLTOVSKOY, E. 1966. Methods for sorting of foraminifers from plankton samples. *Journal of Paleontology*. **40**, (5), 1244-1246.

BRASIER, M. D. 1980. *Microfossils*. Allen and Unwin (London). xii + 193p.

BROOKS, H. K. 1954. A method for removing platy fragments of fossils and minerals from microfossil concentrates. *Micropaleontologist*. **8**, (4), 39-40.

BUZAS, M. A. 1965. Foraminifera from late Pleistocene clay near Waterville, Maine. *Smithsonian Miscellaneous Collection*. **145**, (8), 1-30.

CALLAHAN, J. 1987. A nontoxic heavy liquid and inexpensive filters for separation of mineral grains. *Journal of Sedimentology*. **57**, (4), 765-766.

CARSON, C. M. 1933. A method of concentrating foraminifera. *Journal of Paleontology*. **7**, (4), 439.

CARSON, C. M. 1953. Heavy liquid concentration of foraminifera. *Journal of Paleontology*. **27**, (6), 880-881.

CHARLTON, D. S. 1969. An improved technique for heavy liquid separation of conodonts. *Journal of Paleontology*. **43**, (2), 590-592.

COHEE, G. V. 1937. Inexpensive equipment for reclaiming heavy liquids. *Journal of Sedimentary Petrology*. **7**, (1), 34-35.

COLLINSON, C. 1963. Collecting and preparation of conodonts through mass production techniques. *Illinois State Geological Survey*. Circular 343, 1-16.

DAVIS, L and WEBSTER, G. D. 1985. A modified funnel for heavy mineral separations. *Journal of Paleontology*. **59**, (6), 1505-1506.

FREEMAN, E. F. 1982. Fossil bone recovery from sediment residues by the "interfacial method". *Palaeontology*. **25**, (3), 471-484.

FREY, D. G. 1955. A differential flotation technique for recovering microfossils from inorganic sediments. *New Phytologist*. **54**, (2), 257-258.

GIBSON, T. G and WALKER, W. M. 1967. Flotation methods for obtaining foraminifera from sediment samples. *Journal of Paleontology*. **41**, (5), 1294-1297.

HAUFF, P. L and AIREY, J. 1980. The handling, hazards and maintenance of heavy liquids in the geologic laboratory. *U. S. Geological Survey Circular*. **827**, 24p.

HIGGINS, A. C and SPINNER, E. G. 1969. Techniques for the extraction of selected microfossils. *Geology - Journal of the Association of Teachers in Geology*. **1**, 12-28.

HODSON, F and HODSON, H. 1926. Shortcuts in picking out and sorting foraminifera. *Bulletin of the American Association of Petroleum Geologists*. **10**, (11), 1173-1174.

HOWE, H. V. 1941. Use of soap in the preparation of samples for micropalaeontological study. *Journal of Paleontology*. **15**, (6), 690.

KNOX, A. S. 1942. The use of bromoform in the separation of noncalcareous microfossils. *Science*. **95**, 307-308.

KORNICKER, L. S. 1957. Concentration of ostracodes by alcohol flotation. *Journal of Paleontology*. **31**, (5), 1030.

KRUKOWSKI, S. T. 1988. Sodium metatungstate: a new heavy mineral separation medium for the extraction of conodonts from insoluble residues. *Journal of Paleontology*. **62**, (2), 314-316.

LEHMANN, G and RÖTTGER, R. 1997. Techniques for the concentration of foraminifera from coastal salt meadow sediments. *Journal of Micropalaeontology*. **16**, (1), 144.

LUTZE, G. F. 1968. Jahresgang der foraminiferen-fauna in der Bottsand-Lagune (westliche Ostsee). *Meyniana*. **18**, 13-30.

MERRILL, G. K. 1985. Interfacial alternatives to the use of dangerous heavy liquids in micropalaeontology. *Journal of Paleontology*. **59**, (2), 479-481.

MERRILL, G. K. 1987. Improvements in interfacial methods for conodont separation, pp55-63: In; R. L. Austin (ed), *Conodonts: Investigative Techniques and Applications*. Ellis Horwood Ltd, (Chichester). 1-422.

MILESON, P and JEPPSSON, L. 1983. Heavy liquid and solvent recovery, an economical method. *Journal of Sedimentar Petrology*. **53**, (2), 673-674.

MURRY, P. A and LEZAK, J. 1977. Recovery of vertebrate microfossils with tetrabromoethane. *Curator*. **20**, (1), 15-22.

MURRAY, J. W. 1979. *British nearshore foraminiferids*. The Linnean Society (London) and Estuarine and Brackish-Water Sciences Association, *Synopses of the British Fauna* (New Series), D. M. Kermack and R. S. K. Barnes (eds). No. 16. v + 68p.

SACHS, Jr., K. N, CIFELLI, R and BOWEN, V. T. 1964. Ignition to concentrate shelled organisms in plankton samples. *Deep-Sea Research*. **11**, 621-622.

SAVAGE, N. M. 1988. The use of polytungstate for conodont separations. *Journal of Micropalaeontology*. **7**, (1), 39-40.

SMITH, R. K. 1967. Ignition and filter methods of concentrating shelled organisms. *Journal of Paleontology*. **41**, (5), 1288-1291.

TRAVERSE, A. 1988. *Paleopalynology*. Unwin Hyman (London). xxiii + 600p.

TUCKER, M. E. 1990. *Techniques in Sedimentology*. Blackwell (Oxford). ix + 394p.

TURNER, W. M. 1966. An improved method for recovery of bromoform used in mineral separation. *U. S. Geological Survey Professional Paper*. **550-C**, 224-227.

VON BITTER, P. H and MILLAR-CAMPBELL, C. 1984. The use of "spikes" in monitoring the effectiveness of phosphatic microfossil recovery. *Journal of Paleontology*. **58**, (5), 1193-1195.

VON HOFE GRADY, F. 1979. Some new approaches to microvertebrate collecting and processing. *Newsletter of the Geological Curators Group*. **2**, (7), 439-442.

WRIGHT, R. C and HAY, W. W. 1971. The abundance and distribution of foraminifers in a back-reef environment, Molasses Reef, Florida. *Miami Geological Society*. Memoir 1, 121-174.

YAKOVLEY, N. V and SHTERENBERG, L. YE. 1967. Use of detachable centrifuge test tube for conodont separation. *Paleontological Journal*. **1**, (3), 93-94.

20. THIN SECTION AND SLIDE PREPARATION TECHNIQUES OF MACRO- AND MICROFOSSIL SPECIMENS AND RESIDUES

20.1. INTRODUCTION

The preparation of standard petrographic thin sections (cutting, grinding and polishing) are briefly outlined in this section. This subject is extensively reviewed and discussed by Allman and Lawrence (1972), Hutchison (1974) and Miller (1988). Appendix 2.1 outlines an approach for describing sedimentary rock petrographic thin sections containing fossil plant and animal remains. Randomly orientated fossils and biogenic fragments are frequently observed in thin section, and are particularly well preserved in carbonate and argillaceous sediments. Individual, three-dimensionally preserved fossils, free of matrix, can also be prepared for light and scanning electron microscopy examination.

A brief outline of the principle types of optical light microscopes available in palaeobiology, details of orientating sectioning and specimen mountant's used in petrology and micropalaeontology is also discussed. A specialized technique of preparing larger foraminifera for reflected light examination is also described, in addition to the conventional methods of picking individual microfossil specimens. All these procedures are designed to assist in observing and identifying organisms. For many palaeontological specimens this involves detailed examination of the shell wall and ultrastructural arrangement. Enhancing the contrast of structural components during initial observations, and the production of high quality photographs is of the highest priority, and a variety of procedures to fulfil in this objective are outlined.

The preparation of strew mount slides of assemblages and orientated single specimens prepared from acid insoluble residues is described in detail. Finally, the use of "Finder" graticules to assist in re-locating individual microfossils, fossil fragments or points of interest in thin sections and strew mounts are described.

SAFETY NOTE: the preparation of rock thin sections require the use of high speed rock saws and grinding wheels. During operation some types of machine require the hand holding of specimens and rock chips. These must only be operated by qualified personnel, and trainees must be closely supervised until competent in their use. Operators must wear suitable eye protection, while use of a dust mask reduces the risk of inhaling airborne particles suspended in the fine spray of the coolant. Perspex guards also reduce the risk from fragments and spray. Machinery and guards must be cleaned after use, regularly maintained, and replaced as required (check saw blade rim for wear and alignment, and grinding wheel for the development of a bevelled surface).

20.2. OBSERVING FOSSILS USING OPTICAL LIGHT MICROSCOPES

Three basic types of optical light microscope are used in the observation of fossil material prepared either as thin sections (viewed with transmitted light), or surface examination (viewed with reflected light). Specialized optical facilities can be attached to many microscopes (*e.g.* polarizing, differential interference contrast (DIC), phase contrast, fluorescence), and there are many examples of dedicated variants in use (*e.g.* cathodoluminescence, micro-hardness).

Of these variants, the cathodoluminescence facility is frequently used in palaeobiological studies. The technique is used in the examination of thin sections, grain mounts and prepared (cut, ground and polished) surfaces, complementing conventional transmitted and reflected polarized light observations in detecting impurities present in mineral and crystal zonations not evident in bright field illumination. The method involves bombarding the sample with a beam of electrons from a cold-cathode 30kv electron gun energy source. Minerals respond to the bombardment by emitting various wavelengths of light, with the colour and luminescence observed characteristic of a mineral and the presence or absence of any impurities. Luminescence may take the form of infrared, visible or ultraviolet light (see figure 34.1, ELECTRON MICROSCOPY TECHNIQUES), and phosphorescence may continue from the sample when the energy source is switched off. Results can be quantified by the addition of a spectrophotometer, measuring the spectrum of the emitted luminescence.

Three standard types of microscope used in palaeobiological procedures, their optimum range of magnification and the principle microfossil organisms observed, are listed in table 20.1. Manufactures' new variations of these microscopes and new instrument types are continually under development, and applications to the examination of palaeontological material frequently documented (*e.g.* Baumgartner-Mora and Baumgartner 1994, Hottinger and Mehl 1991, Scott a & b 1989, Scott and Vilks 1991). The scanning light microscope (SLM) has recently been used in viewing and photographing Holocene foraminifera (Scott and Vilks 1991, plate 4). The application of this technique, provide a comparison of SLM views with those recorded under the scanning electron microscope (SEM). The full potential of SLM within micropalaeontology has yet to be explored, although observations and images are comparable with the stereozoom binocular microscope, with the additional benefit of minimizing depth of field problems. Unlike specimens prepared for viewing under the SEM, SLM specimens do not require a metallic coating (see section 34 ELECTRON MICROSCOPY TECHNIQUES), and consequently internal architecture of transparent shells is observable. Further developments using digital video imaging, capturing portions of an image at successive focal planes, and combining them by computer to produce a composite high-contrast image, will assist in eliminating optical aberrations from single plane focusing (Athersuch and Jones 1991).

Specimens prepared for the laser scanning microscope (LSM), as with those for SLM, do not require metallic sputter coating. The LSM is designed to optimize the performance of an optical microscope system (Scott 1989a), with a laser connected to the optical system. The beam is scanned point-to-point over the sample, and a processed image viewed on a cathode ray tube (CRT). The system enables sections to be examined in both transmitted and reflected white and blue (fluorescent) light, and thus combines the advantages of light microscopy with a scanning system. Using a combination of light and scanning facilities, Scott (1989a) provides palynologists and siliceous microfossil workers with a method of observing features critical for taxonomic determination on a single specimen. The technique has successfully been applied to illustrating fossil radiolaria (O'Connor 1996).

Holographic (three-dimensional) microscopy is still in its infancy as far as macro- and micropalaeontological applications are concerned. Empson (1982) suggests there is a potential long term future in the procedure in providing a microscopical data base applicable to taxonomic studies. Applications of this procedure may be forthcoming from vertebrate palaeontologists following development and advances of holographic procedures used in the medical industry in examining internal organs distorted through disease (Blackie 1988).

MICROSCOPE	LIGHT SOURCES (i) principle, (ii) additional	OPTIMUM RANGE OF MAGNIFICATION	MICROFOSSIL GROUPS OBSERVED
Stereozoom binocular	Reflected (incident), transmitted (for use with fixed or rotating stage)	20-500x*	Foraminifera, ostracods, conodonts, scolecodonts, megaspores.
Compound polarizing	Transmitted, reflected	40-1000x	As above, but viewed in thin and polished section, strew slide mount and polished block.
Compound biological	Transmitted, reflected	40-1000x	Mega and miospores, pollen, chitinozoa, scolecodonts, diatoms, dinoflagellates, acritarchs, radiolaria, calcareous nannofossils.

Table 20.1. Comparisons between the standard light microscopes used in palaeobiology laboratories for viewing three-dimensional material, strew mounts, thin and polished sections and blocks. Additional facilities such as fluorescence, DIC and phase contrast can be incorporated on the compound biological microscope. *The addition of magnifying or reducing lenses can increase the range.

The specialized scanning acoustic microscope (SAM) provides another "optical" facility yet to be fully exploited by palaeobiologists, although potential applications within geology have been outlined (Scott 1989b). At its most fundamental it provides a method (by defocusing the "Z" control) of examining both surface and sub-surface properties of homogenous opaque materials. Furthermore, the technique is essentially non-destructive, avoiding unstable transformations introduced during chemical preparation. Specimens are, however, prepared in the form of polished blocks or un-covered polished thin or thick sections.

The acoustic microscope generates and transmits sound waves across the specimen surface. Acoustic properties and microstructural discontinuities within the material, caused by cracks, fractures, crystal and grain boundaries, result in receiving echoes disrupting normal wave propagation, and the acoustic reflection providing a direct, and potentially quantitative measure of the minerals mechanical properties (Somekh 1988). In turn, through a processing system, the reflected wave propagation generates an image with contrast. Magnification is partly controlled by a mineral's composition and the sound frequency used, usually in the range of 0.1-2.0 GHz (gigahertz). Within this frequency range an optimum resolution of 65μm is achievable (Scott 1989b). High frequencies allow for greater magnifications, but impair the depth of penetration, while low frequencies equate with lower magnifications and higher penetration depths.

Many acoustic microscopes accommodate interchangeable rotating heads containing either conventional reflecting light microscope objectives or acoustic sapphire objectives. The sapphire objective has a cavity tip that is immersed in a coupling fluid (usually a water medium) between the lens and the specimen surface, in a similar fashion to an oil immersion lens. The lens can be positioned to within a 10μm accuracy, essential as the specimen remains static while the acoustic lens vibrates. Sound waves are collected through a the fluid filled cavity (acting as a converging lens, refracting the sound to a sharp focus), and transferred by an acoustic transducer, converting the returning pulse height and time interval to a DC signal. The signal is used to modulate the image brightness at a particular point and stored in a computer frame store or a digital frame converter (Briggs 1985).

The organic content of a macerated palynological residue can be viewed by light microscopy. Kerogen concentrates are recognizable in both transmitted and reflected light, and if well-preserved morphological detail is present, identification possible. However, abundant amorphous organic matter (AOM) can, in part, be characterized using quantitative fluorescence microscopy (Smith 1984). Fluorescence also provides a qualitative method of estimating thermal maturity, as particle excitation varies from green → yellow → orange → red with increasing rank. Spectra produced during quantitative analyses must, however, be interpreted with care as mixtures of several different types of AOM can produce similar results (Smith 1984). Further details of equipment and techniques can be found in Phillips (1972) and Ploem and Tanke (1987).

Hands-free fine focusing devices have been described for both the stereozoom binocular and biological light microscopes (Dillé 1983, Whybrow 1982 see section 24 PREPARATION AND CONSERVATION OF VERTEBRATE FOSSILS). This facility is particularly useful while observing delicate preparatory work and fine detail on material with a wide range in depth of field. For a comprehensive introduction to the optical microscope refer to Bradbury (1989). The use of petrological microscopes, and the physics of polarized light is frequently described in optical mineralogy text-books. Three of the most informative texts with chapters on the use, range of applications, care and maintenance of the instrument are provided by Gay (1982), Kerr (1977) and Milner (1962).

20.3. FOSSIL ORIENTATION IN THIN SECTION

A knowledge of shell structure and morphology assists in the identification of many macro-fossils preserved in sedimentary rocks, particularly when used in conjunction with petrographic thin section analysis. Sedimentary rock thin sections are usually prepared parallel or perpendicular to bedding planes, providing views of randomly or preferred orientated shell fragments, identifiable to phylum level or below, and possibly even evidence of their mode of life. Lithified and indurated rocks that cannot be processed to free fossils, can be studied using thin sections prepared in the following way:

1 Cut a parallel-sided rock slice approximately 5mm thick, and trim the sides so that it will fit onto a standard glass slide (75x25mm).

2 Prepare one side of the slice by grinding it on a high speed (2800rpm) horizontal diamond impregnated wheel, removing any saw blade marks.

3 Continue to polish the surface lapping, using either a steel wheel with carborundum powders or glass plates. Lapping is accomplished by using successively finer grades of powder (3F, 600, 800), ensuring the specimen is thoroughly washed between each grade, and to avoid cross contamination between plates, reducing the possibility of scratching the specimen surface.

4 Place the completed chip, prepared surface up, on a hot plate and mount onto a clean glass slide using either an epoxy resin or thermo-plastic cement. The former eliminates the need to "transfer" the completed section to a new slide, and is the most common method by which standard thin sections are prepared today.

5 Once the section has cooled, excess rock can be ground away and the surface lapped. Complete the procedure on a fine grade of carborundum powder. Where the examination of fossils is the primary concern, it may be necessary to grind the section thinner than the standard 30µm to detect detail within fine grained dense structures such as shell walls (Lindholm and Dean 1973). However, if the section requires staining (see section 22 PREPARATION OF CARBONATE STAINED ACETATE PEELS AND THIN SECTIONS), allow sufficient thickness to accommodate etching in acidic solutions.

6 Return the completed (and stained) section to the hot plate, heat to 100°C, place a small drop of Canada Balsam on and apply a coverslip. Expel the excess balsam and bubbles from under the coverslip. Allow the Canada Balsam to "cook" for approximately 5 minutes or until the balsam is brittle when drawn into a strand.

7 Clean, label and examine the slide (see appendix 2.1 for a procedure for describing the features).

Studying microfossils requires a more subtle approach, where only attached encrusting forms occupy a life position. Some large benthic foraminifera (*e.g.* Nummulitacea, Rotaliacea, Soritacea) exhibit little recognizable detail on the test surface useful for taxonomic and morphological studies. However, descriptions of new foraminiferal taxa require both internal and external examinations of the test and wall structure. This is best achieved by viewing and examining orientated thin sections cut through the organisms test. Examination of these sections places considerable emphasis on the recognition of the plane of orientation to assist in identification. A number of sections are useful for this purpose, and are defined below.

(1) **Axial, vertical** or **transverse sections**: are cut parallel to the axis of coiling. This section is required for the identification of the organism, and in well-preserved specimens usually exhibits the septal folds.

(2) **Median, equatorial, horizontal** or **sagittal sections**: are cut across the axis at the equator, through the proloculus in the direction of coiling. The spiral arrangement is usually seen in this view.

(3) **Parallel** or **tangential sections**: are cut parallel to the axis of coiling, but not through the proloculus.

(4) **Oblique sections**: are cut through any plane except the first three, *i.e.* not parallel to either the axis of coiling or the spire.

Careful examination of outer test surfaces of larger benthic foraminifera invariably reveals a smooth surface. Occasionally the surface may exhibit ornamentation of slight undulations, ribbing or small prominent nodules. These features are usually surface expressions of underlying partitions, septal walls or columns. The internal position of these features, together with test wall structure is used in classification and identification of the organism. Hodgkinson and Margerum (1986) have described a rapid and accurate method of obtaining orientated sections of fusiform and discoidal foraminifera from lithified rocks. The procedure, incorporating the use of a high speed grinding wheel, is summarized as follows:

1 Determine the principle orientation of the fossils, and cut the rock into numerous 1cm thick slices parallel to this.

2 Grind both sides flat, removing any saw blade marks.

3 Lap the surfaces on a glass plate using a fine grit powder (600 grade grit or finer).

4 Wash the specimen and examine the prepared surface under a binocular microscope. If the fossils are not readily distinguishable from the surrounding matrix, etch the surface in dilute hydrochloric acid (HCl) for approximately 5 seconds.

5 While examining the surface, circle fossils most likely to make good sections with a china-graph pencil. Permanent markers are not recommended, as ink can often penetrate the rock and stain the fossil.

6 Trim the slice to free the specimen, leaving sufficiently large (approximately 10x10mm) to allow for convenient handling.

7 Decide the angle of the facet to be developed, and carefully begin grinding. Periodically wash and examine the ground surface, and correct the procedure to produce the desired facet. Care, patience and experience are required to avoid the development of a double facet.

8 Polish the surface with a fine grinding powder (800 or 1000 grade grit), thoroughly wash and affix to a glass slide using a thermo-plastic (Lakeside 70C) cement.

9 Continue preparation in the conventional method as detailed above.

This procedure has proved highly successful in the preparation of large numbers of fusiform and planispiral foraminifera where equatorial and axial orientations are required for biometrical work. However, it is less successful with orbitoidal forms where the equatorial layer may be thin and undulose (Hodgkinson and Margerum 1986).

20.4. THIN SECTIONING OF INDIVIDUAL FORAMINIFERA

Once free of matrix, foraminifera can be sectioned following the procedure outlined by Haynes (1981). Specific techniques relating to orientated specimens have been described by Cummings (1950), Hagn (1953), and Kremp (1953). Historically, the sectioning of individual microfossils can be traced back to the use of thermo-plastic (Lakeside 70C) cements (Meyer 1946, Levinson 1950, Emiliani 1951). Thermo-plastic cements have considerable advantages over epoxy resins in preparing orientated microfossil sections, enabling carefully prepared specimens to be ground, turned over and re-ground on the partially prepared surface (Hodgkinson and Margerum 1986, Van Morkhoven 1958). This ensures that the completed section passes through the initial chamber. A composite procedure for sectioning larger foraminifera is as follows:

20.4.1. Larger Foraminifera

This procedure is used in the preparation of axial and equatorial sections of large individual fusiform or planispirally coiled benthic foraminifera (*e.g.* nummulites, fusulinids, alveolinids, soritids, and many rotalids).

1 Grind one side of the specimen down using a high speed horizontal diamond impregnated wheel. Lap the surface on a glass plate using carborundum grits of successively finer grades (3F, 600, 800).

2 Heat a clean microscope slide with a frosted surface facing up, on a hot plate pre-set at a temperature of 120°C.

3 Mount the prepared surface onto the frosted slide using either a thermo-plastic or epoxy resin as mentioned under the standard thin section technique above.

4 Remove the slide from the heat, and allow it to cool.

5 Grind the specimen, initially using a diamond lap and finishing on carborundum grits. Continually examine the section to ensure it is not ground beyond the proloculus (first chamber). Also check that grinding occurs parallel to the glass slide by examining the thickness of the test wall. The equal translucent nature of the shell wall along its circumference indicates success.

6 Wash the surface, or immerse for a couple of minutes in an ultrasonic tank to ensure all carborundum powder is removed from the chambers.

7 Trim round the edge of the section with a sharp razor blade to remove excess mounting medium. If a thermo-plastic mounting medium has been used, then the specimen must be transferred to a clean glass slide. The surface of the section must be painted with an adhesive coat of "Hills Mixture" (50% *Durofix* + 50% Amyl Acetate). Be careful not to cover the mounting media with adhesive.

8 Return the section to the hot plate and apply a thin cover of Canada Balsam, and carefully place on a cover glass. Expel excess mounting media and any air bubbles.

9 Allow the Canada Balsam to "cook", and leave the section to cool. Clean the slide in methylated spirits, and wash in warm water with detergent. Finally dry the slide and label.

Specialized thin sectioning equipment (Arnold 1958, 1965a, Sachs 1965) can assist inexperienced operators in obtaining accurate sections through individual specimens. However, not all thin section laboratories have the resources or personnel to undertake such modifications. In addition some specimens may require impregnating (Arnold 1965b, see below), and serial (multiple) sections may be necessary for specialized taxonomic and morphological studies (Akpati 1969, Hendry *et al.*, 1963, Honjo 1963).

20.4.2. Smaller Foraminifera

The composite procedure of preparing individual specimens described below follows the methods of Wood (1948) and Emiliani (1951), with additional modifications from Van Morkhoven (1958). Recent and hollow specimens may require resin impregnating before sectioning can commence.

1 Place the specimen on a glass slide and immerse in xylene ($C_6H_4(CH_3)_2$) with a drop of Canada Balsam. The Canada Balsam soaks into the chambers as the xylene evaporates.

2 Using double sided sticky tape, mount a small celluloid or mica coverslip onto a glass slide. The use of this type of coverslip will ensure that both sides of the section remain parallel during the two stages of grinding.

3 Place the treated specimen on the coverslip in a drop of Canada Balsam and heat to 120°C until it is set.

4 Grind or file the specimen by using pre-ground glass slides. Frost the surface of the slides on different grades of carborundum (*e.g.* 3F, 600, 800, 1000), but avoid using powders directly on specimens during preparation, as grains become trapped within wall structures (Van Morkhoven 1958).

5 With a microscope check that the correct thickness has been attained. Remove the coverslip with the specimen attached.

6 Invert the coverslip (*i.e.* specimen down) and attach it to a clean slide with a small amount of Canada Balsam. Expel excess mounting medium and bubbles from under the coverslip.

7 Remove the coverslip and grind the surface to the required thickness, constantly checking progress under the microscope.

8 Return the section to the hot plate, place a small drop of Canada Balsam on the section and cover with a glass coverslip.

9 Clean, label and examine the slide.

Emiliani (1951) includes a resin impregnation procedure (stages' 1-4), replacing stages' 1-3 of the above method. Once impregnated, specimens can be processed using conventional sectioning procedures. Emiliani (1951) had considerable success using this method in preparing recent and hollow specimens, while Levinson (1950, 1961) has used similar methods in sectioning individual fragile ostracods.

1 Soak the specimen in a few drops of solvent placed on a glass cavity slide. Experimentation with different solvents by Emiliani (1951) revealed 2-Ethoxyethane ($C_2H_5O.CH_2.CH_2OH$, *Cellosolve*) has suitable properties for penetrating chambers, and evaporates without producing an emulsion. Acetone was found unsuitable, leaving a white emulsion on evaporation, while alcohol required an excessively long time to evaporate. Chloroform evaporated less quickly than acetone, but occasionally left an emulsion. Leave for 1 minute.

2 On a separate glass cavity slide melt a few drops of thermo-plastic (Lakeside 70C).

3 Remove the specimen from the solvent and place it in the heated melt.

4 Keep the slide on the hot plate until the bubbling around the specimen ceases, *i.e.* all the solvent has evaporated (between 30-90 seconds).

5 Place a standard glass slide on the hot plate, and smear the area where specimen is situated with thermo-plastic cement. Place and orientate the specimen in this area. A magnifying light situated near the hot plate will assist during this stage.

6 Emiliani (1951) grinds the slide using a fine carborundum paste (800 or 1000 grade). However, to avoid cross grade contamination, which might result in scratching the specimen surface, chambers becoming infilled with carborundum, or powder embedding in the mounting media, use either pre-ground glass slides or wet-and-dry silicone-carbide paper to prepare the specimen (stage 4 above). Periodically observe

progress under a petrological microscope until the desired plane is reached.

7 Wash and dry the ground slide.

8 Return the slide to hot plate to invert specimen.

9 Continue the grinding procedure as in stage 6 above.

10 Wash, dry and examine the slide. Excess bubbles within the thermo-plastic can be removed by adding a few drops of solvent and gently warming the slide.

11 Before permanently covering the specimen, Emiliani (1951) recommends photographing the preparation. Sections with a high polish provide a better surface for photography at this stage. An immersion oil or glycerol to moisten the surface enhances contrast. Both are non-drying and have a similar RI to the thermo-plastic.

12 Slides should be covered in the standard way, with a glass coverslip and Canada Balsam.

Techniques for both random and preferred orientations of small organisms have been described Pohl and Browne (1973), Kennedy and Zeidler (1976) and Finger and Armstrong (1984). A composite method based on these is as follows:

1 Specimens must be free of all organic contaminants before embedding. Residues or picked individuals should be carefully washed with a detergent (*e.g.* Quaternary O), which in turn must be removed through washing in distilled water (Pohl and Browne 1973).

2 Oven dry at 40-50°C.

3 Brush a layer of water soluble glue (*e.g.* gum tragacanth) onto a $2cm^3$ mica sheet, and allow it to dry.

4 Mark a 1.5cm circle in the centre of the mica sheet.

5 Viewing the residue or specimens under a binocular microscope, and using the micropalaeontological technique of picking with a damp sable hair paint brush (see section 20.7 PICKING MICROFOSSIL RESIDUES IN REFLECTED LIGHT below), randomly orientate specimens within the marked circle on the mica sheet. Finger and Armstrong (1984) suggest using sand grains to prop specimens in the desired position. They also point out that in a population of variably sized individuals, larger specimens should be positioned near the perimeter of the circle, accommodating the concave surface of the epoxy resin.

6 Grind one end of a 1.5cm PVC pipe, cut to a length of approximately 2cm. Using a cyanoacrylate adhesive (*e.g.* "super glue") position and attach this end of the pipe over the specimens onto the mica sheet.

7 Carefully mix, avoiding the introduction of bubbles, a low viscosity two part epoxy (resin plus hardener) and cover the specimens. Transfer to a vacuum oven, and evacuate until the resin "boils". Bleed the vacuum and repeat the procedure, until the specimens are thoroughly impregnated. Add additional resin to a depth of 1cm.

8 Oven cure, and leave overnight to harden.

9 Remove the mica sheet (it should easily snap off), and grind the exposed surface on 600 grade silicone carbide paper. Periodically examine the surface under a binocular microscope to ensure the correct removal of material. Finish grinding on 100 grade paper. Alternatively, the surface can be polished with cloths using chrome oxide and finishing on magnesium oxide (Kennedy and Zeidler 1976).

10 Apply a small amount of mountant to the prepared surface of a frosted glass slide, and carefully attach the epoxy block, adjusting to remove all air bubbles. Place the slides in a mounting press, and cure on a hot plate.

11 Clamp the hardened block in a vice and hacksaw off the excessive resin, leaving a thin (0.5-1mm) slice on the slide.

12 Grind the slide on a series of silicone carbide papers (grades 200, 600, 1000) to the desired thickness (50-30µm). Thinner sections provide better optical resolutions at higher magnifications, however, there is an increased risk in destroying the specimen through excessive grinding (Finger and Armstrong 1984). The completed section can then be covered with a coverslip, or if desired, left uncovered and polished. Sections must be polished if serial photographs are required (Kennedy and Zeidler 1976).

This technique has been successfully applied to both fossil and recent material (Kennedy and Zeidler 1976), while its application in producing high quality sections required for a taxonomic study is clearly illustrated by Finger (1990). The procedure is ideal when a large number of specimens are available for sectioning, and the desired orientation is obtained from their random positioning. If however, only a few specimens are available, then greater care is required during the initial grinding stage when the preferred orientation is determined (*e.g.* Pohl and Browne 1973). Preparation time will inevitably be longer, but fewer oblique sections are required.

20.4.3. Polished sections

Polished thin, thick and double sided polished sections are regularly prepared for petrographic, cathodoluminescence and ore mineralogy light microscope studies

(Hutchison 1974, Lister 1978). Carbon coated polished sections are required for viewing in a scanning electron microscope and micro-probe for analytical work. Optical advantages of polished sections in comparison with conventionally prepared sections include the ability to study both transparent and opaque minerals, supplementing the use of high powered oil immersion objectives, particularly useful in studying fine grained areas. Furthermore, sections can be re-polished when ore minerals tarnish or surface becomes scratched. Initial section preparation is as conventional thin sections, finishing with a series of diamond pastes (6μm, 3μm, 1μm) providing a relief-free polished surface. Completed sections remain uncovered, and should be stored in a desiccator cabinet containing silica gel.

A minimum of three specialized polishing machines, housed in a dust free environment, is required to accommodate all grades during section preparation. Although lapping plates are interchangeable between machines, in practice they must be dedicated to one polishing grade to avoid cross-contamination. The plates are covered with a self-adhesive polishing cloth (a synthetic woven or non-woven fibre, selected on the type of mineral polished). The lap is charged with a small amount of diamond paste of the appropriate grade, and lubricated with an oil spray. Sections are usually polished for 15 minutes, with the machine running at approximately 120rpm. Additional weight can be added to the section if required. Sections must be cleaned in an ultrasonic bath between each diamond paste grade, with a softer synthetic velvet cloth used for the finishing grade.

A series of polished thin sections of ostracods have been used by Kesling and Sohn (1958) to observe hinge line and duplicature development, and reconstruct the internal morphology of a Palaeozoic genus. This approach is in contrast to Sylvester-Bradley (1941), who used fractured surfaces of specimens in reconstruction's, while Nye, Dean and Hinds (1972) and Finger and Armstrong (1984) preferred double polished sections as the method eliminated surface irregularities, and any traces of the grinding compound likely to become embedded in the preparation. Furthermore, double polishing is beneficial in resolving high magnification microstructures.

20.5. PREPARATION OF SECTIONS CONTAINING HARD AND SOFT TISSUES

Of increasing importance in palaeobiological studies is the need to observe the relationship between hard skeletal parts and surrounding soft tissues. Conventional thin section techniques require the destruction of either the hard or soft parts. Biological procedures (microtoming) require the demineralization of hard parts, while geological techniques (as detailed above) concentrate on the sectioning of firm skeletal material or parts hardened or supported through resin impregnation.

Palaeobiological techniques applied to the simultaneous sectioning of hard and soft tissues use a combination of procedures (Williams 1956, 1965, Nye *et al.*, 1972). Initially, specimens are dehydrated through a series of alcohol's of increasing concentrations, and then impregnated with an epoxy resin. The method outlined below by Nye *et al.*, (1972) has been successfully used in the preparation of recent bryozoans, algae, sponges, polychaetes and sipunculids (unsegmented marine worms), with finished sections revealing exceptionally fine histological detail.

1. Trim off excess material with a diamond saw, scalpel or nippers. Specimens containing fixed soft tissue must be dehydrated (by immersion in increasing concentrations of ethyl alcohol to absolute

alcohol, see sections 4 COLLECTING TECHNIQUES FOR MICROFOSSIL AND LIVE FORAMINIFERA SAMPLES and 11 PREPARATION OF RECENT MATERIAL FOR COMPARATIVE STUDIES for details).

2 Place the specimen in a container suitable for thorough impregnation (see section 10 PRESERVATION, CONSOLIDATION AND REPAIR OF UNSTABLE SPECIMENS), and on mixing a sufficient volume of a low viscosity two part epoxy (resin plus hardener) pour over the specimen.

3 Transfer the container to a vacuum oven, evacuating the specimen until it "boils". Release and re-evacuate until thoroughly impregnated.

4 Oven cure, leaving overnight to harden.

5 Cut the block to the desired orientation, and grind the surface using silicone carbide paper, polishing if required. If required, surfaces can be etched and peeled at this stage.

6 Using an epoxy mountant, carefully attach the prepared surface to a glass slide, and allow to cure overnight.

7 Cut excess material away, and grind the surface to the desired thickness using silicone carbide paper (grades 200, 600, 1000).

8 At this stage non-permanent biological stains (*e.g.* Toluidine Blue O) can be introduced.

Nye *et al.*, (1972) concede that these preparations are no substitute for ultra thin microtome sections of decalcified material, but are useful when examined in conjunction, as the relationship between hard and soft parts is evident.

20.6. CONTRAST ENHANCING TECHNIQUES

Shell ultra-structure, evident when viewed in thin section, requires additional enhancement when photomicrographs are required. Preparing double sided polished sections, as mentioned above, are one option. The use of both reflected and transmitted light has been explored by many micropalaeontologists, in particular ostracod workers (Sohn 1961, Sohn *et al.*, 1965, Levinson 1961). Procedures for enhancing the appearance of microfossils are many and varied. Some of the options described below are dealt with in other sections in greater detail. Reference to these will be indicated.

20.6.1. Reflected light studies

Etching of specimens has been briefly described above (see above, 20.3 Fossil orientation in thin section), while the method is detailed in section 23 PREPARATION OF CARBONATE STAINED ACETATE PEELS AND THIN SECTIONS. As a rapidly performed procedure, Ireland (1950), describes a technique of producing and examining "curved surface sections" of calcareous rock fragments. Following acid immersion (left in dilute hydrochloric or acetic acid for approximately 1 hour) and immediate washing, the texture, fabric and replacement features of prepared surfaces can be examined. The smoothest surfaces were obtained from samples in which etching occurred evenly, independent of immersion time or acid concentration. Therefore, before commencing, experiment with a small fragment to determine the correct acid concentration and immersion time required. Ireland (1950) suggests that careful examination of surfaces will provide preliminary information complementing that obtained from thin section.

20.6.1a. Ammonium chloride coating

This technique, originally developed by Bassler and Kellett (1934), involves the coating of specimens with a combination of fumes of concentrated hydrochloric acid (HCl) and ammonia solution (NH_3). A modification by Hessland (1949), uses a small glass pipette tube (diameter 3mm), filled with ammonium chloride powder. The wide end is then plugged and the tube heated. A jet of ammonium chloride is released through the nozzle, and can be directed onto the specimen.

Ammonium chloride coated specimens of calcareous or phosphatic composition must be carefully and thoroughly cleaned once examination is complete. A wide range in humidity in environmental conditions will cause the sublimate to re-hydrate, releasing a hydrochloric acid solution that can etch and corrode calcareous material over time, causing irreversible damage, and ultimately specimen loss. Conodont type specimens damaged in this way have been observed in research collections (Jeppsson *et al.*, 1985). Unfortunately, this procedure is not suitable for very small specimens, as specimens prepared in moderately high humidities result in a large sublimate grain size, masking specimen details (see section 37 PHOTOMACROGRAPHY AND PHOTOMICROGRAPHY TECHNIQUES).

20.6.1b. Magnesium oxide coating

Specimens are coated by passing them through fumes of burning magnesium ribbon. The advantage of this procedure is that it is temporary, and the coating can be removed by lightly brushing the specimen. **N.B. the burning of magnesium causes a bright light that can injure eyesight - wear dark glasses and avoid looking directly at the light source.**

20.6.1c. Silver nitrate coating

This procedure has been outlined in section 21 STAINING TECHNIQUES USED IN MICROPALAEONTOLOGY.

20.6.1d. Staining

All staining procedures and methods are outlined and discussed in sections 21 STAINING TECHNIQUES USED IN MICROPALAEONTOLOGY and 22 PREPARATION OF CARBONATE STAINED ACETATE PEELS AND THIN SECTIONS.

20.6.2. Transmitted light studies

Surface ornamentation, scars and pore canals are ideally studied by using transmitted light. Consequently opaque shells have to be made translucent or transparent, as "wet" specimens show more detail than dry specimens. A number of techniques are available to fulfil this criterion.

20.6.2a. Liquid immersion

Both temporary and permanent mounts can be prepared. Sometimes water may be adequate for viewing in both transmitted and reflected light (Scott 1944). Usually more permanent mounts, using glycerol, immersion oil or canada balsam, result in greater contrast and give better results. Wagner (1957) had success by soaking specimens for several hours in castor oil. Structures present on slabbed samples of friable fine-grained diatomaceous rocks can be significantly enhanced by spraying with a multi-purpose lubricating oil (*e.g. WD-40*). Specimens, cut dry, smoothed with a knife and cleaned with a compressed-air dust gun, reveal maximum contrast approximately 2-4 minutes after spraying and should be viewed and photographed in reflected light at this time (Savrda *et al.*, 1985). This simple procedure can be readily adapted for field use.

20.6.2b. Hydrofluoric acid technique

Safety Note: Hydrofluoric (HF) acid and its fumes can cause severe burns if it comes in contact with skin - ensure all uses wear appropriate safety equipment and protective clothing, and that all procedures are undertaken in a suitable HF fume cupboard. (See appendix 8 for details).

Discovered by accident, while preparing palynological samples, the conversion of calcareous shells to calcium fluoride was confirmed by Grayson (1956). Examples of the calcitic echinoderm *Mellita quinquiesperforata*, the aragonitic molluscs *Arca incongrua* and *Donax variabilis*, and calcitic *Globigerina* sp., foraminifera were tested. X-ray diffraction analyses run before and after the hydrofluoric acid treatment, confirm that the transformation of calcium carbonate into calcium fluorite occurs without any gross structural change. Furthermore, photographic evidence reveals a cleaner specimen, with enhanced morphological features.

Fluorite is more translucent than calcite in both water and glycerol, providing additional optical properties exploited by Sohn (1956) while examining ostracod carapaces. The fluoridation technique developed is summarized as follows:

1 Place the specimen in a platinum dish and cover with water.

2 Add a few drops of 20% (w/v) hydrofluoric acid (HF), sufficient for effervescence to be seen.

3 Leave for between 2-24 hours, until reaction has stopped or specimens have become either transparent or translucent.

4 Neutralize and safely dispose of HF.

5 Once neutral specimens can be safely handled. Use a fine (00) sable hair paint brush to mount the specimen in a glass cavity slide and cover with glycerol or Canada Balsam.

6 Clean and examine slide. Best results are obtained using clean disarticulated valves.

Schallreuter (1982) used HF as a "corroding" agent in the extraction of fossil ostracods from siliceous rocks, again resulting in the transformation of the carapace from calcite to calcium fluorite. Concentrated HF of 38-40% was used in this procedure, which may take several weeks to breakdown a sample crushed to 1-2cm cubes.

A further modification of optical enhancement procedures using HF, is supplied by Ibaraki and Sameshima (1962) in their examination of the stolon structures of *Lepidocyclina* sp., from Central Japan. Initial examinations of the stolon structures using the dye gentiana violet were not successful in enhancing detail. However, thin sections of specimens prepared to a thickness of 100μm, and etched in a 1% HF solution for 10 hours provided the desired results for stereoscopic examination.

20.7. PICKING MICROFOSSIL RESIDUES IN REFLECTED LIGHT

The picking and orientating of microfossils (foraminifera, ostracods, conodonts) from dried residues viewed down a stereozoom binocular microscope, is a labour intensive procedure undertaken by most micropalaeontologists during some stage of a research project. It is frequently considered as the "classical" method of concentrating microfossils from processed residues (Boltovskoy 1966). In short, recognizable microfossils from residues are removed with the aid of a damp paint brush and transferred to a micropalaeontological slide. Separating processed residues into a series of graded fractions assists the examination process (see sections 16 WASHING AND SIEVING TECHNIQUES USED IN MICROPALAEONTOLOGY, 18 MECHANICAL SEPARATION OF MICROFOSSIL RESIDUES and 19 FLOTATION AND LIQUID SEPARATION TECHNIQUES). The picking of large fossil fragments can be assisted by the use of vacuum tweezers. This pen-like hand held instrument enables a hollow probe to be positioned in contact with a specimen, and by using the index finger to cover an extraction vent, allows specimens to be transferred from the residue to a slide or SEM mount. The procedure is as follows:

1 Pour a small amount of the sample onto a picking tray (gridded with a black background) or into a petri-dish or flat bottomed dish. Spread to form a thin layer.

2 Systematically examine each grid under a stereozoom binocular microscope. Commence examination at low power, with the diaphragm partially closed to reduce light intensity and eye-strain. Increase the magnification as required, and in relation to the grade fraction examined. Residues can be gently moved under the microscope with the aid of a mounted needle, or for finer fractions, a fine probe or micro-tool.

3 Microfossils can then be picked from the residue with a dampened fine (OO or OOO, 0.8 to 1.2mm brush width) sable hair brush, or with the aid of vacuum tweezers, and transferred to a micropalaeontological slide. These specialized cavity slides are usually made of cardboard or plastic, and have a black or white background to contrast with the colour of the organism within the residue. Assemblage slides commonly consist of 32 or 64 numbered cells, while single, double or quadruple circular cells (up to 20mm diameter) are useful for single specimens and taxonomic display.

4 Examine each successive fraction until all (or a representative) assemblage has been picked.

This process has come under considerable scrutiny and discussion from micropalaeontologists (see section 4 COLLECTING TECHNIQUES FOR MICROFOSSIL AND LIVE FORAMINIFERA SAMPLES for discussion) citing the disadvantage of the time required in processing each fraction. However, this has to be off-set against the advantage that many of the micro-organisms are undamaged following processing (Boltovskoy 1966).

20.8. SPLITTING WELL PRESERVED TERTIARY *NUMMULITES*

Wrigley (1934) uses, but does not describe in detail, a technique for splitting Eocene *Nummulites* from the Hampshire Basin of southern England. Although similar material can be found in the Paris Basin, inferior preservation means the technique outlined below will not always be successful.

1 Remove fine sediment from the tests by washing or cleaning in an ultrasonic tank.

2 Heat the fossils in a bunsen flame until they are glowing red hot.

3 Immediately drop the glowing fossil into a beaker of cooled (5°C) water.

4 Remove and examine the test. A line of weakness developing along the equatorial plane, as the two halves of test begin to split, should be evident.

5 Repeat the heating and quenching process until the two halves have separated or it is possible to prize them apart using a fine scalpel blade.

Although the technique results in some damage to test outer chambers, the embryonic chamber is usually clearly seen, making identification of microspheric (small proloculus) and megalospheric (large proloculus) forms possible. Mount specimens in a micropalaeontological slide and examine using a stereozoom binocular microscope.

20.9. WALL ULTRASTRUCTURE STUDIES

Specialist techniques used in studies of foraminiferal wall ultrastructure were initiated by Wood (1963), although they had long been used in the microstructural study of ostracod carapace walls (*e.g.* Sylvester-Bradley 1941; Levinson 1951, 1956; Kesling and Sohn 1958; Levinson 1961; Sohn 1961). Examination using cathodoluminescence provides a further facility used in foraminiferal shell ultrastructural studies (see below). Whole specimen analysis was undertaken by Wood (1963) using the following procedure:

1 Clean the specimen surface by immersion in absolute alcohol.

2 Place in a glass cavity slide and cover with xylene.

3 Examine the specimen using a compound petrological microscope under cross polarized light. For optimum results, the field should be brightly illuminated with the condenser down, and a high (x100) magnification.

Crushed specimens, or specimens with a prepared cut surface can be examined using the SEM following the procedure of Hansen and Lykke-Anderson (1976) summarized below:

1 Prepare a normal thin section.

2 Polish surface on a rotary disc using either a diamond or aluminium oxide (Al_2O_3) paste.

3 Thoroughly wash the surface with an oil-free detergent.

4 Etch polished surface with an aqueous EDTA solution buffered to pH 7 by the addition of sodium hydroxide (NaOH) for between 15-60 seconds (time will depend on test wall thickness).

5 Cut specimen from slide, mount on double sided tape and attach to an aluminium SEM stub.

6 Sputter coat with gold to a thickness of 250-500Å, and examine.

Examine fractured surfaces using an adaptation of the technique described above, and outlined by Bellemo (1974), summarized below:

1 Clean specimen by immersion in sodium hypochlorite (NaClO). This removes traces of the organic membranous layer.

2 Etch with a 25% solution of glutaraldehyde buffered to a pH 3.5 for 5-20 minutes.

3 Mount on double sided tape, gold coat and examine under the SEM.

If organic lamellae are examined, the foraminiferan test can be demineralized by immersion in a solution of chromium sulphate ($Cr_2(SO_4)_3 6H_2O$) for 8-30 minutes. Bellemo (1974) cautions that examination of the results following this procedure requires a great deal of skill to differentiate between naturally occurring structures and preparation induced etched surfaces.

The presence of a light emitting activator (impurity) within carbonate grains (*e.g.* bivalent manganese, Mn^{2+}) provides the means by which cathodoluminescence can be employed. However, the presence of bivalent iron (Fe^{2+}) acts as an ion quencher, suppressing any activation from the Mn^{2+}, eliminating the net effects of cathodoluminescence. At low concentrations a linear relationship exists between Mn^{2+} and the photometrically measured cathodoluminescence, but this is not recorded at higher levels. Baumgartner-Mora and Baumgartner (1994) suggest that calcite luminescence may be more sensitive to Mn^{2+} activation than Fe^{2+} quenching.

Thin sections prepared for cathodoluminescence must be prepared to a thickness of 40μm, polished relief-free, and left uncovered. High resolution, well-focused images are more likely to be obtained from highly polished sections. Baumgartner-Mora and Baumgartner (1994) recognized two applications of cathodoluminescence in foraminiferal wall studies. Firstly, ontogenetic (growth) structures are revealed by primary variations in the luminescence of lamella structures, and secondly, in some fossil specimen's cathodoluminescence reveal early diagenetic changes, evident as a contrast difference between the shell wall and the earliest chamber infilling cements. Baumgartner-Mora and Baumgartner (1994) have observed calcite exhibiting similar luminescent properties, replacing and obliterating growth structures and entire sections of test wall, and concluded that calcite precipitation, originating from the pore filling cements, continued during the geological history. They further state that this type of recrystallization is only observable using cathodoluminescence, and is not seen in conventional transmitted light microscopy. In preserving the original optical properties of the fibrous calcite, recrystallization must have been syntaxial, and the preservation of fine growth structures only occurs in specimens that have been penecontemporaneously displaced into deeper water environments, and thus provides a useful biostratigraphical palaeoenvironmental indicator.

Acid insoluble microfossils can also be examined in petrographic thin section. Thin sections of radiolarian cherts examined in this way provide considerably more information about the organism if made to a thickness greater than the standard 30μm (Burma 1965). This ensures that the maximum amount of the 100-2000μm radiolaria test diameter is preserved. Wall structures are further enhanced by the preparation of double sided polished thin sections. These thin sections must have the finished thickness clearly indicated on the label to avoid confusion and misinterpretation when examining under cross polarized light.

20.10. STREW SLIDE MOUNTING MEDIA AND PREPARATION

Larger, opaque three-dimensional material, (*e.g.* scolecodonts, radiolaria) can be permanently preserved by mounting in micropalaeontological slides, where translucent specimens are best mounted in Canada Balsam on glass cavity slides, and examined with a compound microscope using well-condensed transmitted light. The permanent preservation of residues and assemblages of small sized microfossils (palynomorphs, siliceous microfossils, calcareous nannofossils) requires the preparation of strew slides.

20.10.1. Preparation of mounting media

The preparation of slide mounting media suitable for preserving processed palynological (acid insoluble microfossil) residues have been outlined by Barss and Williams (1973) and Barss and Crilley (1976), although debate continues on the long-term preservational effects of many formulae (see section 4 COLLECTING TECHNIQUES FOR MICROFOSSIL AND LIVE FORAMINIFERA SAMPLES for discussion). Descriptions of formulae and preparations are given below.

20.10.1a. *Cellosize*

Barss and Crilley (1976) describe this as an all purpose mounting medium suitable for all micro-organism residues (palynological and nannofossils), coarse and fine fractions. Some mixtures become cloudy during use, a feature Barss and Crilley (1976) attribute to the incorrect preparation of ingredients. This is avoided by carefully following the procedure below, mixing only small amounts, and using immediately.

1 In a 250ml beaker, with a pre-measured mark of 85ml indicated, place 5g of *Cellosize* powder.

2 Add 20ml of methanol to the beaker, and allow this to soak until all the powder is wet

3 Add 100ml of distilled water and thoroughly mix.

4 While continuously stirring (to avoid the mixture sticking to the bottom of the beaker), gently heat the mixture on a hot plate set at 71°C for 20 minutes. The mixture level will drop as residual methanol is driven off.

5 Reduce the hotplate temperature to 60°C. Maintain this heat for between 40-60 minutes, stirring every 2-3 minutes, until the mixture level is within 5ml of the pre-measured 85ml mark.

6 Re-heat the mixture to 71°C, and maintain at this temperature until the mixture is at the pre-measured 85ml mark.

7 While the mixture is still hot, pour it into a plastic funnel lined with a slow qualitative grade filter paper (*e.g.* Whatman grade 5), into a suitable storage bottle. Barss and Crilley (1976) suggest using two filtering apparatuses, and a plastic funnel in preference to glass as the mixture's viscosity increases when cooled. Approximately 70% of the mixture will be recovered at this stage.

8 Add three drops of ethylene glycol mono-methylether to the mixture, and shake well.

9 Secure the top of the bottle and allow the mixture to stand for a few days before use.

10 When using the mixture dispense with a disposable pipette. Two drops are usually sufficient for spreading the residue over the coverslip. An even covering of the residue can be achieved by placing this on the coverslip first, and then adding the mounting medium and spreading the mixture using a toothpick (discarded after each sample) (Barss and Crilley 1976).

When the mixture is dry, the coverslip can be mounted to the slide using *Elvacite* 2044. On exposure to air the viscosity of the mixture increases, and its ability to spread over the coverslip decreases. At this point it should be discarded.

20.10.1b. *Elvacite* 2044

1 Crush 40g of dry *Elvacite* 2044 powder using a pestle and mortar, and pour into a 150ml beaker.

2 Measure 70ml of analytical grade xylene ($C_6H_4(CH_3)_2$) into a measuring cylinder.

3 Gradually add the xylene to the *Elvacite* powder, stirring gently and continuously with a glass rod. The powder dissolves to give a clear thick liquid.

4 Pour the freshly mixed liquid into a glass screw top bottle, and cover with a large inverted beaker. Leave until all the entrapped air bubbles have escaped (approximately 48 hours).

5 The mixture can be stored long term in a glass screw top container.

Use in small quantities, dispensing with a small glass rod. Although the mixture should be discarded when it thickens, and fails to spread smoothly under the coverslip (Barss and Williams 1973), limited regeneration is achieved by the addition of more xylene.

20.10.1c. *Naphrax*

High resolution mounting media are essential for enhancing contrast and microstructural details of transparent siliceous organisms (*e.g.* diatoms, radiolaria). A number of high refractive index media have been successfully employed over the years by siliceous microfossil workers (*e.g. Mikrops* (West 1977), *Aroclor No. 4465* (Schrader 1976, Koizumi 1980), *Hyrax* (Baldauf 1985), *Pleurax* (Koizumi and Tanimura 1985), *Styrax* (Meller 1985), *Dirax, Naphrax*). Simple to use (see below), *Naphrax* is a toluene ($C_6H_5.CH_3$) based mountant. **Use only in a fume hood.** The application of the mounting media is detailed below.

20.10.1d. Glycerol

Glycerol (or glycerine) is commonly used in the slide preparation of biological specimens, but its versatility should not be ignored for applications in palaeobiology.

Advantages include its low RI (1.44, lower than Canada Balsam at 1.52), and it's low melting point of 80°C, coupled with its ability top remain molten when the temperature drops to as little as 50°C. Furthermore, glycerol contains water, and therefore it is not necessary to dehydrate specimens before mounting. When used in slide preparation a small amount of jelly should be placed on a warmed microslide, as warming the slide reduces the possibility of introducing bubbles into the mount. Excess glycerol is readily removed from cooled prepared slides by gently washing in warm water with detergent.

20.10.2. Preparation of mounts and strew slides

Two types of slide preparation can be made:

- **orientated single specimen slides** - are extremely time consuming to prepare, and require specialist equipment for the preparation of slides (Wornardt 1965). Techniques of manipulating single mounts of dinoflagellates have been described by Wilson (1971) and Evitt (1984), while Gocht (1972) details a method of preparing slides containing up to 324 grouped and orientated dinoflagellate cysts. For larger chitinozoa, Jenkins (1967) transferred individuals to a watch-glass containing a 2% aqueous solution of *Cellosize*. Each specimen was then pipetted in a drop of *Cellosize* onto a coverslip. On each cover glass up to 30 drops, each with a specimen, were arranged in rows.

- **strew slides** - can be prepared as either temporary mounts using distilled water, or more permanently using a microscopical mounting media (canada balsam, glycerol jelly, *Hystomount*, *Cellosize*, *Elvacite*, *Naphrax*). Slides are examined using transmitted light with either a biological or petrological microscope. Permanent mounting follows a two stage procedure: (i) depositing the microfossils onto a coverslip, (ii) attaching the coverslip to a clean glass slide. This procedure ensures that the majority of the micro-organisms are secured in a single optic plane. The practice of preparing temporary mounts, especially during chemical processing stages, may be particularly useful in determining if additional cleaning of organisms is required before proceeding with detailed light or electron microscopical studies.

Strew slide preparation has received considerable attention from micropalaeontologists devising quantitative methods of random grain distribution (*e.g.* Bodén 1991, Laws 1983, Moore 1973). Specialized equipment has been used by some workers in preparing slides (*e.g.* Hansen and Martinez Macchiavello 1978, Locker 1996, Wornardt 1965), but this may not be available in all laboratories. Essentially one of two methods (pipette or random settling method) is used in the preparation of strew slides, preference of which depends on the statistical methods employed in analyzing the assemblage. The methods are:

20.10.2a. Pipette method

Preparations and studies of diatom assemblages have shown that this method is biased towards larger particles within the suspension, and therefore unsuitable for quantitative analyses (Bodén 1991, Battarbee 1973). However, it is ideal for reconnaissance work, and suitable for some taxonomic studies.

1 Clean coverslips by immersion in a standard photographic film washing solution. This reduces water surface tension. Place dried coverslips in a flat *Anumbra* petri dish.

2 Gently shake the storage phial containing the residue. Ensure it is all in a suspension of approximately 40ml volume.

3 Using a clean disposable pipette (dispose after each sample), pipette 0.2-0.5ml of the suspension. Place one drop on the coverslip, and use a fresh coverslip to spread the suspension over the surface.

4 Leave to dry at room temperature.

A detailed account of strew section preparation using the pipette method is outlined by Jansonius (1970). Additional modifications to the above procedure, useful in preparing mounts of chitinozoa, radiolaria, small planktonic and juvenile benthic foraminifera, include transferring the coverslip to a hot plate pre-set to a temperature of 55°C, and leaving to dry (approximately 1 hour). The addition of a few drops of xylene may be required to expel air trapped in hollow tests.

20.10.2b. Random settling method

Random settling procedures are employed when quantitative analyses of an assemblage (Schrader and Gersonde 1978), and or a high degree of reproducibility is required (Moore 1973). Strew mounts produced by settling methods are considered to be within acceptable limits of error relative to the original sample (Bodén 1991, Battarbee 1973). Techniques have been developed and applied to the preparation of quantitative radiolarian slides (Locker 1996) and SEM stubs in the study of coccoliths (Andruleit 1996).

1 Clean glass coverslips, and place in a large flat bottom glass basin (the settling basin). Ensure coverslips are centrally placed to aid in viewing the settling procedure, and reduce the drag effects of the container wall on particles.

2 Mix up 0.2g of unflavoured gelatine in 50ml of distilled water. Pipette several drops of the solution onto the coverslips, and allow it to dry. This will assist in holding the particles in place.

3 Fill the basin with water. Ensure the settling basin is positioned on a flat surface, within the proximity of an infrared heat light. **From now until the end of this procedure the settling basin must not be moved.**

4 Gently shake the sample, and pour it into the water. Stir gently with a vertical motion. Rotational movement generates centrifugal

fractionation, and must be avoided (Moore 1973).

5 Allow 30 minutes for the particles to settle, and then gently begin to pipette all but 1-2cm of water.

6 Position an infrared lamp at 5-10cm from the top of the settling basin, and leave switched on until all the water has evaporated.

7 When the coverslip is dry it can be removed using coverslip forceps, and mounted in the preferred mounting media.

8 The remaining dried residue is retained by washing back into the storage container. Additional preservative may be required before sealing the lid. Label as required.

20.10.2c. Slide mounting

Use a cold (*e.g.* *Elvacite*), warm (*e.g.* glycerol) or a hot setting mountant (*e.g.* *Naphrax*, Canada Balsam). The procedure for using *Naphrax* is as follows:

1 Place a cleaned microscope slide on a hotplate set at 160°C, and allow it to warm.

2 Pipette three drops of *Naphrax* onto the slide, and allow it to boil, evaporating the toluene.

3 Before boiling ceases, and using forceps, place the dried coverslip on the section and leave for 30 seconds.

4 Position the coverslip, and with slight pressure expel trapped air bubbles.

5 Remove the slide from the hotplate, and allow it to cool.

6 Gently trim off excess brittle mountant with a mounted blade.

7 Clean the slide, and apply a label before examination.

Canada Balsam slides are prepared in a similar fashion, with the notable difference that the hotplate is set at a temperature of 110-120°C. The Canada Balsam should never boil, and "cooking" is complete when a strand drawn from the slide fractures in a brittle manner. Once cooled excess balsam can be removed with a suitable solvent (see foraminiferal sectioning above).

Slides prepared using a cold setting mounting media are considerably easier to produce:

1 Apply a small amount of mounting medium to the dried cover glass.

2 Using a clean microscope slide, gently lower it onto the coverslip until

contact with the mounting media is made. Carefully invert the slide and attached coverslip and place on a flat surface. Expel any air bubbles that may appear by gently applying pressure to the coverslip.

3 Place to one side to cure overnight. Excess mountant can be trimmed off with a razor blade. Label before examination.

Sarjeant (1974) suggests that between 150 to 500 individual dinoflagellate cysts can be positioned on a slide in an area of 30x20mm, and more for smaller microfossils (acritarchs, silicoflagellates, smaller diatoms), and less for larger (larger diatoms, radiolaria, chitinozoa). The positions of individual specimens can be recorded using a finder slide (Riedel and Foreman 1961, see below).

20.11. USE OF FINDER GRATICULES

Finder slides or graticules are 75x25mm glass slides inscribed with a numbered and lettered grid. They allow the microscopist to record the position of a feature on a slide, and to re-locate the object, using a similar finder slide, on a microscope of the same type fitted with a similar mechanical stage. The position of the point of interest is ascertained by directly reading off the co-ordinates from finder slide. The method is detailed below.

The two most common types of finder slide used are the England Finder (Fig. 20.1a) and the Halton Finder (Fig. 20.1b). Essentially, the grids are similar on both slides, with the difference one of the scale, and consequently the field of view and optimum objective size used with each. The grid lines of the England Finder are situated at 1mm intervals, while on the Halton Finder the interval is at 0.2mm. The England Finder is used on a stage with X-Y movement of 75x25mm, while the Halton Finder registers a maximum grid movement in an X-Y direction of 5x5mm. England Finders are used for low magnification work, while Halton Finders are used on features centrally situated on slides at higher magnifications.

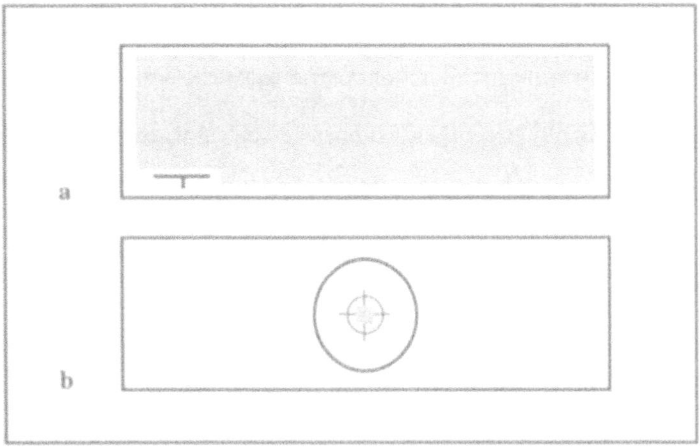

Figure 20.1. (a) The England Finder Slide, has a grid covering the entire slide surface and locating arrows around the edge. (b) the Halton Finder slide has a much smaller grid for use with objectives >x40.

An enlarged section of the finder grid is reproduced in figure 20.2. When correctly positioned and viewed through a compound microscope, reference numbers run horizontally (with vertical rows numbered 1 to 75), and letters A to Z (excluding I) run vertically, forming horizontal rows of the same letter. Each square is then further subdivided into five parts (Fig. 20.3). A central circle (containing the reference number and letter), overlies the four quartiles of the square (numbered left to right 1, 2 and 3, 4).

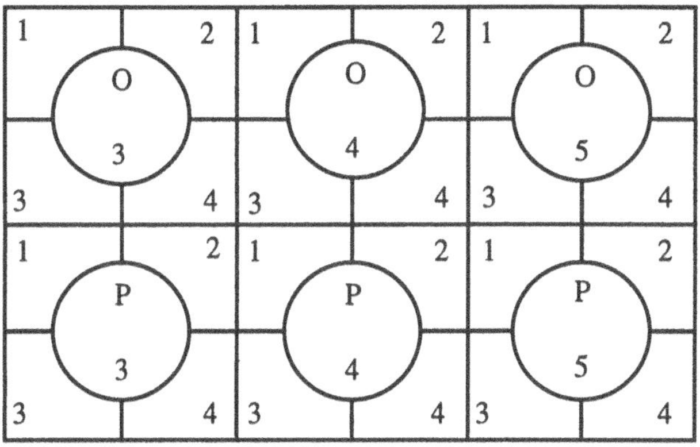

Figure 20.2. An enlarged grid area from a finder slide (see text for explanation).

1 On the slide label of the specimen under examination indicate orientation locating arrows. These marks may be particular to the user, but it is best to employ the same orientation for each slide. This ensures that slide and the finder slide are always orientated together.

2 Bring the point of interest to the centre of field.

3 Ensuring that the position of the stage is not moved, remove the slide and replace with finder slide.

4 Adjust the focus if required until the reference grid of the finder slide can be clearly seen. Record the reference number of the main square (*i.e.* the letter and number in the circle), followed by an oblique stroke and the number of the quartile in which the centre of the field of view lies (Fig. 20.3, point of interest O44/4).

5 To re-locate the point of interest employ the reverse procedure. Place finder slide on stage, and move the stage until the reference point is in the centre of the field of view. Remove the finder slide and replace with the specimen slide, re-focus and examine.

There are two points of note that must be remembered when using this procedure. Firstly, the slide under examination and the finder slide must be referenced from the same point on the mechanical stage. This can be achieved by ensuring that a long and short edge of each slide coincide with the retaining edges of the mechanical stage. Always ensure that the slide is correctly placed within the mechanical stage apparatus, and that the stage is referenced to its maximum left or right and vertical stop

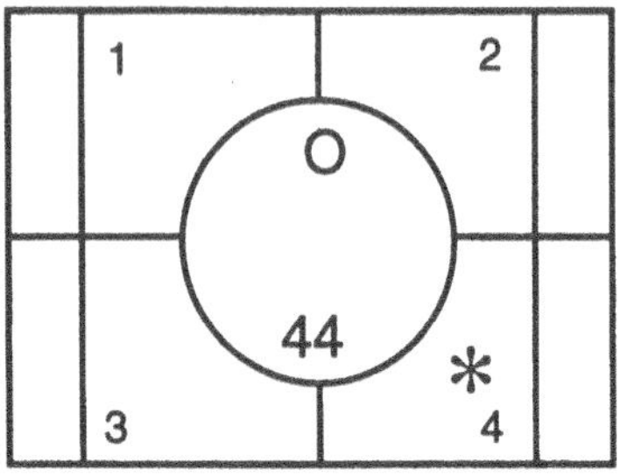

Figure 20.3. The location of a point of interest (* O44/4) on a finder slide grid.

positions. This ensures that the starting point will always be the same regardless of the make of mechanical stage. Should the microscope used not have a mechanical stage with fixed positioning, a fixed stop glass plate can be employed. Available from graticule suppliers, these consist of a 102x51mm glass plate that has four fixed stops, in the form of discs, cemented to the surface. Positioning is achieved by bringing two edges of the slide into contact with three fixed stops of the plate. This is then moved onto the stage until the point of interest is under the cross-wires. The plate is then clamped to the microscope stage using the stage clips. The slide and finder slide are then interchangeable as detailed above. Secondly, only if the specimen slide and finder slide are the same size will the grid references be the same when taken from the left or right. For convenience it may be prudent to take both right and left references, remembering to prefix them in the process.

Micropalaeontological applications in using finder slides are numerous. Traverse (1988) reports success in recording palynomorph locations on strew slides examined on both Leitz and Olympus microscopes. Riedel and Foreman (1961) indexed an important collection of type specimens of North American Palaeozoic radiolaria, and Setty (1966) describes a use in locating diatoms in strew slides. Finder slides are expensive, and should be treated with care. Avoid scratching the surface or damaging the locating edges, and clean with a glass cleaning solution, and polish with a soft cloth.

REFERENCES

AKPATI, B. N. 1969. New technique for serial sectioning of foraminifera. *Journal of Paleontology*. **43**, (4), 1046-1048.

ALLMAN, M and LAWRENCE, D. F. 1972. *Geological Laboratory Techniques*. Blandford, (London). 1-335.

ANDRULEIT, H. 1996. A filtration technique for quantitative studies of coccoliths. *Micropaleontology*. **42**, (4), 403-406.

ARNOLD, Z. M. 1958. A precision sectioning instrument for microfossils. *Micropaleontology*. **4**, (1), 103-112.

ARNOLD, Z. M. 1965a. An improved instrument for sectioning microfossils, pp 232-238: In; B. Kummel and D. Raup (eds), *Handbook of Paleontological Techniques*. W. H. Freeman and Co., (San Francisco). xiii + 852p.

ARNOLD, Z. M. 1965b. The vacuum oven for impregnating microfossils, pp 262-263: In; B. Kummel and D. Raup (eds), *Handbook of Paleontological Techniques*. W. H. Freeman and Co., (San Farncisco). xiii + 852p.

ATHERSUCH, J and JONES, C. 1991. Applications of video image capturing to palaeontology. *Microscopy and Analysis*. No. 24. 7.

BALDAUF, J. G. 1985. Cenozoic diatom biostratigraphy and palaeoecology of the Rockall Plateau region, North Atlantic, DSDP leg 81. *Initial Reports of the Deep Sea Drilling Project*. **86**. 439-478.

BARSS, M. S and CRILLEY, B. 1976. A mounting media for palynological residues. *Geological Survey of Canada*, Paper **76-1B**. 131-132.

BARSS, M. S and WILLIAMS, G. L. 1973. Palynology and nannofossil processing techniques. *Geological Survey of Canada*, Paper **73-26**. 1-25.

BASSLER, R. S and KELLETT, B. 1934. Bibliographic Index of Paleozoic Ostracoda. *Geological Society of America*. Special Paper. **1**. 1-500.

BATTARBEE, R. W. 1973. A new method for the estimation of absolute microfossil numbers, with reference especially to diatoms. *Limnology and Oceanography*. **18**, (4), 647-653.

BAUMGARTNER-MORA, C and BAUMGARTNER, P. O. 1994. Shell structure of fossil foraminifera studied by cathodoluminescence. *Microscopy and Analysis*. No. 40, 35-38.

BELLEMO, S. 1974. Ultrastructures in Recent radial and granular calcareous foraminifera. *Bulletin of the geological Insitute, University of Uppsala*. New Series. 4, 117-122.

BLACKIE, R. A. S. 1988. Three dimensional microscopy using holography. *Microscopy and Analysis.* No. 3. 15-17.

BODÉN, P. 1991. Reproducibility in the random settling method for quantitative diatom analysis. *Micropaleontology.* **37**, (3), 313-319.

BOLTOVSKOY, E. 1966. Methods for sorting of foraminifera from plankton samples. *Journal of Paleontology.* **40**, (5), 1244-1246.

BRADBURY, S. 1989. *An introduction to the optical microscope* (2nd ed). Microscopy Handbooks 1. Royal Microscopical Society, Oxford University Press (Oxford). 86p.

BRIGGS, A. 1985. *An introduction to scanning acoustic microscopy.* Microscopy Handbooks 12. Royal Microscopical Society, Oxford University Press (Oxford). 70p.

BURMA, B. H. 1965. Radiolarians, pp 7-14: In; B. Kummel and D. Raup (eds), *Handbook of Paleontological Techniques.* W. H. Freeman and Co., (San Francisco). xiii + 852.

CUMMINGS, R. H. 1950. Orientated thin-sectioning of microfossils. *Contributions to the Cushman Laboratory for Foraminiferal Research.* **1**, (3-4), 66-67.

DILLÉ, J. E. 1983. Foot-operated fine-focusing device for the light microscope. *Stain Technology.* **58**, (1), 62.

EMILIANI, C. 1951. Notes on thin sectioning of smaller foraminifera. *Journal of Paleontology.* **25**, (4), 531-532.

EMPSON, T. R. 1982. Holographic microscopy of fossil ostracods. Chapter 7, pp 123-146: In; R. H. Bate, E. Robinson and L. M. Sheppard (eds), *Fossil and Recent Ostracods.* Ellis Horwood Ltd., (Chichester). 493p.

EVITT, W. R. 1984. Some techniques for preparing, manipulating and mounting dinoflagellates. *Journal of Micropalaeontology.* **3**, (2), 11-18.

FINGER, K. L and ARMSTRONG, G. L. 1984. A technique to obtain multiple orientations of forminifers in thin section. *Journal of Foraminiferal Research.* **14**, (4), 309-313.

FINGER, K. L. 1990. Atlas of California Neogene Foraminifera. *Cushman Foundation for Foraminiferal Research,* Special Publication, **28**. 1-271.

GAY, P. 1982. *An introduction to crystal optics.* Longman (London). ix + 262p.

GOCHT, H. 1972. "Grouping" preparation of fossil dinoflagellates. *Micropaleontology.* **18**, (2), 235-239.

GRAYSON, J. F. 1956. The conversion of calcite to fluorite. *Micropaleontology.* **2**, (1), 71-78.

HAGN, H. 1953. A new method for preparing orientated thin sections of foraminifera and other small paleontological specimens. *Micropaleontologist.* **7**, (1), 34-43.

HANSEN, H. J and LYKKE-ANDERSON, A-L. 1976. Wall structure and classification of fossil and recent elphidiid and nonionid foraminifera. *Fossils and Strata.* **10**, 1-37.

HANSEN, H. J and MARTINEZ MACCHIAVELLO, J. C. 1978. A new technique for depiction of diatoms. *Micropaleontology.* **24**, (3), 333-335.

HAYNES, J. R. 1981. *Foraminifera.* Macmillan Publishers Ltd, (London). x + 433p.

HENDRY, R. D, ROWELL, A. J and STANLEY, J. W. 1963. A rapid parallel grinding machine for serial sectioning of fossils. *Palaeontology.* **6**, (1), 145-147.

HESSLAND, I. 1949. Investigations of the Lower Ordovician of the Siljan district, Sweden: 1. Lower Ordovician ostracods of the Siljan District, Sweden. *Bulletin of the Geological Institute of Uppsala.* **33**, 97-408.

HODGKINSON, R. L and MARGERUM, R. 1986. On the preparation of oriented sections of larger foraminifera occurring in hard limestones. *Micropaleontology.* **32**, (2), 158-162.

HONJO, S. 1963. New serial micropeel technique. *Bulletin of the State Geological Survey of Kansas.* **165**, (6), 1-16.

HOTTINGER, L and MEHL, J. 1991. X-ray microscopy in micropalaeontology. *Microscopy and Analysis.* No. 21, 21-23.

HUTCHISON, C. S. 1974. *Laboratory handbook of Petrographic Techniques.* Wiley-Interscience (New York). xiii + 527p.

IBARAKI, M and SAMESHIMA, T. 1962. Fine structures of *Lepidocyclina nipponica* from Shimo-shiraiwa, Izu, Central Japan. *Reports of Liberal Arts and Science Faculty, Shizuoka University.* Section Natural Science. **3**, (3), 147-148.

IRELAND, H. A. 1950. Curved surface sections for microscopic study of calcareous rocks. *Bulletin of the American Association of Petroleum Geologists.* **34**, (8), 1737-1739.

JANSONIUS, J. 1970. Classification and stratigraphic application of Chitinozoa. *Proceedings of the North American Paleontological Convention,* **G**. 789-808.

JENKINS, W. A. M. 1967. Ordovician Chitinozoa from Shropshire (techniques pp 439-442). *Palaeontology.* **10**, (3), 436-488.

JEPPSSON, L, FREDHOLM, and MATTIASSON, B. 1985. Acetic acid and phosphatic fossils - a warning. *Journal of Paleontology.* **59**, (4), 952-956.

KENNEDY, C and ZEIDLER, W. 1976. The preparation of oriented thin sections in micropaleontology: an improved method for revealing the internal morphology of foraminifera and other microfossils. *Micropaleontology.* **22**, (1), 104-107.

KERR, P. F. 1977. *Optical Mineralogy.* (4th ed). McGraw-Hill (New York). xvi + 492p.

KESLING, R. V and SOHN, I. G. 1958. The paleozoic ostracode genus *Alanella Boucek,* 1936. *Journal of Paleontology.* **32**, (3), 517-524.

KOIZUMI, I. 1980. Neogene diatoms from the Emperor seamount chain, Leg 55, Deep Sea Drilling Project. *Initial Reports of the Deep Sea Drilling Project,* **55**, 387-407.

KOIZUMI, I and TANIMURA, Y. 1985. Neogene diatom biostratigraphy of the middle latitude Western North Pacific, DSDP leg 86. *Initial Reports of the Deep Sea Drilling Project,* **86**, 289-300.

KREMP, G. 1953. Preparation of orientated thin sections of microfossils. Micropaleontologist. **7**, (1), 29-33.

LAWS, R. A. 1983. Preparing strewn slides for quantitative microscopical analysis: a test using calibrated microspheres. *Micropalaeontology.* **29**, (1), 60-65.

LEVINSON, S. A. 1950. A technique for sectioning microfossils. *Science.* **111**, 60.

LEVINSON, S. A. 1951. Thin sections of Palaeozoic Ostracoda and their bearing on taxonomy and morphology. *Journal of Paleontology.* **25**, (5), 553-560.

LEVINSON, S. A. 1956. The identification of ostracod genera and species in thin sections. (Abstract only). *Journal of Paleontology.* **30**, (4), 1010.

LEVINSON, S. A. 1961. Identification of fossil ostracodes in thin section, pp Q70-Q73: In; R. C. Moore (ed), *Treatise on Invertebrate Palaeontology, Part Q, Arthropoda 3.* Geological Society of America and the University of Kansas (Lawrence, Kansas). xxiii + 442.

LINDHOLM, R. C and DEAN, D. A. 1973. Ultra-thin sections in carbonate petrology: a valuable tool. *Journal of Sedimentary Petrology.* **43**, (1), 295-297.

LISTER, B. 1978. *The preparation of polished sections.* HMSO (London), Institute of Geological Sciences, Report 78/27. 20p.
LOCKER, S. 1996. Quantitative radiolarian slides prepared from soft marine sediments. *Micropaleontology.* **42**, (4), 407-411.
MELLER, A. Einschlußmittel mit hohem Brechungsindex für Diatomeen. *Mikrokosmos.* **74**, 55-60.
MEYER, C. 1946. Notes on cutting and polishing thin sections. *Economic Geology.* **41**, 166-172.
MILLER, J. 1988. Microscopical techniques: 1. Slices, slides, stains and peels, pp 86-107: In; M. Tucker (ed), *Techniques in Sedimentology.* Blackwell Scientific Publications. ix + 394p.
MILNER, H. B. 1962. *Sedimentary Petrography. Volume 1, methods in sedimentary petrography* (4th edition). Allen and Unwin (London). 643p.
MOORE, Jr., T. C. 1973. Method of randomly distributing grains for microscopic examination. *Journal of Sedimentary Petrology.* **43**, (3), 904-906.
NYE, Jr., O. B, DEAN, D. A and HINDS, R. W. 1972. Improved thin section techniques for fossil and recent organisms. *Journal of Paleontology.* **46**, (2), 271-275.
O'CONNOR, B. 1996. Confocal laser scanning microscopy: a new technique for investigating and illustrating fossil radiolaria. *Micropaleontology.* **42**, (4), 395-402.
PHILLIPS, L. 1972. An application of fluorescence microscopy to the problem of derived pollen in British Pleistocene deposits. *New Phytologist.* **71**, 755-762.
PLOEM, J. S and TANKE, H. J. 1987. *Introduction to fluorescence microscopy.* Microscopy Handbooks 10. Royal Microscopical Society, Oxford University Press (Oxford). 56p.
POHL, E. R and BROWNE, R. G. 1973. A thin section method for ultrasmall paleontological specimens. *Journal of Paleontology.* **47**, (2), 221-225.
RIEDEL, W. R and FOREMAN, H. P. 1961. Type specimens of North American Paleozoic radiolaria. *Journal of Paleontology.* **35**, (3), 628-632.
SACHS, K, N. 1965. Preparing paired thin sections from single larger foraminifera, pp 238-241: In; B. Kummel and D. Raup (eds), *Handbook of Paleontological Techniques.* W. H. Freeman and Co., (San Francisco). xiii + 852p.
SARJEANT, W. A. S. 1974. *Fossil and living Dinoflagellates,* (techniques, appendix A, pp 128-133). Academic Press (London). vii + 182.
SAVRDA, C. E, BOTTJER, D. J and GORSLINE, D. S. 1985. An image-enhancing oil technique for friable, diatomaceous rocks. *Journal of Sedimentary Petrology.* **55**, (4), 604-605.
SCHALLREUTER, R. 1982. Extraction of ostracods from siliceous rocks, Chapter 10, pp 169-176: In; R. H. Bate, E. Robinson and L. M. Sheppard (eds), *Fossil and Recent Ostracods.* Ellis Horwood Ltd., (Chichester). 493p.
SCHRADER, H-J. 1976. Cenozoic planktonic diatom biostratigraphy of the Southern Pacific Ocean. *Initial Reports of the Deep Sea Drilling Project.* **35**, 605-672.
SCHRADER, H-J and GERSONDE, R. 1978. Diatoms and silicoflagellates. *Utrecht Micropalaeontological Bulletins.* **17**, 129-176.
SCOTT, A. C. 1989a. Geological applications of laser scanning microscopy. *Microscopy and Analysis.* No. 10, 17-19.
SCOTT, A. C. 1989b. Geological applications of scanning acoustic microscopy. *Microscopy and Analysis.* No. 11, 39-41.
SCOTT, H. W. 1944. Muscle scar patterns on some upper Paleozoic ostracodes. *Journal of Paleontology.* **18**, (2), 162-171.
SCOTT, D. B and VILKS, G. 1991. Benthic foraminifera in the surface sediments of the deep-sea Arctic Ocean. *Journal of Foraminiferal Research.* **21**, (1), 20-38.
SETTY, M. G. P. A. 1966. Preparation and method of study of fossil diatoms. *Micropaleontology.* **12**, (4), 511-514.
SMITH, P. M. R. 1984. The use of fluorescence microscopy in the characterisation of amorphous organic matter. *Organic Geochemistry.* **6**, 839-845.
SOHN, I. G. 1956. The transformation of opaque calcium carbonate to translucent calcium fluoride in fossil ostracoda. *Journal of Paleontology.* **30**, (1), 113-114.
SOHN, I. G. 1961. Techniques for the preparation and study of fossil ostracodes, pp Q64-Q70: In; R. C. Moore (ed), *Treatise on Invertebrate Palaeontology, Part Q, Arthropoda 3.* Geological Society of America and the University of Kansas (Lawrence, Kansas). xxiii + 442.
SOHN, I. G, BERDAN, J. M and PECK, R. E. 1965. Ostracods, pp 75-89: In; B. Kummel and D. Raup (eds), *Handbook of Paleontological Techniques.* W. H. Freeman and Co., (San Francisco). xiii + 852p.
SOMEKH, M. G. 1988. Scanning acoustic microscopy. *Microscopy and Analysis.* No. 6, 7-9.
SYLVESTER-BRADLEY, P. C. 1941. The shell structure of the Ostracoda and its application to their palaeontological investigation. *Annals and Magazine of Natural History.* Series 11. **8**, (43), 1-33.
TRAVERSE, A. 1988. *Paleopalynology.* Unwin Hyman (London). xxiii + 600p.
VAN MORKHOVEN, F. P. C. M. 1958. A simplified method for grinding foraminifera. *Micropaleontology.* **4**, (2), 209-210.
WAGNER, C. W. 1957. *Sur les ostracodes du Quaternaire Récent des Pays-bas et leur utilisation dans l'étude géologique des dépôts Holocenes.* Mouton & Co., ('s-Gravenhage). 1-250.
WEST, R. G. 1977. *Pleistocene geology and biology* (2nd Edition). Longman, (London). xi + 440p.
WHYBROW, P. J. 1978. Fibre optics and foot operated microscope control equipment to assist in the preparation of fossils. *Newsletter of the Geological Curators Group.* **2**, (3), 141-143.
WILLIAMS, A. 1956. The calcareous shell of the brachiopod and its importance to their classification. *Biological Reviews.* **31**, 243-287.
WILLIAMS, A. 1965. Techniques for the preparation of fossil and living brachiopods, pp H251-H256: In; R. C. Moore (ed), *Treatise on Invertebrate Palaeontology, Part H, Brachiopoda (1).* Geological Society of America and the University of Kansas (Lawrence, Kansas). xxxii + 521.
WILSON, G. J. 1971. A chemical method for palynological processing of chalk. *The Mercian Geologist.* **4**, (1), 29-36.
WOOD, A. 1948. Sectioning smaller foraminifera. *Journal of Paleontology.* **22**, (4), 530.
WOOD, A. 1963. Wall structure of foraminifera in polarised light. *Micropaleontology.* **9**, (4), 432.
WORNARDT, W. W. 1965. Mechanical finger for preparation of diatom slides, pp 283-287: In; B. Kummel and D. Raup (eds), *Handbook of Paleontological Techniques.* W. H. Freeman and Co., (San Francisco). xiii + 852p.
WRIGLEY, A. 1934. A Lutetian fauna of Southampton Docks. *Proceedings of the Geologists' Association.* **45**, (1), 1-16.

21. STAINING TECHNIQUES USED IN MICROPALAEONTOLOGY

21.1. INTRODUCTION

In this section staining techniques used in the recognition of live material, and as an aid to the differentiation of microfossils within a processed residue will be described. The many applications of stains and dyes in petrology, palaeontology, mineral recognition and field techniques are discussed elsewhere (Green 1993), a review that contains an extensive bibliography.

21.2. DETERMINATION OF LIVE MATERIAL FROM STAINING

With the increase in studies on foraminifera ecology, particularly quantitative work on taxa distribution and their relationship within the total foraminiferal population, it became essential to differentiate live and recently living foraminifera from those of older dead generations but present within the same sample. The first systematic approach at describing methods was by Walton (1952). In this paper he compared four stains that determined the presence of live foraminifera. Two, previously described methods, involved chemical tests for the presence of protein, while the other two methods, one of which had been described, involved the use of selective biological stains.

21.2.1. Chemical Tests

These are based on the colour reactions between the chemical reagents and one or more of the constituent radicals present within the complex protein molecule. Different colour reactions result from each method as each reagent acts on the presence of differing groups of radicals present.

21.2.1a Millon's Reagent (Phleger 1952)

This reagent can only be used on agglutinated foraminifera, and is therefore of limited use for ecological work involving mixed assemblages. Prepared as follows:

1 One part (by weight) of mercury (Hg) with two parts (by weight) of nitric acid (NH_3, S.G. 1.42).

2 Dilute the solution with two volumes of water.

3 Allow the precipitate that forms to settle out. The resulting supernatant liquid (a solution of mercuric nitrate in nitric acid with some nitrous acid) is used.

4 Thoroughly wash the sample to be tested over a 63μm mesh sieve, to remove all the organic salts and fixative solution. Failure to do this will result in the mercury in the reagent precipitating out and rendering the solution inert.

5 Place a small amount of the residue in a petri dish with a few drops of distilled water.

6 Add a few drops of the reagent, and gently heat the solution under a high intensity lamp for 4-6 hours.

7 The solution will give a red colour to the proteins present within the foraminifera. This is a result of the hydroxyphenyl group (C_6H_4OH) being present in the protein molecule.

21.2.1b. Biuret Reagent (Phleger 1945, 1951)

The solutions are prepared as follows:

1 Thoroughly wash the sample over a 63µm mesh sieve, place in 500ml glass beaker and cover with distilled water.

2 Add an equal volume of 10% sodium hydroxide (NaOH) and mix thoroughly.

3 Add to this, one drop at a time, 0.5% copper sulphate solution ($CuSO_4$), constantly agitating the mixture in the process until a blue solution is produced. Do not add an excessive amount of copper sulphate solution, as copper hydroxide crystals may be precipitated

4 After about 20-30 minutes a blue pink colour is obtained on any protein material present within the residue. This coloration is as a result of the reaction between the reagents in the solution and two or more linked peptide groups (-CONH-).

Unfortunately, the stained specimens are unstable, and if left in contact with the specimens the sodium hydroxide will hydrolyse the protein and other organic substances on the test surface. If required for further study, the specimens have to be washed free of reagent. A stable Biuret solution can be made as follows:

1 Prepare a 10% potassium hydroxide (KOH) solution.

2 Dissolve 0.75g of copper sulphate in a small amount of water, and while agitating the potassium hydroxide solution, slowly add it a few drops at a time. It may be necessary to cool the potassium hydroxide solution during this stage in a constant running water bath.

3 Immerse the specimens in the solution for approximately 20-30 minutes.

The resulting solution, with an indefinite shelf life, will be dark blue in colour. It can be added directly to the washed sample. In both solutions described above the coloration is often weak, and best detectable on broken tests. Moreover, it is usually of a temporary nature.

21.2.2. Biological Stains

21.2.2a. Methyl green-eosin (Rhumbler 1935)

The solution is prepared as follows:

1 Mix 50ml of 1% methyl green with 0.8g of eosin dissolved in 50ml of a 50% alcohol solution, and 50ml of absolute alcohol.

2 Wash the residue, and immerse in the stain solution for half to three-quarters of an hour.

The solution should differentially stain the detrital material (green), faecal material (green) and organic material (red), not very satisfactory for the rapid determination of living foraminifera.

21.2.2b. Rose Bengal (Walton 1952 - protein stain)

The solution is prepared as follows:

1 Dissolve 1g of rose Bengal in 1 litre of distilled water. The concentration of the stain is not critical since it alters only the intensity not the effectiveness.

2 Wash the sample as above to remove sea water, salts and fixative.

3 Immerse the residue retained on the sieve in the stain solution, and leave for approximately 10 minutes.

4 Re-wash the stained residue to remove the excess stain, and then examine the material.

Protoplasm present within foraminiferal tests stains a deep rose red colour. The technique works best on smaller hyaline foraminifera, particularly if protoplasm is present in the final few chambers, as this will preferentially stain first. The stain is obscured in agglutinated foraminifera, and calcareous hyaline and porcelaneous forms with thick tests, particularly in dried residues (Murray 1991).

Walton (1952) concluded that the rose Bengal technique was "the most reliable and efficient stain for the recognition of living Foraminifera". However, he also points out "the only completely accurate method of determining the true state of foraminiferal protoplasm is by observation of pseudopodial action or cytological sectioning and staining of the nuclei". Although sectioning is impractical for palaeontologists undertaking ecological studies, rose Bengal staining coupled with the optical observations of the movement of pseudopodia, protoplasmic colour and casts following decalcification (Myers 1942), provide complementary methods in determining live protozoa.

The inconclusive nature of the rose Bengal technique was questioned by Martin and Steinker (1973). Following Walton's (1952) technique, their observations showed inconsistencies of stain intensity within species, even when the time the residue was in contact with the stain was increased (in some cases up to 24 hours), and partial staining of empty tests, ostracod carapaces, molluscan shell material, ooids, pellets and siliceous sand grains. Their conclusion was, that although time-consuming, "the only reliable methods ... for the determination of living individuals involve direct observation in an attempt to recognise signs of life". Staining by rose Bengal thus recognises the presence of protoplasm, that is then directly, and sometimes incorrectly, interpreted as representing live material.

21.2.2c. Sudan black B (Walker *et al.*, 1974, lipid stain)

Further criticism against the rose Bengal technique was outlined by Walker *et al.* (1973, 1974). A number of protoplasmic stains were tested by Walker and Schafer (1974), and Walker *et al.* (1974), and compared with rose Bengal. The most convincing results were obtained from acetylated or saturated Sudan black B, with stain penetration into the test improved if the solutions were heated and specimens had been fixed prior to staining.

Preparation of acetylated Sudan black B is as follows:

1 Dissolve 1g of Sudan black B in 100ml of diethyl ether.

2 Filter the solution through Whatman No.1 grade paper.

3 Boil the solution.

4 Add 0.5ml of acetic anhydride in 20ml of ether. Boil the solution for 20 minutes by means of a reflux system, i.e. maintaining the solutions boiling point without loss by evaporation.

5 Cool and filter the solution.

6 Transfer the solution to a separating funnel, and with the addition of cold water decant off until the aqueous layer is no longer coloured, or acidic. Test with universal indicator paper to confirm the solution is neutral.

7 Decant the final solution into an evaporating dish and allow the ether to evaporate.

8 Scrape the metallic black crystals of acetylated Sudan black B from the dish, and prepare a 1% solution in 70% ethanol (should make approximately 100ml of working stain solution).

9 Thoroughly wash the fixed samples in distilled water.

10 Add pre heated (40°C) acetylated Sudan black B.

11 Place the sample, immersed in the stain solution, in a constant temperature water bath set at 40°C for about 30 minutes.

12 Decant off the excess stain solution, then wash the sample two or three times in 70% ethanol.

13 Allow the sample to dry before examining it.

If protoplasm is present in the final six or seven chambers it will stain dark blue-black. Observations by Walker *et al.*, (1974), who examined dissected material, revealed that test linings and outer test surfaces remained unstained, and that washing and heating of material removed algae adhering to the test wall.

Unfortunately the above procedure takes many hours to complete, only a small amount of working stain can be made at one time, and specialized glass ware is required. With only a small decrease in stain accuracy, a saturated solution of Sudan black B can be prepared in minutes in the following way:

1 Add approximately 10g of stain to 1 litre of 70% ethanol, i.e. enough to saturate the solution.

Add to the sample as above following stages' 9-13.

Despite the evidence that acetylated Sudan black B is a more reliable stain than rose Bengal, particularly when used in the laboratory, it offers no advantage for use in the field if material cannot be fixed. Satisfactory results (i.e. live material taking up the stain) can be obtained by the addition of rose Bengal either pre-mixed in sea water, or added directly to the sample, but again care must be taken in interpreting the results. Conclusive determinations between stained recent live material and tests of recent dead foraminifera, requires careful optical examination, although the uncertainty of recognizing recently dead tests still remains.

This problem has been partly overcome by the experimental work of Bernhard (1988), in contrasting rose Bengal and Sudan black B staining with the Adenosine-5'-triphosphate (ATP) assay procedures, commonly used in bacteriological determination. ATP is present in all living organisms, but rapidly degrades immediately following death. Comparisons of three identical samples collected from McMurdo Sound, Antarctica, were undertaken. The foraminiferal assemblages in each sample were known to be living at the time of collection, but subsequently killed in the laboratory by storing in flasks of fresh sea water, capped and left at 20°C for 5 hours. The experimental work was undertaken over a four week period. Results indicated the percentages of live material from the three experiments to calculate as, 30% from the ATP assay, 47.2% from the rose Bengal assemblage, and 18.5% from the Sudan black B assemblage. Furthermore, they indicated that foraminifera that had been dead for up to four weeks continued to stain with both rose Bengal and Sudan black B. Moreover, the intensity of the stain indicated that protoplasm degradation was a slow process. It is not clear if similar results would be obtained from live material collected from warmer climates, in which protoplasm degradation would occur at a faster rate. Comparative experimental work is required to verify this. However, these results confirm the unreliability of rose Bengal, indicating a 57.3% margin of error by the staining of recently dead foraminifera still containing protoplasm, and the underestimation of live material within an assemblage by an error of 38.3% when using Sudan black B.

21.3. STAINING MICROFOSSIL RESIDUES

The staining of individual microfossils picked from residues is often used to enhance the structural details of the organism during optical examination. A number of techniques specific to foraminifera, ostracods and organic walled microfossils, using food colourings, are briefly outlined in Allman and Lawrence (1972). As most of these stains are non-selective it is important to ensure that all the matrix has been removed, as if it is of calcareous composition it too will stain. The procedure is as follows:

1. Clean specimens are immersed in stains mixed in distilled water, alcohol or ether.
2. Both the time of immersion and stain concentration control the final intensity of the stain colour.
3. On removal from the stain, use the capillary action of a fine paint brush or tissue paper to draw off excess stain, and allow the specimen to dry before mounting in a cavity slide and examining.

Unfortunately, all but one of the food colourings (a red/pink powder H4938) used by Allman and Lawrence (1972) is now not manufactured by Bush Boake Allen Ltd., (London). Furthermore, the food colourings are only supplied in minimum quantities of 20 or 25kg cans, and a minimum value order is required. Smaller quantities of suitable powdered food colourings may be available from local suppliers (particularly suppliers to Asian food retailers). Quantifying the long-term effects of these additives (if any) on calcium carbonate tests and carapaces is currently under investigation.

The type and colour of the stain are a matter of personal preference and availability, however as results vary from different specimens it is best to compare the effects of several stains. Brotzen (1936) describes a method that takes advantage of the non-selectivity of these stains, using methylene blue to differentiate calcareous particles (shell fragments and microfossils) from argillaceous and siliceous material. The procedure is as follows:

1. Place the sample in a petri dish or watch glass containing 0.5ml of distilled water.
2. Add one drop of methylene blue solution (0.25g to 100ml of distilled water).
3. Argillaceous and siliceous material stains within 2 minutes, while calcareous material usually takes a little longer. It is therefore important to examine the material around two minutes to ensure it is not over stained, thus inhibiting particle differentiation.
4. For larger quantities of residue use more water to cover the sample, and then add one drop of methylene blue for every 0.5ml.
5. After staining decant off the solution and allow the material to dry and then examine.

A similar method has been described by Herman and Metz (1972) using a standard Alizarin Red-S solution similar to that used in carbonate staining. Immerse an oven dried residue in the stain solution for between 30 seconds and 1 minute. Excess stain is decanted or washed off over a 63µm screen. The residue is then dried and examined. This method can also be applied to the detection of ostracods in an argillaceous rich residue. A method employing silver nitrate and used specifically on single ostracod carapaces has been documented by Triebel (1947, in German, translated to English by Levinson 1951), and Artusy and Artusy (1956). The procedure is as follows:

1. Place clean specimens on a platinum or aluminium support, and heat with a bunsen for 3-5 seconds. This treatment will have the effect of darkening the specimen.

2. Allow the specimen to cool to room temperature, then immerse in a 5% solution of silver nitrate (5g $AgNO_3$ to 100ml distilled water) and allow it to stand for 30 seconds. **N.B. The solution must be stored in a dark bottle.**

3. Carefully remove the specimen and place on filter paper to remove the excess solution.

4. Re-heat the specimen over the bunsen for approximately 1 minute.

5. Allow the specimen to cool then mount in a normal micropalaeontological slide. The specimen will appear with a dark metallic surface lustre.

As pointed out by Allman and Lawrence (1972, p 98), the pore canals become obscured by this procedure, and furthermore the resulting semi-opaque nature, limits the usefulness of prepared specimens in transmitted light microscopy observations. However, the methods of Henbest (1931, using malachite green dissolved in alcohol) and Artusy and Artusy (1956, using food stains) overcome these problems with reversible procedures. A summary of the procedures is as follows:

1. Place clean specimens in a watch glass or petri dish.

2. While observing under the binocular microscope apply a small amount of the stain with the aid of a paint brush. As the microfossils are liable to float, ensure that they become immersed with the aid of the brush.

3. After the specimen has absorbed enough of the stain, remove from the solution and allow it to dry. The stain tends to gravitate to the depressions of the fossil, resulting in sharp contrast differences in the relief and the enhanced appearance of the raised areas.

4. The stain can be removed by immersing the specimen in alcohol and giving it a gentle wash with the paint brush.

The stains most commonly used to enhance the appearance of organic walled microfossils are (i) methylene blue; (ii) methyl violet; (iii) basic fuchsin; and (iv) ruthenium red. These stains are usually added to prepared cleaned residues, temporarily stored in distilled water, or for longer term storage, alcohol (see section 26 EXTRACTION TECHNIQUES FOR ACID INSOLUBLE MICROFOSSILS). Both methylene blue and ruthenium red have been successfully used to stain cherts containing microfossils. Hand specimens are immersed in a 20°C stain solution in a 1:1 alcohol/distilled water mixture, for between 4-12 hours (methylene blue), or 30 minutes (ruthenium red), depending on which dye is used. This procedure can also be used for the staining of chert in thin section.

REFERENCES

ALLMAN, M and LAWRENCE, D. F. 1972. *Geological Laboratory Techniques*. Blandford, (London). 1-335p.
ARTUSY, R. L and ARTUSY, J. C. 1956. The use of food colouring as a new technique for staining microfossils. *Journal of Paleontology*. **30**, (4), 969-970.
BERNHARD, J. M. 1988. Postmortem vital staining in benthic foraminifera: duration and importance in population and distributional studies. *Journal of Foraminiferal Research*. **18**, (2), 143-146.
BROTZEN, F. 1936. Foraminiferen aus dem schwedischen untersten Seon von Erikskal in Schonen Sweden. *Sveriges geologiska undersokning. Afhandlingar*. Serie C, **30**, (396), 1-206.
GREEN, O. R. 1993. The use and applications of stains and dyes in geological techniques. *Science Technology*. **28**, 2-26.
HENBEST, L. G. 1931. The use of selective stains in paleontology. *Journal of Paleontology*. **5**, (4), 355-364.
HERMAN, Y and METZ, M. C. 1972. Staining technique for recent and fossil calcareous invertebrates. *Journal of Paleontology*. **46**, (1), 152.
LEVINSON, S. A. 1951. The Triebel technique for staining ostracodes. *Micropaleontologist*. **5**, (2), 27.
MARTIN, R. E and STEINKER, D. C. 1973. Evaluation of techniques for recognition of living foraminifera. *Compass of Sigma Gamma and Epsilon*. **50**, (4), 26-30.
MURRAY, J. W. 1991. *Ecology and Palaeoecology of Benthic Foraminifera*. Longman Scientific. x + 397p.
MYERS, E. H. 1942. A quantitative study of the productivity of the foraminifera in the sea. *Proceedings of the American Philosophical Society*. **85**, 325-342.
PHLEGER, F. B. 1945. Vertical distribution of pelagic foraminifera. *American Journal of Science*. **243**, 377-383.
PHLEGER, F. B. 1951. Ecology of Foraminifera, northwest Gulf of Mexico. Part 1, Foraminifera Distribution. *Memoir of the Geological Society of America*. **46**, 1-88.
PHLEGER, F. B. 1952. Foraminifera ecology off Portsmouth, New Hampshire. *Bulletin of the Museum of Comparative Zoology (Harvard)*. **106**, (8), 315-390.
RHUMBLER, L. 1935. Rhizopoden der Kieler Bucht, gesammelt durch R. Remane 1 Teil. *Schriften des naturwissenschaftlichen Vereins fur Schleswig-Holstein*. Band **21**, H2, 143-195.
TRIEBEL, E. 1947. Methodische und Technische Fragen der mikropalaontologie. *Senckenberg Buch, 19 W Kramer, (Frankfurt)*. 47p.
WALKER, D. A, LINETON, A. E and SCHAFER, C. T. 1974. Sudan Black B: a superior stain to rose Bengal for distinguishing living from non-living foraminifera. *Journal of Foraminiferal Research*. **4**, (4), 205-215.
WALKER, D. A and SCHAFER, C. T. 1973. An investigation of staining methods to differentiate living and non-living foraminifera. (Abstract only). *Geological Survey of Canada, Part B*. Paper **73-1B**, 131.
WALKER, D. A and SCHAFER, C. T. 1974. Sudan Black-B: a stain for quantitative determination of living foraminifera. (Abstract only). *Geological Survey of Canada, Part B*. Paper **74-1B**, 131.
WALTON, W. R. 1952. Techniques for the recognition of living foraminifera. *Contributions from the Cushman Foundation for Foraminiferal Research*. **3**, (2), 56-64.

22. PREPARATION OF CARBONATE STAINED ACETATE PEELS AND THIN SECTIONS

22.1. INTRODUCTION

Many stains and staining schemes have been developed to aid the identification of carbonate minerals. Variations in mineralogy and rock fabric result in differential solution rates, while stain concentration and active shelf life influence the reaction time and absorption rates. Minerals immersed for excessive time periods are very dark, and features are obscured. Likewise, if a specimen is removed from the stain too early, stain effect will be insufficient. In both cases, minerals, fabrics and textures are indistinguishable.

Carbonate thin section staining and peel slab procedures combine the techniques of (1) mineral differentiation and (2) gross analysis of the rock texture and fabric. Staining and peel techniques are straight forward to lean, and with practice easy to execute. Furthermore, they have the advantage of being relatively cheap and repeatable. The following procedures are outlined:

- Preparation of a slab
- Etching the slab
- Staining a slab and thin section
- Preparation of a peel
- Common staining tests to differentiate aragonite, calcite and dolomite

A technique for peeling individual microfossils is outlined in section 32 SPECIALIST TECHNIQUES USED IN THE PREPARATION OF INDIVIDUAL MICROFOSSIL SPECIMENS, while non-selective staining of microfossil specimens and assemblages are detailed in procedures within section 21 STAINING TECHNIQUES USED IN MICROPALAEONTOLOGY.

22.2. SLAB PREPARATION

1. Prepare specimen by cutting a 1cm thick slab. Remember to cut perpendicular to fabric or bedding. If peels are to be taken trim the slab to fit the glass size. Mark one side of the slab with the specimen number.

2. Grind the unmarked side of the slab to remove the saw blade marks. Use a selection of carborundum powders, progressing to finer grades, *e.g.* 220, 120, 3F, 600, 800. The final polish should be on grade 800 or 1000. Remember to thoroughly wash the specimen between grades to avoid contamination. Use a different glass plate for each grade.

3. Wash, dry and examine the polished surface. If scratches or saw blade marks are still visible repeat stage two. If the specimen is noticeably uneven then grind the surface using first a diamond wheel and then a steel lapping wheel with carborundum powder before proceeding to the glass plates.

Failing to spend adequate time in preparing the surface of the specimen will result in a poor peel quality, with a failure to reproduce fine microstructural differences (Wilson and Palmer 1989).

22.3. ETCHING THE SLAB

Etching is the most important single step in the preparation of successful good quality peels. **UNDERTAKE THE FOLLOWING WORK IN A WELL VENTILATED AREA.**

1 Prepare a 10% HCl solution by mixing 100ml commercial grade HCl with 900ml distilled water. Use of a stronger acid will destroy microstructural detail (Wilson and Palmer 1989).

2 By holding the polished surface face down, and at a slight angle, immerse it in the acid solution for twenty seconds to one minute (depending on specimen solubility).

In specimens where differential etching of the carbonate minerals with HCl results in excessive relief, a 0.1M solution of ethylenediametetra-acetic acid (EDTA), buffered to pH 10 with the addition of sodium hydroxide (NaOH) can be used as an alternative (Mandado and Tena 1986). Etching times may vary between 3-7 minutes (increasing if multiple specimens are etched), depending on specimen lithology, mineralogy and texture. Conventional staining procedures (outlined below) have been used in conjunction with this process, although when preparing a peel the acetate sheet can be removed after 5 minutes (Mandado and Tena 1986).

3 Remove the specimen and wash the surface with running water. This will stop the reaction. Allow the surface to dry.

4 Examine the etched surface of the specimen using a hand lens or binocular microscope. The specimen surface will appear 3-D, indicating well-pronounced relief between the different carbonate minerals. Ensure the surface is not handled, as any grease from fingers can obscure detail.

22.4. STAINING THE SLAB AND THIN SECTION

Stain mixtures and recipes will be described in detail below under section 5, staining tests.

22.4.1. Slab

1 Immerse the slab in the stain solution for 2-3 minutes, examine during immersion to decide the optimum concentration required. Experimentation may be required. The intensity of the stain will increase the longer the specimen is immersed. N.B. Remember that the stain intensity will be reduced if a peel is taken.

2 Remove specimen and either gently wash under a stream of running water (some of the stain will be removed as it is water soluble), or allow specimen to air dry (stain will become more intense).

22.4.2. Thin Section

Carbonate thin sections, prepared in the standard way, but without a cover slip, can be stained using the following technique. However, great care must be taken as prepared thin sections will usually be at a thickness of 30µm, which can easily be etched away. If possible have prepared sections left at a thickness between 35 to 40µm to allow for etching and staining. It is customary to stain only half of the section, which represents the bulk composition of the rock.

1 Etch half the section in a 0.2-0.5% HCl solution for about 10 seconds. The acid concentration and time of immersion are dependent on the slide thickness and rate of reaction. Determined by trial an error, experience provides the preparator with the best guide. However, observing the voracity of effervescence provides a means of assessing the time required.

2 Wash under a stream of running water and dry section.

3 Immerse the etched side in a 50-75% diluted mixture of stain solution for 20-30 seconds. The reason for using a dilute mixture is two-fold. Firstly, the stains are mixed in acid, and will therefore continue to etch the surface of the section, and secondly, with a dilute stain mixture there is more control over the intensity of the stain taken up by the specimen. The problems of stain concentration and time of immersion are similar to those encountered during etching. Weaker stain solutions allow for longer immersion times, providing the preparator with greater control in determining stain intensity.

4 Remove section from stain and gently wash under running water. Remember that the stain is water soluble, so do not touch the stained area.

5 When the thin section is dry apply a cover slip in the usual way.

22.5. PEEL PREPARATION

1 Set the slab on a flat surface, either in a sand bath or supported with plasticine, with the prepared surface uppermost.

2 Cut a piece of acetate sheet to a size slightly larger than the specimen surface. Be careful with the acetate sheet. It will mark easily. Only handle the edges.

Acetate sheet can be purchased in a variety of thicknesses. A thickness of 180µm is usually sufficient for most work, although where specimen relief is low (rocks with micritic compositions), and less acetone is applied to the surface, thinner (125µm, 95µm) sheets can be used. Using too thin a sheet can result in a wrinkled peel (Wilson and Palmer 1989).

3 Flood the surface of the specimen with acetone. Insufficient acetone results in a poor reproduction of detail (Wilson and Palmer 1989).

4 Immediately apply a piece of acetate sheet on the etched stained surface before the acetone evaporates. This is achieved by bending the acetate sheet and placing it at the centre of the specimen and rolling it out over the surface. Sufficient acetone should remain on the specimen during this stage, so that as the acetate is rolled over the surface excess acetone and trapped air is expelled.

A major problem in preparing good quality peels is the presence of bubbles. This may result because the sediment is so porous that the acetone seeps from the surface through the rock before the acetate sheet adheres (Wilson and Palmer 1989). It can also result from the acetone evaporating from the specimen surface, either as a result of excessively high temperatures in the working environment (avoid doing this in direct sunlight or by an oven or hot-plate), or taking too long between flooding the surface and applying the acetate sheet (practice makes perfect). There should be no need to gently "smooth" the sheet to expel excess bubbles. Even with "dry fingers" the acetate sheet can exhibit finger-prints, and furthermore, once the sheet has come into contact with the acetone it is structurally weakened, and even the smallest indentation will permanently mark the surface. Porous rock specimens should be resin impregnated before the peel is attempted (see section 10 PRESERVATION, CONSOLIDATION AND REPAIR OF UNSTABLE SPECIMENS).

5 Leave the acetate sheet to dry on the specimen for a minimum of 15 minutes. Attempting to remove the peel before the acetone has dried frequently results in tearing.

6 Carefully remove the peel starting at one end of the specimen.

7 Once the peel has been removed from the specimen quickly trim the excess acetate sheet from the edges, cutting as close as possible to the edge of the peel.

8 Mount the peel between two clean glass plates, hinged down the edge like a book. Seal the remaining edges of the slide with *Sellotape*, which is preferable to masking tape, particularly if the peel fills the view of the slide. As an alternative method of storage, particularly for smaller peels, consider mounting them between 35mm slide transparency holders. This has the advantage in that they can be projected onto a large screen for examination.

9 Label the slide.

22.6. STAINING TESTS

22.6.1. Determination of Aragonite using Feigl's solution

This test was first described by Feigl (1937). It is dependent on the slightly differing solubility's of calcite and aragonite in water, and is related to their differing crystallographic structure. The solution is made as follows:

i Dissolve 11.8g of manganese sulphate ($MnSO_4.7H_2O$) in 100ml of water.

ii To this solution add 1g of silver sulphate (Ag_2SO_4).

iii Boil and allow to cool.

iv Filter the suspension

v Add two drops of 10% sodium hydroxide (NaOH) solution.

vi Allow the solution to stand for two hours.

vii A precipitate will form, filter the solution.

viii Use the solution immediately or store in a dark bottle.

N.B. If the solution is stored for any time, it should be tested on a specimen known to be aragonite to determine its effectiveness. The staining procedure is as follows:

1 The prepared surface is etched in 10% HCl as outlined above.

2 Immerse the specimen in the Feigl's solution for 2-4 minutes.

3 Careful agitation of the specimen is usually required to ensure all the surface is in contact with fresh stain.

4 Remove the sample carefully and wash using distilled water.

5 Allow surface to dry. Peel if required.

Results

Calcite	:	remains unstained.
Aragonite	:	is stained black.
Dolomite	:	remains unstained.

22.6.2. Determination of calcite and dolomite using Alizarin Red-S and Potassium Ferricyanide

22.6.2a. Alizarin Red-S (Sodium Alizarin Sulphonate)

First put forward by Hügi (1945) and applied to carbonates by Friedman (1959). Determines the presence or absence of magnesium differentiating calcite/aragonite from dolomite. Carbonate determinations by the staining procedures of Friedman (1959), and Warne (1962) are illustrated in figures' 22.1 and 22.2, while the combination staining and etching procedure of Dickson (1965) is illustrated in figure 22.3. The solution is made up as follows:

i Dissolve 0.1g of alizarin red-S in 100ml of 0.2% HCl and mix thoroughly.

This stain can be stored in a dark bottle for a limited time. Temperatures greater than 20°C will shorten the shelf life. The staining procedure is as follows:

1 Prepare and etch the surface as described above.

2 Place the specimen in the stain for approximately 4 minutes, remember that calcite will stain intensely in 2-3 minutes.

3 Remove specimen from stain and lightly wash under running water.

Results

Calcite	:	stained pink or red, depending on crystal orientation.
Aragonite	:	stained pink or red, depending on crystal orientation.
Dolomite	:	remains unstained.

22.6.2b. Potassium Ferricyanide

The test was originally proposed by Heeger (1913), as a positive test for the presence of iron in carbonates, i.e. the stain intensity increases with Fe content, thus differentiating between the ferroan phases in calcites and dolomites, however it cannot distinguish between ferrous and ferric iron. The solution is made up as follows:

i Dissolve 2g of potassium ferricyanide in 100ml of 1.5% HCl. This stain is unstable and must be used immediately on the day of preparation.

The staining procedure is as follows:

1 Immerse the specimen in the stain for 4 minutes.

2 Remove and lightly wash under running water.

PROCEDURE I

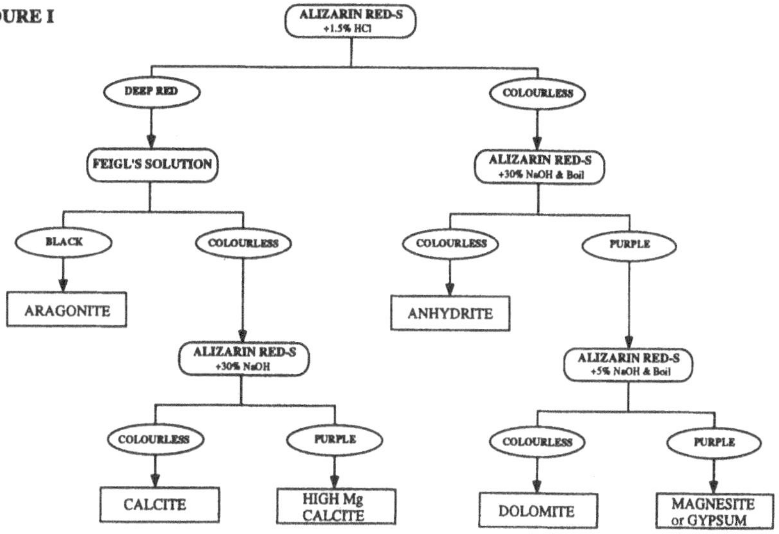

PROCEDURE II

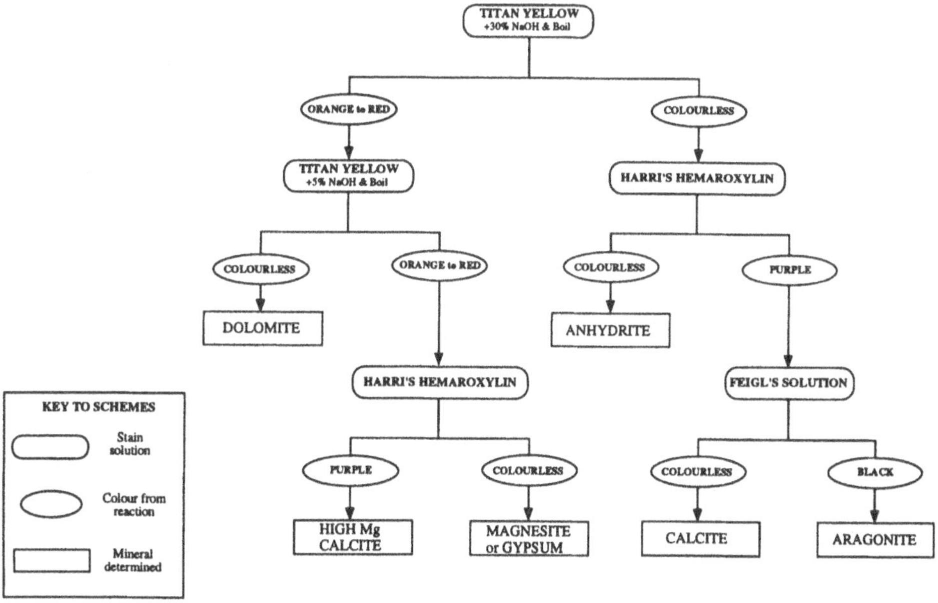

Figure 22.1. Friedman's (1959) staining procedures for the determination of the most commonly occurring carbonate minerals.

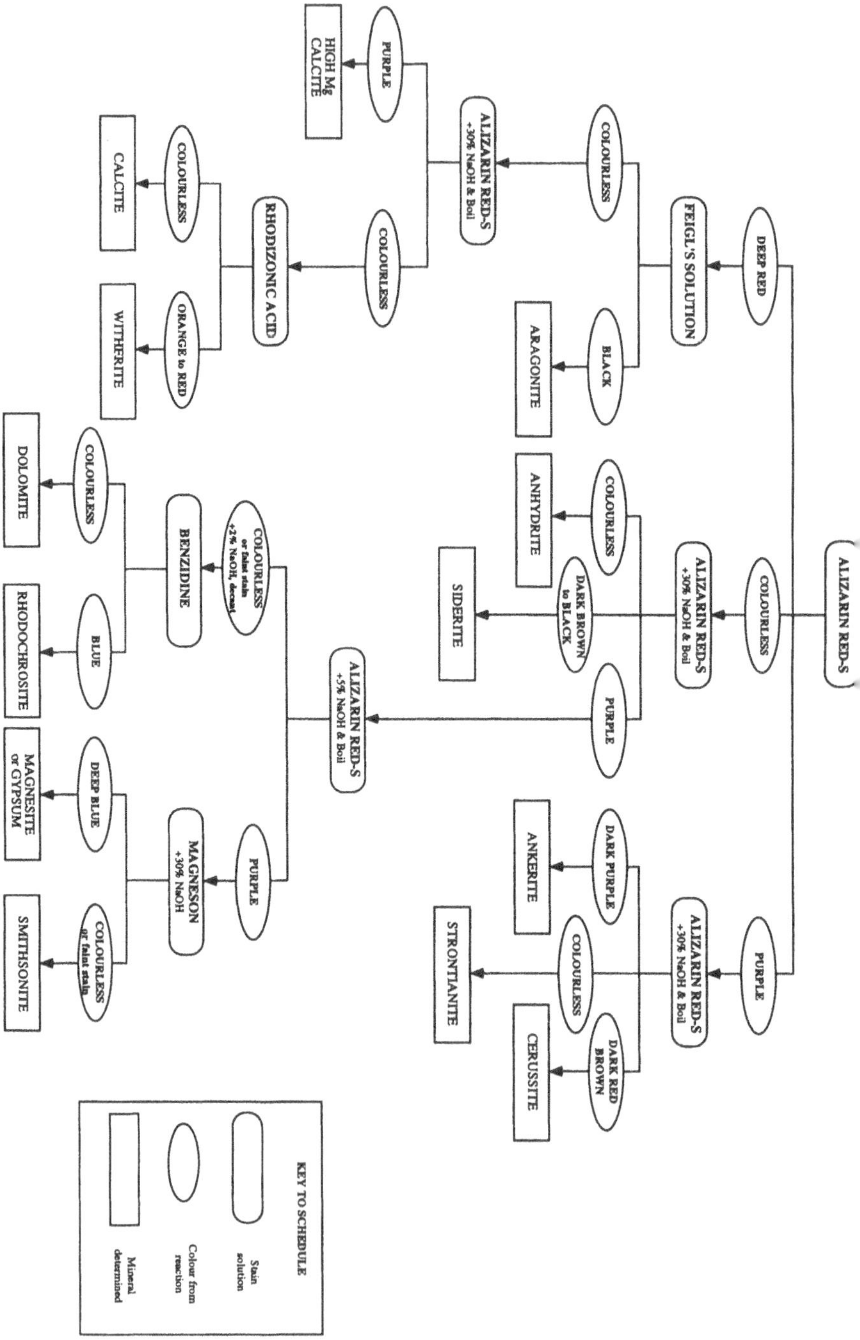

Figure 22.2. Warne's (1962) modified staining procedures for the determination of commonly occurring carbonate minerals.

Results

Iron free calcite	:	remains unstained.
Iron poor calcite	:	stained light blue.
Iron rich calcite (siderite)	:	stained dark blue.
Iron free dolomite	:	remains unstained.
Ferroan dolomite (FeO)	:	stained light blue.
Iron rich dolomite (ankerite)	:	stained dark blue.

As the iron content of the carbonate increases so will the intensity of the stain. Iron free dolomite and siderite are usually only seen when the solution is heated, the reaction time is quicker for siderite.

22.6.2c. Alizarin Red-S and Potassium Ferricyanide

Dickson (1966) described this procedure, a combined stain for determining iron (Fe) and Magnesium (Mg). Stains are prepared as follows:

Alizarin Red-S

i Dissolve 0.3g of alizarin red-S in 150ml of 1.5% HCl.

Potassium Ferricyanide

i Dissolve 2g of potassium ferricyanide in 100ml of 1.5% HCl.

Mix the two solutions in a ratio of 3:2 (alizarin red-S : potassium ferricyanide). The staining procedure is as follows:

1 Immerse the specimen in the stain for 4 minutes.

2 Remove specimen from stain, lightly wash and dry.

3 Peel if required.

Results

Iron free calcite	:	stained pink or red.
Iron poor calcite	:	stained mauve.
Iron rich calcite	:	stained royal blue or purple.
Iron free dolomite	:	remains unstained.
Iron poor dolomite	:	stained light turquoise.
Iron rich dolomite	:	stained dark turquoise.
Aragonite	:	stained pink or red.

The relationship between the carbonate group of minerals and the stains imparted on them by using a combined alizarin red-S and potassium ferricyanide solution is diagramatically interpreted in figure 22.4.

22.6.3. Determination of Mg-calcite using Titan Yellow

A non permanent method of staining using Titan yellow (Clayton yellow) to detect the presence of Mg-rich calcite was described by Friedman (1959), and modified by Winland (1971). An improved method was described by Choquette and Trusell (1978), in which the stain was made permanent. The method can be applied to slabs or thin sections, but the latter must be mounted in an epoxy resin, as both Lakeside and Canada Balsam are soluble in the stain solution. Cover slips must also be epoxy mounted. The solutions are prepared as follows:

Titan yellow

i Dissolve 1g of Titan yellow powder, 8g of sodium hydroxide (NaOH) pellets and 4g of ethylenediaminetetra-acetic acid disodium salt (diNaEDTA) in 1 litre of distilled water.

N.B. This solution has a shelf life of approximately two years if stored in a dark bottle.

Stain fixer

i Slowly dissolve 200g of NaOH pellets in 1 litre of distilled water. **CAUTION: HEAT AND FUMES ARE EVOLVED DURING THIS REACTION, AND THE SOLUTION IS CORROSIVE.**

N.B. This solution has an indefinite shelf life, but must be stored in polythene bottles as it will etch glass.

The staining procedure is as follows:

1 Etch slabs in a 5% acetic acid solution for 30 seconds.

2 Dry the surfaces in a low temperature oven or stream of warm air.

3 Immerse the specimen in the stain solution for about 20 minutes.

4 Dry the surface in warm air.

5 Immerse the stained surface in the fixer solution for 30 seconds.

6 Dry surface in warm air, cover thin sections (using an epoxy mounting media), or peel slabs.

Results

Calcite with 5-8% $MgCO_3$: pink to pale red.

PROCEDURE		ETCHING TIME	CARBONATE	RESULT
Stage I	Etch in 1.5% HCl	10 - 15 seconds	CALCITE FERROAN CALCITE	Considerable etch
			DOLOMITE FERROAN DOLOMITE	Negligible etch
Stage II	Stain combination of Alizarin Red-S and Potassium Ferricyanide, mixed in 1.5% HCl. Ratio of ARS to PF : 3 : 2	30 - 45 seconds	CALCITE	Pale-pink red
			FERROAN CALCITE	Very pale pink-red and pale blue-dark blue Together give mauve-purple-royal blue
			DOLOMITE	Colourless
			FERROAN DOLOMITE	Pale deep turquoise developing with ferrous content
Stage III	Alizarin Red-S in 1.5% HCl	10 - 15 seconds	CALCITE FERROAN CALCITE	Very pale pink-red
			DOLOMITE FERROAN DOLOMITE	Colourless

Figure 22.3. Dickson's (1965) thin section determination of ferroan-rich and ferroran poor calcite and dolomite using staining and etching characteristics.

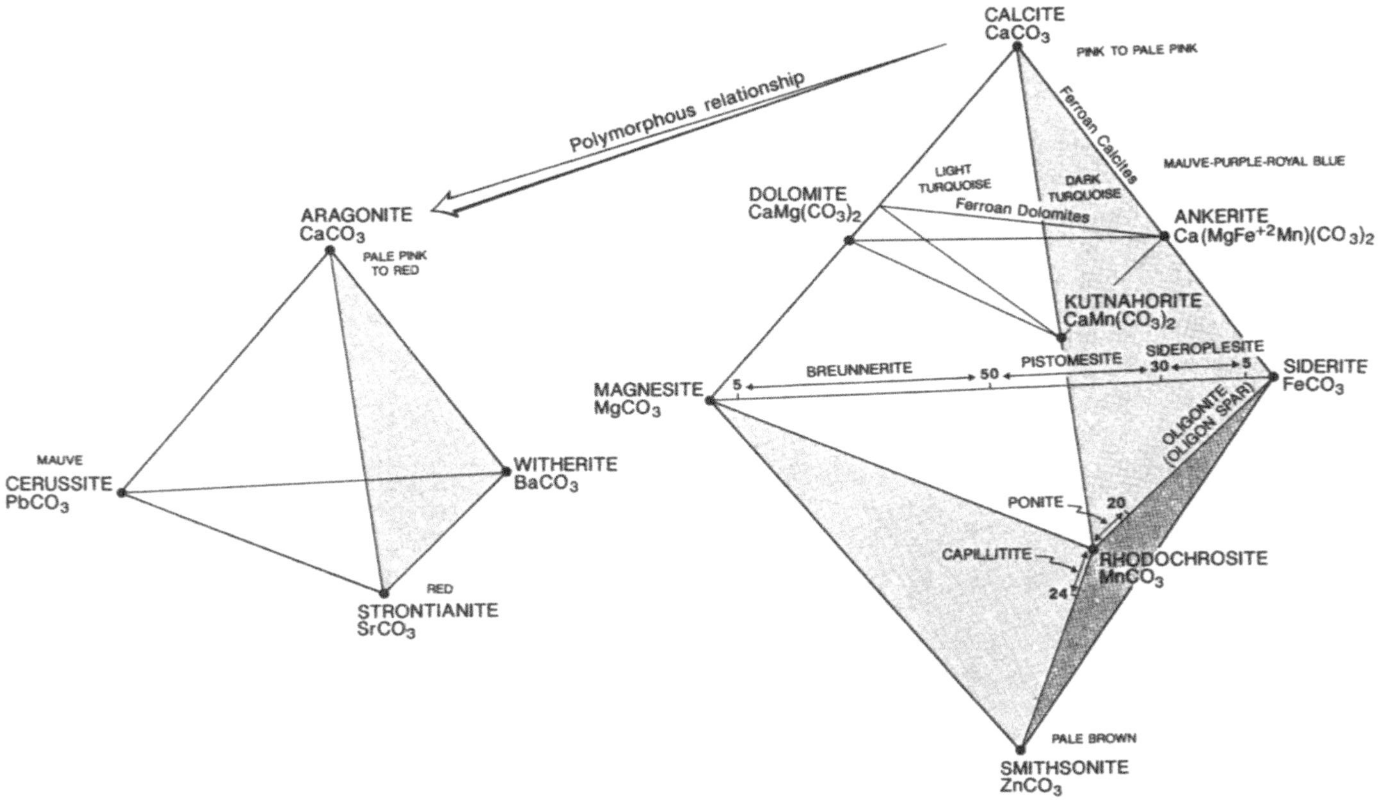

Figure 22.4. A diagrammatic interpretation of the solid solution series relationships between the end-members of the carbonate group of minerals. Superimposed are the dye colours resulting from an Alizarin Red-S and Potassium Ferricyanide (3:2) combination. A visual relationship between calcium-iron-magnesium substitution and the stain combination (following Dickson's 1965 procedure, see figure 22.3) can be seen in the Ferroan calcite and Ferroan dolomite series (diagram from Green 1993).

The stain is sensitive to Mg-calcite levels of 3% or more $MgCO_3$. With increasing Mg content the calcite takes on a deep red colour, but this is also dependent on the crystal orientation and particle size; c-axis normal sections through crystals stain more vividly, as do finer components (micritic cements and pellets) than parallel sections.

22.6.4. Double staining technique for carbonate thin sections

The combination of two staining schemes on the same slide has the advantage of allowing for two independent determinations of carbonate components within the slide (Green 1993). The two staining schemes used in this study are the combination of (i) alizarin red-S and potassium ferricyanide (Dickson 1966) for determining Fe/Mg carbonates; and (ii) the Titan yellow stain for the determination of Mg-rich calcite following the method of Choquette and Trussell (1978), in which the stain was made permanent. Thin sections must be mounted in an epoxy resin, as both Lakeside and Canada Balsam are soluble in the Titan yellow stain solution. Cover slips must also be epoxy or polyester resin mounted.

Carbonate thin sections should be prepared in the standard way, minus the cover slip. However, great care must be taken as thin sections will usually be at a thickness of 30μm, and can easily be etched away. If possible have thin sections prepared to a thickness of between 35 to 40μm to allow for etching and staining.

1 Etch one third of the section in a 0.2-0.5% HCl solution for about 10 seconds (see section 22.4.2 above for explanation).

2 Wash in and dry section.

3 First stain one third of the slide with the combination of alizarin red-S and potassium ferricyanide (3:2 ratio)

Prepare stains as follows:

Alizarin Red-S

i Dissolve 0.3g of alizarin red-S in 150ml of 15% HCl.

Potassium Ferricyanide

i Dissolve 2g of potassium ferricyanide in 100ml of 15% HCl.

4 Mix the two solutions in a ratio of 3:2 (alizarin red-S : potassium ferricyanide).

5 Decant a volume of the solution to the staining vessel. Dilute the mixture with the addition of an equal volume of distilled water.

The staining procedure is as follows:

6 Immerse the etched third in the stain solution for 20-30 seconds. See comments in section 22.4.2 above for guidance on time of immersion and stain concentration.

7 Remove specimen from stain, lightly wash under running water, and allow to air dry. Remember that stain is water soluble, so avoid touching stained area.

Stain the other end of the slide (one third) using Titan yellow.

The solutions are prepared as follows:

Titan yellow

i Dissolve 1g of Titan yellow powder, 8g of sodium hydroxide (NaOH) pellets and 4g of ethylenediaminetetra-acetic acid disodium salt (diNaEDTA) in 1 litre of distilled water.

N.B. This solution has a shelf life of approximately two years if stored in a dark bottle.

Stain fixer

i Slowly dissolve 200g of NaOH pellets in 1 litre of distilled water. **CAUTION: HEAT AND FUMES ARE EVOLVED DURING THIS REACTION, AND THE SOLUTION IS CORROSIVE.**

N.B. This solution has an indefinite shelf life, but must be stored in polythene bottles as it will etch glass.

The staining procedure is as follows:

8 Etch thin section in a 5% acetic acid ($CH_3.COOH$) solution for 30 seconds. Allow surface to dry.

9 Immerse section in stain solution for about 20 minutes.

10 Dry surface in warm air.

11 Immerse stained surface in fixer solution for 30 seconds.

12 Dry surface in warm air, cover thin sections immediately, using an epoxy mounting media or a polystyrene resin mountant.

Results

An unstained portion should remain in the middle of the slide, with stained thirds to each side. The calcite of the area stained with Titan Yellow will be pink to pale red, depending on the concentration of $MgCO_3$ in the 5-8% range. The stain is sensitive to Mg-calcite levels of 3% or more $MgCO_3$. With increasing Mg content the calcite takes on a deep red colour, but this is also dependent on the crystal orientation and particle size; c-axis normal sections through crystals stain more vividly, as do finer components (*e.g.* micritic cements and pellets) than parallel sections.

The third of the section stained with the combination of alizarin red-S and potassium ferricyanide will appear mauve, purple or red, depending on the amount of iron and magnesium present. An increase in the iron content within the calcites is evident in a colour change from mauve to royal blue and purple. In dolomites an increase in the iron content is evident by a colour change from light to dark turquoise.

A summary of the results from both stained portions is given below:

Fe free/Mg rich calcite	:	stained pink or red.
Iron poor calcite	:	stained mauve.
Iron rich calcite	:	stained royal blue or purple.
Iron free dolomite	:	remains unstained.
Iron poor dolomite	:	stained light turquoise.
Iron rich dolomite	:	stained dark turquoise.
Aragonite	:	stained pink or red.

REFERENCES

CHOQUETTE, P. W and TRUSELL, F. C. 1978. A procedure for making the Titan-yellow stain for Mg-calcite permanent. *Journal of Sedimentary Petrology*. **48**, (2), 639-641.

DICKSON, J. A. D. 1965. A modified staining technique for carbonates in thin section. *Nature*. **205**, 587.

DICKSON, J. A. D. 1966. Carbonate identification and genesis as revealed by staining. *Journal of Sedimentary Petrology*. **36**, (2), 491-505.

FEIGL, G. 1937. *Qualitative analysis by spot tests*. Nordemann Publishing Co., (New York). 400p.

FRIEDMAN, G. M. 1959. Identification of carbonate minerals by staining methods. *Journal of Sedimentary Petrology*. **29**, (1), 87-97.

FRIEDMAN, G. M. 1971. Staining, pp 511-530: In; R. E. Carver (ed), *Procedures in Sedimentary Petrology*. Wiley-Interscience (New York). xiii +653p.

GREEN, O. R. 1993. The use and applications of stains and dyes in geological techniques. *Science Technology*. **28**, 2-26.

HEEGER, J. E. 1913. Ueber die mikrochemische Untersuchung fein verteiler Carbonate im Gesteinsschliff. *Centralblatt fur Mineralogie, Geologie und Paläontologie*. 44-51.

HÜGI, TH. 1945. Gesteinsbildend wichtige Karbonate und deren Nachweis mittels Fäbmethoden. *Schweizerische Mineralogische und Petrograhische Mitteilungen*. **25**, 114-140.

MANDADO, J. A and TENA, J. M. 1986. A peel technique for sulfate (and carbonate) rocks. *Journal of Sedimentary Petrology*. **56**, (4), 548-549.

MILLER, J. 1988. Microscopical techniques: 1. Slices, slides, stains and peels, pp 86-107: In; M. Tucker (eds), *Techniques in Sedimentology*. Blackwell Scientific Publications. ix + 394p.

REID, W. P. 1969. Mineral staining tests. *Mineral Industrial Bulletin, Colorado School of Mines*. **12**, (3), 1-20.

WARNE, S. S. J. 1962. A quick field or laboratory staining scheme for the identification of the major carbonate minerals. *Journal of Sedimentary Petrology*. **32**, (1), 29-38.

WILSON, M. A and PALMER, T. J. 1989. Preparation of acetate peels, pp 142-145: In; R. M. Feldman, R. E. Chapman and J. T. Hannibal (eds), *Paleotechniques*. The Paleontological Society Special Publication. **4**. iv + 358p.

WINLAND, H. D. 1971. Non-skeletal deposition of high-Mg calcite in the marine environment and its role in the retention of textures: In; O. P. Bricker (ed), *Carbonate Cements*. John Hopkins University Studies in Geology No. 19, (Baltimore). 278-284.

WOLF, K. H, EASTON, A. J and WARNE, S. 1960. Techniques of examination and analysis sedimentary carbonate skeletons, minerals and rocks, pp 253-341: In; G. V. Chilingar, H. J. Bissell and R. W. Fairbridge (eds), *Carbonate rocks: physical and chemical aspects. Developments in sedimentology 9B*. Elsevier Publishing Co. 413p. (Staining techniques pp 259-272).

23. PREPARATION OF AMBER SPECIMENS CONTAINING FOSSILS

23.1. INTRODUCTION

Fossilized tree resins range from the Carboniferous to Recent times. Deposits of fossil amber are known from the Lower Cretaceous to Early Tertiary times, and are generally products derived from the extinct conifer *Pinus succinifera*. The most famous deposits are from the Baltic, in which a renewed commercial interest has been shown following the opening-up of the Eastern Europe market and commercial access to localities of derived (allochthonous) deposits. The palaeobiological significance of amber deposits as a media for the exceptional preservation of a fossil assemblage (*Fossil-Lagerstätten*) has been briefly reviewed by Schlüter (1990), and appraised in popular fashion by Poinar and Poinar (1994).

23.2. ORIGIN AND FORMATION

Resins stratigraphically younger than one million years old are generally known as copal, of which New Zealand kauri deposits, exuded by the conifer *Agathis australis* are an excellent example (Thomas 1969). Even younger resin deposits (a few hundred years) are known as Gum. However, despite the long history commercial exploitation of fossil resins, remarkably little is known on the chemistry and geochemical transformations from gums and resins to amber (Poinar 1992). Geochemical similarities to modern plant resins, from which it originates, are recognised. Furthermore, resins are not restricted to coniferous plants, but occur in a wide range of flowering plants (angiosperms).

Although secreted by the plant's parenchyma cells (thin walled cells, typical of the cortex and pith), a single tree may produce several types of resin. Thomas (1969) recognised five different types of resin from the kauri tree of New Zealand, each chemically different and therefore possessing different properties in relation to rates of decay. Fresh resins (oleoresins) are a complex mixture of terpenoid hydrocarbon compounds, acids and alcohols (Poinar 1992). These "essential oils" are very volatile, and at normal environmental temperatures are lost as the resin ages. Naturally exuded resins are normally broken down by physical and biological processes and disintegrate. The conversion of resin to amber requires unique preservational conditions, where sufficient oxidation and polymerisation, the process by which monomers (small molecules) chemically combine to produce polymers (large molecules), results in the resin becoming harder (copal). Poinar (1992) distinguishes amber and copal on their physical properties (Table 23.1).

If the hardened copal is in a suitable environment, protected from extensive oxidation, it may, after a few million years, transform into amber. The rate of transformation varies, depending upon the prevailing physical and biological conditions. All known amber beds are, or have been, associated with a marine incursion and related deposits (Poinar 1992). It has been suggested that sea-water maybe essential for amber formation, altering the oxygen content and lowering temperature, or in providing suitable salts that at the correct temperature and pressure, catalyse the polymerization process (Poinar 1992). The processes that turn wood into coal may closely parallel those that turn resin into amber, particularly as many amber deposits occur in association with low grades of coal.

INCREASING AGE (oxidation and polymerization) →			
Physical properties	**Resins, gums & saps**	**Copal**	**Amber**
Temperature range (°C melting)	Viscous at room temperature	<150	200-380
Strength	Moldable	Non-moldable	Non-moldable
Solubility (i) water (ii) acetone	Soluble	Soluble (sticky surface)	Insoluble Insoluble
R.I.		1.5-1.6	1.542±0.003
S.G. (increases with age)		1.03-1.08	1.04-1.10
Fracture			Conchoidal, no cleavage
Hardness			2-3
Tenacity			Brittle
Lustre			Resinous to greasy
Fluorescence (fresh, broken surface)		Faint, light sheen	Distinctly bluish
Ignition		Sputting flame, white smoke	Steady flame, black smoke
Odour (when burning)		Sweet, lemon, resinous	Acrid, burnt, resinous
Piercing with hot needle		Melts around entry, white smoke	Does not melt, becomes granular or gelatinous, dark smoke

Table 23.1. Comparison of the physical properties of copal and amber (data from Poinar 1992). Although amber occurs in a range of colours (yellow, brown, reddish or whitish), transparency ranges from transparent to translucent, while the streak is always white. Crystalographically amber and gum products are amorphous, as they have neither a regular structural arrangement nor characteristic shape, and their properties are the same in all directions.

Because the complex, and little understood, process of formation, the age of amber cannot be determined by direct analysis. Dating is achieved by examining the enclosed fossils, or those present in the related marine deposits. An extensive range of biological inclusions in amber has been reported (Poinar 1992, Poinar and Poinar 1994, Rice 1980). These have included examples from the plant kingdom (bacteria, algae (slime and moulds), fungi, bryophytes, pteridophytes, gymnosperms, angiosperms), and animal kingdom (protozoa, annelids, mollusca, arthropods (crustacea, insects, arachnids), vertebrates (amphibians, [frogs], reptiles [lizards], bird feathers and small mammals)). The exceptional preservation and rapid entombment of specimens provide palaeobiologists with excellent examples of palaeosymbiosis, and palaeontologists with unique examples of microevolution at the species level.

Other products that resemble Baltic amber, but with a slightly different chemical composition, are known from Romania (rumanite), Sicily (simetite) and Myanmar (formerly Burma, burmite). Amber from Upper Cretaceous sediments of Hrádek (Hrebec) in the Czech Republic is known as valchovite.

23.3. PREPARATION PROCEDURES

The semi-precious nature of amber means that preparation procedures are not widely published, as it is not in the interests of commercial jewellers to divulge or even discuss techniques employed. However, many of these procedures are directly related to the gemological appearance of specimens, and understanding them is not essential to the palaeontological conservation of specimens (Rice 1980). The brief outline that follows is centred on amber specimens conservation and preparation to enhance their qualities for photographic, optical observations and a technique for extracting well-preserved fossilized inclusions from amber.

23.4. CUTTING AND POLISHING

Cutting and polishing procedures employed on amber specimens are similar to those used in the preparation of acrylic plastics. Amber is not sectile, and therefore can not be cut into slices or shavings (Poinar 1992). Specimens can be cut dry using a junior hacksaw or jewellers saw with a very fine blade. Slow and even strokes should be used. Files and sandpaper can also be used in surface preparation, although the former have to be continually cleaned with a stiff metal or nylon brush, as they continually clog (Rice 1980). Softer copal can be prepared using a water-proof abrasive paper. This aids in keeping the surface cool, particularly as friction generated during polishing can cause the surface to become sticky. Generally, however, the soft nature of copal means that it does not polish well.

Mechanical cutting and grinding procedures are not recommended, and should be avoided. The use of a high revolution blade can induce a number of catastrophic and irreversible problems to the specimen. High-speed friction imparts excessive strain in the surface layers of the specimen that can result in it shattering, or overheating and the generation of a rippled "orange-peel" effect (Rice 1980). Continued action results in further weakening of surface layers as the melting point is approached and volatile compounds escape. The generation of a negative electric charge results in a build-up of static electricity, which can give the preparator a mild shock. Excessive friction, from both cutting and polishing, can be recognised by the characteristic odour generated by the procedure. If a mechanical cutting or polishing wheel is used, the motor speed should not exceed 100rpm.

Surface polishing of amber should be undertaken using an aluminium oxide abrasive and a leather chamois. Diamond pastes should be avoided, as these cause the surface to become dull and cloudy (Rice 1980). Use only those felt pads specifically dedicated to amber preparation, thus avoiding diamond paste contamination of specimen surfaces.

23.5. ARRESTING SURFACE DEGRADATION

Natural degradation of the outer surface of specimens occurs when they are exposed to air. Oxidation to a "garnet red" colour is evident in specimens incorrectly stored (Lichter 1993). This may be accompanied by surface decomposition, initially evident by the development of a network of fine hairline cracks. If development continues unabated the outer layer may begin to crumble, and the red colouring intensifies. Correct storage is essential to avoid this development, although colour enhancement of specimens is possible and frequently undertaken.

Translucent, milky or cloudy amber is the result of fine microscopic bubbles present within the amber. Specimens can be made transparent by immersion in a clear oily solution of benzoin ($C_{14}H_{12}O_2$) (Lichter 1993). **Extreme care must be taken when using this chemical, procedures should be undertaken in a fume cupboard, or well-ventilated area, as fumes can cause breathing problems.** A much safer procedure is to clarify the specimen by immersing it in a rape seed oil, and slowly raising the temperature to its boiling point, and then allowing it to cool. The hot oil penetrates the fine cracks and pores of the specimen, replacing the air and thereby making it transparent (Poinar 1992).

A technique for enhancing the clarity of specimens has been suggested by Azar (1997), and is summarised as follows:

1. Wash amber specimens in a mixture of water and sodium hypochlorite ($NaClO$) to remove clay matrix. Allow specimens to air dry.

2. Immerse specimens in glycerol ($CH_2OH.CHOH.CH_2OH$) or a glycerol (40%) alcohol (60%) solution, and examine under a stereozoom binocular microscope. Alcohol penetrates any micro-fractures, resulting in greater clarity during optical examination (Azar 1997).

3. Suitable pieces can then be ground on a series of carborundum impregnated cloths, until the amber is transparent.

4. Coat the specimen surface with acetone varnish or Canada Balsam. This will further enhance surface details, allowing for more accurate identifications to be performed.

5. For the final optical examination and photography, immerse specimens in oil of cedarwood to enhance detail (Azar 1997).

23.6. COLOUR ALTERATIONS

The examination and photographing of the entrapped organisms within amber can often be enhanced by modifying the natural colour or its general appearance. Two main procedures can be employed, and will either lighten or darken the amber.

23.6.1. Colour reduction

Reducing the colour intensity in amber can be achieved by unheated autoclaving of specimens. This effectively drives residual oil out into pockets of more porous material. Specimens should then be hardened by heating. Initially the heat softens the amber, and gas inclusions expand forming small discoidal fractures. Heating increases the oxidation process, usually in a concentric manner from the surface to the specimen core. Consequently, outer layers usually have a deeper colour than inner layers (Rice 1980). Exposure of specimens to a strong source of ultraviolet light can also cause a colour reduction (Lichter 1993), but has the disadvantage in turning surface layers into a white powder.

23.6.2. Colour enhancement

Increasing the colour intensity can be achieved by placing the specimens in a pure sand base within an iron pot, and heating for 30-40 hours. This will darken the amber producing a rich brown colour, and deep red if left. The colour is characteristic of older naturally oxidized specimens. Many prepared specimens are artificially aged in this manner, and care must be taken when purchasing material from a previously unheard of locality.

23.7. CLEANING AND STORING SPECIMENS

Specimens should be cleaned using a soft cloth, dampened with luke-warm water. Avoid solvents, even on samples with a surface coating of varnish, and any jewellery cleaning solutions. Never use ammonia based cleaning solutions, soap or even the mildest detergent.

Controlling the microenvironment of amber specimens is vital for their preservational longevity. Long-term storage of specimens should be in air-tight containers, or in mineral oil or glycerol (Poinar 1992). These should be periodically checked to ensure there has been no leakage or evaporation, and topped up as required. Small specimens can be embedded or coated in clear polyester resin (Lichter 1993). Epoxy resins, with their mild yellow-tints, may mask the natural colour of amber. Specimen surfaces can also be coated with an air-resistant wax or a clear polyurethane varnish.

23.8. EXTRACTING ORGANISMS FROM AMBER

Galippe (1920) recognized that the immersion of amber specimens, for up to 96 hours, in diethyl ether ($(C_2H_5)_2O$) usually resulted in the softening of the amber, making it easier to cut. Partial dissolution of the amber using alcohol, was attempted by Larsson (1978), but frequently resulted in the destruction of the fossil. Recently, Azar (1997) has described a method of extracting organic remains from Lebanese amber by dissolving them in chloroform ($CHCl_3$). The procedure is summarized below:

1 Treat specimens individually by placing them in containers with approximately 20ml of chloroform (sufficient to cover the specimen). Seal the container with an air tight lid to avoid solvent loss through evaporation.

2 Monitor the reaction, as organisms are usually freed from the amber within 2 hours.

3 Freed components are as subtle as freshly collected material (Azar 1997), although their translucency can be enhanced, before permanent mounting, by a 30-60 minute immersion in sodium hypochlorite.

The above method proved to be more satisfactory than heating methods, which frequently resulted in specimens cracking and burning (Azar 1997). Both ethanol (C_2H_5OH) and butan-1-ol ($CH_3.(CH_2)_3.OH$) proved ineffective, while acetone ($(CH_3)_2CO$), toluene ($C_6H_5.CH_3$) and 1,1,1-trichloroethane ($CH_3.CCl_3$) succeeded only in softening the amber (Azar 1997).

Azar (1997) reports considerable success with this method in freeing exceptionally well preserved plant fragments and insect structures, and speculates on the possible use of this material in ancient DNA studies as contamination with other inclusions is minimised. Testing of the method on amber from other localities is now required, and must follow Azar (1997) by experimenting with less valuable and imperfect specimens to avoid the accidental destruction of good quality material.

23.9. PALAEOBIOLOGICAL APPLICATIONS

The unique preservational qualities of amber provide palaeontologists with organisms of a quality not found by any other means of fossilization (Henwood 1992). Nature has essentially used its own methods of fixation and dehydration (Poinar and Poinar 1994), although an element of good fortune has ensured that the amber has not been destroyed during its subsequent geological history. If sufficient quantities (and quality) of amber entombed organisms can be recovered from an area, then extensive bio- and genetic diversity studies can be pursued. Biodiversity studies provide data for palaeoecological reconstruction and habitat organisation, particularly from high organism diversity ancient rain forests, similar to those in existence today (containing over half the plant and animal species in existence). In turn, supporting evidence for climate change and palaeogeographical distribution is provided.

Poinar and Poinar (1994) recognized the association between biodiversity and genetic variation. Defining this as the genetic make-up of a species accumulated within similar breeding organisms, genetic diversity is reduced when a population ceases to gain new genetic material from neighbouring members. In turn this could initiate a slow (or rapid) decline to extinction via increased predation, habitat loss or climate change. A broad range of genetic material is required to maintain a stable population, and as Poinar and Poinar (1994) indicate "essential to the long-term survival of life".

Early attempts at TEM examination of amber enclosed specimens resulted in labour intensive preparations of sections, and results of variable quality that invariably ended with damage to the specimen and contamination of the microscope column as specimens degassed (see Poinar and Poinar 1994 for review). Overcoming these problems enabled the finer details of organism ultra-structure to be studied (Poinar and Hess 1982, Poinar and Poinar 1985), and through differentiating structures and cell organelles within the nucleus initiated studies of ancient DNA (deoxyribonucleic acid). Careful laboratory procedures and improved methodology have reduced problems of sample contamination, and improved the reliability of extracted DNA (Pääbo 1993, Poinar *et al.*, 1994).

Sequencing (the procedure for determining the sequence of a nucleic acid), cloning (procedures by which a self replicating DNA molecule can accept an inserted DNA sequence and multiply), and amplification (making multiple copies of DNA sequences) of very small amounts of DNA have enabled techniques to be applied to minute fragments of fossilized material. Consequently, it is now possible to use the procedure to accurately identify an organism. This is particularly if morphological features are obscured, missing or cannot be recognized (Canno *et al.*, 1992a & b, DeSalle *et al.*, 1992, Pääbo 1993, Poinar *et al.*, 1993). The procedure can also be used to confirm the phylogenetic affinities, determining the relationship between an existing organism and its present day descendants. Furthermore, if the rate of change is known, ancient DNA sequences can be used as "molecular clocks" in estimating the time necessary for species separation, providing additional evidence to support any palaeogeographical reconstruction (Poinar and Poinar 1994). Advances in microbiological research will doubtless govern the applications of ancient DNA studies in palaeobiology. The palaeobiologist has to ensure that material studied and passed to the microbiologist for analysis is genuine, with original fossils enclosed in genuine amber, and collected from a known amber occurring locality. With the full potential and unique preservational nature of amber enclosed fossils now recognised, only specimens with a known pedigree and well-documented history of preparation can provide reliable analytical data.

23.10. AMBER IMITATIONS

Although colour alterations can be employed to falsify the original geological history of a specimen, outright imitations are a more serious problem. Coloured plastics, sold and authenticated as amber by "locals" in amber producing areas, can even have experts fooled until laboratory verification is undertaken (*e.g.* Poinar and Poinar 1994). A simple test for distinguishing fake amber uses a saline solution, and is prepared as follows:

1 Dissolve 2.5 tablespoons of salt in a cup of water.

2 Carefully place the amber fragments in the solution. If genuine the fragments will float. Imitation material will sink to the bottom of the cup.

Copal is frequently passed off as amber by unknowing dealers. Copal can easily be moulded around insects and used an embedding media. Chemical treatment can leave specimens transparent, and after spraying with shellac, with properties (*e.g.* specific gravity 1.07, refractive index 1.54) very similar to that of natural amber.

Some amber specimens are devoid of any fossil inclusions. However, in some specimens it has been shown that an artificial cavity, now filled with copal, was drilled, and a small insect inserted before the hole was plugged. Only careful optical examination can detect the modification (Poinar 1982). Occasionally, small pieces of amber cut from prepared specimens are reheated under increased pressure and pressed or moulded into a form known as ambroid.

A wide range of inclusions in amber effect its opacity in transmitted light, from completely transparent to virtually opaque. Inclusions of carbon and organic debris within the amber can give the appearance of it being black. However, amber never naturally occurs black, but ranges in colour from light yellow to deep red-brown. Specimens sold as "black amber" are more than likely examples of jet a resinous hard variety of lignite (fossilized wood), capable of taking a high polish.

REFERENCES

AZAR, D. 1997. A new method for extracting plant and insect fossils from Lebanese amber. *Palaeontology*. **40**, (4), 1027-1029.

CANNO, R, POINAR, H, ROUBIK, D. W and POINAR, Jr., G. O. 1992a. Enzymatic amplification and nucleotide sequencing of portions of the 18s rRNA gene of the bee *Proplebeia dominicana* (Apidae: Hymenoptera) in 25-40 million year old Dominican amber. *Medical Science Research*. **20**, (7), 249-251.

CANNO, R, POINAR, H. N and POINAR, Jr., G. O. 1992b. Isolation and partial characterization of DNA from the bee *Proplebeia dominicana* (Apidae: Hymenoptera) in 25-40 million year old amber. *Medical Science Research*. **20**, (17), 619-623.

DESALLE, R, GATESY, J, WHEELER, W and GRIMALDI, D. 1992. DNA sequences from a fossil termite in Oligo-Miocene amber and phylogenetic implications. *Science*. **257**, 1933-1936.

GALIPPE, V. 1920. Recherches sur la résistance des microzymas à l'action du temps et sur leur survivance dans l'ambre. *Comptes Rendus de l'Académe des Sciences, Paris*. **170**. 856-868.

HENWOOD, A. 1992. Exceptional preservation of dipteran flight muscle and taphonomy of insects in amber. *Palaios*. **7**, (2), 203-212.

LARSSON, S. G. 1978. Baltic amber - a palaeobiological study. *Entomonograph*. **1**, 1-192.

LICHTER, G. 1993. *Fossil Collectors Handbook*. Sterling (New York), 160p.

PÄÄBO, S. 1993. Ancient DNA. *Scientific American*. (November), 60-66.

POINAR, Jr., G. O. 1982. Amber - true or false? *Gems and Minerals*. (April, No. 534), 80-84.

POINAR, Jr., G. O. 1992. *Life in Amber*. Stanford University Press (California). xiii + 350p.

POINAR, Jr., G. O and HESS, R. 1982. Ultrastructure of 40-million-year-old insect tissue. *Science*. **215**. 1241-1242.

POINAR, Jr., G. O and POINAR, R. 1985. Preservative qualities of recent and fossil resins: electron micrograph studies on tissue in Baltic amber. *Journal for Baltic Studies*. **16**, 222-230.

POINAR, Jr., G. O and POINAR, R. 1994. *The quest for life in Amber*. Addison-Wesley (Reading, Massachusetts). xiii + 219p.

POINAR, Jr., G. O, POINAR, H. N and CANNO, R. 1994. In; B Herrmann and S. Hummel (eds), *Ancient DNA*. Springer-Verlag (New York).

POINAR, H. N, POINAR, Jr., G. O and CANNO, R. 1993. DNA from an extinct plant. *Nature*. **363**, 677.

RICE, P. C. 1980. *Amber: the golden gem of the ages*. The Kosciuszko Foundation Inc., (New York). x + 289p.

SCHLÜTER, T. 1990. Baltic amber, pp 294-297: In; D. E. G. Briggs and P. R. Crowther (eds), *Palaeobiology - A synthesis*. Blackwell Scientific Publications (Oxford). xii + 583p.

THOMAS, B. R. 1969. Kauri resins - modern and fossil, pp 599-618: In; G. Eglinton and M. T. J. Murphy (eds), *Organic Geochemistry*. Springer-Verlag (New York). 828p.

24. PREPARATION AND CONSERVATION OF VERTEBRATE FOSSILS

24.1. INTRODUCTION

This section outlines techniques specific for the preparation of vertebrate fossils. The objectives in preparing invertebrate and vertebrate fossils are frequently similar, and so are the mechanical and chemical methods used in preparation. Differences result in the size of material under preparation, and scale of the procedures employed. There are also differences in the laboratory equipment used in monitoring and observing the effects and results of preparation.

A considerable amount of thought has gone into describing and explaining the modern techniques used in preparing vertebrate fossils (*e.g.* Converse 1989, Leggi and May (eds) 1995). Consequently, many consider this branch of palaeobiological preparation a specialized science in itself. Certainly it is one in which objectives are not be satisfactorily attained without gaining considerable experience from persistent and repetitive practice. The following account of methods and procedures does not attempt to compete with the excellent dedicated texts of Converse (1989) and Leggi and May (eds, 1995), but complement them by providing a comprehensive introduction. From this it will become evident that many procedures are adaptable to (and have been adapted from) invertebrate and micropalaeontological techniques.

Techniques outlined within this section detail the specialized field and laboratory methods used in preparing vertebrate bones. Particular reference is made to the preparation of skulls, partial and complete skeletons (including fish), and late Tertiary age sub-fossilized bones. Both mechanical and chemical methods are referred to, and although repetition of methods with other sections is kept to a minimum, some is inevitable for the continuity of the techniques under discussion. Reference is made to other sections where procedures are explained in greater detail.

24.2. FIELD COLLECTING OF VERTEBRATE SKELETONS

The success of prospecting and collecting for large vertebrate fossils relies in part on the meticulous site recording of previous generations of geologists, particularly the Victorian collectors. These have enabled modern collectors to return, recollect and re-examine sites and outcrops. Major technological advances (*e.g.* mechanical diggers and excavators) have been introduced to assist in the collecting of material (Greenwald 1989).

The careful plotting of float specimens recovered from scree slopes may result in finding *in-situ* material. However, complete skeletons are extremely rare, and their removal requires considerable skill and patience, and occasionally the use of explosives (Greenwald 1989). Physical strength may also be required, even when used in conjunction with a pulley systems and heavy lifting gear.

Of course not all vertebrate material is of macro-size. Stamina and strength will also be required for processing large amounts of gravel and sand size sediment for micro-invertebrate material (Frizzell 1965, Novacek and Clemens 1977, Von Hofe Grady 1979). In addition a great deal of patience and a well-trained eye is necessary in detecting the rare bones and teeth.

Multiple boned skeletons cannot, and should not, be excavated in haste. Field collecting, packing and transportation procedures must be approached methodically. Although these have been described in an earlier section for individual fossils (see part II FIELD TECHNIQUES), skeletons or articulated assemblages of bones require careful preparation.

24.2.1. Pre-preparation

1 Construct a detailed and accurate plan of the site, and number each component.

2 Photograph and illustrate each of the components *in-situ*.

3 If individual bones cannot be extracted on a bone-by-bone basis, then jacketed blocks will have to be prepared (Rixon 1976, see also section 5 CONSOLIDATION, STABILIZATION AND REPLICATION OF SPECIMENS).

24.2.2. Skulls, pelvic & pectoral girdles

Skulls are essentially a collection of semi-fused bones enclosing a cavity. Consequently, and in common with the wide flat bones of pelvic and pectoral areas, they frequently suffer from crushing induced by the sediment overburden. Extraction and development of these bones, in both the field and laboratory, is an exceedingly long and laborious process, requiring delicate mechanical development, strengthening and consolidation, and replication (moulding and casting) (*e.g.* Whybrow 1982, see below, Gower and Sennikov 1996).

Standard consolidation methods are usually not sufficient to strengthen these types of bones. Alternative methods involve the use of a viscous poly(vinyl butyral) adhesive (*e.g. Butvar* B76) dissolved in acetone and supported by a translucent light woven glass fibre strand mat (Rixon 1976). The method is as follows:

1 Clean the bone surface of all loose material, and treat with a low viscosity poly(vinyl butyral) (*e.g. Butvar* B98). This solution soaks into the bone and aids in strengthening. Allow the consolidant to harden.

2 Cut strips of strand glass fibre mat to suitable lengths and widths.

3 Starting at the edge of the bone, coat the bone with a solution of B76 dissolved in acetone. Although this solution is viscous it is mobile enough to impregnate the mat and allow for positioning. Place a second strip on top of this and liberally coat with the solution. Stipple the cloth lightly down to the bone surface. A small border should be carried over to all edges of the bone.

4 Repeat the process, ensuring a slight overlap of the strips, until the entire specimen is covered.

5 Allow the covering to dry hard.

Once consolidated large bones can be encased in a plaster jacket (see section 5 CONSOLIDATION, STABILIZATION AND REPLICATION OF SPECIMENS), labelled top and bottom and transported to the laboratory.

Laboratory development of the cranium from the holotype specimen of *Archaeopteryx lithographica* by Whybrow (1982) required a variety of specialized procedures, specifically developed for the preparation of this priceless specimen. The main concern in preparing a fossil of this type, is protection. Fragile and delicate areas were covered with tissue paper and foam rubber.

Conventional chemical methods of development (detailed below) could not be used for fear of destroying the delicate feather impressions and causing structural damage severely weakening the bones. The procedure employed by Whybrow (1982) can be broken down into five clearly defined stages:

1 **Defining the bone:matrix boundary** - an x-ray of the main slab revealed no additional bones concealed by matrix surrounding the cranium.

2 **High stock removal** - initially intended to remove the cranium by excavating through 60mm of limestone matrix. Subsequently decided the simplest and safest method was to cut around and underneath the cranium, removing it as a block from the main slab.

3 **Mechanical development** - removal of delicate matrix was performed under a stereozoom binocular microscope at x40 magnification, modified so that foot operated switches controlled focus and magnification (Whybrow 1978). A sharpened tungsten carbide rod mounted in a dental percussion mallet and hand held carbon steel mounted needles were used to expose the bone.

4 **Additional specimen support** - prepared areas were replicated by casting in epoxy resin from RTV silicone rubber moulds. The areas were then embedded in a water soluble polyethylene glycol wax to provide additional support, and enable other areas to be developed.

5 **Consolidation on completion of preparation** - soluble wax was removed by immersion in cold water. Dried bones were then coated with a cyanoacrylate hardener.

This delicate preparation has provided a virtually uncrushed skull (Whybrow 1982; pp 190-191, fig 4a-c) which can be used for comparison with other specimens to aid in studies on ancestry (Whybrow 1982).

24.3. FIELD COLLECTING OF MICROVERTEBRATE MATERIAL

Of increasing importance in the study of vertebrate fossils is the recovery of small specimens, including bone fragments, teeth and otoliths (ear bones). Specialized collecting and processing procedures have to be employed. Adapting methods used in the mining industry, "screen-washing" is the process by which large quantities of material are washed, dried and examined for fossil fragments. Ideally processing should be carried out in the laboratory. However, the remoteness of the field area combined with the cost of transporting many hundreds of kilograms of sediment to the laboratory may result in preparation at or near the point of sampling.

The procedure was formally described by Hibbard (1949), who recognised that slightly different procedures are required when collecting from either new (working) or old (disused) quarries. Spoil heaps in old quarries can be systematically dry sieved using appropriate mesh sizes (*e.g.* 16mm, 8mm, 4mm *etc.*, see section 16 WASHING AND SIEVING TECHNIQUES USED IN MICROPALAEONTOLOGY). The effectiveness of separating fossil fragments in unconsolidated sand, silt and marls is increased when the sediment is dry (Miller 1989). Sampling in new quarries enables freshly exposed bedding planes to be examined for fragments, enabling bulk processing to be targeted at fossil rich horizons. However, the absence of fossil fragments may not necessarily indicate that the horizon is barren. Subsequent processing may yield a diverse semblage of both vertebrates and invertebrates (Hibbard 1949).

The procedure has remained essentially unchanged, although labour intensive field processing (McKenna 1962, 1965, Novacek and Clemens 1977), has through necessity been modified for smaller working groups using nylon bags, while retaining the high recovery rates of fossil material (Von Hofe Grady 1979). If the fossils are rare, large quantities of material (50-600kg) can be processed to leave residues of 8 to 12kg for laboratory examination (Frizzell 1965, Von Hofe Grady 1979). If materials are available washing boxes can be constructed on-site with wood and wire mesh or galvanised screen mesh bases of appropriate size (Hibbard 1949, McKenna 1962, 1965, Miller 1989). Lightweight nylon bags provide an effective alternative and, with a mesh size of approximately 0.08mm, can increase the recovery rate (Von Hofe Grady 1979). A composite wet sieving or screen washing method is as follows:

1. Bulk samples are soaked in water without reagent (in the laboratory used distilled water). The addition of deflocculant bases (*e.g.* sodium carbonate, sodium bicarbonate) will deteriorate the organic fibres (Frizzell 1965).

2. Fill the bags (or washing boxes) with sediment. Suspend the bags in a moderately flowing river, and gently agitate. Avoid an excessive scrubbing action of the sediment against the mesh as this may result in damage to the microfossils (Converse 1989). In the laboratory wet sieve through a nest of sieves (see section 16 WASHING AND SIEVING TECHNIQUES USED IN MICROPALAEONTOLOGY) of appropriate mesh size to retain the smallest fragments.

3. When clay fraction material ceases to wash out the sediment can be dried. The wet concentrate should be washed into a pan or onto absorbent cloth if the screen is required for supplementary samples. To avoid cross sample contamination thoroughly rinse the mesh bag or clean sieve screens before reuse.

4. Coarse elements of the fraction can be examined and picked with the unaided eye or hand lens. Finer fractions must be examined under a

stereozoom binocular microscope and picked with a fine brush or micro-dissection forceps. Some fragments (*e.g.* otoliths) are best picked when the residue is wet (Frizzell 1965).

5 Back in the laboratory, fossils can be further concentrated from the matrix by using liquid separation techniques (see section 19 FLOTATION AND LIQUID SEPARATION TECHNIQUES), while individual specimens can be further prepared. Additional cleaning can be undertaken by washing in water and brushing with a fine brush. Individual specimens can be placed in an ultrasonic bath in water with detergent (see section 12 MECHANICAL METHODS FOR PREPARING FOSSIL SPECIMENS). Acid preparation work (see below) can also be performed.

6 Small specimens can be stored in micropalaeontological slides, and larger specimens in small plastic trays with perspex lids.

24.4. LABORATORY PREPARATION PROCEDURES

Many of the laboratory procedures used in preparing vertebrate fossils have been described in detail in other sections. The procedures will not be repeated in this section, although methods pertinent to vertebrate preparation will be described.

24.4.1. Removing specimen jackets

Specimens and bones supported by expanded polyurethane foams can be removed with the aid of a sharp knife or heated wire. Trimming must, however, be undertaken in a well-ventilated area (see section 5 CONSOLIDATION, STABILIZATION AND REPLICATION TECHNIQUES). The ease at which a plaster jacket can be removed is very much dependent on the thickness of plaster covering the specimen and the success of consolidating procedures carried out during field preparation. Furthermore, a considerable amount of time may have passed since the specimen was jacketed and the initiation of laboratory preparation. Consequently, the maximum amount of information provided on the jacket label (*e.g.* orientation of fossil, thickness of paper and plaster) will assist the preparator during jacket removal.

One, or a combination, of two methods is used in cutting the jacket open (Converse 1989). Probably still the most common method is the use of a small hand held keyhole saw. Considerable care is required to ensure the cut penetrates only the plaster and the underlying paper covering the fossil. Hand held electric saws are common place in most preparation laboratories, and can be used in cutting the jacket. However, some electric tools may not provide the manual sensitivity provided by a hand saw. Converse (1989) describes the use of a *Stryker* electric cast cutter. This tool, designed for use in the medical industry for the removing of casts, has a 5cm diameter blade. It does not, however, operate in the conventional rotary manner, but by a rocking back and forth over short distances. This provides the operator with greater control, as the blade can be held without fear of being cut. Thus it is considerably safer than a rotary tool, and furthermore, generates less airborne dust particles.

No matter what tools, or combination of tools, are used in cutting, extreme care and good judgement are required by the preparator when opening the jacket. Very thick plaster jackets (4-6cms) may require removing with the aid of masonry bits and high-powered pneumatic saws. The appropriate protective safety equipment (*e.g.* eye protection, gloves, particulate dust mask) must be used by the preparator (see sections 9 SAFETY IN THE LABORATORY, 12 MECHANICAL METHODS FOR PREPARING FOSSIL SPECIMENS and Appendix 5 - RESPIRATOR AND FILTER CARTRIDGE COLOUR CODING), and great care taken not to damage the fossil.

Once the plaster jacket has been opened, the processes of development (matrix removal) and preparation can begin. This will involve both mechanical and chemical methods (Lörcher and Keller 1984). The underside of the specimen must be prepared first (refer to label for correct orientation, see section 5 CONSOLIDATION, STABILIZATION & REPLICATION OF SPECIMENS). Soft matrix can be loosened and removed with the aid of dental picks and soft brushes. Harder matrix will require the use of hand-held power tools (*e.g.* ultrasonic, airbrassive and pneumatic pens), uses of which are outlined in section 12 MECHANICAL METHODS FOR PREPARING FOSSIL SPECIMENS. Avoid excessive vibration of the specimen, particularly when more of the bone becomes exposed.

The removal of matrix must be closely observed so that as little damage as possible is inflicted on the specimen (Converse 1989). Small cracks easily develop within bone. These must be filled with a dilute solvent based consolidant (*e.g. Butvar* B76, *Paraloid* B72), avoiding contact with the matrix, and allowed to dry before proceeding with the preparation. Do not soften and wash clay grade sediment by the addition of water. This can result in bone fragments expanding and cracking. Furthermore, water is immiscible with solvent based consolidants, resulting in poor bonds between breaks, and turning the consolidant milky white (although this can be removed by applying an acetone wash to the affected area). Wet specimens must be dry before preparation proceeds.

Fragmented specimens should only be glued together when all the matrix has been removed (Converse 1989). Reassembling fragmented specimens may require reference to field notes, plans and photographs of the excavation site (see earlier). Careful reference to these notes, and preparing each fragment in numerical order, will avoid problems and reduce the potential for losing small individual pieces. This will provide some measure of control to the preparator.

24.4.2. Specimen consolidation and repair

Consolidants used in the repair of specimens and bones in the field are also extensively used in the laboratory. Bones are particularly susceptible to decay, and can deteriorate rapidly when surrounding matrix is removed and they begin to dry. Few, if any fossils, require no consolidation or repair during preparation (Converse 1989).

The poly(vinyl butyral) solutions of *Butvar* B98 (mixed with propan-2-ol (isopropyl) alcohol, $(CH_3)_2CHOH$) and *Butvar* B76 (mixed with acetone or alcohol, can be used as consolidants and adhesives respectively. Faster drying of the specimen is achieved when acetone mixed solutions are used. Several *Butvar* formulas exist providing resins with slightly different properties (Table 24.1). Of these B76 is the most versatile for palaeontological use.

Converse (1989) reminds preparators of the importance of mixing poly(vinyl butyral) solutions correctly. *Butvar* is manufactured in a crystal flake form, and will not fully dissolve if the solvent is poured onto a large volume of crystals. This is caused by the crystals rapidly developing a sticky gel coat, inhibiting crystals at the base of the container from completely going into solution.

Poly(vinyl butyral) resin System (Butvar)	**Mixed with & soluble in:**	**Viscosity* (cp)**	**Molecular weight (in thousands)**
B-72	Alcohol	7,000 - 14,000	170-250
B-74	Alcohol	3,000 - 7,000	120-150
B-76	Alcohol/acetone	500 - 1,000	90-120
B-79	Alcohol/acetone	100 - 400	50-80
B-90	Alcohol/acetone	600 - 1,200	70-100
B-98	Alcohol/acetone	200 - 400	40-70

Table 24.1. Poly(vinyl butyral) (PVB) - *Butvar*, technical data (from manufacturers' data). * viscosity of a 15% solution by weight of the polymer in 60:40 toluene:ethanol at 25°C.

24.4.3. Acid development (Harris 1971, Howie 1974, Lindsay 1987)

The preparation of calcium phosphatic vertebrate fossils in acidic solutions has been exploited by palaeontologists for the past 70 years. Principally this has involved the immersion of specimens in dilute aqueous acidic solutions (acetic or formic acid), dissolving any calcium carbonate and liberating the fossil from matrix. A number of advantages, in comparison with mechanical preparation, have been suggested (Lindsay 1987):

- removes the risk of mechanical collapse of the specimen resulting from the use of chisels and saws.
- provides a three-dimensional fossil which can be studied directly, and would be destroyed if serial sectioning or peel sections were made.
- provides a means of matrix removal from areas where mechanical tools cannot reach.
- exploits differential solution between the matrix and fossil to a finer extent than mechanical preparation.
- is less labour intensive than mechanical preparation, allowing the preparator to work on other projects while the acid slowly dissolves the matrix.

However, the advantages have to be clearly considered and weighed against the disadvantages. Acid vapours and fumes, even in weak concentrations, can present a health hazard to the preparator. **All procedures using acids must be performed in a fume cupboard.** Furthermore the action of the acid is uncontrollable, and can seriously weaken the structural integrity of the specimen, although detailed casting, particularly of cranial cavities, can be performed before individual elements become detached (Schaeffer 1965, Whybrow 1982). High quality preparations can be extremely fragile, requiring extensive consolidation and careful expert mounting to display (see section 38 ILLUSTRATION AND EXHIBITING FOR DISPLAY AND PUBLICATION).

The acetic acid procedure developed by Toombs (1948) and Rixon (1949), and successfully applied as part of the modified procedure in the technique known as the Transfer Method (Toombs and Rixon 1950). The transfer method involves attaching exposed areas of the fossil to a supportive medium, and then carefully dissolving the matrix. The procedure has been applied to bryozoans, graptolites, fossil plants and fish skeletons (Lindsay 1987). Where material cannot be transferred, or direct observation of the surface or both sides is required, the acid procedure (in combination with mechanical development) must be employed. A composite method of the acid technique is as follows:

24.4.3a. Pre-preparation

1 Remove loose matrix by brush or needle, or gentle mechanical preparation. Clean freshly broken surfaces of the fossil and re-apply fragments.

2 Fill any cracks with coloured filler, or temporarily with plasticine. Consolidate these and all areas of exposed bone or original fossil material. This stage of restoration will be repeated several times during the preparation.

These two procedures alone may take many hours to perform, and should be undertaken in a methodical manner.

24.4.3b. Acid immersion

Acetic ($CH_3.COOH$) and *Formic* acid ($H.COOH$)

Rixon (1976) suggests six categories for dealing with vertebrate remains (see section 27 EXTRACTION TECHNIQUES FOR PHOSPHATIC FOSSILS), depending upon the skeleton part under preparation and the completeness of the skeleton. Particular care must be used when working on material from fissures and cave breccias. Phosphatic bones from these deposits are commonly replaced or partially replaced by calcium carbonate that suffers partial or complete dissolution in acetic acid. Either acetic acid ($CH_3.COOH$, <15% solutions) or formic acid ($H.COOH$, <10% solutions) can be used, although the solubility of calcium phosphate is greater in the latter, even in concentrations as low as 2% (Lindsay 1987).

1 Immerse the specimen in a 10-15% solution of acetic acid ($CH_3.COOH$). Unless ionised by dilution, the acid will not react. Ensure the specimen remains completely covered during development. This will inhibit the formation of calcium acetate crystals if

evaporation of the solution occurs. These are extremely difficult to remove during washing stages, and can damage delicate structures and fabrics (Converse 1989). Consider using buffered acetic acid solutions on fragile specimens (see section 27 EXTRACTION TECHNIQUES FOR PHOSPHATIC FOSSILS).

2 Frequent visual checks on the reaction and its effect on the fossil must be undertaken. Monitor the development of micro-fissures on the specimen surface, and gently brush away any surface residue with a fine soft brush, without applying too much pressure (Lörcher and Keller 1984). At any time the reaction can be stopped by removing the specimen from the acid, and neutralizing by immersing it in ammonia. Water alone does not stop the reaction immediately, and will initially increase etching activity (Converse 1989).

3 Place the specimen in a sink, and flush with running water, or frequently change the water if a water bath is used.

4 Remove the specimen, and air dry. If a calcium acetate salt deposit appears on the surface during drying, immediately continue with washing. In many respects washing is the single most critical stage of the procedure, as salts resulting from evaporation not only disfigure the surface, but will disrupt the internal structure of the specimen, particularly if it is stored in an un-controlled environment with fluctuating humidity.

5 Once dry carefully examine the specimen with an magnifying lamp or under a stereozoom binocular microscope. Remove softened matrix with the aid of a mounted needle and soft brush. The contrast between the bone and matrix can be enhanced by flooding the area with heavy distillate (considerably safer than alcohol or acetone) when examining with a microscope (Schaeffer 1965).

6 Fill fissures and cracks to prevent further internal damage, and consolidate.

7 Before re-immersion of the specimen, cover all exposed areas of bone with a protective layer of poly(butyl methacrylate). Pay particular attention to newly exposed areas, and ensure they are protected.

8 When the sealant has dried a repeat immersion of the specimen in a fresh acid solution can commence.

Fossil development is a slow process, which must be undertaken in a methodical way. Lörcher and Keller (1984) preparing *Ichthyosaur* fragments, immersed specimens in acid for a total of 91.5 hours. Washing residual traces of acid from the bones required a further 250 hours. This computed to an average dissolution rate of 7-8mm/hour, corresponding to a continuous rate of 0.07mm/hour, with seven interruptions for washing, drying and consolidation. Control of the procedure cannot hastened by use of a stronger acetic or formic acid solution. The immersion time of specimens within acid must be kept to a minimum (Lindsay 1987, Lörcher and Keller 1984), although with time the dissolution of calcium carbonate (increases the pH) suppresses the solubility of calcium phosphate (Lindsay 1987). Acetic acid concentrations of 5% (restricting the number of H^+ ions in solution) can be just as effective as 15-20% solutions.

The recognition of mineralized dinosaur soft tissue (*e.g.* Briggs *et al.*, 1997, Martill 1991), although extremely rare, reinforces the need for careful examination of material prior to acid development. Haste in releasing skeletal material from the enclosing matrix could result in the loss of important taphonomic information (Briggs *et al.*, 1997).

Mercaptoacetic (Thioglycollic) acid ($CH_2(SH).COOH$)

Specimens enclosed in a hematite matrix or held by an iron oxide cement (Fe_2O_3) can be prepared by immersion in a solution of mercaptoacetic acid and calcium hydrogen phosphate following the procedure of Howie (1974). The acid converts ferric ions to the ferrous state, which then forms a soluble ferrous thiogycollate salt. The procedure is summarized as follows:

1 Immerse the specimen in a 5% solution of mercaptoacetic acid ($CH_2(SH).COOH$) and 0.9% calcium hydrogen phosphate ($CaHPO_4$). Mercaptoacetic acid dissolves calcium phosphate, at a greater rate than acetic acid (Lindsay 1987). However, the addition of calcium hydrogen phosphate eliminates this.

2 Decant the spent acid solution, and immerse the specimen in a 5% solution of ammonium hydroxide (NH_4OH) to neutralize the acid, and then thoroughly wash with a detergent and water. Ferro-thiogycollate ions will oxidize during washing, leaving a brown stain on the surface of the matrix and bone. To avoid this use a detergent during the initial washing stages.

Matrix softened during the acid immersion stage can be gently removed by brushing or mechanically with the use of an *Airbrassive* pen charged with a soft abrasive powder (*e.g.* sodium bicarbonate) (Howie 1974).

Ammonium chloride (NH_4Cl)

Standard acid procedures described above are ineffectual on encrustation's of calcium sulphate (gypsum, $CaSO_4{\cdot}2H_2O$). Specimens can be heated, dehydrating the gypsum and reducing it to a powder, but gypsum filled cracks may enlarge and fracture the specimen (Schaeffer 1965). Harris (1971) describes a chemical procedure for the removal of gypsum using a 10-20% solution of ammonium chloride (NH_4Cl), without deteriorating calcium phosphatic bone or teeth fragments. The procedure is as follows:

1 Completely immerse the specimen in a dilute solution of ammonium chloride. The container used in preparing material must be sufficiently large enough to hold four to five times the specimen volume.

2 Occasionally stir the solution, and change when evidence of dissolution has ceased. However, the reaction proceeds without effervescence, and must be periodically checked to ensure that the solution is effective.

With a slower reaction rate than ammonium chloride, glycerol ($CH_2OH.CHOH.CH_2OH$) may also is dissolve gypsum, without dissolving calcium phosphate (Harris 1971).

24.4.3c. Sub-fossil bone (Doyle 1983, 1987)

Doyle (1983, 1987) describes a method, and the construction of apparatus, used in conserving fragile Pleistocene and Holocene bird remains from New Zealand. Most of the specimens had suffered from poor long term storage, and becoming excessively dry. The apparatus enabled a continuous dispersion of an aqueous poly(vinyl acetate) (PVA) emulsion to be administered. Aqueous solutions of PVA are extensively used in geological and archaeological conservation, applied by a number of mechanisms, including vacuum impregnation, partial and total immersion of the specimen, or brushing and dropping (Rixon 1976). However, sub-fossil bone retaining collagen or other protein decomposition products does not respond to these methods of application. Brushing, even with very dilute solutions, fails to provide sufficient penetration, while immersion or vacuum impregnation can result in strain on the bone by causing it to swell. Furthermore, uncontrolled drying of wet material can cause shrinkage, splitting and irreparable damage to the specimen (Doyle 1987).

The method involves the application of two different concentrations of PVA: (i) a concentrated solution to seal the bone ends, followed by (ii) a dilute solution to penetrate through as much bone as possible. The procedure is:

1 Re-attach loose or damaged specimen labels using a water-proof varnish (*e.g.* 30 % solution poly(butyl methacrylate) dissolved in acetone or ethyl acetate).

2 Re-attach loose fragments with a concentrated PVA solution. Large cracks must be filled with PVA, and if possible closed with a clamp. Seal limb bones at either the proximal or distal ends with concentrated PVA by brushing on. On other bones seal all damaged surfaces. The purpose of sealing all porous areas is to prevent the dilute PVA from seeping through during the main stage.

3 Allow bones to dry for about 2 hours or until the PVA becomes transparent.

4 Construct a scaffold to support the bones, which are suspended using clamps so that they hang vertically. Doyle's system (1983 fig 1, 1987 fig 1) allows flexibility in both size, shape and number of bones that can be treated. Place a trough under the bones to collect excess PVA that flows through the saturated bones.

5 A 5l reservoir container for dilute PVA is situated above the bones. Dilute PVA is thus gravity fed through individual silicone tubes fixed in position 1cm above the bones.

6 A continuous cycle pump returns used PVA from the collecting trough situated beneath the fossils, back to the reservoir tank. Wire mesh over the opening of the return tube will prevent any small bone fragments contaminating the reservoir.

7 Ensure all clamps are closed when filling the reservoir tank 2/3 full with dilute PVA (diluted approximately 1:10 with distilled water, giving a solids content of 5%, Doyle 1983). Doyle (1983) undertook experimental runs to ensure the PVA concentration did not air dry leaving a gloss appearance.

Rixon (1976) suggests an emulsion for consolidating bone that should satisfy the following requirements:

- high plasticiser content (not less than 20%)
- small particle size
- negatively charged
- homopolymer (single polymer) emulsion
- matt or semi-matt finish

PVA is an ideal medium for the growth of moulds and fungi, particularly on sub-fossil bone (Doyle 1983, 1987). Development can be prevented by the addition of a 0.1% fungicide solution (*e.g. Parmetol* K40) to the reservoir tank (Doyle 1983).

8 By trial and error the flow of PVA can be regulated to suit the absorption rate of each bone. The optimum flow rate ensures that PVA spreads evenly throughout the fossil before dripping out of the bottom, and without flowing down the outer surface of the fossil (Doyle 1983). At approximately 3 hour intervals the tubes should be moved about the entry surface of the fossil, ensuring total saturation and complete impregnation (Doyle 1983). As the solids are gradually absorbed into the bone, the PVA must be discarded and replaced after approximately 6hrs, depending on the condition of the bone.

9 Complete saturation of the bones is indicated when it becomes soft and pliable. At this stage it must be removed from the clamps (acetone can be used as a PVA solvent if they are stuck). Any splits or cracks must be bound with strips of polythene sheet secured with masking tape and 'G' clamps. Allow bones to air-dry thoroughly, preferably supported on metal "zig-zags" (Doyle 1983). Do not oven dry, as this may result in further splitting. Specimens developing cracks at this stage should be placed in a relative humidity tent of Rh 85%, which is reduced gradually by opening the tent to allow the bones to dry out slowly (Doyle 1987).

10 When the bones are thoroughly dry, remove the polythene strips, and wipe all bone surfaces with lint-free tissue soaked in acetone. This removes excess PVA pooled in recesses at the sealed ends, and reduces surface shine.

11 Clean the equipment by flushing tubes with water and detergent to prevent clogging.

REFERENCES

BAIRD, D. 1980. The burnt dope technique and other intertidal ploys from America. *Newsletter of the Geological Curators Group*. 2, (8), 519-520.

BRIGGS, D. E. G, , WILBY, P. R, PÉREZ-MORENO, B. P, SANZ, J. L, and FREGENAL-MARTÍNEZ, M. 1997. The mineralization of dinosaur soft tissue in the Lower Cretaceous of Las Hoyas, Spain. *Journal of the Geological Society, London*. 154, (4), 587-588.

CONVERSE, H. H. 1989. *Handbook of paleo-preparation techniques* (2nd edition, revised by R. W. McCarty). Florida Paleontological Society. v + 109p.

DOYLE, A. M. 1983. The conservation of sub-fossil bird bone. *Geological Curator*. 3, (7), 447-451.

DOYLE, A. M. 1987. The conservation of sub-fossil bone, pp 463-465: In; P. R. Crowther and C. J. Collins (eds), The Conservation of Geological Material. *The Geological Curator*. 4, (7), 375-474.

FREEMAN, E. F. 1982. Fossil bone recovery from sediment residues by the "interfacial method". *Palaeontology*. 25, (3), 471-484.

FRIZZELL, D. L. 1965. Otoliths, pp 125-127: In; B. Kummel and D. Raup (eds), *Handbook of Paleontological Techniques*. W. H. Freeman and Co., (New York). xiii + 852p.

GENTRY, A. W. 1979. Curation of fossil vertebrates. *Special Papers in Palaeontology*. 22, 87-95.

GOWER, D. J and SENNIKOV, A. G. 1996. Endocranial casts of early archosaurian reptiles. *Paläontologische Zeitschrift*. 70, (3/4), 579-589.

GREENWALD, M. T. 1989. Techniques for collecting large vertebrate fossils, pp 264-274: In; R. M. Feldman, R. E. Chapman and J. T. Hannibal (eds) *Paleotechniques*. The Paleontological Society Special Publication No. 4. iv + 358p.

HARRIS, J. M. 1971. Chemical preparation of gypsum encrusted vertebrate material. *Journal of Paleontology*. 45, (2), 350.

HIBBARD, C. W. 1949. Techniques of collecting microvertebrate fossils. *Contributions from the Museum of Paleontology University of Michigan*. 8, (2), 7-19.

HOWIE, F. M. P. 1974. Introduction of thioglycollic acid in preparation of vertebrate fossils. *Curator*. 17, 159-166.

LEIGGI, P and MAY, P. J. (eds). 1995. *Vertebrate Paleontological Techniques*, Volume 1. Cambridge University Press. 368p.

LINDSAY, W. 1987. The acid technique in vertebrate palaeontology: a review, pp 455-461: In; P. R. Crowther and C. J. Collins (eds), The Conservation of Geological Material. *The Geological Curator*. 4, (7), 375-474.

LÖRCHER, F and KELLER, T. 1985 (for 1984). Preparation techniques for material from the Posidonienschiefer (Lias Epsilon, Upper Liassic) of Germany. *The Geological Curator*. 4, (3), 164-168.

MARTILL, D. M. 1991. Organically preserved dinosaur skin: taphonomic and biological implications. *Modern Geology*. 16, 61-68.

McKENNA, M. C. 1962. Collecting small fossils by washing and screening. *Curator*. 5, (3), 221-235.

McKENNA, M. C. 1965. Collecting microvertebrate fossils by washing and screening, pp 193-203: In; B. Kummel and D. Raup (eds), *Handbook of Paleontological Techniques*. W. H. Freeman and Co., (San Francisco). xiii + 852p.

MILLER, B. B. 1989. Screen-washing unconsolidated sediments for small macrofossils, pp 260-263: In; R. M. Feldman, R. E. Chapman and J. T. Hannibal (eds) *Paleotechniques*. The Paleontological Society Special Publication No. 4. iv + 358p.

NOVACEK, M and CLEMENS, W. A. 1977. Aspects of intrageneric variation and evolution of *Mesodma* (multituberculata, Mammalia). *Journal of Paleontology*. 51, (4), 701-717.

RIXON, A. E. 1949. The use of acetic and formic acids in the preparation of fossil vertebrates. *Museum Journal*. 49, 116-117.

RIXON, A. E. 1976. *Fossil animal remains: their preparation and conservation*. Athlone Press, (London). viii + 304p.

SCHAEFFER, B. 1965. Fishes, pp 110-118: In; B. Kummel and D. Raup (eds), *Handbook of Paleontological Techniques*. W. H. Freeman and Co., (New York). xiii + 852p.

SLAUGHTER, B. H. 1968. The neglected angle in descriptive and interpretive vertebrate paleontology. *Journal of Paleontology*. 42, (5), 1311-1312.

TOOMBS, H. A. 1948. The use of acetic acid in the development of vertebrate fossils. *Museum Journal*. 48, 54-55.

TOOMBS, H. A and RIXON, A. E. 1950. The use of plastics in the transfer method of preparing fossils. *Museum Journal*. 50, 105-107.

TOOMBS, H. A and RIXON, A. E. 1959. The use of acids in the preparation of vertebrate fossils. *Curator*. 2, 304-312.

VON HOFE GRADY, F. 1979. Some new approaches to microvertebrate collecting and processing. *Newsletter of the Geological Curators Group*. 2, (7), 439-442.

WHYBROW, P. J. 1978. Fibre optics and foot operated microscope control equipment to assist in the preparation of fossils. *Newsletter of the Geological Curators Group*. 2, (3), 141-143.

WHYBROW, P. J. 1982. Preparation of the cranium of the holotype of *Archaeopteryx lithographica* from the collections of the British Museum (Natural History). *Neues Jahrbuch für Mineralogie, Geologie und Paläontologie, Monatshefte*. H3, 184-192.

PART III

LABORATORY TECHNIQUES

C. CHEMICAL PROCEDURES

25. Extraction techniques for palaeobotanical and palynological material.

26. Extraction techniques for acid insoluble microfossils.

27. Extraction techniques for phosphatic fossils.

28. Extraction techniques for uncrushed graptolites.

29. Extraction techniques for calcareous microfossils from argillaceous sediments.

30. Extraction techniques for calcareous microfossils from carbonate sediments.

31. Extraction techniques for agglutinated foraminifera from calcareous sediments.

32. Specialist techniques used in the preparation of individual microfossil specimens.

33. Extraction techniques for calcareous nannofossils.

25. EXTRACTION TECHNIQUES FOR PALAEOBOTANICAL AND PALYNOLOGICAL MATERIAL

TECHNIQUES USED IN PALYNOLOGY ARE PARTICULARLY HAZARDOUS. NEW AND INEXPERIENCED USERS SHOULD BE MONITORED BY EXPERIENCED LABORATORY PERSONNEL. VAPOUR AND FUMES OF CONCENTRATED ACIDS, HYDROFLUORIC, NITRIC AND HYDROCHLORIC, ARE EXTREMELY DANGEROUS. ALL WORK MUST BE UNDERTAKEN IN AN APPROPRIATE FUME CUPBOARD OR WELL VENTILATED AREA WITH THE OPERATOR WEARING ACID RESISTANT PROTECTIVE CLOTHING. SUITABLE MEDICAL EQUIPMENT, NEUTRALIZING CHEMICALS AND SPILLAGE ABSORPTION GRANULES SHOULD BE CLOSE AT HAND IN CASE OF ACCIDENT OR SPILLAGE. HF WORK SHOULD ONLY BE UNDERTAKEN IN A LABORATORY WITH A QUALIFIED HF FIRST AIDER PRESENT.

25.1. INTRODUCTION

Extraction procedures used in palynology are complex and time consuming, adaptable to rock type and composition (Aldridge 1990). Applied methods in plant palaeoecology are extensively covered by DiMichele and Wing (eds, 1988), and will not be expanded in this volume. A successful residue preparation invariably requires individual steps to be tailored to the rock type processed. Many comprehensive accounts of procedures have been outlined (*e.g.* Barss and Williams 1973, Colbath 1985, Doher 1980, Gray 1965a, Lennie 1968, Litwin and Traverse 1989, Moore *et al.*, 1991, Norem 1956, Phipps and Playford 1984, Sittler 1955, Traverse 1988), with all procedures involving the use of HF traced back to the technique of Assarsson and Granlund (1924). The main stages of palynological processing are illustrated in the flow-chart figure 25.1. As with all micropalaeontological and palynological work, care must be employed if quantitative work is the primary objective on a faunal and floral assemblage concentrated within a residue. Biasing raw data, even following stringent processing procedures, can effect the frequency of a particular component (Farley 1988), consequently "in processing techniques where grains might be lost, it is advisable to check the "superfluous, non-palynomorph" fraction to insure nothing of interest is in it" (Farley 1988, p 127).

Generalized procedures of maceration usually involve the following procedures:

- Physical breakdown
- Removal of calcareous material
- Removal of siliceous material
- Removal of other inorganic material
- Oxidation
- Sieving, cleaning and concentrating the organic rich residue
- Staining and slide preparation.

Not all these steps will be required for every sample. Furthermore, some of the stages (*e.g.* removal of other inorganic material and oxidation) may be inappropriate if palynofacies analysis is undertaken. Sequences can be modified according to the sediment type processed, although procedural changes must be documented and tested for their influence on results. However, carbonate removal must always precede silica removal. This is discussed in greater detail below when describing individual processing methods (see 25.2.2. Chemical treatment).

During preparation, samples must always be kept covered to eliminate airborne contaminants. During all stages of processing be alert for the possible loss and damage of palynomorphs, particularly when decanting after centrifuging, as a froth may have formed around the top of the tube. Unless indicated, centrifuge at speeds of 2500-3000 rpm for 2-3 minutes using 10-15ml tubes, and wash all residues using distilled water.

Safety Note: All palynological processing should be carried out in a purpose-built laboratory, and in an efficient fume-cupboard. Full protective clothing (rubber apron, full-face mask, gloves and laboratory coat sleeve protectors) should be worn when working with HF. Ensure adequate neutralization and acid disposal facilities are available.

25.2. POLLEN AND SPORE EXTRACTION FROM ARGILLATES AND SILICATES

25.2.1. Pre-treatment

25.2.1a. Sample washing

Thoroughly clean rock samples by washing and scrubbing in distilled water. If necessary etch surfaces in concentrated hydrochloric acid (HCl) or nitric acid (NHO_3), to remove recent debris, before proceeding with crushing. However, it is important to remember that acidic cleaning can result in the oxidation of palyno-debris near the sample surface. If the bulk sample is large enough mechanically remove weathered surfaces or areas contaminated with recent organic material.

25.2.1b. Rock crushing

1 Wrap sample (20-50 grams) in paper or a plastic bag to avoid contamination. If using a rock crusher ensure the jaws and sediment collecting bin are thoroughly cleaned between each sample.

2 Break sample into 1-2mm fragments (do not crush or powder rock), with a hammer on a metal plate.

3 If required continue the mechanical breakdown of the sample by using a pestle and mortar until the particles are of coarse grit size (0.5-1mm).

The Quaternary O method (Hills and Sweet 1972), avoids excessive mechanical crushing of the sample. The procedure is as follows:

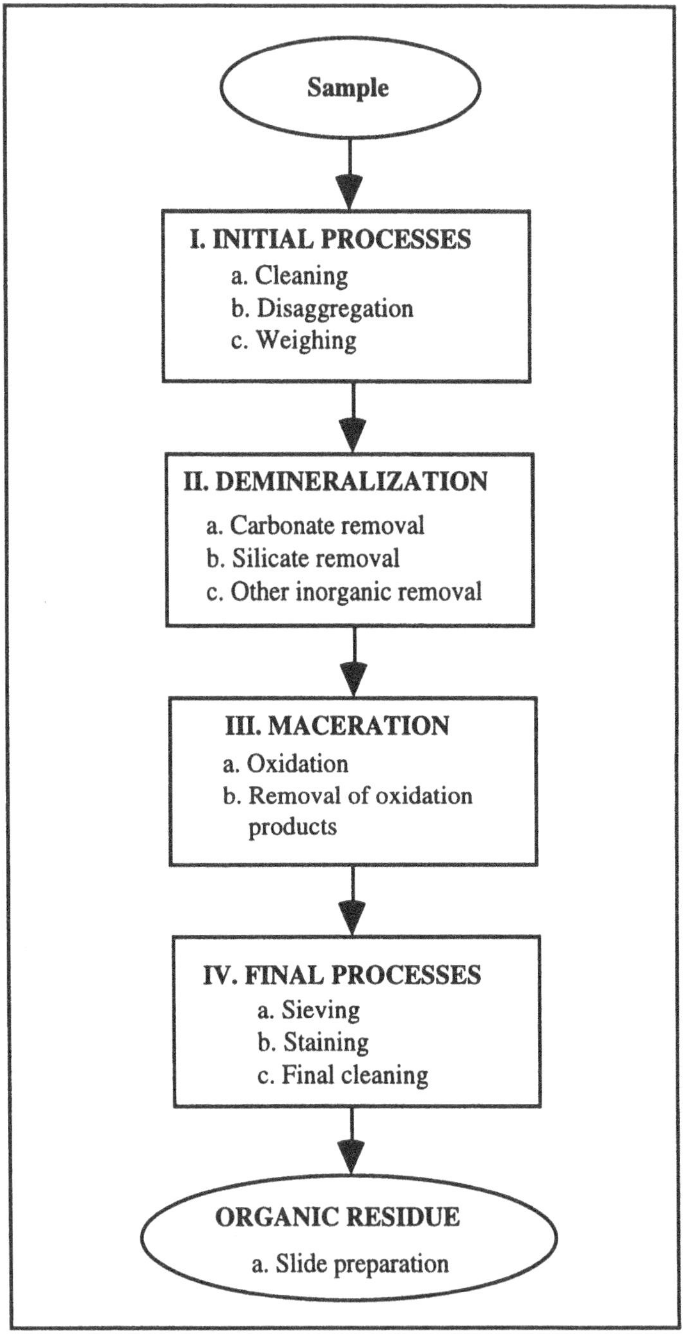

Figure 25.1. A basic plan showing the main stages for palynological processing of sedimentary rocks. Coals and peat's require a different sequence of operations (from Traverse 1988).

1 Soak fragments (0.5-1cm^3) in water from several hours to one week.

2 Mix a solution of 1 part Quaternary O to two parts distilled water, and add one volume of the solution to 3 volumes of the sediment. **N.B. Undiluted Quaternary O is suitable for strongly indurated or slightly calcareous rocks.**

3 Boil for 1-3 hours.

4 Thoroughly wash the residue, and repeat the procedure until the desired breakdown fraction is attained.

Some samples may disaggregate completely with just one treatment while others require repeated boiling. This procedure was found to be less effective on limestones and coals unless they contained a high percentage of argillaceous matter, while for carbonaceous sediments the use of Quaternary O accelerated the breakdown, particularly if alternated with the use of Schultze's solution. Once disaggregated standard chemical procedures can be followed. It may be prudent to crush more than the amount required and then sub-sample 20 grams for processing.

25.2.2. Chemical treatment

25.2.2a. Hydrochloric acid (HCl) (carbonate removal)

The two stage method advocated by Reissinger (1950) is preferred, although his reagent concentration of 6% is ineffective on many samples and extends the processing time. Test for the presence of carbonates using a dropper bottle of 10% HCl. If positive proceed as follows.

1 Place the samples into labelled clear plastic or polypropylene beakers. These should be sterilized with chromic acid, followed by a small amount of cold 10% HCl.

2 Add approximately 100ml of 20% concentrated, or 10% warmed HCl. The addition of concentrated HCl limits the formation of insoluble chlorides when HF is added. Add more acid as required, and leave for 12 hours (overnight), or until the reaction has ceased. If the reaction becomes violent spray the surface of the acid with 90% alcohol, reducing surface tension and liberating trapped CO_2.

3 Dilute the acid with distilled water and decant the supernate. Repeat the decantation procedure three or four times.

Although this stage removes the carbonate component and organisms such as foraminifera, pseudochitinous remains are often preserved and, in the case of some foraminifera, recognized in the macerated palynological residue (Tappan and Loeblich 1965, Wilson and Hoffmeister 1952). Laboratory work by Cohen and Guber (1968) demonstrated that "microforaminifera" can be produced from "normal-sized" foraminifera during standard palynological preparations, although this mode of formation is not mutually exclusive of any taphonomic changes. Cohen and Guber (1968) identified remains of rotaliid foraminifera, between 50-200μm in diameter, in which the early inner chambers in comparison to late outer chambers, had thicker pseudochitinous layers.

25.2.2b. Hydrofluoric acid (HF) (silicate removal)

Both hot and cold HF techniques have been documented (*e.g.* Herngreen 1983), preference deemed by the preparator and any time constraints imposed during sample processing. Investigations into the quality of HF used (general purpose or analytical grade) is in preparation (L. Ross *pers. com*), with initial results indicating a preference for analytical grade chemicals, reducing the risk of introducing contaminants of organic debris present in the acid.

1 Add approximately 50ml of cold concentrated HF (40% w/v) to each sample. Stir the liquid with a plastic or teflon rod to test reaction. After initial reaction has subsided, top up with HF to 100ml. Experienced operators may speed up the process by immersing containers in a warm water bath. Swirl 2-3 times daily. Swirling is preferable to stirring as there is less danger of damaging the palynomorphs.

2 Add additional HF to replace any lost by evaporation. Gentle use of a plastic or teflon coated rod can be used to detect the presence of "grit" at the base of the beaker. Repeat the HF stage if the siliceous component is not completely digested (Norem 1956).

3 Decant HF into large polypropylene beakers containing a neutralizing agent of orthoboric acid (H_3BO_3) or calcium carbonate ($CaCO_3$). Fill beakers containing the residue with distilled water. Repeat until the residue solution is neutral (test with pH indicator paper). Flushing the sample plus acid solution in a dialysis bag with continuously running water reduces the time of this stage (Jackson *et al.*, 1974, McKee 1977, Batten and Morrison 1983). This modification has the additional advantage of reducing the risk of damage or loss of delicate light-weight microfossils. If quantitative studies are undertaken, it is at this stage residues should be spiked with a specific amount of exotic *Lycopodium* spores.

4 Sieve the neutralized samples through a 20μm nylon monofilament mesh. For large volumes of residue use a 20μm and 10μm mesh to separate into two fractions. Smaller mesh size ensures all palyno-debris is retained.

5 Transfer the residue into a glass sinter funnel (P100, 100-400μm), and filter into a 1 litre buchner flask. Use a "blow-ball" to free blocked funnel pores.

6 Inspect the residue by examining temporary slide mounts for the presence of calcium fluoride (CaF_2) crystals. Follow procedures outlined below for their removal.

25.2.2c. Boiling in concentrated hydrochloric acid (calcium fluoride precipitate removal)

1 Boil the residue in 100ml of 40-50% HCl for 1 minute. Wash at least four times in warm water. Add 10% HCl to the last wash to discourage flocculation.

2 Pour the residue into a large beaker, and sieve through a 20µm monofilament nylon mesh.

3 Decant the residue into a plastic phial, add a few drops of formaldehyde or HCl to act as a preservative, and label. N.B. if HCl is added to stained residues (see below) for long term storage, re-staining will be required before permanent slides can be made.

For palynofacies studies it is advisable to keep the amount of processing to a minimum, as certain treatments may alter the nature of the palynomorphs and organic debris within the residue (L. Ross *pers. com*).

25.2.3. Heavy mineral separation

There are two methods that can be used, depending on time and finances available, for the removal of heavy minerals such as rutile, tourmaline and zircon.

25.2.3a. Swirling

A simple particle size sorting procedure, however, there is a danger of sample biasing and the differential loss of some palynomorphs (Traverse 1988), particularly if palynomorphs are bound with pyrite (L. Ross *pers. comm*).

1 Transfer the residue onto a very large watch-glass, by using a rotary action swirling concentrates large minerals and particles in the centre of the watch glass. Palynomorphs remain in suspension and are removed by pipette. Similar results can be achieved by using two beakers, mixing the residue in one and then decanting the palynomorph rich 'float' into the second.

25.2.3b. Heavy liquid

1 Decant as much water as possible from each residue tube.

2 Mix each residue slurry with zinc bromide ($ZnBr_2$, S.G. 2.1). Pour into a 'U' shaped rubber tube inserted in a 100ml centrifuge tube, and top with additional zinc bromide. Acidify the solution with the addition of a few drops of HCl to prevent precipitation.

3 Prepare four centrifuge tubes (of equal weight) and place in a centrifuge. Centrifuge for 20 minutes at 2500 rpm.

4 The palynomorph rich light fraction accumulates at the top, while the barren heavy fraction is concentrated at the bottom. Retain the heavy fraction for later examination. Wash the light fraction into a beaker with dilute (10%) HCl to dissolve the zinc bromide.

5 Centrifuge this fraction for a further 2-3 minutes at 2500 rpm.

6 Pour the residue into a large beaker of water diluting any remaining acid, and re-sieve through a 20µm mesh to recover the sample. Store in plastic phials.

Other methods of gravity separation using the liquids bromoform and alcohol, zinc chloride in water and Thoulet solution, have been used by various palynologists and are outlined by Herngreen (1983). Centrifuging may cause the destruction of some organic debris, which could bias palynofacies analysis (L. Ross *pers. com*).

25.2.4. Oxidation (lignin removal)

Samples requiring oxidation (i.e. the removal of pyrite and lightening of palynomorph exines) can be treated with nitric acid (HNO_3), while fine organic debris may be removed by alkali treatment with 5% potassium hydroxide (KOH).

1 Place the residue in a beaker containing 50ml of 10% nitric acid, and leave in a fume cupboard until the reaction has ceased (approximately 10 minutes). If the reaction is not violent top to the 100ml mark with 10% nitric acid. If oxidation is insufficient with nitric acid, use Schulze's solution (Chlor-zinc-iodine) in a similar manner. Some bituminous clay residues may require longer periods of acid immersion. Heating the acidic solution is not recommended, as in some residues this may result in biased assemblages (*e.g.* Hughes *et al.*, 1964).

2 Wash the solution into a plastic beaker with distilled water and sieve through a 20µm mesh screen. Decanting and centrifuging twice can be undertaken as an alternative (Norem 1956).

N.B. Care must be taken during oxidation to avoid the destruction or weakening of palynomorphs, and was omitted by Dean (1989) when macerating Lower Devonian low-grade metamorphic grits from Southwest England.

3 Add 25ml of 10% sodium hydroxide. After 20 minutes the solution should turn from black to brown in colour. If not repeat the procedure using Schulze's solution. Top up with distilled water and centrifuge. Repeat and wash until the water is clear.

4 Finally, wash the residue in 95% alcohol before staining and transferring to a storage phial.

Neves and Dale (1963) used a reversible flow system when filtering oxidized residues, thus ensuring that the fine meshed screen did not become blocked.

25.2.5. Ultrasonic treatment

The long term storage of residue suspensions results in palynomorphs clumping together at the base of the phial. They can be disaggregated, without damaging the palynomorphs, by placing the phials in an ultrasonic bath for 10-30 seconds, although in some cases 5 seconds may be sufficient (Caratini 1980, Peppers 1970). This also breaks down organic matter aggregates, particles of which can be removed by re-sieving the sample. Smith and Butterworth (1967) report that successful separations can be achieved on residues that have been slightly under-macerated with Schulze's solution before alkali and ultrasonic treatment. Reduced chemical contact time, resulting in the under-maceration of samples, is necessary for techniques using ultrasonic disaggregation methods, where combinations of the two damage poorly preserved or brittle palynomorphs (McIntyre and Norris 1964).

25.2.6. Sieving and filtration

An alternative to concentrating palynomorphs by centrifuging and decanting is to employ a method of sieving or filtration. This may be as simple as using a nest of sieves with either metal or nylon mesh (*e.g.* Cwynar, Burden and McAndrews 1979, Eagar and Sarjeant 1963, Ediger 1986, Kidson and Williams 1969, Neves and Dale 1963), or a more elaborate method using specialized filtration apparatus (*e.g.* Raine and Tremain 1992, Vidal 1988). In some cases sieving is combined with an ultrasonic stage (Caratini 1980, McIntyre and Norris 1964).

A 53μm mesh sieve is capable of retaining chitinozoans and large palynomorphs, while a 5-7μm mesh is required to retain the smallest palynomorphs. Sintered glass funnels were preferred by Neves and Dale (1963) and Ediger (1986), while Kidson and Williams (1969) and Caratini (1980) used more expensive wire mesh sieves. More cost effective, durable, and disposable between samples is the use of mono-filament nylon or polyester mesh (Cwynar *et al.*, 1979, Raine and Tremain 1992, Vidal 1988).

The apparatus described by Vidal (1988) and Raine and Tremain (1992; *Satorius Membrane Filter, SM 165 10*) is well suited for the standardized mass production of residues in laboratories where sample turn-over is high. Furthermore, by replacing the screen between samples the risk of cross residue contamination, that can result from conventional screening when sieves are not adequately cleaned, is reduced. However, during use the filter can become clogged, but is quickly cleaned by reversing the water flow through the unit (Raine and Tremain 1992). Vidal (1988) employed the system between the stages of silicate (HF) and fluorocarbonate (boiling in 30% HCl) removal, and again after oxidation (HNO_3). It was particularly successful in concentrating assemblages from poorly fossiliferous samples.

25.2.7. Staining

If the specimens are pale in colour they can be stained with an aqueous solution of 0.1% Safranine-O, Malachite green or Bismarck brown. The addition of a few drops of 2% potassium hydroxide (KOH) is required if the stain is not readily absorbed. To stain a residue:

1 Add a small amount of the stain to the mounting media, and mix it thoroughly until an even colour has been obtained.

2 Pipette a few drops of the residue into the solution, and mount onto a glass slide. If glycerol is used to mix the stain, and to mount the residue, the excess can be stored in a phial without deteriorating the palynomorphs. Gentle heating reduces the viscosity of the glycerol.

N.B. It is always better to under-stain than over-stain.

25.2.8. Slide and stub production

Palynological slide preparation requires a careful systematic approach to ensure contamination does not occur, and if it does, it is instantly recognized. The use of control slides is preferred by some workers (*e.g.* Norem 1956).

The concentrate is finally strew-mounted onto slides, using glycerol jelly for temporary mounts or *Elvacite* 2044 for permanent mounts (Barss and Williams 1973). The procedure involves:

1 Pipette a small amount of residue onto a No. 0 cover slip. For quantitative studies material must be evenly distrubuted to avoid bias during counting.

2 Leave for the water to evaporate. Place a small drop of *Elvacite* on a clean glass slide.

3 Position the dried cover glass with sample on the glass slide, label and allow to dry.

4 Once dry, the slide can then be examined using a biological microscope.

Peppers (1970) stored all processed cleaned fractions. Material coarser than 210µm was stored in an alcohol/glycerol mix, while finer material suitable for section mounting, was stored in 50% absolute alcohol/xylene mix with Canada Balsam. The stability of this mixture over a long time has yet to be thoroughly evaluated. Phials should be carefully stored and continually monitored.

Other suitable mounting media include glycerol, Canada Balsam and epoxy resin (Traverse 1988). Silicone oil was preferred by Faegri and Iversen (1989), a modified procedure of which is described by Traverse (1988, p 455). In addition to *Elvacite*, Barss and Williams (1973) describe the use of *Clearcol* and *Cellosize*. However, the unavailability of *Clearcol* required a revision of the *Cellosize* (Barss and Crilley 1976). The preferred permanent mounting medium of many is *Elvacite* as it does not cause organic debris swelling and offers low fluorescence (L. Ross *pers. comm*).

A systematic idealized mounting procedure for light microscopy slides and scanning electron microscopy stubs has been advocated by Watson and Sincock (1992) in preparing plant cuticle material. They readily admit however, it is not always possible to follow, but it is worth while reiterating for the small number of examples to which it can be applied. Ideally, a small piece of cuticle should be cut into three. One piece is stained (if required), and mounted either way up for LM. The two remaining pieces are studied under the SEM, and mounted on stubs to exhibit both adaxial and abaxial surfaces, joined if possible. Specimens are positioned on the stub using double sided tape, and mounted so that the two inner and two outer surfaces can be observed (*e.g.* Watson 1977).

The SEM examination of palynomorph residues requires the attachment of strew covered glass cover slips to aluminium stubs, secured using double sided tape. Following gold sputter coating material can be examined by electron microscope (Dean 1989, see section 34 ELECTRON MICROSCOPY TECHNIQUES for details).

25.2.9. Recovering residue and cuticle mounts from damaged slides

Wilson (1972) describes procedures employed in the reclamation and re-mounting of residues, prepared in glycerol or some other non-permanent mounting media, from old or damaged palynological slides. Furthermore, additional processing, cleaning and staining of residues can be undertaken if desired.

1. Gently clean the glass slide with cotton wool moistened with alcohol. If the slides have been sealed with a lacquer this should be removed with solvent.

2. Glycerol mounts should be soaked in a beaker of distilled water placed on a hot plate at 100°C, until the cover slip and residue has sunk to the bottom of the beaker, and the glycerol melts. Canada balsam mounts require soaking in methanol at room temperature - **not** on a hot plate.

3. Once all the residue has been washed from the slide (use a fine sable hair paint brush), remove the slide, clean for re-use or discard if broken.

4. Transfer the residue to a 15ml tube and thoroughly wash in a solvent followed by distilled water.

5. Residues can then be remounted, or if further processing is required, (i.e. acetylation to remove cellulose, additional staining, or decreasing the stain intensity by washing in 20% HCl), it can be undertaken.

A technique of enzymatic hydrolysis was successfully employed by Watson and Sincock (1992) in salvaging cuticle material from slides 20 to 75 years old.

1. Individually immerse slides in a beaker or petri dish containing a 10% solution of Trypsin at 20°C for 24 hours. Older slides with stubborn glycerol may require longer.

2. The cover slip loosens and falls off, and complete hydrolysis of the underlying glycerol releases the cuticle specimen. In older slides excess jelly may have to be cut away using a surgical knife.

3. Clean the slide for re-use (if unbroken).

4 Remount the specimen.

It is not essential to remove all fragments of glycerol, as small pieces can be incorporated within the fresh jelly of the new slide mount. A great advantage of this procedure, as opposed to that of Wilson (1972) described above, is that the slide labels are largely unaffected by the restoration process.

25.3. POLLEN AND SPORE EXTRACTION FROM CARBONATES

The processing of chalk samples for palynological specimens has been extensively detailed by Wilson (1971). Processing of carbonate rich samples usually results in the complete or partial destruction of any calcareous organisms. If calcareous faunal elements are required for study, divide the sample in two and process for either acid insoluble or calcareous microfossils. The procedures used for the removal of silicates, oxidation, acetylation and liquid separation are similar to those described above, and will only be briefly outlined in this section at the appropriate stage of use.

25.3.1. Pre-treatment

1 Clean the surface of the sample. A sample size of 250g of pure chalk or 150g of impure chalk is usually sufficient.

2 Crush into small $1cm^3$ fragments, and place in a 2 litre plastic beaker.

25.3.2. Chemical treatment

25.3.2a. Hydrochloric acid (HCl) (carbonate removal)

1 To reduce a violent reaction resulting from the addition of HCl, moisten the sediment with 50% acetone. Covering the sample with 100-200ml of kerosene suppresses froth and foam, allowing for the use of stronger acid concentrations (Nørgaard *et al.*, 1991).

2 Slowly add cold concentrated HCl, and thoroughly stir the sediment.

3 The vigorous reaction must be constantly monitored, and sprayed with acetone if it becomes violent.

4 Top up with fresh HCl, and re-charge the solution until the reaction has ceased.

5 If the sample is oil rich, then a surfactant (*e.g.* MP 10, Triton X-100) can be added drop by drop. This aids the dissolution of heavy hydrocarbon particles, liberating it from the carbonate, and assisting acid digestion (Nørgaard *et al.*, 1991).

It is important to ensure that the sediment is thoroughly washed at the end of this stage to avoid the formation of calcium fluoride. Samples rich in hydrocarbons require filtering with lukewarm water through a 10μm mesh to remove the acid and hydrocarbon waste products. The addition of a small amount of detergent, and the gentle massaging of the mesh under running water is enough to thoroughly clean the residue.

6 Wash the cleaned residue into plastic 15ml centrifuge tubes.

25.3.2b. Hydrofluoric acid (HF) (silicate removal)

1 Add concentrated HF to the sediment, mixing occasionally and allowing to stand for 24 hours.

2 Digestion can be accelerated by gently heating the tubes within a detergent filled (*e.g. I.C.I.'s Lissapol*, Lennie 1968, Wilson 1971) temperature bath set at 120°C. Water filled baths must not be used.

3 Decant and neutralize the solution, then centrifuge the residue at 2500 rpm for 3 minutes.

25.3.2c. Boiling in concentrated hydrochloric acid (calcium fluoride precipitate removal)

1 Wash residues into a glass beaker, cover with concentrated HCl and boil.

2 Re-centrifuge or wash through a 10μm monofilament nylon mesh.

25.3.3. Oxidation (lignin removal)

1 Add 5ml of concentrated (95%) fuming nitric acid (HNO_3) to the residue in each 15ml centrifuge tube (Wilson 1971).

2 Filter through a glass sinter funnel (P40, 16-40μm).

25.3.4. Acetylation (cellulose removal)

The removal of unwanted cellulose follows standard palynological procedures.

1 Wash the residue in glacial acetic acid.

2 Carefully add 5ml of a 9:1 mixture of acetic anhydride and sulphuric acid. Stir the mixture and keep at a constant temperature (90°C) for 10 minutes.

3 Wash the residue in distilled water.

25.3.5. Heavy mineral separation

Further treatment may be required if the residue contains heavy minerals.

1 Separate using zinc bromide ($ZnBr_2$) dissolved in 10% HCl, and prepared to a specific gravity of 2.0.

Additional cleaning and residue concentration can be achieved by short centrifuging. Deflocculation of the palynomorph rich residue may require ultrasonic treatment (see section 25.2 EXTRACTION OF POLLEN AND SPORES FROM ARGILLACEOUS AND SILICEOUS RICH SEDIMENTS for details).

25.4. POLLEN AND SPORE EXTRACTION FROM EVAPORITES

A variation on the palynomorph extraction procedure from carbonate rocks, applicable to calcium sulphate (gypsum) rich sediments, has been described by Clay-Poole (1990). Although conventional HCl and HF stages dissolve calcium carbonate and siliceous minerals respectively in evaporites, a high concentration of calcium ions may remain in gypsum rich residues. These can re-precipitate as calcium fluoride, obscuring palynomorphs that may be present in very low concentrations. Boiling in ethylenediaminetetra-acetic acid (ethylenedinitrilotetra-acetic acid (EDTA, $[CH_2.N(CH_2.COOH)_2]_2$) eliminates this problem by dissolving both calcium carbonate and calcium sulphate (Clay-Poole 1990). Following the initial disaggregation of the sample, the modified chemical treatment is as follows:

25.4.1. Chemical treatment

25.4.1a. Ethylenediaminetetra-acetic acid (EDTA) (carbonate and sulphate removal)

1 Equally divide the sample by weight into a series of 10-15ml volume centrifuge tubes.

2 Add EDTA in a proportion three parts EDTA solution to one part sediment. A stock solution of EDTA is prepared by dissolving 74.45g of technical grade EDTA in one litre of distilled water, buffered to a maximum pH 10 with the addition of sodium hydroxide (Clay-Poole 1990). Increased alkalinity, although accelerating the dissolution process, has the potential of destroying the palynomorphs. Mix the solution with the sediment and, while occasionally stirring, heat for 15-30 minutes.

3 Centrifuge the residue at 2500-3000rpm for 2-3 minutes, and decant the supernate. Wash the sample twice in distilled water.

25.4.1b. *tetra*-Sodium pyrophosphate (clay deflocculation and humic acid removal)

1 For the deflocculation of clay and clay grade particles, prepare a 0.1 M solution of *tetra*-sodium pyrophosphate ($Na_4P_2O_7.10H_2O$, 44.6g/lt). Add one part of the residue to three parts solution. Heat and stir for 15

minutes. This stage dissolves humic acid and reduces the potential of a violent reaction when HF encounters the residue.

2 Centrifuge the residue and decant the supernate. Wash the sample twice in distilled water.

25.4.1c. Hydrochloric acid (HCl) (carbonate removal)

1 As a precaution to ensure the complete removal of carbonate material, add cold dilute HCl, and thoroughly stir the residue (Clay-Poole 1990).

2 Centrifuge the residue and decant the supernate. Wash the sample twice in distilled water.

25.4.1d. Hydrofluoric acid (HF) (silicate removal)

1 Add concentrated HF to the sediment, mixing occasionally and allowing to stand for 24 hours.

2 Centrifuge the residue and decant the supernate. Wash the sample twice in distilled water.

25.4.2. Acetylation (cellulose removal)

1 Wash the residue in glacial acetic acid.

2 Carefully add 5ml of a 9:1 mixture of acetic anhydride and sulphuric acid. Stir the mixture and keep at a constant temperature (90°C) for 10 minutes.

3 Re-wash the residue in glacial acetic acid.

4 Centrifuge the residue and decant the supernate. Wash the sample twice in distilled water and store the residue in labelled phials.

25.5. POLLEN AND SPORE EXTRACTION FROM HOLOCENE PEAT

By following the procedure of West (1977), in preparing temporary glycerol mounts of residues after each processing stage, cleaned residues suitable for permanent mounts are readily determined, and unnecessary processing stages avoided. A modification in the treatment of Holocene clay-rich lacustrine sediments has been described by Bates *et al.*, (1978), and is outlined below.

25.5.1. Pre-treatment

1 Place 1-2g of sample in a labelled beaker and dry in a laboratory oven pre-set at 60°C.

2 Crumble the sample between fingers. If more than one sample is being processed wear disposable gloves to avoid cross-sample contamination of material trapped under fingernails.

25.5.2. Unsaturated soil colloid removal

1 Cover with a 10% solution of sodium hydroxide (NaOH) or 5% potassium hydroxide (KOH). Stir with a glass rod.

2 Place sample in a pre-heated (100°C) water bath until the solution has become brown, and the peat is thoroughly broken down (usually around 10 minutes, but may take up to an hour).

3 Decant off the excess solution.

4 Wash the residue in distilled water, leaving for no less than 2 hours between decanting the supernate. Repeat the washing procedure two or three times.

5 After the final decant, flush the residue into 10 or 15ml centrifuge tubes and centrifuge at 2000 rpm for 1-2 minutes. Fresh pollen samples will not settle in 50ml tubes (West 1977), but if larger tubes are used centrifuge times must be increased to 4-5 minutes.

25.5.3. Oxidation (lignin removal)

Pollen grains are easily destroyed during the process of oxidation. Reactions should be stopped as soon as the sample is bleached. In alkaline solutions lignin becomes soluble, and can be removed as follows:

1 Add 8ml glacial acetic acid (CH_3COOH) and 4.5ml sodium chlorate ($NaClO_3$) (100g to 200ml distilled water) to the residue and stir with a glass rod.

2 Add 1ml of concentrated sulphuric acid (H_2SO_4) a few drops at a time. Leave until the liquid becomes yellow or orange in colour (approximately 10 minutes). If oxidation is slow, it can be accelerated with the addition of a few drops of concentrated HCl. However, if the reaction is rapid avoid using Erdtman's method of acetolysis, as the pollen grains may become swollen and rupture.

3 Centrifuge and decant off the excess liquid.

N.B. Do not allow water to come into contact with the sample.

25.5.4. Silicate removal

If present, silicates can be removed at this stage by a number of methods. Flotation and gravity separation is one option, while West (1977) describes the following chemical method:

1 Flush with 7% HCl. Transfer 0.5ml of the sample to a 10ml platinum crucible.

2 Add 40% HF until the crucible is two-thirds full.

3 Gently boil for 2-3 minutes on a sand bath.

4 Allow the mixture to cool, then pour into a beaker containing 20ml 7% HCl, and transfer equal volumes to centrifuge tubes.

5 Centrifuge and decant the excess liquid, wash the residue and repeat.

6 Heat residue in a 7% HCl solution to remove colloidal silicon dioxide and silicofluorides.

7 Before the final wash neutralize the solution by adding a few drops of sodium hydroxide (NaOH). Centrifuge and decant.

8 Examine the residue under a microscope. If clumps of inorganic material are still present, repeat with a further stage of hot HCl.

For clay rich sediments, the time required for silicate removal is significantly reduced by treating the sample with a deflocculant (*e.g.* 0.1 M *tetra*-sodium pyrophosphate), warming the solution for 10-20 minutes until all the clay particles are in suspension, and then centrifuging at 3000 rpm for 5 minutes (Bates *et al.*, 1978). The clay fraction remains in suspension, while the palynomorphs are concentrated at the base of the centrifuge tube. This treatment may have to be repeated several times, before the HF procedure.

25.5.5. Dehydration

1 Flush the sample into a 10 or 15ml centrifuge tube.

2 Top up with glacial acetic acid ($CH_3.COOH$), stir or shake, centrifuge then decant off excess liquid. If the residue does not disperse, dehydration is incomplete, and must be repeated, although washing with either alcohol or acetone assists at this stage.

N.B. Do not allow water to come into contact with the sample.

25.5.6. Acetylation (cellulose removal)

Reagents must be freshly mixed before to use, although acetylation or acid hydrolysis may not be required if small amounts of cellulose are present in the sample. Mix 0.5ml of concentrated sulphuric acid (H_2SO_4) with 10ml of glacial acetic acid.

1 Add the reagent to the sample. **N.B. this reaction is exothermic. Do not allow water to come into contact with the solution.**

2 Place in a pre-heated (100°C) water bath, stirring occasionally until the solution turns brown (up to 25 minutes).

3 Centrifuge the sample.

4 Decant off the excess liquid.

5 Wash the residue in glacial acetic acid.

6 Centrifuge and decant off excess liquid.

7 Wash and decant the residue in distilled water, centrifuge between each washing stage until a clear liquid remains.

25.5.6a. Erdtman's Acetolysis

First described by Erdtman (1943), this violent method of acetolysis is effective in removing organic material that cannot otherwise be destroyed. It must not be used on poorly preserved specimens, as these may be destroyed or swollen beyond recognition. Well-preserved grains become enlarged and yellow-brown in colour.

1 The acetolysis mixture is prepared by mixing 1ml of concentrated sulphuric acid (H_2SO_4) with 9ml of acetic anhydride ($(CH_3.CO)_2O$).

2 Mix the dehydrated residue with the acetolysis mixture in a 10ml centrifuge tube.

3 Place in a water bath set at 100°C and leave for 1-2 minutes. Stir the contents during heating, but take care not to boil the residues.

4 Centrifuge for 1-2 minutes then decant excess acid.

5 Wash with glacial acetic acid, centrifuge and decant.

6 Wash and centrifuge twice with distilled water. Add a few drops of dilute sodium hydroxide (NaOH) to the first washing.

7 Flush residue into a labelled vial, stain with Safranine O if required and prepare sections (see section 25.2 EXTRACTION OF POLLEN AND SPORES FROM ARGILLACEOUS AND SILICEOUS RICH SEDIMENTS).

25.6. POLLEN AND SPORE EXTRACTION FROM BITUMEN

A successful technique for the extraction of Pleistocene pollen from tar deposits has been outlined by Faruqi and Copley (1966), and is as follows.

1 Place 5g of sample in a 150ml glass test tube.

2 Add 20ml of naphtha, and stir with a glass rod until the tar is in suspension.

3 Pour 10ml of the saturated tar-in-naphtha solution into separate 50ml centrifuge tubes. To each add 5ml of acetic acid ($CH_3.COOH$) and 5ml of methanol (CH_3OH). When the mixture is thoroughly stirred with a glass rod, dissolved tar separates and rises leaving a clear layer of methanol and acetic acid.

4 Centrifuge for 5 minutes at 2000 rpm. Decant the supernate.

5 Add a further 10ml of naphtha, 5ml of acetic acid and 5ml of methanol to each centrifuge tube.

6 Centrifuge and decant as in stage 4. Repeat this procedure until the upper portion of the solution is translucent.

7 Wash the residue with alcohol and then distilled water.

8 Oxidize the residue by adding 10ml of a saturated solution of potassium chlorate ($KClO_3$), and agitate for a few minutes.

9 Centrifuge and decant.

10 Wash the residue in distilled water, centrifuge and decant. Repeat this stage to ensure the residue is thoroughly clean.

11 Wash the residue into a labelled phial.

12 Stain the residue with Safranine O and prepare sections as outlined under the techniques in section 25.2, EXTRACTION OF POLLEN AND SPORES FROM ARGILLACEOUS AND SILICEOUS RICH SEDIMENTS.

25.7. MEGASPORE EXTRACTION FROM HIGH RANK COALS (>50% carbonaceous matter)

Methods used in the processing coals are adaptable for chemical and petrological compositional variations within samples (Peppers 1970). Techniques for macerating coals of different ranks have been outlined by Smith and Butterworth (1967), concluding that different spore types are susceptible to varying degrees of destruction during processing, particularly in oxidation and alkali treatments. Smith and Butterworth (1967) have demonstrated that over-maceration of a sample can result in the increased representation of resistant spore types, and thus generate misleading quantitative data. They further conclude (p 105) that "comparative work should always be based on material prepared in the same way".

25.7.1. Pre-treatment

1 Crush 10g of sample until fragments are approximately 5mm in diameter. Most of the sediment processed should pass through a 4.25μm mesh sieve, as finer material contains many broken exines. Approximately 1g of coal is required for processing.

25.7.2. Chemical treatment

1 Place the sample in a large (1 litre) labelled plastic beaker.

Continue with the oxidation procedures outlined below.

25.7.3. Oxidation (lignin removal)

25.7.3a. Methods using Schulze's reagent and alkali (Schulze 1855)

Schulze's solution, a commercially available reagent that is a mixture of potassium chlorate and concentrated nitric acid has been successfully used as an oxidant in two methods of coal maceration (Smith and Butterworth 1967). Two variations of the method are used.

Wet method

One part saturated potassium chlorate ($KClO_3$) solution to two parts concentrated nitric acid (HNO_3). Then follow the procedure outlined below.

Dry method (Raistrick 1934)

One part dry $KClO_3$ (1g) to an equal amount of sample (1g), and then add a drop at a time, 10ml of concentrated (70%) nitric acid HNO_3.

The dry method is suitable for all except for weathered coal samples, as it is quicker and spores are less liable to suffer damage by excessive contact with the oxidizing solution. Highly weathered, or core samples with a relatively high sulphur content, oxidize naturally if exposed to air. However, a partly oxidized residue must be re-treated (using the wet method, *e.g.* Peppers 1970), or a non-representative palynomorph assemblage results because pollen and spores are not randomly distributed within a coal, and certain lithological types differentially oxidize (Smith and Butterworth 1967).

1 Add one part Schulze's solution to one part residue. Add a few drops at a time of concentrated (70%) nitric acid (HNO_3). For the optimum reaction the volume of nitric acid must be sufficient to dissolve the potassium chlorate (larger volumes slow the reaction). If the proportion of potassium chlorate to coal is increased the reaction is quicker, but there is an increased possibility of damaging the palynomorphs.

2 Leave for 16 hours, or overnight, or until swelling occurs, stirring at intervals if possible. As a rule weathered (surface) coals need less time for oxidation (*e.g.* a "few minutes", Peppers 1970), while "fresh" (subsurface) samples react slowly in Schulze's solution and may require 72 hours. A maceration of a high rank coal sample may take a week, while a lignite requires only 48 hours. Higgins and Spinner (1969) suggest testing a small amount of the sample in 5-10% potassium hydroxide (KOH). If the mixture turns dark brown in colour, the coal particles have broken down leaving only isolated spores and other plant debris, then the process involving Schulze's solution can cease. A less reliable test, for use by more experienced preparators, is that the coal "feels soft" when stirred. The preparation and examination of wet mounts are used to check how successful organic matter has separated.

3 If during this stage little exothermic reaction or effervescence from the solution is observed, slowly add a little solid potassium chlorate ($KClO_3$). Should heating begin, cool immediately in a water bath, and if the mixture is going to boil over, add distilled water. Leave for one week.

4 Dilute the solution with the addition of distilled water. Allow it to stand for one hour and decant off the supernatant liquid.

5 Add a few drops of cold 10% hydrochloric acid (HCl). **N.B. an exothermic reaction results during this stage.**

6 Flush the residue into a series of 10-15ml centrifuge tubes with distilled water. Centrifuge the residues at 6000 rpm for 4-5 minutes. Repeat this stage to separate the palynomorphs.

Alkali treatment

1 Add 20-30ml of 10% potassium hydroxide (KOH), and gently stir, converting the suspension into a solution and dissolving the humic matter. **N.B. exothermic reaction results.** Other alkalis tested

by Smith and Butterworth (1967) are potassium carbonate (K_2CO_3) and ammonium hydroxide (NH_4OH).

2 Leave for 2 hours or until the solution becomes translucent. Repeat the stage if the breakdown is incomplete, and ensure the complete removal of all dissolved humic matter.

3 Decant off the excess solution and centrifuge the residue at 6000 rpm.

If any silicates are present they can be removed after maceration by treating the residues in a 20-30ml of 40% HF for at least 24 hours.

25.7.3b. Method using fuming nitric acid (Zetzsche and Kälin 1932a)

1 Add 20-30ml of fuming nitric acid to 0.5-1g of sample in a plastic beaker.

2 For low and medium rank bituminous coals leave for 16 hours. Higher rank coals may break down in shorter periods of time. Top up the acid as required during this time.

3 Filter the material through a sintered glass funnel.

4 Wash the residue in fuming nitric acid, then concentrated followed by dilute nitric acid, and finally wash with distilled water.

Most coals macerate within 24 hours, particularly if the specific gravity of the acid is greater than 1.5. The advantage of this method is that it does not require the use of an alkali stage, with its damaging effect on the spore walls. Furthermore, it appears that the oxidation period appears to be less critical, thus reproducible assemblages are consistently obtained. The palynomorphs are stained yellow in appearance, in contrast to the pale or colourless exines resulting from the method using Schulze's solution.

25.7.3c. Method using hydrogen peroxide (Zetzsche and Kälin 1932b)

A dilute solution can be used for macerating lignites, but for high rank coals a concentrated solution yields better results (in some cases better than the two methods described previously).

1 Gently boil 1g of sample with 50ml 30% (w/v) H_2O_2 for 4-6 hours, or until the coal has oxidized. During evaporation maintain the volume of solution with additional hydrogen peroxide. **N.B. take care if there is a large amount of pyrite in the sample as the reaction may become violent.**

2 As the colour of the solution clears examine the residue. Maceration times vary between coals of different rank.

3 The milky solution formed is cleared by the addition of a few drops of 10% HCl.

4 Centrifuge the residue, and wash with distilled water.

25. 7.3d. Other methods

For coals of high rank, Raistrick (1934) discovered that they responded better to treatment if, before the addition of Schulze's reagent, they were soaked in cold pyridine (C_5H_5N) for 24 hours, or even boiled in the solution for 16 hours. The pre-treated coal residue is then mixed with an equal weight (1g) of potassium chlorate and 25ml of fuming nitric acid. Maceration usually takes about 16 hours. Smith and Butterworth (1967, p 102) report that this method gives better results with higher rank coals than the potassium chlorate and concentrated nitric acid method, but is not as efficient as the hydrogen peroxide method.

Hoffmeister (1960) successfully used sodium hypochlorite, while Funkhouser and Evitt (1959) used *aqua regia* (concentrated nitric acid saturated with potassium dichromate). A mixture of solid periodic acid (H_5IO_6) and 50% perchloric acid ($HClO_4$) was used by Spielholtz *et al.*, (1962) in their maceration technique. Smith and Butterworth (1967) consider these inferior to the methods detailed above.

25.7.4. Washing and storing residue

1 Sieve and wash the coarse fractions, allowing material that passes through a 0.5mm sieve at least one hour to settle out before decanting off excess liquid.

2 Store residues in distilled water with a few drops of mould inhibitor. Microscopical examination of successfully macerated residue reveals an abundance of megaspores, cuticle and other plant debris. Megaspores yellow-brown in colour are preferable however, if they are dark brown or black in appearance further oxidation may be required. If the water turns brown during examination, add a few drops of the potassium hydroxide (KOH) solution, however, be cautious as excess alkali may result in spores swelling and rupturing.

3 Individual megaspores can be isolated from the residue while examining in water in a petri dish under a binocular microscope, by using a fine (00) sable hair brush or mounted needle.

4 Mount and stain if required. Examination is usually by transmitted light on a biological microscope.

A summary of palynological processing procedures required for specimens of different lithologies is presented in table 25.1.

	Clay	Unconsolidated sand	Argillaceous sediments	Calcareous rich and carbonate sediments	Shale and impure coal	Coals	Lignite	Peat	Oil, tar, asphalt and bitumens	Amber	Evaporites
Water ±surfactant			1								2
EDTA											1
Elutriation ±wet sieving		1									
Benzene-alcohol-alcohol mixtures									1		
Dispersing agent	1										
H_2SO_4 + washing in $CH_3.COOH$										1	
HCl (carbonate removal)	2	2	2	1	1						4
NaOH or KOH (soil colloid removal)	4	4				2	1	1			3
HF (silicate removal)	3	3	3	2	2			3			5
$CH_3.COOH$ (dehydration)								4			6
Acetolysis (cellulose removal)	5	5		5			2	5		2	
Gravity (liquid) separation	6	6	4	3							
Oxidation (lignin removal)			5	4	3	1	3	2	2		
Sieving/filtration by alcohol/water	•	•	•	•	•	•	•	•	•	•	•
Staining	7	7	6	6	4	3	4	6	3	3	7
Slide preparation	8	8	7	7	5	4	5	7	4	4	8

Table 25.1. Summary of procedures and order in which they should be carried out for palynological processing. It should be noted that elutriation techniques have not been described in detail as these involve specialized equipment not available in every laboratory. Sieving, filtration, centrifuging, etc., are procedures required after a number of processing stages, and are indicated by the symbol •. See text for details. (Modified after Herngren 1983).

25.8. MACERATION OF COALIFIED PLANT CUTICLE MATERIAL

Modern bulk maceration techniques for plant cuticle material in sedimentary rocks were first described by Harris (1926) and reviewed in detail by Wesley (1954). The procedure requires the removal of coaly structureless tissues that optically obscures features of the cuticle. Harris' (1926) procedure is essentially an oxidizing process, and forms the basis of the technique still used by palaeobotanists today. Modifications in separating cuticle from argillaceous matrix have been described by Andrews (1961) and Watson and Sincock (1992).

1 Place the specimen in a labelled beaker, and clean in warm 20% HCl for 24-48 hours, removing any calcareous material that would be transformed to form insoluble calcium fluoride.

2 Thoroughly wash specimen in distilled water.

3 Immerse in 48% HF for 48 hours. Some specimens may require longer to remove all siliceous material.

The original procedure of Harris (1926) involved the oxidation of material **before** it was immersed in HF. This part of the procedure was reversed by Andrews (1961) who dissolved the rock matrix in HF before nitric acid immersion. This enables plant material to be removed intact, and allows for greater control during the oxidation stage.

4 Decant and neutralize the liquid.

Oxidation is by one of three methods. In descending order of severity these are: (i) Schulze's solution (concentrated nitric acid (HNO_3) containing 5% potassium chlorate ($KClO_3$)) for rapid reactions; (ii) 20% chromium (VI) oxide (chromium trioxide, CrO_3) for slow gentle reactions; (iii) sodium hypochlorite.

5 Cover with 10% nitric acid (HNO_3) containing 5% potassium chlorate ($KClO_3$) or 5% sodium hydroxide (NaOH).

6 Leave for at least 24 hours.

In their study of English Wealden Bennettitales Watson and Sincock (1992), noted a preservational difference in similar material collected from the United Kingdom and Germany. Specimens from the UK required between 10 minutes and 48 hours immersion in Schulze's solution, while the more robust German material could be left for 4-7 days.

7 Decant off the solution, and thoroughly wash the sample in distilled water.

8 Immerse the residue in a very dilute (<4%) ammonium hydroxide solution, (the smell of ammonia is just detectable). Stronger solutions

cause cuticle fragments to curl, and appear to inhibit the "humic acid" solution (Watson and Sincock 1992). This process may take between a few minutes or many hours, with the residue safely left in the solution overnight. Wash the cuticle surface with a sable hair paint brush to remove stubborn matrix adhering to the surface.

Ultrasonic treatment for short time periods (10 seconds) can be applied, but may result in the destruction of adaxial cuticle (Watson and Sincock 1992).

9 Wet sieve the cuticle through a fine (38μm) mesh to remove all traces of the ammonia. Failure to do this adversely effects glycerol mounts, causing the mountant to irreversibly set. Processed residues consist of cuticle fragments that rarely exceed 2cm in length, plus any large palynomorphs retained on the sieve mesh.

For a rapid reconnaissance of cuticle material, the following quick procedure can be performed:

1 Detach small fragments of cuticle from the sample, and place in a shallow dish, and immerse in concentrated nitric acid.

2 Continue to examine the specimen every fifteen minutes using a binocular microscope. Specimens, initially be very dark brown, black or opaque will gradually lighten.

3 When the internal structure of the specimen is evident, remove the cuticle from the acid and wash in distilled water. Detailed microscopical examination can then proceed.

Where plant material is small and forms only a thin layer on a sample, it can be removed while still attached to a thin layer of matrix. The fragments are then placed on a fine nylon mesh screen for immersion in an HF bath. This modification was first outlined by Stubblefield and Banks (1978), and further modified by Doran (1982) to accommodate larger specimens, and facilitate the recovery of large well-preserved plant stems without the need to remove individual overlying branches.

Prior to mounting for LM observation, specimens can be stained with Safranine O (see 25.2 EXTRACTION OF POLLEN AND SPORES FROM ARGILLACEOUS AND SILICEOUS RICH SEDIMENTS). Techniques for the recovery and restoration of old and damaged slides are covered in the above section under slide preparation. Long term conservation problems and procedures of LM slides and SEM stub material have been discussed by Collinson (1987).

Ideally both LM and SEM preparations should be performed on the same specimen, although in practice this is seldom possible (Watson and Sincock 1992). The "peel" procedure described by Jacobson and Schopf (1979) provides a method of recovering smaller palynomorphs that have been attached to SEM stubs without the use of a cement, and thinly coated (<23nm) with a metallic electrolyte. With these criteria met, transferred material permits transmitted light observations. A single drop of peel solution (5 parts amyl acetate, 1 part nitro-cellulose) applied to the stub and allowed to dry, can be carefully removed and mounted on a glass microscope slide in a permanent mounting medium with a coverslip affixed. To avoid replicating the concentric milling marks on the metallic stub, original SEM mounting must be on a base of glass, mica or celluloid film (see section 34 ELECTRON MICROSCOPY TECHNIQUES for details), although they may aid in the relocation of specimens (Jacobson and Schopf 1979). Experience is required in matching SEM to LM observations, as considerably more internal detail is seen in incident light, furthermore the metallic coating imparts a slight colouring to the specimen, similar to that produced by staining and enhancing image contrast (Jacobson and Schopf 1979).

25.9. PALAEOBOTANICAL PROCEDURES

Some palaeobotanical studies can involve the comparison of recent and fossil material. Careful preparation of both types of material is required to reveal histological features. To identify and examine all the relevant features, surfaces and sections of three oriented planes perpendicular to one another (Fig. 25.2) must be viewed by either light or electron microscopy. Preparation of recent and sub-fossil wood is influenced by the type of tree examined (hard or softwood), and the type of wood (bark, sapwood or heartwood).

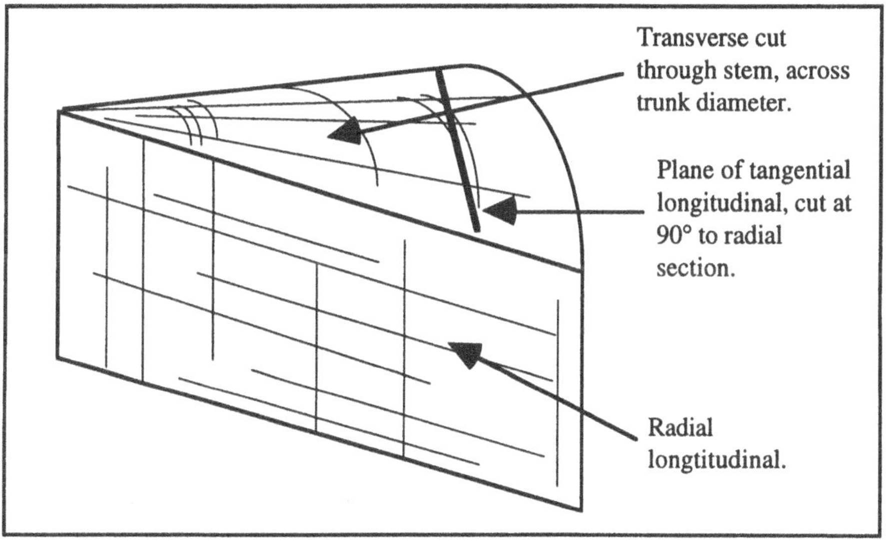

Figure 25.2. Diagrammatic illustration through a portion of tree stem illustrating the respective positions of the oriented planes of section required to obtain the optimum anatomical detail required for identification.

25.9.1. Cubes and blocks for histological examination

Sample from living trees with a diameter of 7.5cm or greater, and at least 1.5m above ground level. The sampling area must be free from knots and any other obvious distortions. Ultimately, the sample size should be large enough to produce a finished cube or block with three oriented faces (Fig. 25.2) of approximately 1-1.5cm diameter (Abbott *et al.*, 1982). Sectioning can be undertaken using either a sharp razor blade, or if greater accuracy is required a microtome. However, depending on the preservational state of the sample, either softening or hardening (resin embedding) may be required. Abbott *et al.*, (1982) provide details on a number of softening (and preserving) procedures applicable to recent woods. These procedures are summarized and tabulated below (Table 25.2). Increased optical contrast on prepared surfaces is obtainable through staining for a few seconds in a bath of 1:1000 aqueous solution of acridine orange ($C_{17}H_{20}ClN_3.ZnCl_2$) dye. Stained surfaces fluoresce under short wave light, revealing greater anatomical detail.

The SEM observation of Recent wood, and particularly beach, has revealed some unusual artefacts peculiar to preparation (Kucera 1986). Tangential split sections of modern wood specimens were compared with naturally split sections exhibiting structures devoid of cell wall deformations induced from cutting. However, species of beach observed under the SEM frequently show screw-shaped strips, formed from transverse tensile stresses invoked by the splitting process. They are not recognized in radial preparations of beach or other species, or evident on cut specimens (Kucera 1981). Furthermore, there does not appear to be any evidence of these artefacts in fossil specimens, but their appearance cannot be ruled out when preparing tangential preparations of sub-fossil beach wood specimens.

25.9.2. Histological examination of stained recent wood

Abbott *et al.*, (1982) describe staining procedures suitable for sections from softwoods (coniferous) and hardwoods.

25.9.2a. Softwoods

1 Gently agitate the cut section in a series of ethyl alcohol solutions (50, 30, 20, 10%) for two minutes, pipetting supernatant fluid between each 'wash'. Finally, wash in distilled water for two minutes.

2 Add 2-3% ammonium iron(III) sulphate *12-hydrate* (iron alum, $NH_4Fe(SO_4)_2.12H_2O$) for 20-30 minutes.

3 Flood the section with distilled water, and add 2-3 drops of 0.5% haematoxylin ($C_{16}H_{14}O_6.H_2O$). Leave until the section takes on a dark blue (not black) colour.

4 Wash in several changes of distilled water.

5 Cover the section with a 2% Safranine O ($C_{20}H_{19}ClN_4$) solution. Immersion may require only a few seconds, or up to 24 hours to achieve the desired intensity of colour. If the section is over stained, wash in distilled water to remove excess.

6 Wash in a series of ethyl alcohol solutions (10, 20, 30, 50, 70, 95, 100%) for two minutes at each concentration.

7 Immersion in xylene ($C_6H_4(CH_3)_2$) prevents the gradual fading of the stain, stabilizing it for permanent slide mounting (see 25.2 POLLEN

PROCEDURE	METHOD	RESULTS AND EFFECTS
Waterlogging	Gentle boiling	Block sinks and becomes waterlogged.
Glycerol:alcohol	50:50 volume	Softens and lubricates. May require a few days or many months.
Acetic acid methods (a) Vinegar + H_2O	(i) Waterlog and boil for 30 minutes in 50:50 volume. (ii) Rinse in distilled water. (iii) Boil for 30 minutes in 50% NH_3 solution. (iv) Wash and dry.	Further sections can be cut from re-waterlogged blocks.
(b) Formalin + acetic + ethyl alcohol	Store cubes in a 5:5:90 volume	Blocks can be kept for many years, requiring no further softening before sectioning.
(c) H_2O_2 + glacial acetic	50:50 volume Weaker solution (1 part H_2O_2 + 2 parts acid) at room temperature softens wood more gradually.	Placed in plastic bottles with loosely fitting caps, and oven heated at 50°C. Rapidly softens cubes. N.B. Extended treatment of 2-3 days can result in total maceration and delignification. Shorter time periods can cause over softening of outer layers, which may require trimming before sectioning.
Glycols (a) Poly(ethylene glycol) (commercial antifreeze)	Boil pre-waterlogged specimens	Store (wet or dry) or section immediately. Addition of 10-15ml of acetic acid or vinegar to PEG may be required to harden some specimens (*e.g.* oak).
(b) Triethylene glycol (for conifers and permeable hardwoods)	Immerse section in 50-100ml of triethylene glycol, and heat at 120-150°C for 20-30 minutes. Waterlog specimens if required. For hardwoods add 0.2-0.5% (by weight) of phenosulphonic or paratoluenesolphonic acid.	Softened woods can be stored dry for many months. N.B. acidity of glycol increases with re-use.
Ethylenediamine and hydrofluoric acid	Immerse sections in a 4% ethylenediamine solution. Evacuate with vacuum pump, repeating the procedure to ensure complete ethyl penetration. In a fresh solution heat to 70-75°C. Remove and dry with paper towel. Immerse transverse sections in a 50% HF solution, and leave for 16 hours. Neutralize before removal.	Hardwood specimens may require soaking in HF for many months before suitable sections can be obtained. N.B treat radial sections separately from transverse sections, omitting the HF stage. **CAUTION: Use HF with extreme care (see notes at the beginning of this chapter and appendix 8 for advice on care and handling).**

Table 25.2. Preserving and softening procedures of recent wood (data from Abbott *et al.*, 1982).

AND SPORE EXTRACTION FROM ARGILLATES AND SILICATES).

25.9.2b. Hardwoods

1 Wash the section in a 95% ethyl alcohol solution for at least 2 minutes. If the section has been stored in glycerol, commence the procedure with a 70% solution.

2 Immerse the section in a 2% solution of Safranine O (stains the lignified walls, *e.g.* vessels, fibres, tracheids).

3 By gentle agitation wash in two or more changes of 95% ethyl alcohol for 1 minute. Keep changing the solutions until the final wash remains clear.

4 Flood the section, and gently agitate with Fast green FCF (C Green No. 3, $C_{37}H_{34}N_2Na_2O_{10}S_3$, stains parenchyma and tension wood fibres of cellulose walls) for about 15 seconds. Excess time results in a combination of the green and red stains turning the section blue.

5 Wash section in 100% ethyl alcohol for 10 seconds.

6 Stabilize the stain by immersing the section in xylene, and permanently mount (see 25.2 POLLEN AND SPORE EXTRACTION FROM ARGILLATES AND SILICATES).

25.9.3. Preparation of peels

Microstructural studies of pyritized and silicified wood are undertaken by two methods. Firstly, by standard and polished thin sections studied under transmitted and reflected light, and secondly, by preparing peels and examining them in section in transmitted light. Both are standard procedures that have been employed by palaeobotanists for over 70 years (*e.g.* thin sections Heard 1927, Matten 1968; peels Walton 1928, 1930, Joy *et al.*, 1956, Jennings 1972, Abbott *et al.*, 1982). Thin sectioning procedures are outlined in section 20 THIN SECTION AND SLIDE PREPARATION TECHNIQUES OF MACRO AND MICROFOSSIL SPECIMENS, and will not be referred to further, while standard carbonate peel techniques are described in the section PREPARATION OF CARBONATE STAINED ACETATE PEELS AND THIN SECTIONS, however, subtle differences employed by palaeobotanists in preparing pyritized material are outlined below.

A historical review of procedures for calcareous coal-balls, including the preparation of liquid plastic sheets, is found in Stewart and Taylor (1965). The procedure below, based on Jennings (1972) outlines a peel alternative using poly(vinyl chloride), in contrast to examining cut and polished pyritized material in thin section using reflected light.

1 Cut and prepare a flat polished surface, finishing with 1000 grit carborundum powder.

2 Etch the prepared surface for 1-1.5 minutes in a concentrated nitric acid (HNO_3) solution at 70°C. **N.B. this procedure, and the following chemical stages must be undertaken in a well-ventilated fume cupboard.**

3 Neutralize the acid by immersing the specimen in an aqueous solution of sodium hydroxide (NaOH) for 1-2 minutes.

4 Brown hydroxides may form on the surface, however these can be removed by immersion of the surface in a concentrated HCl solution.

5 Neutralize the acid by immersing the surface in a dilute 5-10% ammonium hydroxide (NH_4OH) solution and acidify in a saturated oxalic acid ($(COOH)_2.2H_2O$) solution.

6 Oven dry the specimen at 60°C.

7 Mix a saturated solution of poly(vinyl chloride) dissolved in butanone (ethyl methyl ketone, $C_2H_5.CO.CH_2$), and pour onto the surface of the specimen.

8 Leave for 24 hours, then carefully remove the peel with the aid of a razor blade. Do not touch the surface of the peel that has been in contact with the specimen.

9 Complete drying of the surface may take 3-4 days. The presence of any adhering iron sulphide particles can be removed by immersing the peel in a bath of nitric acid at room temperature. The concentration of the acid does not effect the quality of the poly(vinyl chloride) peel.

Examination and photography of the peel is undertaken in transmitted light. Jennings (1972) illustrated a significant improvement in contrast and detail observed in comparison with polished sections viewed in reflected light.

REFERENCES

ABBOTT, M, BAREFOOT, A. C and HANKINS, F. W. 1982. Appendix 2, Specimen preparation, pp 165-172: In; A. C. Barefoot and F. W. Hankins *Identification of modern and Tertiary Woods.* Clarendon Press, Oxford. vii + 189.

ALDRIDGE, R. J. 1990. Extraction of Microfossils, pp 502-504: In; D. E. G. Briggs and P. R. Crowther (eds), *Palaeobiology - A synthesis.* Blackwell Scientific Publications (Oxford). xiii + 583p.

ANDREWS, H. N. Jr. 1961. *Studies in Paleobotany.* John Wiley & Sons Inc. (New York). 487p.

ASSARSSON, G and GRANLUND, E. 1924. En metod för pollenanalys av minerogena jordarter. *Geologiska Föreingen Förhandlinger.* Band. **46**, 76-82.

BARSS, M. S and CRILLEY, B. 1976. A mounting medium for palynological residues. *Geological Survey of Canada,* Paper **76-1B**. 131-132.

BARSS, M. S and WILLIAMS, G. L. 1973. Palynology and nannofossil processing techniques. *Geological Survey of Canada,* Paper **73-26**. 1-26.

BATES, C. D, COXON, P and GIBBARD, P. L. 1978. A new method for the preparation of clay-rich sediment samples for palynological investigation. *New Phytologist.* **81**, 459-463.

BATTEN, D. J and MORRISON, L. 1983. Methods of palynological preparation for palaeoenvironmental, source potential and organic maturation studies. *Norwegian Petroleum Directorate-Bulletin 2 (Oljedirektoratet).* 35-53.

CARATINI, C. 1980. Ultrasonic sieving to improve palynological processing of sediments: a new device. *International Commission for Palynology Newsletter.* **3**, (1), 4.

CLAY-POOLE, S. T. 1990. Calcium dissolution and clay flocculant removal in pollen extractions from sediments of the arid southwest, USA. *Journal of Paleontology.* **64**, (3), 483-484.

COHEN, A. D and GUBER, A. L. 1968. Production of pollen-sized "microforaminifera" from "normal" foraminifera. *Micropaleontology.* **14**, (3), 361-362.

COLBATH, G. C. 1985. A comparison of palynological extraction techniques using samples from the Silurian Bainbridge Formation, Missouri, USA. *Review of Palaeobotany and Palynology*. **44**, 153-164.
COLLINSON, M. E. 1987. Special problems in the conservation of palaeobotanical material, pp 439-445; In: P. R. Crowther and C. J. Collins (eds), The Conservation of Geological Material. *The Geological Curator*. **4**, (7), 375-474.
CWYNAR, L. C, BURDEN, E, McANDREWS, J. H. 1979. An inexpensive sieving method for concentrating pollen and spores from fine-grained sediments. *Canadian Journal of Earth Sciences*. **16**, (5), 1115-1120.
DEAN, A. 1989. Palynomorphs from deformed low-grade metamorphic rocks: an SEM-based technique. *Journal of the Geological Society, London*. **146**, (4), 597-599.
DIMICHELE, W. A and WING, S. L. (eds). 1988. Methods and Applications of Plant Paleoecology. *Paleontological Society Special Publication*. **3**, iv + 1-171.
DOHER, I. L. 1980. Palynomorph preparation procedures currently used in the paleontology and stratigraphy laboratories, U.S. Geological Survey. *U. S. Geological Survey Circular*. **830**. 1-29.
DORAN, J. B. 1982. A modification of the bulk maceration technique for paleobotany. *Journal of Paleontology*. **56**, (3), 823-825.
EAGAR, S. H and SARJEANT, W. A. S. 1963. Fossil hystrichospheres concentrated by sieving techniques. *Nature*. **198**, 81.
EDIGER, V. S. 1986. Sieving techniques in palynological sample processing with special reference to the MRA system. *Micropaleontology*. **32**, (3), 256-270.
ERDTMAN, G. 1943. *An introduction to pollen analysis*. Plant Science Books, vol. 12, and Chronica Botanica. Waltham, Massachusetts and New York. 239p.
FAEGRI, K and IVERSEN, J. 1989. *Textbook of pollen analysis* (4th edition). Revised by K. Faegri, P. E. Kaland and K. Krzywinski. John Wiley and Sons (Chichester).
FARLEY, M. B. 1988. Environmental variation, palynofloras, and paleoecological interpretation, pp 126-146: In; W. A. DiMichele and S. L. Wing (eds) Methods and Applications of Plant Paleoecology. *Paleontological Society Special Publication*. **3**, iv + 1-171.
FARUQI, S. A and COPLEY, A. J. 1966. A technique for separating pollen and spores from tar. *Micropaleontology*. **12**, (3), 370.
FUNKHOUSER, J. W and EVITT, W. R. 1959. Preparation techniques for acid-insoluble microfossils. *Micropaleontology*. **5**, (3), 369-375.
GRAY, J. 1965a. Part III, Techniques in palynology (coordinator), pp 469-706: In; B. Kummel and D. Raup (eds), *Handbook of Paleontological Techniques*. W. H. Freeman and Co., (San Francisco). xiii + 852p.
HARRIS, T. M. 1926. Note on a new method for the investigation of fossil plants. *New Phytologist*. **25**, 58-60.
HEARD, A. 1927. On Old Red Sandstone plants showing structure, from Brecon (South Wales). *Quarterly Journal of the Geological Society, London*. **83**, 195-209.
HERNGREEN, G. F. W. 1983. Palynological preparation techniques. *Norwegian Petroleum Directorate-Bulletin 2 (Oljedirektoratet)*. 13-34
HIGGINS, A. C and SPINNER, E. G. 1969. Techniques for the extraction of selected microfossils. *Geology - Journal of the Association of Teachers in Geology*. **1**, 12-28.
HILLS, L. V and SWEET, A. R. 1971. The use of "Quaternary O" in megaspore palynological preparations. *Review of Palaeobotany and Palynology*. **13**, 229-231.
HOFFMEISTER, W. S. 1960. Sodium hypochlorite, a new oxidising agent for the preparation of microfossils. *Oklahoma Geological Notes*. **20**, 34-35.
HUGHES, N. F, DE JEKOWSKY, B and SMITH, A. H. V. 1964. Extraction of spores and other organic microfossils from Paleozoic clastic sediments and coals. *C. R. Sième Congress of International Stratigraphy and Géologie, Carbonifère*. **3**, 1095-1109.
JACKSON, T. A, GERMS, A and MOORMAN, M. 1974. An improved method for the chemical maceration of sedimentary rocks. *Journal of Paleontology*. **48**, (4), 844-845.
JACOBSON, S. R and SCHOPF, J. M. 1979. Preservation of palynological specimens examined by SEM for later microscopical study and reference. *Journal of Paleontology*. **53**, (3), 744-746.
JENNINGS, J. R. 1972. A polyvinyl chloride peel technique for iron sulphide petrifications. *Journal of Paleontology*. **46**, (1), 70-71.
JOY, K. W, WILLIS, A. J and LACEY, W. S. 1956. A rapid cellulose peel technique in paleobotany. *Annals of Botany*. **20**, 635-637.
KIDSON, E. J and WILLIAMS, G. L. 1969. Concentration of palynomorphs by use of sieves. *Oklahoma Geological Notes*. **29**, 117-119.
KUCERA, L. J. 1981. Cutting wood specimens for observation in the scanning electron microscope. *Journal of Microscopy*. **124**, (3), 319-325.
KUCERA, L. J. 1986. Splitting wood specimens for observation in the scanning electron microscope. *Journal of Microscopy*. **142**, (1), 71-77.
LENNIE, C. R. 1968. Palynological techniques used in New Zealand. *New Zealand Journal of Geology and Geophysics*. **11**, (5), 1211-1221.
LITWIN, R. J and TRAVERSE, A. 1989. Basic guidelines for palynomorph extraction and preparation from sedimentary rocks, pp 87-98; In: R. M. Feldman, R. E. Chapman and J. T. Hannibal (eds) *Paleotechniques*. The Paleontological Society Special Publication No. 4. iv + 358.
MATTEN, L. C. 1968. *Actinoxylon banksii* gen. et. sp. nov.: A progymnosperm from the Middle Devonian of New York. *American Journal of Botany*. **55**, 773-782.
McINTYRE, D. J and NORRIS, G. 1964. Effect of ultrasound on recent spores and pollen. *New Zealand Journal of Science*. **7**, (2), 242-257.
MCKEE, K. 1977. A dialysis unit in palynological sample preparation. *The British Micropalaeontologist*. **5**, 12-13.
MOORE, P. D, WEBB, J. A and COLLINSON, M. E. 1991. *Pollen Analysis*. (2nd edition). Blackwell Scientific Publications (Oxford). viii + 216p.
NEVES, R and DALE, B. 1963. A modified filtration system for palynological reparations. *Nature*. **198**, 775-776.
NOREM, W. L. 1956. An improved method for separating fossil spores and pollen from siliceous rocks. *Journal of Paleontology*. **30**, (5), 1258-1274.
NØRGAARD, I, RASMUSSEN, A-M, SCHIØLER, P and STOUGE, S. 1991. A new method for bulk palynological processing of oil-contaminated chalks. *Journal of Micropalaeontology*. **10**, (2), 202.

PEPPERS, R. A. 1970. Correlation and palynology of coals in the Carbondale and Spoon Formations (Pennsylvanian) of the Northeastern Part of the Illinois Basin. *Bulletin of the Illinois State Geological Survey*. **93**, 1-173.

PHIPPS, D and PLAYFORD, G. 1984. Laboratory techniques for extraction of palynomorphs from sediments. *Papers, Department of Geology, Univesity of Queensland*. **11**, 1-23.

RAINE, J. I and TREMAIN, R. 1992. Apparatus for rapid sieving of palynological residues. *New Zealand Geological Survey*. Report **PAL 151**. 14p.

RAISTRICK, A. 1934-1935. The correlation of coal-seams by microspore-content: Part 1. The seams of Northumberland. *Transactions of the Institute of Mining Engineers (London)*. **88**, 142-153; 259-264.

REISSINGER, A. 1950. Die "Pollenanalyse" ausgedehnt auf alle Sedimentgesteine der geologischen Vergangenheit. Part II: *Palaeontographica Abteilung B*. Band **90**, 99-126.

SCHULZE, F. 1855. Über des vorkommen wohlerhaltener cellulose in braunkohle und steinkohle. *Bericht uber die zur Bekanntmachung geeigneten Verhandlungen Koniglich Preussischen Akademie der Wissenschaften zu Berlin*. **21**, 676-678.

SITTLER, C. 1955. Méthodes et techniues physicochimiques de préparation sédiments envue de leur analyse pollinique. *Revue de l'Institut Français du Pétrole et Annales des Combustibles Liquides*. **10**, (2), 103-114.

SMITH, A. H. V and BUTTERWORTH, M. A. 1967. Miospores in the coal seams of the Carboniferous of Great Britain. *Special Papers in Palaeontology*. **1**, x + 324p.

SPIELHOLTZ, G, THOMAS, L. A and DIEHL, H. 1962. Isolation of spores by wet oxidation. *Micropaleontology*. **8**, 109-110.

STEWART, W. N and TAYLOR, T. N. 1965. The peel technique, pp 224-232: In; B. Kummel and D. Raup (eds), *Handbook of Paleontological Techniques*. W. H. Freeman and Co., (San Francisco). xiii + 852p.

STUBBLEFIELD, S and BANKS, H. P. 1978. The cuticle of *Drepanophycus spinaeformis*, a long-ranging Devonian lycopod from New York and eastern Canada. *American Journal of Botany*. **65**, (1), 110-118.

TAPPAN, H and LOEBLICH, Jr., A. R. 1965. Foraminiferal remains in palynological preparations. *Revue de Micropaléontologie*. **8**, (2), 61-63.

TRAVERSE, A. 1988. *Paleopalynology*. Unwin Hyman (London). xxiii + 600p.

VIDAL, G. 1988. A palynological preparation method. *Palynology*. **12**, 215-220.

WALTON, J. 1928. A method of preparing sections of fossil plants contained in coal-balls or in other types of petrifications. *Nature*. **122**, 571-572.

WALTON, J. 1928. Improvements in the peel method. *Nature*. **125**, 413-414.

WATSON, J. 1977. Some Lower Cretaceous conifers of the Cheirolepidiaceae from the USA and England. *Palaeontology*. **20**, (4), 715-749.

WATSON, J and SINCOCK, C. A. 1992. *Bennettitales of the English Wealden*. Monograph of the Palaeontolographical Society (London). 1-228, pls 1-23. (Publ. No. 588, part of Vol. **145** for 1991).

WESLEY, A. 1954. A short synopsis of some microscopical methods in Palaeobotoany. *Proceedings of the Leeds Philosophical and Literary Society*. Science Section. **6**, 168-179.

WEST, R. G. 1977. *Pleistocene geology and biology* (2nd edition). Longman (London). xi + 440p.

WILSON, G. J. 1971. A chemical method for the palynological processing of chalk. *The Mercian Geologist*. **4**, (1), 29-36.

WILSON, G. J. 1972. A method for the recovery of mounted palynological residues. *The Mercian Geologist*. **4**, (2), 139-141.

WILSON, L. R and HOFFMEISTER, W. S. 1952. Small foraminifera. *Micropaleontologist*. **6**, (2), 26-28.

ZETZSCHE, F and KÄLIN, O. 1932a. Untersuchungen über die membranen der sporen und pollen VII. Eine methode zur isolierung des polymerbitumens (sporenmembranen, kutikulen usw) aus kohlen. *Braunkohle*. **31**, 345-351 and 363-366.

ZETZSCHE, F and KÄLIN, O. 1932b. Untersuchungen über die membranen der sporen und pollen VIII. *Helv. chim. acta*. **15**.

26. EXTRACTION TECHNIQUES FOR ACID INSOLUBLE MICROFOSSILS

26.1. INTRODUCTION

A variety of techniques used in the extraction of acid resistant microfossils are described in this section, although many are similar to those outlined in section 25 EXTRACTION TECHNIQUES FOR PALAEOBOTANICAL AND PALYNOLOGICAL MATERIAL. In the following account microfossil groups not specifically mentioned under previous headings are emphasised, although many of these may be present in assemblages as "by-products" of processing for another microfossil group. Many of the extraction procedures described are readily modified to macrofossils of similar biomineral compositions (*e.g.* phosphatic and siliceous fossils). The preparation of strew mount slides is detailed in section 20 THIN SECTIONING AND SLIDE PREPARATION TECHNIQUES OF MACRO AND MICROFOSSIL SPECIMENS. Methods are described under the headings of three common biomineral types, each to some extent an acid resistant group:

i Phosphatic
ii Siliceous
iii Organic walled (excluding pollen and spores)

SAFETY PROCEDURES AND LABORATORY PRACTICES OUTLINED IN SECTION 25 EXTRACTION TECHNIQUES FOR PALAEOBOTANICAL AND PALYNOLOGICAL MATERIAL MUST BE FOLLOWED - THEY WILL NOT BE REPEATED IN THIS SECTION.

Hydrofluoric acid (HF) is used in some of the procedures described below. It is extremely dangerous and must only be used in laboratories with facilities capable of handling disposal and potential spillage. HF should be used following training or under supervision, and informing other laboratory workers, adhering to chemical suppliers' recommendations and state legislation. Consult the safety notes of Appendix 8 - Care and handling of HF, and Appendix 9 - Spillage and disposal of unwanted chemicals.

26.2. PHOSPHATIC

This group of microfossils includes conodonts, hyoliths and phosphatic ostracod-like arthropoda (Phosphatocopina), together with two groups usually studied by macropalaeontologists, micro-vertebrate remains and microscopic phosphatized brachiopods (Müller 1978). For details of the specialist extraction procedures of phosphatized fossils refer to 27 EXTRACTION TECHNIQUES FOR PHOSPHATIC FOSSILS.

26.3. SILICEOUS

Four groups of microfossils have siliceous skeletons. These are:

i Radiolaria
ii Diatoms
iii Silicoflagellates
iv Ebridians

Sediments are frequently processed exclusively for both radiolaria and diatoms, and as a consequence specific techniques have been developed for the extraction of these from different lithologies suffering varying degrees of induration. Usually occurring together, silicoflagellates and ebridians are planktonic marine organisms, often seen in residues prepared for other siliceous organisms.

26.3.1. Radiolaria

Range from 100-2000μm in diameter, although colonial associations can attain a size of 250mm. Living radiolaria are composed of amorphous (opaline) silica ($SiO_2.nH_2O$). This unstable form of silica is frequently replaced in pre-Tertiary fossils, following the removal of water, to chalcedony a related variety of silica. Skeleton replacement by calcite and pyrite is also known. A closely related group to the radiolarians, the acantharians, have skeletons composed of celestite (strontium sulphate, $SrSO_4$), but these are not preserved in the fossil record, while fresh-water heliozoans are another group that bear a superficial resemblance to the radiolarians however, they too do not produce a fossilized skeleton (Kling 1978). Radiolaria have a continuous stratigraphic range from the Cambrian to Recent, although evidence suggesting they may range back into the Precambrian has yet to be authenticated, and many of these accounts appear to be describing acritarchs (Kling 1978, Brasier 1980).

As radiolaria are susceptible to diagenetic alteration, a far wider range of collecting and preparation techniques are used to extract assemblages from Mesozoic in comparison to Cenozoic aged sediments. The evolution of procedures, many of which can be performed in a field laboratory (De Wever *et al.*, 1979, Baumgartner *et al.*, 1981), reflect the micropalaeontological value applied to non-siliceous preserved radiolaria, and the targeting of samples that generate high yields when processed on return to the laboratory. Unfortunately, the most productive rocks (argillaceous shales and marls) are difficult to test in the field, but routine processing procedures are relatively quick, so little time is wasted in processing barren samples. Carbonate and siliceous rocks can be tested in the field, by etching a surface in hydrochloric or hydrofluoric acid respectively (see section 4 COLLECTING TECHNIQUES FOR MICROFOSSIL AND LIVE FORAMINIFERA SAMPLES for details). Careful examination of the dried etched surfaces with a hand lens may reveal the presence of radiolaria (that have remained unetched, and stand proud of the surface). In this way only the most productive horizons are sampled for laboratory processing.

Composite, modified extraction procedures, based on Burma (1965), Pessagno (1977b, c), Riedel and Sanfilippo (1977), Sanfilippo and Riedel (1985) and Sanfilippo, Westberg-Smith and Riedel (1985), collectively summarized in Dyer and Copestake (1990), are outlined below.

26.3.1a. Argillaceous rocks

Radiolaria most commonly occur in sediments of argillaceous (mudstones, shales and marls) composition. Approximately 25-50g of sample, although less for pelagic sediments, is required for processing.

Hydrogen Peroxide + MP 10 (a multi-purpose cleaner and degreaser)

1 Place the sample in a beaker and oven dry at 100°C. Loosely disaggregate the sample to 2cm^3 fragments.

2 Sub-sample and weigh out approximately 25g of sediment. Place in a glass beaker. When the sample is completely dry, remove from the oven and place on a warm laboratory hot plate (set at 70°C).

3 Cover the sediment with approximately 10ml of 15% hydrogen peroxide (H_2O_2), and an equal volume of a multi-purpose surfactant or degreaser (*e.g.* MP 10). MP 10 (see Appendix 1, supplier's list) is a water based alkali (pH 12-12.5) solution containing both anionic and non-ionic surfactants and a glycol ether solvent, and assists in the removal of oil based drilling mud. **N.B. the reaction is exothermic - can be very violent if excess MP 10 is added.**

4 Gently stir the sediment until all the aggregates are dispersed.

5 When the initial reaction has ceased additional hydrogen peroxide and MP 10 can be added to the residue, until approximately 50ml of each solution have been added.

6 Leave the beaker for 15-20 minutes.

7 Wash the sediment through a nest of stainless steel sieves, terminating with a 38μm sieve, as coarser screens retain adult forms only (Kling 1978). Retreat coarse material (>1.7mm, stages 2-6) if additional fine residue is required. Thoroughly wash the residue to remove all the mud fraction, and if necessary use additional MP 10 during the washing process to remove persistent contamination from oily drilling fluids.

8 Any carbonate component present can be removed by the addition of a 10% solution of hydrochloric acid (HCl), although the presence of a weak siliceous cement may inhibit this process. Brief treatment with dilute (10%) hydrofluoric acid (HF), or boiling in a weak solution (5%) of sodium hydroxide (NaOH) assists in releasing the radiolaria.

9 Rinse the cleaned residue into an evaporating dish, or back into the beaker, and oven dry the residue at 60°C.

10 Dry sieve residues into three or four fractions to aid in optical examination. Bag and label the dried residues into "poly-grip" sample bags.

Petroleum spirit

1 Break down the sample into 2cm^3 fragments, place in a beaker and oven dry at 100°C.

2 Soak in petroleum spirit for 2 hours. Pessagno (1977c) suggests the use of a *Waring Blender* will reduce the sediment to a mud in 30-60 seconds. Due to the volatile nature of petroleum spirit **do not** heat or boil the solution.

3 Decant and filter the excess petroleum spirit for re-use, and immediately cover the sediment with boiling water. Stir and allow the sediment to stand until disintegration has ceased (usually 5-30 minutes).

4 Wet sieve the sediment over a 38µm sieve to remove the clay size fraction.

5 Add 10% hydrochloric acid (HCl) to remove the carbonates.

6 Further cleaning of the specimens can be achieved by either re-boiling the residue with a dispersant, or by immersion in an ultrasonic tank filled with a distilled water and detergent solution for a few seconds. This aids in dispersing any fine organic matter (Funkhouser and Evitt 1959). If small quartz grains adhere to the fossil surface, treating residues with a hot solution of sodium hydroxide (NaOH) before ultrasonic treatment, loosens the particles. However, care must be taken with these procedures, as ultrasonic vibrations destroy fragile tests, and prolonged exposure to sodium hydroxide (NaOH) may result in shell surface dissolution.

7 Wet sieve and oven dry the residue as above.

8 The radiolaria are concentrated using the Franz Isodynamic Separator, which separates the magnetic clay fraction from the non-magnetic radiolaria (ostracod and foraminifera) fraction. A pure residue is obtainable using this method (Eckert *et al.*, 1961).

Hydrofluoric acid (Pyritized organisms)

Safety note: Hydrofluoric acid (HF) is extremely dangerous, consult the notes in section 25 EXTRACTION TECHNIQUES FOR PALAEOBOTANICAL AND PALYNOLOGICAL MATERIAL, Appendix 8 - Care and handling of HF, and Appendix 9 - Spillage and disposal of unwanted chemicals for conditions of use and procedures to be followed.

The extraction of pyritized organisms such as radiolaria and diatoms from Jurassic-Cretaceous and Palaeocene argillaceous rich sediments of the North Sea, illustrates many of the problems in following a standardized processing technique. Conventional micropalaeontological techniques result in residues in which recovered organisms are often "dirty", with surfaces obscured with a fine layer of clay-grade silica-rich matrix. Standard cleaning procedures, such as ultrasonic and boiling with a deflocculant, frequently result in specimen damage (S. Packer *pers. comm*).

Obtaining a residue requires a processing procedure that cleans the individual specimens and assists in matrix removal, causing the least damage to the fragile elements and shell processes. Furthermore, the removal of contaminants within a residue processed from well-cuttings, impregnated with a mica-rich drilling mud, must also be successful. The preferred method of concentrating pyritized radiolaria is with heavy liquids, as the use of the magnetic separator results in a biased assemblage. The two principle reasons for un-reliability result firstly, from the partial pyritization of the assemblage, where some of the microfossil shells exhibit replacement and others only partial replacement. In many assemblages it is possible to observe both complete and partially pyritized microfossils.. Secondly, smaller radiolaria are concentrated in a fine grade size fraction, and therefore not suitable for a successful electro-magnetic separation.

Processing, including a hydrofluoric acid (HF) chemical stage provides a means of obtaining a clean, undamaged, assemblage, however, it must be noted that HF processing may be of limited use, particularly if the organisms present are not completely pyritized and replaced from their original siliceous composition.

The modified method below follows that of Pessagno (1977a, b), Sanfilippo and Riedel (1985), and Dyer and Copestake (1990).

1. Clean the sample of drilling mud and recent floral debris. Follow steps 1-8 of the hydrogen peroxide plus MP 10 method.

2. Wash the residue into a plastic polythene screw top bottle (with a volumetric capacity of approximately 250ml).

3. Oven dry the residue at 60°C.

4. In an HF fume cupboard - cover the sample with approximately 50ml of 5% HF. Leave for 4 hours.

5. Decant and neutralize the excess HF into a beaker of orthoboric acid (H_3BO_3), **or** fill the beaker with distilled water and allow to stand for 1 hour before decanting.

6. Add more distilled water, leave for 4-6 hours, decant and neutralize. Repeat this stage at least four times.

7. Once the solution is neutral (test with pH paper or pH solution), wash the residue over a 38μm stainless steel sieve.

8. Oven dry the residue at 60°C.

Microfossil tests composed of calcite, present in argillaceous sediments, can be transformed to calcium fluoride by immersion in HF for 24-48 hours (Pessagno 1977a). Pyritized fossils remain undamaged. Refining the technique, by quantifying the strength of HF required, and the contact time between the acid and sediment, forms a current experimental study (S. Packer *pers. comm*).

26.3.1b Calcareous rocks

Hydrochloric acid (HCl)

1 Place the sample in a beaker and oven dry at 100°C. Loosely disaggregate the sample to $2cm^3$ fragments.

2 Sub-sample and weigh out approximately 25g of sediment. Place in a glass beaker and slowly add concentrated (20%) HCl. If a vigorous reaction results add methanol (CH_3OH) before continuing. Formic acid (H.COOH) is a suitable alternative to hydrochloric acid, particularly if the reaction cannot be adequately controlled. Test the process on a sub-sample before proceeding.

3 When the reaction has ceased, neutralize and decant the acid by the addition of distilled water.

4 Wet sieve the sample over a 38µm sieve.

5 Return the residue to the large beaker and cover with a 30% hydrogen peroxide solution. In some samples this reaction may generate large amounts of heat, while for others the beaker requires heating on a hot-plate. Excess matrix becomes dislodged during this stage, although individual radiolaria may require more than one treatment.

6 Wet sieve the residue through a nest of sieves (180µm, 63µm onto a 38µm), and oven dry at 60°C.

7 Concentrate the radiolaria using the carbon tetrachloride (CCl_4) flotation method (see section 19 FLOTATION AND LIQUID SEPARATION TECHNIQUES). Baumgartner *et al.*, (1981) exploit the difference in settling velocities between the radiolaria and other grains, and illustrate apparatus providing an inverted water flow, although they note that no separating procedure is 100% successful, and that all fractions must be carefully examined. Dry sieve, examine the grade fractions and pick residues.

Hydrogen Peroxide (H_2O_2)

1 Calcareous marls, deep-sea clays, silts and oozes are best processed by immersing in a 30% solution of hydrogen peroxide (H_2O_2), with the addition of 5g of *tetra*-sodium pyrophosphate ($Na_4P_2O_7.10H_2O$) per

150ml of solution. Heat the mixture until all the pyrophosphate has dissolved. Hydrogen peroxide (H_2O_2) assists in the physical breakdown of the sample, while the effective removal of organic material is achieved with the addition of a 5% sodium hypochlorite (NaClO) solution, preferably after sediment disaggregation and wet sieving.

2 Add approximately 5-10ml of sediment to the solution, and boil for 2-20 minutes. Repeat the stage for the disaggregation of larger fragments. Unconsolidated sediments disaggregate in approximately 2-3 minutes.

3 Wet sieve the sediment over a 38µm sieve, and retain the coarse fraction, returning it to the beaker and repeating stages 1 and 2 if further residue is required.

4 Add 10% HCl to remove the carbonate component.

5 Decant and wash the residue several times, and finally wet sieve. Additional cleaning and removal of clay-grade material, if required, involves treating with hydrogen peroxide + *tetra*-sodium pyrophosphate and brief ultrasonic immersion. Any encrusting iron oxides are removed by immersion in 1% oxalic acid ($(COOH)_2.2H_2O$) and 2% ammonium oxalate ($(COONH_4)_2.H_2O$) solutions (Sanfilippo and Riedel 1985). Store residue dry or in distilled water in a glass or plastic phial.

26.3.1c. Siliceous rocks

Safety note: Hydrofluoric acid (HF) is extremely dangerous. Read safety notes in Appendix 8 - Care and handling of HF, Appendix 9 - Spillage and disposal of unwanted chemicals, and section 25 EXTRACTION TECHNIQUES FOR PALAEOBOTANY AND PALYNOLOGICAL MATERIAL for conditions and procedures of use.

Hydrofluoric acid (HF)

The use of HF to extract radiolaria from cherts has been outlined by Dumitrica (1970) and Pessagno and Newport (1972). The fundamental success of this technique requires well-preserved radiolaria showing no sign of recrystallization, although replacement of original opaline silica by the more stable chalcedony may have occurred. This is confirmed by thin section examination, where radiolarian tests are sharply defined in contrast to the matrix, and a fragile meshwork of spines is clearly visible. The differential dissolution between the silica of the matrix (average grain size 3µm) and specimen (average grain size 0.3µm) is fundamental to the successful release of radiolaria specimens using this method.

1 Etch the chert fragments in concentrated (52-55%) HF for 3-5 minutes.

2 Remove from the acid and gently wash the surface using distilled water.

3 Examine the surface under a binocular microscope. This provides the best indicator of the method's success. Well-preserved radiolaria stand proud of the sample surface, and with further controlled etching can be released from the rock.

4 Crush the sample into $2cm^3$ fragments.

5 Place in a plastic beaker and add a 10% HF solution. Leave for 24 hours. Depending on the sample reaction it may be necessary to alter the acid concentration, effectively increasing or decreasing the length of reaction time. Test with a small sample before proceeding with the bulk. If there is a danger of the fragments crushing freed skeletons, suspend the sample in a nylon mesh during HF immersion (Sanfilippo and Riedel 1985). Use of a coarse mesh (250μm) enables freed radiolaria to rain down to the bottom of the beaker.

6 Neutralize the acid by decanting and the addition of orthoboric acid (H_3BO_3). Safely dispose of the neutralized solution. Repeat the process until the residue is neutral (test with pH paper).

7 Wash and wet sieve the residue, then oven dry. To reduce skeleton damage on metallic mesh surfaces Baumgartner *et al.*, (1981) suggest the use of nylon mesh cloths suspended in plastic funnels. This has a number of advantages over the conventional metallic sieve screens. The material is resistant to the acidic and oxidizing solutions used in residue generation, and is relatively cheap in comparison with fine mesh sieves, and therefore replaceable with each sample processed. This is advisable, reducing the possibility of cross-sample contamination, as the spiny radiolaria skeletons readily adhere to the mesh surface, resisting the gentle washing required to remove them.

Depending on their mode of preservation, radiolaria can be viewed in reflected (stereozoom binocular microscope) or transmitted light (biological or petrological microscope). Large single specimens are picked directly from dried residues, examined under a stereozoom binocular microscope, and mounted on standard multi or single celled micropalaeontological slides. Finer, smaller specimen assemblages are prepared as strew slides, and examined with a conventional light microscope or, with well-preserved siliceous fossils, a Laser Scanning Microscope (O'Connor 1996, see section 20 THIN SECTION AND SLIDE PREPARATION TECHNIQUES OF MACRO- AND MICROFOSSIL SPECIMENS). Cherts prepared as polished thick sections (40-60μm) must be viewed in transmitted light with the condenser well down (Brasier 1980). Extraction procedures are summarized on the accompanying flow chart (Fig. 26.1).

26.3.2. Diatoms

Diatoms range in size from 5-2000μm, although the usual size range is between 20-200μm. Living and recent diatoms are collected by scrapping the green scum from

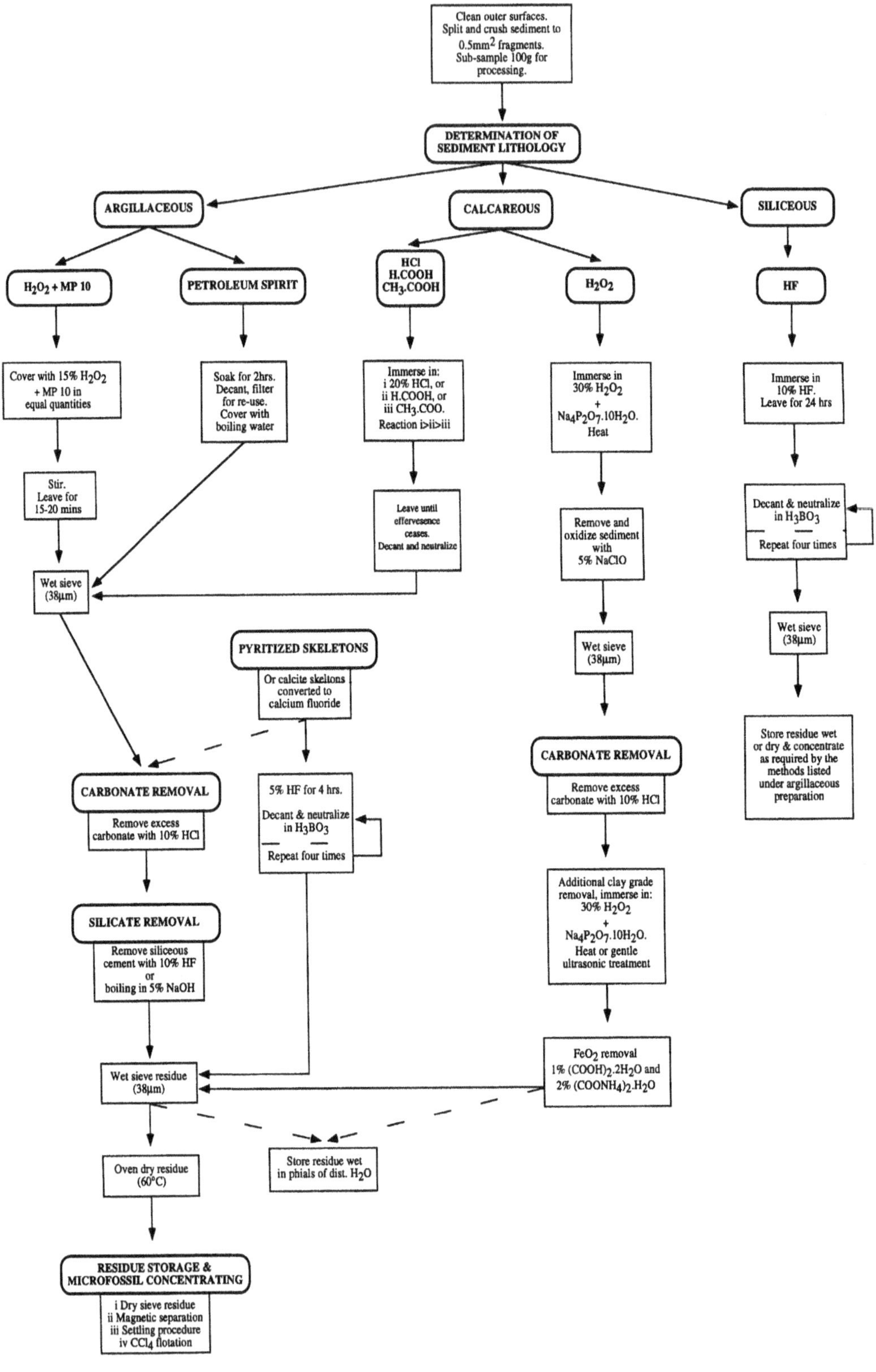

Figure 26.1. Composite flow chart illustrating radiolarian processing procedures for sediments of argillaceous, calcareous and siliceous composition.

pond floors, or the surfaces of mud, pebbles, shells, and marine and fresh-water vegetation in shallow rock pools.

Fossil diatoms studied from fresh-water or marine diatomites present few problems in preparation, and clean residues are relatively easy to generate (see 26.3.2a below). The cell walls of diatoms are hyaline perforate, composed of either a single or double layer of silica, and best viewed in transmitted light. Diatoms known range is from the Mid-Cretaceous to Recent, but they may extend back into the Jurassic and even the late Palaeozoic (Burckle 1978).

26.3.2a. Recent unconsolidated sediments

Preparation and frustule cleaning of Recent material have been briefly outlined by Hendey (1974) and West (1977), and a composite procedure is as follows:

1 Remove the coarse gravel fragments by screening over a 2mm sieve. Collect the fines in a small conical flask or beaker. Approximately 0.1g of sample is required for processing.

2 Cover with a small quantity (25ml) of 30% hydrogen peroxide (H_2O_2).

3 When the sample has dispersed, thoroughly mix in 10ml of potassium dichromate ($K_2Cr_2O_7$).

4 The addition of 10% hydrochloric acid (HCl) removes the carbonate component. Allow the mixture to stand for one hour or until the effervescence has ceased. Top with distilled water, allow to stand until clear, and decant. Repeat washing until the solution is neutral.

5 Oxidize the organic matter by adding 2ml of concentrated sulphuric acid (H_2SO_4), followed by 10-15ml of potassium permanganate ($KMnO_4$). Periodically gently agitate the mixture, and allow it to stand for 16-24 hours. If the solution shows a tendency to turn brown add more permanganate until a purple colour persists.

6 Add 10ml of oxalic acid ($(COOH)_2.2H_2O$) solution to remove the manganese oxide precipitate and excess permanganate. Gently warm the beaker until a faint effervescence results. On commencement of the reaction add more oxalic acid until all the colour has dispelled. Overheating causes the solution to boil vigorously, and can result in sample loss. Allow the solution to cool.

7 Wash and centrifuge the residue until the liquid is clear.

8 The addition of 1-2ml of ammonia solution (NH_3) prevents clumping of the cleaned diatoms in the residue. The presence of iron is indicated when the liquid develops a faint orange-brown discoloration. This is removed by the addition of one or two drops of hydrochloric acid. Further centrifuging is then required to clean the residue of iron

hydroxide. Fine clays (if present) are dispersed with the addition of a small amount of 5% sodium hexametaphosphate solution ($(NaPO_3)_6$, *Calgon*) and gently warmed at 40°C for 30 minutes.

9 When the solution is clear, flush the residue into storage phials and add a few drops of formalin. Alternatively store the residue in alcohol.

10 Transfer a drop of the suspension to a cover slip and prepare temporary or permanent slides. West (1977) suggests *Mikrops* as a mounting medium (R.I. 1.63), which melts at a temperature of 110°C.

In preference to centrifuging, Hendey (1974) advocated gravity settling and decanting for sample washing, reporting that rapid washing often failed to remove all residual acid from the residue. Trapped acidic solution may leach out in permanent mounts (and presumably stored residues), destroying the mountant and impeding optical resolution. Ruined diatoms result from a hasty cleaning procedure!

26.3.2b. Partially consolidated sediments

Setty (1966) describes three methods of concentrating fossil diatoms from marine and fresh water sediments. Two of the methods involve chemical concentration, while the third method is chemico-mechanical concentration. Additionally, descriptions of strew slide preparation, frustule staining, and use of the England Finder Slide are also comprehensively described by Setty (1966). Chemical concentration methods are more suited for partially consolidated fresh water, lacustrine or estuarine samples, that are rich in organic matter. Marine samples, low in organic matter, but with a high percentage of clay grade material, are successfully processed using the chemico-mechanical method. A fourth method (Schrader 1973) has proved particularly successful in the processing of ODP and DSDP samples, especially as it is not essential to dry sediments before commencing with the procedure.

Method 1. Takes approximately 10.5 days to obtain a residue.

1 Oven dry the sample at 125°C for 12 hours.

2 Allow the sample to cool to room temperature. Separate approximately 200g of sediment, and place it in a 250ml glass beaker.

3 Cover the sample with a 15% solution of hydrogen peroxide (H_2O_2), and boil for 20 minutes to disaggregate the sample.

4 Fill the beaker with distilled water and allow it to stand for 24 hours.

5 Decant the supernatant liquid.

6 Cover the sample with concentrated hydrochloric acid (30% HCl), and boil for 15 minutes.

7 Allow the sample to cool, top with distilled water and leave for 24 hours.

8 Decant the supernate liquid and top with distilled water. Repeat until neutral.

9 Following the final decanting of distilled water, cover the sample with a 25% nitric acid (HNO_3) solution. Top with distilled water and allow to stand for 24 hours.

10 Decant the liquid and transfer the residue to a 1.5 litre beaker. Top with distilled water, leave for 24 hours, then carefully decant.

11 Repeat stage 10 at least four times or until neutral.

The result is a clean residue, consisting of a concentration of acid insoluble microfossils with finer clastic material, free of all organic matter and soluble minerals. The clastic fines can be separated to leave a microfossil rich residue by centrifuging, using sodium hexametaphosphate ($(NaPO_3)_6$), *Calgon*) as a dispersing agent.

Method 2. Takes up to 8 weeks to process a sample.

Follow steps 1 to 5 of method 1, then;

6 Add (30%) concentrated hydrochloric acid (HCl) to the residue, warm (not boil) the mixture for 15 minutes, then allow to stand for 72 hours.

7 Decant the acid, top with distilled water and allow the solution to settle for 24 hours.

8 Repeat this procedure 8 to 10 times, until the final wash is neutral.

9 Following the final decantation, add chromium (VI) oxide (CrO_3) and allow the solution to stand for 72 hours.

10 Decant and repeat this procedure 5 times.

11 After the final decantation of chromium (VI) oxide top with distilled water and allow to stand for 24 hours.

12 Repeat this procedure 6 to 8 times.

13 Decant and store the residue in 70% butan-1-ol (butyl alcohol). Pipette the residue to make strew slides.

Method 3. Takes approximately 14 days to process a sample.

Follow steps 1 to 9 in method 1, then;

10 Fill the beaker with dilute (1%) ammonia solution (NH_3) and allow to stand for 6 hours.

11 Decant the supernate, and repeat this stage 15 to 20 times. This eliminates colloidal clay from the residue.

12 Add distilled water to the residue and allow to stand for 24 hours.

13 Decant suspension, and repeat this stage at least four times. After the final decantation, for short periods of time, store residues in distilled water, however, if slides are not prepared immediately, use butyl alcohol for long term storage.

Method 4. Takes approximately 5 days to process a sample.

The procedures of Setty (1966), although successful, are excessively long, and in many cases impractical for the time allocated for many research projects. The technique of Schrader (1973) is by far the easiest to adapt to partially lithified samples, although for dolomites Lagle (1984) provides a comprehensive two stage procedure (see 26.3.2d below for details). The most important factors in a successful procedure are the removal of carbonate, organic matter and the mud grade size fraction (<20μm) from the residue, components capable of impairing the optical properties of the diatom frustules.

Schrader (1973) processed wet and dry samples from core sediments, proposing a standardized method for cleaning diatoms. For dry samples approximately 5 grams of sediment, placed in an oven set at 60°C for 24 hours, is required, while wet samples must remain sealed in plastic poly-grip bags refrigerated at 6°C, until processing commences. However, there is some evidence that diatom frustules corrode when stored in sea-water over time (Lewin 1961). A similar disaggregation procedure is then applied to both wet and oven-dried samples, as follows.

1 Place the sediment in a 200ml beaker.

2 Add 50ml of analytical grade acetic acid ($CH_3.COOH$) and 50ml of analytical grade hydrogen peroxide (H_2O_2). The acetic acid dissolves carbonate components, while the hydrogen peroxide physically disaggregates the sediment.

3 For violent reactions, and the formation of a foamy froth, spray the surface of the solution with ethanol.

4 Heat the mixture for 20 minutes, until all the peroxide has evaporated.

5 Once active boiling has ceased add 25ml of distilled water, then pour the mixture into 50ml glass centrifuge tubes.

6 Centrifuge at 1200-1300 rpm for 2 minutes. Decant the supernate containing the suspension of clay minerals.

7 Re-fill the centrifuge tube with distilled water and repeat the centrifuging and decanting process at least 7 times.

8 Add a solution of 0.5% *tetra*-sodium pyrophosphate ($Na_4P_2O_7.10H_2O$) to the residue, and re-centrifuge. This deflocculant ensures that the very fine clay grade components remain in suspension, and are decanted. Repeat this procedure at least 4 times.

9 Wash the residue into a 50ml glass or plastic vial with distilled water, with the addition of two drops of neutralized formaldehyde (HCHO) to prevent the growth of algae.

The above method has been successfully used by Schrader (1976) and Koizumi (1980) in processing DSDP sediments. However, some workers (*e.g.* Lohman 1972; Ciesielski 1983) prefer the use of hydrochloric acid (HCl) to remove the carbonate component, while Koizumi and Tanimura (1985), undertaking quantitative work, dried wet sediment samples, and using a slightly longer processing procedure did not boil or centrifuge residues in order not to destroy the more delicate diatom frustules.

26.3.2c. Lithified clastic sediments

Extracting fossil diatoms from lithified rocks is a much longer and complicated procedure (*e.g.* Lohman 1972, Schrader 1973, Lagle 1984, Hinchey and Green 1994). The conventional diatom processing techniques of Lohman (1972) and Schrader (1973), successful on partially consolidated sediments, prove ineffective on lithified clastic sediments. A composite processing procedure, incorporating (i) physio-mechanical/chemico-mechanical sediment disaggregation; (ii) chemical treatment assisting in the breakdown and removal of any carbonate component; (iii) oxidation to remove unwanted organics and disperse the clay-grade material, and (iv) mechanical separation of the clay-grade fraction from the diatom frustules has been outlined by Hinchey and Green (1994). The amended procedure is summarized below:

Mechanical breakdown

1 Coarse crush the sample to fragments of approximately $1cm^3$ in size.

2 Oven dry sample at 60°C for 24 hours.

3 Weigh the dried sample, and sub-sample approximately 0.5g.

4 Soak the sediment in distilled water for 12 hours. Sediment that forms a muddy slurry can be treated in stage 8.

5 Remove any large fragments and place them on a $6cm^2$ of aluminium foil, and cover with distilled water. Repeat until the sediment is saturated.

6 Wrap the sample tightly in the foil preventing it from leaking. For additional security include a label with the sediment, and mark the sample number on the foil package.

7 Using tongs place the aluminium envelope into a flask of liquid nitrogen. Leave for 2 minutes and then remove, and immediately transfer the envelope into a beaker of recently boiled water. Repeat the freeze-thaw process until the sediment forms a muddy paste. Fragments that do not disaggregate by this process should be transferred to a mortar, and gently crushed with a pestle (avoid using a grinding action).

Physio-chemical breakdown

8 Open the foil envelope and empty the sediment into a clean 250ml beaker. Continue disaggregation with the addition of a 20ml 30% hydrogen peroxide (H_2O_2) solution, and gently heat for 2 hours on a hot plate at 90°C. Do not allow to the solution to boil. Alternatively, following the freeze-thawing process, argillaceous sediments are treated using the petroleum spirit procedure (see section 29 EXTRACTION TECHNIQUES FOR CALCAREOUS MICROFOSSILS FROM ARGILLACEOUS SEDIMENTS for details) (Abelmann 1988).

9 Fill the beaker with distilled water and leave for 12 hours.

Carbonate removal

10 Decant the supernate and add 30ml 0.5% hydrochloric acid solution (HCl). Warm gently on a hot plate at 40°C for 2 hours then leave for a further 4 hours. Allow the beaker to cool, fill with distilled water and leave for a further 24 hours. This process of acid etching assists in the break down of the sediment by exploiting weaknesses along fissures containing a calcium carbonate cement.

11 Decant the supernate and add 20ml 15% HCl and warm gently until the reaction ceases or the liquid has turned a lemon yellow colour. Top with distilled water, leave to settle and decant.

12 Repeat the washing, settling and decanting process until the solution is neutral (test with pH indicator paper).

Organic matter removal

This optional procedure is required for the removal of larger quantities of organic matter. Samples are treated as follows:

13 Add 20ml of 25% nitric acid solution (HNO_3) to the residue. Leave for 6 hours to settle and then decant. Add distilled water, allow to settle and decant.

14 Repeat washing, settling and decanting until the solution is neutral.

Stubborn organic concentrations can be dealt with in the following manner:

15 Add 10ml of potassium permanganate ($KMnO_4$) and 0.5ml of 10% sulphuric acid (H_2SO_4) to the residue, and simmer for 1 hour. If the solution shows a tendency to turn brown, add more permanganate until a purple colour persists.

16 Add 1-2g of powdered *di*-sodium disulphite ($Na_2S_2O_5$) to the solution to remove the black manganese dioxide (MnO_2) residue. Alternatively, add 20ml of concentrated (30%) sulphuric acid, boil at 150°C until white fumes are liberated, then add a few crystals of sodium nitrite ($NaNO_3$) to oxidize the organic matter.

17 Top with distilled water, allow to stand before decanting. Repeat washing until the solution is neutral.

18 Further breakdown is achieved by short ultrasonic treatment. Ensure the sample is in suspension in distilled water with a small amount of industrial detergent, then immerse for about 2 seconds.

Clay dispersal and removal

19 Decant the excess supernate, then add 100ml of 5% sodium hexametaphosphate solution ($(NaPO_3)_6$, *Calgon*), mixed with 38g/100ml of sodium carbonate (Na_2CO_3) and gently warm at 40°C for 30 minutes.

20 On cooling, gently agitate the sample to suspend the residue, then centrifuge at 800 rpm for 4 minutes. Decant the supernate of suspended clay particles.

21 Top the centrifuge tubes with 5% $(NaPO_3)_6$, and repeat six times.

22 Finally, centrifuge the residue three times with distilled water.

23 Use wet mounts to check the sample for colloidal clays or remaining organic matter. If the sample is clean then proceed with permanent slide mounts. However, if the sample contains large amounts of clay aggregates then undertake the following procedure.

24 Add 20ml of 1% ammonia solution (NH_3) to the residue, and allow it to stand for 4 hours. Decant the supernate, and repeat the procedure 6-8 times.

25 After the final decantation, top with distilled water, allow to settle and decant. Repeat washing until the solution is neutral.

The residue now contains a concentrate of clean siliceous material, suitable for the preparation of permanent slides or SEM stub mounts. Strew slide preparation is detailed in section 20 THIN SECTION AND SLIDE PREPARATION TECHNIQUES OF MACRO- AND MICROFOSSIL SPECIMENS.

26.3.2d. Lithified carbonate sediments

Lagle (1984), in processing dolomites, favoured a longer controlled procedure of extraction, using a two stage acid etching and digestion technique followed by oxidation. Following the initial physical disaggregation of the sample (with jaw crusher, hammer or pestle and mortar) to fragments approximately 2mm in length, separate 5 grams into a 500ml beaker.

1 Etch by covering the fragments with a 0.5% hydrochloric acid solution, and warming on a hot plate pre-set at 40°C.

2 Repeat this stage until all signs of an effervescent reaction has ceased.

3 Swirl the contents in the beaker, then decant the suspended residue into 50ml centrifuge tubes, and centrifuge at 1200 rpm for 2 minutes.

4 Decant the clear supernate, and repeat the procedure until the wash is neutral.

5 Return the sediment to a 500ml beaker for continued thorough acid digestion. Carefully immerse the aggregates in 11.5% hydrochloric acid, cover the beaker, and leave until effervescence has ceased (this process may take many hours).

6 Wash and centrifuge the residue.

7 Cover the residue with 15% hydrogen peroxide, and leave in the fume cupboard (Lagle 1984, suggests this may require up to 14 days).

8 When all signs of the reaction have ceased, swirl the sediment and decant only the suspended material.

9 Wash, centrifuge and decant, discarding the supernatant liquid.

10 After the final washing, cover the residue in a 0.5% solution of sodium pyrophosphate ($Na_4P_2O_7.10H_2O$). Leave for two minutes then re-centrifuge.

11 Wash and centrifuge in distilled water at least five times.

12 After the final wash flush the residue into a storage phial and add two drops of buffered (37%) formaldehyde.

The inclusion of the acid etching stage is particularly useful when processing dolomites, as it weakens the rock matrix along fissures developed around the delicate diatom frustules, and establishes a greater surface area of matrix for the more violent, and potentially more destructive, acid digestion stage that follows. Processing diatomites generates a jelly-like organic-rich residue containing abundant well-preserved microfossils (Lagle 1984). A comprehensive flow-chart indicating the main stages in diatom processing are illustrated in figure 26.2.

26.3.3. Silicoflagellates

Silicoflagellates are unicellular marine flagellates, with an opaline silica skeleton, ranging from the early Cretaceous to Recent. They range in size from 20-50μm in diameter, and usually occur in association with smaller (3-25μm) chrysomonad cysts and ebridians in marine diatom rich assemblages. Consequently, they are prepared and studied together in a manner similar to diatoms.

26.3.4. Ebridians

Ebridians are similar to silicoflagellates in size, however the living ebridian cell has two unequal flagella, and an internal skeleton of solid construction, unlike the tubular construction of the silicoflagellates. Ebridians range from the Palaeocene to Recent, but are most numerous in Eocene and Miocene marine sediments.

26.4. ORGANIC-WALLED MICROFOSSILS

Five groups of microfossils have organic walls. These are:

- i Acritarchs
- ii Tasmanitids
- iii Chitinozoa
- iv Dinoflagellates
- v Scolecodonts

A note on the term 'Hystrichospheres': this term is now no longer used, but will be encountered in literature over thirty-five years old (Evitt 1963). At one point all spiny organic walled microplankton were assigned to the 'class' hystrichospheres' or 'hystrichosphaerids' (Sarjeant 1961). However, it soon recognized that many of the organisms assigned to this heading were fossil dinoflagellates (Evitt 1961, Evitt 1963). Subsequently, all the fossil forms have now been attributed to either the dinoflagellates, acritarchs or tasmanitids (Williams 1978).

26.4.1. Acritarchs

Acritarchs form a polyphyletic group placed in a class within the algae, some forms may represent dinoflagellate cysts, although the affinities of most are uncertain. They range in age from Precambrian to Recent, and in size from 3-250μm in diameter, but the majority are less than 100μm, and thus rarely visible in hand specimen. The exact chemistry of the organic wall remains under investigation, but is compositionally similar to sporopollenin (Mendelson 1993). Sampling is based on lithological criteria, however, they occur in abundance in fine-grained dark carbonaceous shales, mudstones

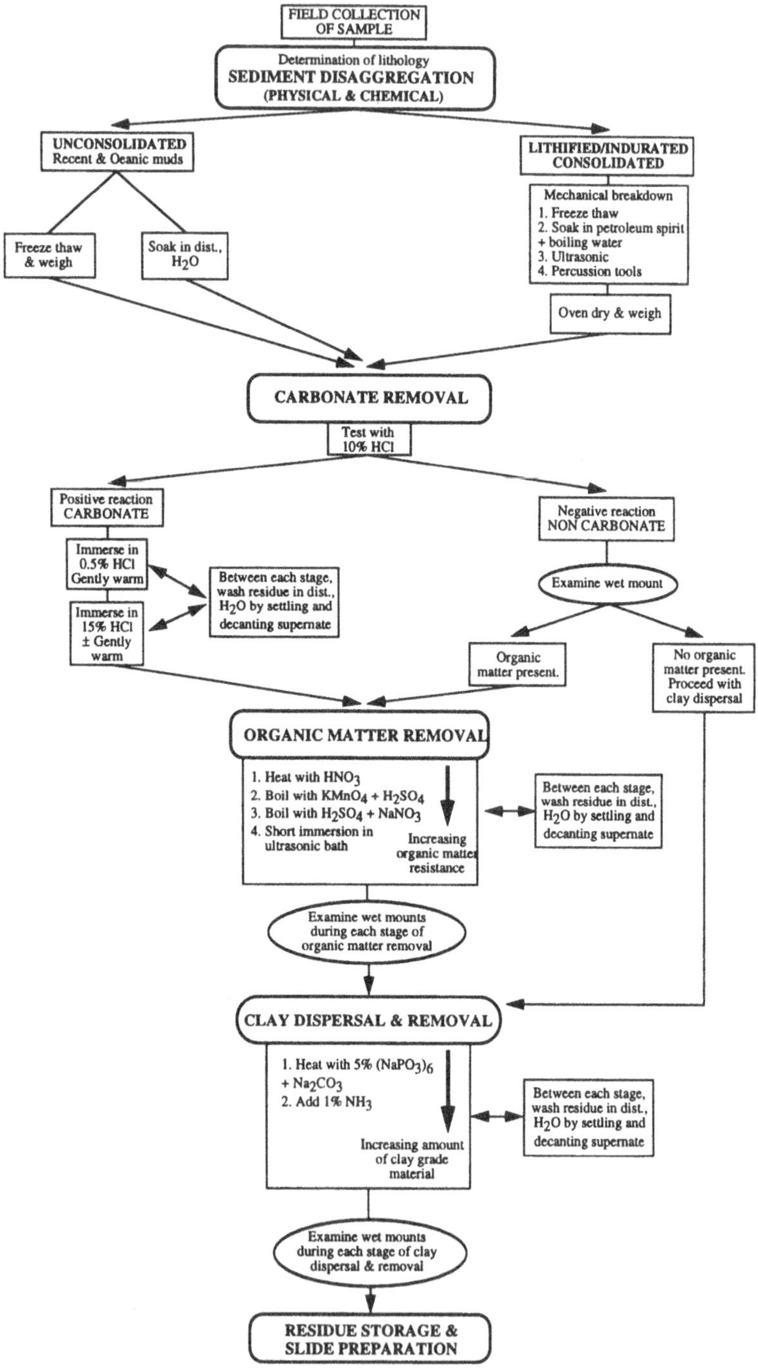

Figure 26.2. Flow diagram illustrating the main stages of diatom processing (modified and re-drawn from Hinchey and Green 1994).

and clays. For sediments of unknown yield approximately 100 grams is required to provide a representative assemblage, although larger amounts are necessary for coarser grained rocks. Generally, however, a 10-25 gram sample size is sufficient to generate a representative assemblage suitable for biostratigraphical and sedimentological studies (Grey 1993), although a sample size as small as 2 grams may be sufficient (Playford and Dring 1981). Well-sorted sands, oxidized sediments and recrystallized limestones are usually barren.

Preparation techniques are critical, as the fragile brittle nature of the organism, particularly in large specimens, is liable to destruction during some stages of conventional palynological processing. The minimum amount of crushing must be employed, while centrifuging stages, used in a standard palynological preparation, are best omitted (Grey 1993). Some Mesozoic and Cenozoic samples readily disaggregate by pulverisation, scrubbing the surface, soaking in solvent or sodium hypochlorite (NaClO), and boiling in sodium carbonate (Na_2CO_3). A modified processing procedure for more indurated sediments has been outlined by Grey (1993), as follows:

1 Clean the outer surface of recent plant debris.

2 Split sediment along bedding planes.

3 Crush the sediment to an average size of 0.5mm diameter, and place a representative sample in a polythene bottle with a screw top.

4 Cover with 10% HCl. Leave for 24 hours.

5 Top with distilled water, allow to settle and decant. Repeat this stage at least four times, until the sample is neutral.

6 Cover with 40% HF, and leave for 24 hours.

7 Top with distilled water, allow to settle and decant into orthoboric acid (H_3BO_3) crystals to neutralize. Repeat this stage at least six times, until the solution is neutral.

8 Boil with concentrated (30%) HCl.

9 Top with distilled water, allow to settle and decant. Repeat this stage at least four times to ensure the complete removal of any gel-like fluorides.

10 Add concentrated (95%) nitric acid (HNO_3), and leave to macerate for 1-2 minutes. Oxidation is less extreme using nitric acid than Schulze's reagent, providing the preparator with grater control in monitoring the reactions effects. Colloid removal, using either 5% sodium hydroxide (NaOH) of potassium hydroxide (KOH), is optional, but residues must be thoroughly washed in distilled water before proceeding (Playford and Dring 1981).

11 Top with distilled water, allow to settle and decant. Repeat this stage at least four times.

12 Wash and filter the residue, using a fine sinter glass funnel (P40, 16-40μm, Higgins and Spinner 1969), or 10μm and 25μm monofilament mesh (Grey 1993) until neutral (test with pH indicator paper).

13 Acritarchs are concentrated by using carbon tetrachloride (CCl_4) and decanting (*e.g.* the organic walled fossil variation (K) of Brasier 1980). Heavy liquid separation is un-successful for pyrite infilled material (Grey 1993).

Both acritarchs and dinoflagellates are transparent, and lack any prominent contrasting body colour. However, light microscopical examination is significantly improved by staining specimens during the final washing and filtering stage. During the final wash, with the residue retained in the sinter funnel, add a few drops of Safranine O and allow the stain solution to filter through the residue. Prepare a temporary slide mount by pipetting a few drops, and examine the effects of stain intensity. For increased stain intensity repeat the procedure. The amount of stain absorbed varies considerably within individual species, so do not expect all specimens to stain uniformly.

Staplin (1977) reports gradational colour changes from pale yellowish-green to grey-black in acritarchs present in diagenetically altered rocks, where temperatures of up too 250°C have been achieved. Beyond this temperature acritarchs are usually destroyed, although they may be preserved in some slates, and recognizable in thin sections of phyllite and schist (Burmann 1968). A comprehensive flow chart outlining the non-routine palynological processing for organic walled microfossils is presented in figure 26.3.

26.4.2. Tasmanitids

Tasmanitids are hollow spheres with a thick punctate sporopollenin wall, that is yellow to reddish-brown or dark brown in colour. The two or three layered wall is 5-20μm thick, but the outer layer is rarely preserved. The middle layer forms the bulk of the wall, and is characterized by the presence of radial canals, surrounding a spongy or fibrous inner layer. They range in size from 100μm to over 600μm in diameter, while the stratigraphic range is from Cambrian to Miocene, although a related living representative known to be the cyst of pelagic chlorophyllous green-algae, extends the range to the present (Williams 1978).

26.4.3. Chitinozoa

Chitinozoa are an extinct group of marine micro-organisms ranging from the Upper Cambrian to the Lower Carboniferous, and comprise a hollow flask-shaped test with a pseudochitinous wall, open at one end. A considerable variation in test shape and ornamentation is evident, but all have a basic radial symmetry around a central elongate axis, and occur either singularly or in chains. The affinities of the group are in dispute, regarded by some as protozoan, it is also suggested that they are remains of metazoan egg capsules. Benthonic, substrate attached or possibly free-swimming nektonic, they range in size from 50μm-2mm in length, and 200-300μm in diameter.

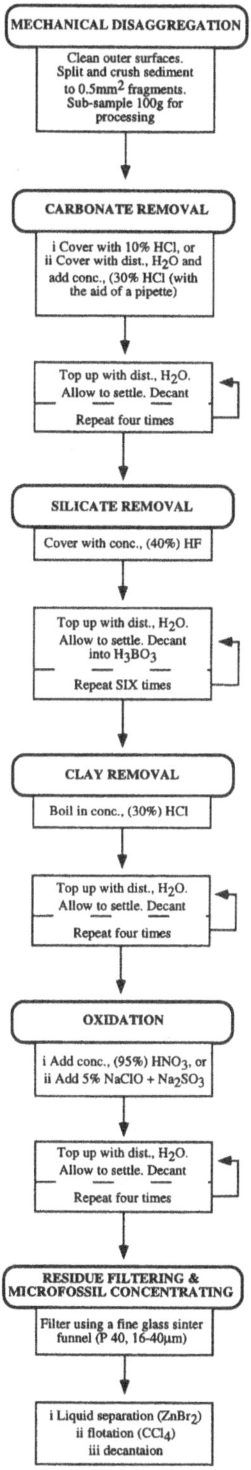

Figure 26.3. Non-routine palynological processing (excluding centrifuging stages), suitable for acritarchs, tasmanitids, chitinozoa and non-siliceous scolecodonts.

Chitinozoa have very fragile pseudochitinous tests, consisting of single, double or triple layered walls. However, the most common wall type consists of two layers, but as with tasmanitids this may reflect preferential preservation. Well-preserved material is usually translucent or amber coloured, although tests recovered from metamorphosed sediments are opaque, black and extremely brittle (Jansonius and Jenkins 1978).

Preparation techniques are similar to those used for acritarchs and other organic walled microfossils, but must avoid excessive use of mechanical processes that might damage the brittle, fragile tests. Jansonius and Jenkins (1978) report that 20 tests per gram of residue is considered a good yield, however, it may be less than one test per gram. Compared with other organic-walled microfossils this is very low (acritarchs and spores are usually in the thousands per gram). Consequently, when undertaking pilot studies larger sub-samples (500 grams) must be processed to yield a representative chitinozoan assemblage. Smaller sub-samples (250 grams) can be processed from sediments known to contain chitinozoa (Jenkins 1967). Chitinozoa are commonly found in calcareous shales, and extracted using the following procedure:

1. Break the rock into 0.5-1cm^3 fragments with a pestle and mortar.

2. Dry sieve through a 38µm screen. Discard the fine material generated during the initial breakdown, particularly if quantitative studies are undertaken, as the chitinozoan tests are small, unidentifiable, and likely to have suffered damage.

3. Using a sediment sample splitter divide the sample into two. Retain half for later use.

4. Place the other half of the sample in a 1 litre beaker and cover with 500ml of distilled water.

5. Remove the carbonate by gently adding drops of concentrated HCl.

6. Neutralize the solution by decanting and the addition of distilled water. Avoid agitating the partially broken down sample (this may result in additional damage to the fragile tests). Ensure the finer residue is not lost during decanting.

7. Transfer the partially digested residue to 250 ml polypropylene screw-top bottles, and add 100ml of cold concentrated (40%) HCl. Control the reaction with the addition of distilled water, and adding the acid a few drops at a time over a period of minutes before increasing the concentration. Secure the lids, allowing for vapour to vent. Leave the sample for a few days, agitating by gently tilting the bottles.

8. When the reaction has ceased neutralize by decanting with distilled water.

9. Once neutral, wash the residue over a 38µm sieve by immersing the mesh in water, and gently moving the sieve in an up-and-down motion to within 0.5cm of the top, this action avoids damage to the

chitinozoa. Do not employ the directed jet of water method (see section 16 WASHING AND SIEVING TECHNIQUES USED IN MICROPALAEONTOLOGY). This process is time consuming as only small amounts of residue can be washed at any one time, and several changes of water are required. Continue until no fines pass through the mesh.

10 Examine the residue while it is wet by pouring a small amount into an evaporating dish with a white background, and view under a stereozoom binocular microscope using reflected light. The brown-black appearance of the chitinozoa contrasts with the evaporating dish, and their absence is likely to indicate a barren sample than a short-coming in the processing procedure.

Several bleaching methods (Schulze's solution, fuming nitric acid, concentrated nitric acid, sodium hypochlorite and hydrogen peroxide) were tested by Jenkins (1967), who concluded that success depended upon the specimens preservational state. Where tests were strongly carbonized no reagent would bleach them, and prolonged treatment resulted in disintegration of the chitinozoa. With less carbonization any of the reagents could be used, however the reaction involving sodium hypochlorite was much easier to control than those involving acidic solutions.

11 Opaque specimens are bleached by placing them in an evaporating dish (white bottomed) with a small amount of distilled water. While observing individual chitinozoa under a stereozoom binocular microscope, add a few drops of sodium hypochlorite solution (NaClO). When individual specimens are sufficiently translucent, stop the reaction with the addition of a few drops of sodium sulphite solution (Na_2SO_3). Finally wash and decant the residue with distilled water to remove the sodium sulphite solution.

12 Oven dry the cleaned residue at 60°C.

13 Using a fine sable hair brush pick individual specimens, place in a micropalaeontological slide, or a glass cavity slide with either glycerol or Canada Balsam. Slide mounting procedures are described in section 20 THIN SECTION AND SLIDE PREPARATION TECHNIQUES OF MACRO AND MICROFOSSIL SPECIMENS.

26.4.4. Dinoflagellates

Dinoflagellates are a major class of unicellular biflagellate algae, first recorded in the Silurian, becoming common with a continuous fossil record from the Upper Triassic to present. They commonly occur in dark grey or black argillaceous marine rocks, but a few freshwater occurrences are known from the Eocene. Most fossilized cysts are composed of sporopollenin, a complex carbon, hydrogen, oxygen and nitrogen polymer, although some, composed of cellulose, are less liable to be preserved (Edwards 1993). A few cysts contain silica, while a small group is composed of calcite (calcispheres), and can only be extracted using nannofossil techniques (see section 33 EXTRACTION

TECHNIQUES FOR CALCAREOUS NANNOFOSSILS). They range in size from 5µm to 2mm.

Approximately 10 grams of clay, silt, mudstone, sandstone, limestone or shale is sufficient to yield an assemblage, although concentrations are higher in carbonaceous and argillaceous rocks compared with arenaceous and highly calcareous rocks. A 10 gram sample of shale may result in a yield of several hundred thousand, compared to 100 grams of chalk or sandstone that yields far fewer (Sarjeant 1974). The robust nature of the organism allows for the use of standard palynological preparation procedures. Extraction techniques have been outlined by Gray (1965), Barss and Williams (1973), Sarjeant (1974) and Evitt (1984). A composite procedure is as follows:

Safety note: Hydrofluoric acid (HF) and concentrated hydrochloric acid (HCl) are extremely dangerous, consult the notes in section 25 EXTRACTION TECHNIQUES FOR PALAEOBOTANICAL AND PALYNOLOGICAL MATERIAL, Appendix 8 - Care and handling of HF, and Appendix 9 - Spillage and disposal of unwanted chemicals for conditions of use and procedures to be followed.

Processing

1. Divide the sample as a precaution against accidents and for reference or descriptive purposes. Clean the surface of indurated rocks in distilled water to remove residual weathered surfaces, and break the sample into fragments of approximately $3mm^3$. For argillaceous sediments a sample size of 25 grams is usually sufficient to yield a representative assemblage, while other lithologies, yielding fewer specimens require the processing of a 400g sample size. Place the fragments in a glass or polythene beaker.

2. Remove carbonates by covering the sample with 10-20% HCl. Should the reaction result in the formation of a froth, spray the surface with acetone to reduce the solutions surface tension. Conversely the reaction is increased by placing the glass beaker on a hot plate and gently stirring the mixture. When the reaction has ceased, add distilled water, allow the residue to settle and decant.

3. Repeat the washing, settling and decanting until the liquid is neutral (test with pH paper).

4. Transfer the residue to a 250 ml screw-top polypropylene container. For silicate removal slowly add 100-200ml of cold concentrated HF. If the reaction is violent spray the acid with a mist of distilled water, and continue the addition of HF using a 10% concentration. Leave for 24 hours, stirring occasionally. Decant the acid, neutralize with orthoboric acid and safely dispose.

5. Top with distilled water, allow to settle and decant. Repeat the washing process until the solution is neutral.

6 Boil in concentrated (40%) HCl to remove fluoro-silicates.

Removal of Heavy Minerals

At this stage of the procedure it is useful to examine a small amount of the residue, to assess cyst concentration and cleanliness, by preparing temporary slide mounts (Sarjeant 1974). Prepare slides by pipetting a small amount of residue onto a glass microslide, allow the water to evaporate, and examine under transmitted light. Liquid separation procedures are employed to remove heavy mineral concentrations, a solution of zinc bromide ($ZnBr_2$) and 10% HCl is used as follows:

7 Transfer the residue into 15ml centrifuge tubes and decant the excess distilled water.

8 Add approximately 7ml of zinc bromide ($ZnBr_2$) to each of the tubes and thoroughly mix.

9 Centrifuge the mixture for 3 minutes at 3000 rpm.

10 Pipette the organic rich float fraction into a clean 100ml beaker containing 20ml of 10% HCl.

11 With the residue in suspension, pour equal amounts in clean tubes and centrifuge. Decant, retaining the supernate, and repeat the washing process until the mixture is neutral.

12 Wash, neutralize and examine the heavy fraction retained in the centrifuge tube, as this may contain organic walled microfossils trapped between grains, and thus require further treatment. When no organic matter is evident discard this fraction.

Oxidation

Funkhouser and Evitt (1959) consider this stage the most precarious of all operations when processing for organic-walled microfossils. It can result in damage or, if insufficient care is taken, complete or partial destruction invalidating the use of the organism in statistical work. In general, coalified material is acted upon faster than the more resistant organic-walled microfossils. The amount and type of oxidation varies from each sample, and therefore it is imperative to continually monitor the residue during the process. Two types of reaction can be seen:

i dissolves the organic matter directly,

ii converts it into a form (humic acids) that is soluble in a base.

In practice the two reactions occur simultaneously, with their relative importance determined by the composition of the sample

13 Heat to near boiling in nitric acid (HNO_3) to remove pyrite and other large unwanted organic fragments.

One of two procedures is then employed. Evitt (1984) suggests the following:

14 Add 25ml of 5% sodium hypochlorite (NaClO) to the centrifuged residue, thoroughly mix the solution and pour into a beaker. When the reaction subsides alternately add drops of concentrated hydrochloric acid (HCl) and sodium hypochlorite.

Evitt (1984) comments on the reaction, "which may follow a delay of one to five minutes and vary in intensity with the sample from nearly imperceptible to surprisingly - on occasion, startlingly! - vigorous,..". A colour change from blackish to brownish, plus the generation of heat also accompanies this reaction, which is stopped by saturating the mixture with excess sodium hypochlorite. This highly effective treatment substitutes for Schulze's solution, with the advantage that prolonged immersion does not result in damage to the cysts.

Sarjeant (1974, p 131-132) illustrates and describes an alternative method using 95% fuming nitric (HNO_3) acid, in a combined process of oxidation and filtration using a glass sinter funnel.

15 Short centrifuge the residue with detergent solution, and decant to remove the clay grade fraction.

Acetylation

Acetolysis provides two useful functions. Firstly, this procedure removes unwanted cellulose from the residue, and secondly it helps darken cysts, enhancing their appearance for microscopical examination.

16 Wash the residue in acetic acid ($CH_3.COOH$).

17 Prepare a 5ml mixture of acetic anhydride ($(CH_3.CO_2)_2O$) and sulphuric acid (H_2SO_4), mixed in a ratio of 9:1.

18 Carefully add the two mixtures together. The reaction is exothermic, and they should be combined and mixed while still hot. The colourless solution turns a rich reddish brown.

19 Leave the mixture to cool for 30 minutes.

20 Centrifuge the residue-solution mixture, and neutralize by washing with distilled water. Washing, centrifuging and decanting must be repeated to ensure the residue is neutral.

An alternative to this method of acetylation is the addition of concentrated (95%) fuming nitric acid (HNO_3), or a 20% sodium hydroxide (NaOH) solution to the residue.

However, the breakdown of any residual organic matter requires continual monitoring to establish a suitable immersion time for the residue (Sarjeant 1974).

A further process for removing fine organic debris combines the use of gravity liquid separation and centrifuging, with residues mixed in a 4% solution of *Darvan No. 4* Sarjeant (1974). The unwanted fine material floats, while the coarser dinoflagellate rich material sinks.

Staining

Although acetylation enhances the colourless cysts by increasing the transparency, the contrast of wall structures and processes is improved further through staining. Although staining with the biological stains basic fuchsin and Bismarck brown is successful, Safranine O, which stains the dinoflagellates red, is the most satisfactory.

21 Add a small amount of stain to the residue in the centrifuge tube, top with distilled water, and mix until the stain dissolves. Additional stain or distilled water will increase or reduce the stain intensity as required.

22 Excess stain can be removed by centrifuging and decanting the solution. Repeat the process until the liquid is clear.

Permanent slide preparation is outlined in section 20 THIN SECTION AND SLIDE PREPARATION TECHNIQUES OF MACRO- AND MICROFOSSIL SPECIMENS.

26.4.5. Scolecodonts

Scolecodonts form a morpho group consisting of known fossilized teeth or jaws of polychaete worms (annelids), either bottom dwelling or nektonic organisms, separating them from the superficially similar conodonts. Composed of chitinous, horny, or siliceous material, jet black in colour with a high vitreous lustre, scolecodonts are shallow water marine organisms. Scolecodonts range from the Cambrian to Recent, although most fossil forms occur in Palaeozoic Ordovician to Devonian rocks. The majority of these fossils are smaller than conodonts, ranging in length from 0.1 to 1.5mm. Processing is by the organic-walled (non-routine) palynological procedure, omitting the hydrofluoric (HF) acid stage for scolecodonts of known siliceous composition.

Taugourdeau (1971), extracting scolecodonts from Frasnian (Devonian) calcareous siltstones for systematic studies, successfully used methods involving both hydrochloric (HCl) and hydrofluoric (HF) acids.

Tasch and Shaffer (1961) describe a method of oxidizing scolecodonts, decreasing the denticle opacity, making it suitable for study by transmitted light microscopy, and enhancing their photographic appearance. The method involves placing individual specimens on a glass cavity slide and the application of a drop of *Clorox*. The transformation is observed using the light microscope, with the solution renewed as required, and stopped by removing and washing the scolecodonts when denticles appear amber coloured and translucent in appearance.

The treatment time depends on the size and thickness of the specimen, and the amount of organic material present. There are however, two distinct disadvantages to the procedure. Firstly, the evaporation of *Clorox* around the specimen results in the formation of sodium hypochlorite crystals, a potential source of damage to the specimen in addition to obscuring photographic details. Continued observation, under high power, of the oxidizing reaction, and replenishing of the solution avoids this problem. Secondly, oxidation results in a structural weakening of the scolecodont denticle and, on completion of the procedure, specimens must be permanently protected by a mountant and cover glass. Transferring to a micropalaeontological slide or SEM stub must be avoided for fear of damaging the structurally weakened denticle. However, Scanning electron microscopy is an important and essential tool in examining acid insoluble specimens, particularly where surface sculpture is an important feature in distinguishing taxa. Specimen mounting procedures are described in section 34 ELECTRON MICROSCOPY TECHNIQUES.

REFERENCES

ABELMANN, A. 1988. Freeze-drying simplifies the preparation of microfossils. *Micropaleontology*. **34**, (4), 361.
BARSS, M. S and WILLIAMS, G. L. 1973. Palynology and nannofossil processing techniques. *Geological Survey of Canada*, Paper **73-26**. 1-25.
BAUMGARTNER, P. O, BJØRKLUND, K. R, CAULET, J-P, DE WEVER, P, KELLOGG, D, LABRACHERIE, M, NAKASEKO, K, NISHIMURA, A, SCHAAF, A, SCHMIDT-EFFING, R and YAO, A. 1981. EURORAD II, 1980 - Second European meeting of radiolarian paleontologists: current research on Cenozoic and Mesozoic radiolarians. *Eclogae geologicae Helvetiae*. **74**, (3), 1027-1061.
BRASIER, M. D. 1980. *Microfossils*. Allen and Unwin (London). xii + 193p.
BURCKLE, L. H. 1978. Marine Diatoms, pp 245-266: In; B. U. Haq and A. Boersma (eds), *Introduction to Marine Micropaleontology*. Elsevier Biomedical (New York). viii + 376.
BURMA, B. H. 1965. Radiolarians, pp 7-14: In; B. Kummel and D. Raup (eds), *Handbook of Paleontological Techniques*. W. H. Freeman and Co., (San Francisco). xiii + 852p.
BURMANN, G. 1968. Diacrodien aus dem Unteren Ordovizium. *Paläontolgische Abhandlungen* . Band 2, **4**, 635-793.
CIESIELSKI, P. F. 1983. The Neogene and Quaternary diatom biostratigraphy of subantarctic sediments, deep sea drilling project Leg 71. *Initial Reports of the Deep Sea Drilling Project*. 71, 635-665.
DE WEVER, P, RIEDEL, W. R, BAUMGARTNER, P. O, DUMITRICA, P, BJØRKLUND, K. R, CAULET, J-P, DROBNE, K, GRANLUND, A, KOCHER, R and SCHAAF, A. 1979. Recherches actuelles sur les Radiolaires en Europe. *Annales de la Societé geologique du Nord, Lille*. **98**, 205-222.
DUMITRICA, P. 1970. Cryptocephalic and crytothoracic Nassellaria in some Mesozoic deposits of Romania. *Revue roumaine de geologie, geophysique et geographie, Serie de Geologie*. **14**, (1), 1-124.
DYER, R and COPESTAKE, P. 1990. A review of Late Jurassic to earliest Cretaceous radiolaria and their biostratigraphic potential to petroleum exploration in the North Sea, pp 215-235; In: D. J. Batten and M. C. Keen (eds) *Northwest European micropalaeontology and palynology*. Ellis Horwood.
ECKERT, R, HAY, W. W, LORENZ, G and VOGT, P. 1961. The magnetic separator as a tool in micropaleontology. *Journal of Paleontology*. **35**, (5), 876-877.
EDWARDS, L. E. 1993. Dinoflagellates, pp 105-129: In; J. H. Lipps (ed) *Fossil Prokaryotes and Protists*. Blackwell Scientific Publications (Oxford). ix + 342p.
EVITT, W. R. 1961. Observations of fossil dinoflagellates. *Micropaleontology*. **7**, (4), 385-420.
EVITT, W. R. 1963. A discussion and proposals concerning fossil dinoflagellates, hystrichospheres and acritarchs, 1. *Proceedings of the National Academy of Science (USA)*. **49**, (2), 158-164.
EVITT, W. R. 1984. Some techniques for preparing, manipulating and mounting dinoflagellates. *Journal of Micropalaeontology*. **3**, (2), 11-18.
FUNKHOUSER, J. W and EVITT, W. R. 1959. Preparation techniques for acid-insoluble microfossils. *Micropaleontology*. **5**, (3), 369-375.
GRAY, J. 1965a. Part III, Techniques in palynology (co-ordinator), pp 469-706: In; B. Kummel and D. Raup (eds), *Handbook of Paleontological Techniques*. W. H. Freeman & Co., (San Francisco). xiii + 852p.
GREY, K. 1993. Appendix F: Acritarch biostratigraphy of the Ediacarian of the Centralian Superbasin, pp 109-115: In; R. J. F. Jenkins, J. F. Lindsay and M. R. Walker (eds), *Field Guide to the Adelaide Geosynclinal and Amadeus Basin*. Record No. 1993/35. Australian Geological Survey Organisation. 1-133.
HENDEY, N. I. 1974. The permanganate method for cleaning freshly gathered diatoms. *Microscopy*. **32**, 423-426.
HIGGINS, A. C and SPINNER, E. G. 1969. Techniques for the extraction of selected microfossils. *Geology - Journal of the Association of Teachers in Geology*. **1**, 12-28.
HINCHEY, J. V and GREEN, O. R. 1994. A guide to the extraction of fossil diatoms from lithified or partially consolidated sediments. *Micropaleontology*. **40**, (4), 368-372.
JASONIUS, J and JENKINS, W. A. M. 1978. Chitinozoa, pp 341-357: In; B. U. Haq and A. Boersma (eds), *Introduction to Marine Micropaleontology*. Elsevier Biomedical (New York). viii + 376p.
JENKINS, W. A. M. 1967. Ordovician Chitinozoa from Shropshire (techniques pp 439-442). *Palaeontology*. **10**, (3), 436-488.
KLING, S. A. 1978. Radiolaria, pp 203-244: In; B. U. Haq and A. Boersma (eds), *Introduction to Marine Micropaleontology*. Elsevier Biomedical (New York). viii + 376p.
KOIZUMI, I. 1980. Neogene diatoms from the Emperor seamount chain, Leg 55, Deep Sea Drilling Project. *Initial Reports of the Deep Sea Drilling Project*. **55**, 387-407.
KOIZUMI, I and TANIMURA, Y. 1985. Neogene diatom biostratigraphy of the middle latitude Western North Pacific, DSDP leg 86. *Initial Reports of the Deep Sea Drilling Project*. **86**, 289-300.

LAGLE, C. W. 1984. Recovery of siliceous microfossils by the disaggregation of dolomite, pp 185-194: In; R. E. Garrison, M. Kastner and D. H. Zenger (eds), *Dolomites of the Monterey Formation and other organic rich units*. Pacific Section, Society of Economic Petrologists and Mineralogists (Los Angeles, California). **41**, 1-215.
LEWIN, J. C. 1961. The dissolution of silica from diatom walls. *Geochimica et Cosmochimica Acta*. **21**, 182-198.
LOHMAN, K. E. 1972. A procedure for the microscopical study of diatomaceous sediments; In: R. Simonsen (Ed), Symposium on recent and fossil diatoms. *Nova Hedwegia*. **39**, 267-285.
MENDLESON, C. V. 1993. Acritarchs and prasinophytes, pp 77-104: In: Lipps, J. H. (eds), *Fossil Prokaryotes and Protists*. Blackwell Scientific Publications (Oxford). ix + 342p.
MÜLLER, K. J. 1978. Conodonts and other phosphatic microfossils, pp 277-291: In; B. U. Haq and A. Boersma (eds), *Introduction to Marine Micropaleontology*. Elsevier (New York). viii + 376p.
O'CONNOR, B. 1996. Confocal laser scanning microscopy: a new technique for investigating and illustrating fossil radiolaria. *Micropaleontology*. **42**, (4), 395-402.
PESSAGNO, Jr, E. A. 1977a. Upper Jurassic Radiolaria and radiolarian biostratigraphy of the California Coast Ranges. *Micropalaeontology*. **23**, (1), 56-113.
PESSAGNO, Jr, E. A. 1977b. Lower Cretaceous radiolaria biostratigraphy of the Great Valley Sequence and Franciscan Complex, California Coast Ranges. *Cushman Foundation for Foraminiferal Research*. Special Publication. **15**. 1-87.
PESSAGNO, Jr, E. A. 1977c. Radiolaria in Mesozoic stratigraphy, pp 913-950 (techniques 918-920): In; A. T. S. Ramsay (ed), *Oceanic Micropalaeontology* (2 Volumes). Academic Press (London). xii + 1453p.
PESSAGNO, Jr, E. A and NEWPORT, R. L. 1972. A technique for extracting Radiolaria from radiolarian cherts. *Micropaleontology*. **18**, (2), 231-234.
PLAYFORD, G and DRING, R. S. 1981. Late Devonian Acritarchs from the Carnarvon Basin, Western Australia. *Special Papers in Palaeontology*. **27**, 1-78.
RIEDEL, W. R and SANFILIPPO, A. 1974. Radiolaria from the southern Indian Ocean, DSDP Leg 26. *Initial Reports of Deep Sea Drilling Project*. **26**, 771-813.
RIEDEL, W. R and SANFILIPPO, A. 1977. Cainozoic Radiolaria, pp 847-912 (techniques 852-858): In; A. T. S. Ramsay (eds), *Oceanic Micropalaeontology* (2 Volumes). Academic Press (London). xii + 1453p.
SANFILIPPO, A and RIEDEL, W. R. 1985. Cretaceous radiolaria, pp 573-630: In; H. M. Bolli, J. B. Saunders and K. Perch-Nielsen, (eds) *Plankton Stratigraphy*. Cambridge University Press (Cambridge). viii + 1032p.
SANFILIPPO, A, WESTBERG-SMITH, M. J and RIEDEL, W. R. 1985. Cenozoic radiolaria, pp 631-712: In; H. M. Bolli, J. B. Saunders and K. Perch-Nielsen (eds), *Plankton Stratigraphy*. Cambridge University Press (Cambridge). viii + 1032p.
SARJEANT, W. A. S. 1961. The hystrichospheres - a review and discussion. *Grana Palynologica*. **2**, 101-111.
SARJEANT, W. A. S. 1974. *Fossil and living Dinoflagellates*, (techniques, appendix A, pp 128-133). Academic Press (London). vii + 182p.
SCHRADER, H-J. 1973. Proposal for a standardized method of cleaning diatom-bearing deep-sea and land-exposed marine sediments. *Nova Hedwegia*. **45**, 403-409.
SCHRADER, H-J. 1976. Cenozoic planktonic diatom biostratigraphy of the Southern Pacific Ocean. *Initial Reports of the Deep Sea Drilling Project*. **35**, 605-672.
SETTY, M. G. P. A. 1966. Preparation and method of study of fossil diatoms. *Micropaleontology*. **12**, (4), 511-514.
STAPLIN, F. L. 1977. Interpretation of thermal history from color of particulate organic matter - a review. *Palynology*. **1**, 9-18.
TASCH, P and SHAFFER, B. L. 1961. Study of scolecodonts by transmitted light. *Micropaleontology*. **7**, (3), 369-371.
TAUGOURDEAU, P. 1971. Scolecodontes du Frasnien du Boulonnais. *Revista Española de Micropaleontologia*. **3**, (1), 71-108.
WEST, R. G. 1977. *Pleistocene geology and biology* (2nd Edition). Longman (London). xi + 440p.
WILLIAMS, G. L. 1978. Dinoflagellates, Acritarchs and Tasmanitids, pp 293-326: In; B. U. Haq and A. Boersma (eds), *Introduction to Marine Micropaleontology*. Elsevier Biomedical (New York). viii + 376p.

27. EXTRACTION TECHNIQUES FOR PHOSPHATIC FOSSILS

27.1. INTRODUCTION

The extraction and preparation of phosphatic preserved macro and microfossil material requires the careful use of acidic solutions in controlled etching techniques. In the area of micropalaeontology the application is widely known in relation to the extraction of conodont elements (*e.g.* Collinson 1963, 1965, Stone 1987) and other phosphatic microfossils (Bengtson 1977). Acids used in the extraction of conodonts are familiar to vertebrate and invertebrate palaeontologists in the preparation of phosphatized bones and shell material respectively. Their use in the extraction of large fossils creates additional problems in that exposed surfaces must be protected from contact with fresh acid (Converse 1989).

In common with all acid preparation procedures, the best results are obtained when the chemical composition between the fossil and matrix differ. Success is also enhanced when the acid dissolves the matrix at a faster rate than the fossil. This action of controlled etching requires considerable skill and experience on behalf of the operator, particularly in micropalaeontological preparations where the material is examined in an SEM. Such procedures are not new to micro or macro-palaeontologists (see Cooper and Whittington 1965, for a brief history, and Stone 1987 for a review of conodont extraction procedures).

27.2. MACROFOSSILS

27.2.1. Pre-treatment

1 Scrub the sample with a bristle brush in a sink. This will remove any loose debris.

2 Oven dry at 120°C for 1 hour.

3 Remove excess matrix surrounding the fossil by using a hammer and chisel, or one of a variety of pneumatic or electrical rotary and reciprocal tools (see section 12 MECHANICAL METHODS OF PREPARING FOSSIL SAMPLES).

The use of acetic and formic acids in the development of calcium phosphatic and chitino phosphatic fossils were formalised over a ten year period commencing during the late 1940's (*e.g.* Bell 1948, Toombs 1948, Rixon 1949, Ewing 1950, Toombs and Rixon 1959; summarised and reviewed in Cooper and Whittington 1965, and Rixon 1976). Methods were principally used in the development of vertebrate material, although adopted and successfully used in the development of invertebrate fossils, in particular phosphatized brachiopods (Bell 1948), conodonts (Beckmann 1952) and partially silicified brachiopods (Muir-Wood and Owen 1952).

Safety Note: All procedures should be undertaken in a fume cupboard. Where acid preparations have to be continually observed, controlled environments (*e.g.* Sass 1963) reduce equipment corrosion or fumes reaching the preparator. Ensure such equipment conforms with legal requirements.

27.2.2. Phosphate determination

Before embarking on the chemical preparation of either macro and micro material, it may be prudent to determine the presence of phosphate within the sample. A rapid procedure involving the application of an acidified vanadomolybdate solution, and observing the precipitate, achieves this (see section 6 FIELD STAINING TECHNIQUES FOR DETERMINING CALCITE, DOLOMITE AND PHOSPHATE).

27.2.3. Chemical extraction

Where mechanical development is not sufficient in complete matrix removal, chemical techniques have to be employed. Essentially, these will completely free the fossil from all matrix, and disarticulate adjoining bones (*e.g.* jaw and limb mechanisms), enabling detailed examination of all faces and facets. When considering the use of acid baths, a preparator must remain with the sample until processing is complete, failure to monitor the reaction could result in irreversible damage to the fossil.

Prior to commencing acid bath work, test a small amount of matrix in acids of different strengths to determine the most suitable concentration. Also strengthen and protect exposed areas of the fossil by painting on a layer of poly(butyl methacrylate). This will inhibit etching of the exposed bone.

1 Completely immerse the specimen in a 10-15% solution of acetic acid ($CH_3.COOH$). If part of the specimen is free of acid, calcium acetate crystals will form. These are very difficult to remove, and can cause damage to the microfabric of delicate vertebrate material (Converse 1989). A buffered acetic acid solution may be desirable for extractions of smaller and delicate fossils, eliminating the etching effects of the acid (see below under chemical extractions of microfossils) which are to be examined in an SEM. Muir-Wood and Owen (1952) successfully extracted partially silicified brachiopods, in which the silica was in a colloidal or amorphous form, and not cryptocrystalline, using an acetic acid solution buffered to pH 4-6 with a 20% ammonium acetate ($CH_3.COONH_4$) solution

2 Frequently monitor the speed of the reaction, and its effect on the fossil material. Prolonged exposure (5 days+) of fossil material to the acid, may result in its destruction (Griffith 1954). The reaction can be stopped at any time by removing the specimen from the acid and neutralizing by immersing it in ammonia. Water will not immediately stop the reaction, and may initially increase etching activity (Converse 1989). The controlled etching and continued visual monitoring of

material ensures that the minimum damage is inflicted on delicate specimens (Sass 1963).

3 Place the specimen in a sink, and flush with running water, or frequently change the water if a water bath is used.

4 Remove and allow specimen to air dry. If any calcium acetate salt deposits are seen on the surface, continue with the washing stage.

5 Once dry the specimen can be examined with a magnifying lamp or under a stereozoom binocular microscope, and the softened matrix removed with the aid of a mounted needle.

6 If the specimen requires further acid treatment, the newly exposed areas should be protected by covering with a protective lacquer [poly(butyl methacrylate)]. This will both strengthen the specimen and seal delicate areas from further etching.

7 Only when the sealant has dried should the specimen be re-immersed in a fresh acid solution. Extra vigilance is required in monitoring the etching process to ensure no damage to the fossil.

At advanced stages of the procedure large delicate vertebrate specimens may require additional support as the matrix is removed. Latex rubber and polyester resin supports can be moulded to encase the specimen and limit fracturing while the matrix is being removed (Converse 1989). Rixon (1976) recognised six categories within which vertebrate remains can be assigned, and procedures tailored for successful extraction:

(1) Isolated bones
(2) Isolated skulls (± jaw mechanisms)
(3) Isolated small lower jaws and teeth
(4) Isolated, articulated joints
(5) Complete or partial skeletons
(6) Bone beds, fissure fillings and cave breccias

Small, delicate fossils such as plants and teleost (fish) skeletons, which naturally split through the bone centres revealing the internal anatomy, can be successfully prepared using the "transfer method" (Toombs and Rixon 1950). This procedure enables the external morphology and anatomy to be studied. The fossil is rigidly held in place and supported by a backing, commonly polyester resin. The embedding procedure is detailed in section 10 PRESERVATION, CONSOLIDATION AND REPAIR OF UNSTABLE SPECIMENS). Polyester resins are preferable to epoxy resins, which tend to have a slight yellow tint and cure at an exothermic temperature that may distort or crack thin fragile specimens.

Specific and detailed procedures relating to vertebrate fossil preparation are presented in Rixon (1976), Converse (1989) and Leiggi and May (1995), see also section 24 PREPARATION AND CONSERVATION OF VERTEBRATE FOSSILS.

27.3. MICROFOSSILS (CONODONTS AND SMALL SHELLY FOSSILS - SSF's)

Conodonts range in size from 0.1mm to 4mm, consequently they are rarely seen in the field. When visible to the naked eye, the elements occur in abundance, usually on bedding planes. Resistant to chemical weathering, they often occur as unaltered fossils, translucent amber-brown in colour, with a waxy lustre. Alteration results in a range of colours from translucent grey through opaque white and grey to opaque black.

27.3.1. Pre-treatment

A sample size of 500-1000 grams is usually sufficient to yield a representative fauna from limestones (10-100 conodonts/kg, Collinson 1965). Bioclastic limestones are generally more productive than micritic limestones, while oolitic limestones may contain rounded and worn specimens. Yields from dolomites are lower (@50/kg), even if formic acid is used in processing, possibly reflecting the insoluble nature of the rock.

Conodont yields from clastic sediments are extremely variable. Shales can yield in excess of 1000/kg (Collinson 1965), but this is rare, and despite being easy to process are generally less reliable than limestones. Likewise with sandstones and siltstones, results are poor as sediment composition inhibits disaggregation.

1 Scrub the sample with a bristle brush in a sink. This will remove any loose debris.

2 Oven dry fragments at 120°C for 1 hour.

3 Break the sample, using a hammer or jaw-crusher, into fragments approximately 2 cm^3, discard the fine powder. Place approximately 500g in a labelled 2000ml glass or plastic beaker (smaller amounts can probably be processed more quickly (100 grams of sample in 1000ml beaker). Two or three beakers of the same sample can be processed simultaneously).

A 2kg sample should be sufficient, but samples between 500 grams and 1kg can be used. From Palaeozoic rocks a yield of between 10-20 conodonts can be expected. Varker (1967) collected a 3kg of sample and then process a standard 1kg sub-sample. Pre-weighing samples before processing is essential in faunal studies requiring a quantitative approach. Extraction yields can be improved considerably by employing buffered acidic solutions during the chemical extraction stages (see details below). Furthermore, less damage will be inflicted upon the structurally weaker biomineralogical and diagenetically damaged elements of the fauna.

27.3.2. Chemical extraction

Safety Note: All procedures should be undertaken in a fume cupboard.

Previously described techniques, concentrating on conodont extraction, outline methods applicable, in the main, to limestones or calcareous cemented rocks (Allman and

Lawrence 1972, Bell 1948, Brasier 1980, Collinson 1965, Gunnell and Morey 1932, Higgins and Spinner 1969, Simes 1973, 1977).

Methods outlined below involve the use of acids and heavy liquids that should only be used under guidance from a qualified person and in a well ventilated working area. **HANDLE WITH CARE.** Notices of use should be well displayed and containers used during processing well labelled.

27.3.2a. Calcareous rocks

Formic acid (H.COOH)

Until quite recently traditional methods of extraction using this procedure (*e.g.* Brasier 1980, Aldridge 1990) suggest using a 10-15% acid solution covering a crushed sample, and topped with hot (80°C) water. Leaving for 6-24 hours, decanting, wet sieving and repeating the process on the course material. Recent experimental work (Jeppsson and Anehus 1995) shows that this procedure must be abandoned, and illustrates the importance of using double buffered formic acid solutions for conodont extraction. The acid must be buffered with both calcium carbonate ($CaCO_3$) and *tri*-calcium phosphate ($Ca_3(PO_4)_2$)(*tert-calcium phosphate*). Jeppsson and Anehus (1995) demonstrated that solutions buffered with only calcium carbonate or *tri*-calcium phosphate destroy phosphatic fossils. The revised extraction process is as follows:

Follow stages 1 - 3 of the pre-treatment method.

4 Mix up a (10%) solution of formic acid (H.COOH). To every litre of acid solution add 20-30g of calcium carbonate and 1.2-0.7g *tri*-calcium phosphate (Jeppsson and Anehus (1995) recommend 20 and 1.2g respectively for total dolomite removal). The pH should not exceed 3.6. It is essential that all carbonate is removed from the residue if density separation procedures employ sodium polytungstate, (see below and section FLOTATION AND LIQUID SEPARATION TECHNIQUES).

5 To avoid solution evaporation, cover with a lid and place in a well-ventilated fume cupboard to allow gas generated from the process to escape.

6 After 6-24 hours, when most of the carbonate should have dissolved and effervescence ceased, wash the residue by half-filling the bucket with water and gently stirring.

7 Allow the coarse material to settle (approximately 5 minutes), and decant off the liquid.

8 Very coarse or large fragments can be treated with a fresh batch of buffered acid.

9 Thoroughly wet sieve the residue over a 63μm sieve.

The effectiveness of the dissolution/disaggregation process can be further monitored by the addition of a "spike", a measured quantity of a substance (*e.g.* apatite) of similar specific gravity to the microfossils being recovered, to the crushed residue (von Bitter and Millar-Campbell 1984). Spikes can also be used to monitor the microfossil recovery during the liquid separation stage of the process.

Acetic (glacial) acid ($CH_3.COOH$)

Conodont element destruction using unbuffered acetic acid solutions of 5%, 10% and 15% on different volumes of limestone was illustrated by Jeppsson *et al.*, (1985). Buffering the acetic acid with calcium acetate [$(CH_3.COO)_2Ca$] ensured improved recovery and less etching of uncorroded conodont elements. Jeppsson *et al.*, (1985) used a 10% fresh acetic acid solution, mixed with a filtered acetic acid solution, an "acetate soup", used on previous samples. A mixture of 70% of the first to 30% of the second was found to be effective in dissolving the rock while not damaging the conodont elements. However, this may not be suitable in laboratories where the procedure is not performed regularly, and stocks of used acid are not accumulated or can be stored. The addition of sodium acetate ($CH_3.COONa$) to acetic acid, as a substitute for filtered calcium acetate has proved satisfactory. A revised method is as follows:

Follow stages 1 - 3 of the pre-treatment method.

4 Mix a 10% acetic acid solution, and to every 100ml add 3-4g of sodium acetate ($CH_3.COONa$) or calcium acetate [$(CH_3.COO)_2Ca$], sufficient to reduce the pH to approximately 3.4-3.5.

5 Change the acid by decantation every 8 hours, examining the washed fines for evidence of dissolution. This process may take several weeks to dissolve a large quantity of sample. The rock is best treated in smaller quantities than in the formic acid method. It is important to change the acid regularly and ensure that the sample is completely immersed. The production of calcium acetate at the acid-air interface during the reaction can result in problems. This salt is only slightly soluble in water, and on forming a coat over the sample inhibits and retards the reaction.

6 When all effervescence has ceased, wash the residue by half-filling the bucket with water and gently stirring.

Complete the procedure by following stages' 7-9 of the formic acid method.

Monochloroacetic acid ($CH_2Cl.COOH$)

The method is similar to that outlined for both formic and acetic acids. The use of this acid results in faster reaction times compared with acetic acid (Beckmann 1952). However, as with both formic and acetic, the unmonitored use of this acid has been shown to destroy conodont elements (Ziegler *et al.*, 1971). Until more detailed experimental work determines suitable buffered concentrations and pH's that do not destroy phosphatic elements, its use on material should be carefully monitored, and if possible avoided.

Other acids can be used to dissolve limestones, such as tartaric and citric acid, but like formic and chloroacetic acid, they are more expensive than glacial acetic acid.

27.3.2b. Argillaceous rocks

Conodonts in rocks of a predominantly argillaceous composition can be extracted using a two-stage chemical processing technique. Initial processing follows one of three chemical methods listed below, and detailed in section 29 EXTRACTION TECHNIQUES FOR CALCAREOUS MICROFOSSILS FROM ARGILLACEOUS SEDIMENTS.

Petroleum spirit

Follow pre-treatment stages outlined above.

4 Place 500g of fragments in a large beaker, and cover with petroleum or white spirit. Allow the fragments to soak for 24 hours. This is one of the most effective methods of reducing shales to a fine mud.

5 Drain off and filter the spirit for re-use.

6 Cover the sample with hot water, containing a few drops of concentrated detergent (this will help clean the specimens). Boil until the rock disintegrates and a mud is produced.

Hydrogen Peroxide (H_2O_2)

Soak the fragments in a 15% solution of H_2O_2. Immerse the sample for 24 hours, then boil the solution for 30 minutes. This technique is effective on most samples except black shales.

Sodium hypochlorite (NaClO)

Most argillaceous rocks will break down in a 10-15% solution of sodium hypochlorite. This process may, however, take three or four weeks, after which some of the conodonts may be bleached white, although they remain identifiable. After soaking residues must be boiled in water, with a few drops of detergent for about 30 minutes. Following this stage continue the processing with either the formic or acetic acid procedure as detailed above.

Norby (1972) successfully used both the hydrogen peroxide and sodium hypochlorite methods in examining slabs of black shales containing conodont assemblages. In addition to freeing the assemblages, the three dimensional position of elements was recorded. This was achieved by initial x-ray photography of the slab to determine the position of the elements, applying a protective coat to all but one surface prior to chemical preparation. Between each chemical stage the slab was photographed recording the position of each element.

27.3.2c. Siliceous rocks

Safety Note: Hydrofluoric acid (HF) is extremely dangerous. Read safety notes under section EXTRACTION TECHNIQUES FOR PALAEOBOTANY AND PALYNOLOGICAL MATERIAL for conditions of use and procedures to be followed.

With only rare occurrences of conodonts in silicified rocks, this method is not widely used. However, a renewed interest by Ethington and Austin (1993) in studying Ordovician cherts containing conodonts has resulted in a re-examination and modification to the method outlined by Collinson (1965).

Initial preparation follows the pre-treatment stages outlined above.

4 Place sample in a plastic bucket and cover with 10% (by volume) hydrofluoric acid (HF).

5 Allow the sample to stand for 24 hours, or until all chemical reaction has ceased.

6 Decant off and neutralize the HF solution with the addition of sodium carbonate (Na_2CO_3).

7 Wet sieve the residue over a 38µm sieve onto filter paper. Catching all the fines ensures that the smallest conodont fragments are retained. These should be transferred to micropalaeontological slides before the filter paper has dried, reducing loss by blowing away, or breakage by becoming entangled within the dried fibres of the filter paper (Ethington and Austin 1993).

8 Repeat stages' 4-7 on the fraction >38µm if insufficient residue has been produced, or fragments have not completely disaggregated.

Ethington and Austin (1993) report that after seven repetitions of the acid dissolution stage, the original 50g of sample was reduced by 19g (38%), freeing eight specimens. However, in common with Collinson (1965), specimens recovered using this method are brittle. The delicate nature of elements invariably results in some being lost, particularly when trying to remove them from the filter paper. Furthermore, if initial preservation is poor the freed elements may amount nothing more than a collection of phosphatic grains of sand grade size.

The etching procedure described by Ford and Lee (1997) attempts to provide an alternative, less destructive method of examining and photographing conodont elements. Unfortunately, the results are not permanent, and there is no guarantee that poorly preserved material will not be destroyed. The procedure was employed on poorly fossiliferous Early Permian cherts from the South Island of New Zealand, providing precise faunal correlation between Permian sequences. The method is as follows:

Follow pre-treatment stages 1 and 2 outlined above.

3 Chert specimens were split along bedding planes (surfaces most likely to reveal conodonts) to produce thin slabs of approximately $5cm^2$.

4 Specimens, with the prepared surface uppermost, are then placed in a polypropylene beaker and covered with 35% HF.

5 **Avoid breathing in the fumes generated during etching** - carefully monitor the colour change to white of the specimen surfaces over a 15-30 minute period. During this time observation will reveal the formation of gel forming over the surface.

6 Decant, neutralize and safely dispose of HF. Cover samples with water, but **do not** pour the water directly onto the specimen surface. Care must be taken to ensure that the gel covering is not dislodged. Avoid washing the surface with a brush.

7 Repeat the process until the sample can be handled (it is advisable to wear vinyl gloves), transferred to a petri dish, immersed in water and examined under a binocular microscope for the presence of conodonts.

8 Samples yielding conodonts should be photographed immediately, or within 24 hours if stored in water. The translucency of the gel coat is impaired as it undergoes crystallization and begins to break up, a process that is initiated when acid neutralization commences. Although proceeding at a slower rate for samples kept wet, the maximum storage time recommended prior to photography is 24 hours (Ford and Lee 1997).

9 Re-etching of samples containing conodont specimens can be undertaken as required. On completion of the process the gel coat can be removed by gently brushing with a stiff bristled paint or tooth brush. Samples should be soaked in orthoboric acid (H_3BO_3) or sodium carbonate (Na_2CO_3) to ensure they are thoroughly neutralized, and can be safely handled without gloves, prior to storage.

Ford and Lee (1997) have demonstrated with this procedure how highly micro-fractured conodont elements can be successfully reconstructed, identified and used in age determinations and correlation's within poorly fossiliferous sediments. This technique is particularly useful where it proves impossible to extract identifiable elements by conventional means.

27.4. TREATMENT OF RESIDUES

27.4.1. Sieving and separating grade size fractions

Initial treatment of the residue involves wet sieving to produce grade size fractions that may concentrate the conodonts (see section 16 WASHING AND SIEVING TECHNIQUES). This is undertaken using a nest of sieves within the size ranges of

4mm to 63µm. Wash the residue with a fine jet of water for approximately 10 minutes, this will remove any trace of the maceration acid and fine clay sized particles. The contents of each of the sieves should then be washed into an evaporating dish and oven dried at 60°C. Initial stereozoom binocular examination of the fractions will reveal if further separating and concentrating procedures are required.

27.4.2. Heavy liquids

Safety Note: Many of the organic heavy liquids (*e.g.* Bromoform, Tetrabromoethane, Methylene iodine) used in residue separation are carcinogenic or toxic. They should be handled with care, and only used within a fume cupboard or well-ventilated area, with the preparator wearing suitable protective clothing.

Separations of the dried residues to concentrate the conodonts using heavy liquids are extensively discussed in section 19 FLOTATION AND LIQUID SEPARATION TECHNIQUES. The use of Bromoform ($CHBr_3$) and Tetrabromoethane ($CHBr_2.CHBr_2$) have been very successful. However, the potential health hazard related to these chemicals has resulted in the suggestion of alternatives. One such method, although not widely accepted, uses water soluble sodium polytungstate (Savage 1988).

The method is essentially similar to that of the more traditionally known procedure, using separating funnels and filter paper to remove the heavy and light fractions (see section 19 FLOTATION AND LIQUID SEPARATION TECHNIQUES).

If a large number of liquid separations are to be undertaken, the procedure can be very labour intensive. Charlton (1969) modification the tedious and time consuming stirring stages by constructing a series of weighted plastic paddles, and immersing them in the top funnel. Motion was achieved with the aid of a compressed air jet, which when activated ensured the residue was kept in constant motion within the liquid.

Comparing manual and automatic methods under controlled experimental conditions, Charlton (1969) demonstrated the efficiency of the modifications, reducing a 2hr manual method to 30 minutes for 25g of residue. Furthermore, it was possible to demonstrate that automatic stirring contributed to the efficient of conodont recovery, although smaller conodonts proved more difficult to remove through liquid separation than larger conodonts.

27.4.3. Magnetic separation

Residue separation using the Frantz Isodynamic Magnetic Separator has been described by Dow (1960, 1965). The procedure is detailed in section 18 MECHANICAL SEPARATION OF MICROFOSSIL RESIDUES. The procedure can be time consuming, with residues requiring many runs to concentrate a sufficient faunal collection. Between each run examine the residue (the conodonts should be concentrated in the non-magnetic fraction) optically, and adjust the speed of feed, forward and side slope and amperage accordingly.

During successive runs, a fine powder may collect in the fractions. This can be removed by dry sieving (see section 16 WASHING AND SIEVING TECHNIQUES USED IN MICROPALAEONTOLOGY). Many samples will, however, only require one pass through the separator, the degree of separation being at the operator's discretion or determined by the lithology of the residue.

Further concentration of conodonts by the removal of quartz and dolomite grains (the two most common minerals of the residues), are outlined by Dow (1960, 1965), and detailed in section 18 MECHANICAL SEPARATION OF MICROFOSSIL RESIDUES.

27.5. RESIDUE PICKING AND MOUNTING

A fine (00 or OOO) sable hair brush is adequate when used under a low power (x20) binocular microscope. Picking is done in the conventional way for microfossils (see section 20 THIN SECTION AND SLIDE PREPARATION TECHNIQUES OF MACRO- AND MICROFOSSIL SPECIMENS). For examination the residue should be placed in a picking tray, preferably with a light back ground.

Permanent mounting can be done using either the single or multi-celled cavity slides with a water soluble gum. This will enable the specimens to be moved or re-orientated if required later. Slides with a white background tend to be more effective than black for the examination of the dark conodonts.

The main stages of phosphatic fossil development are summarised in the flow chart illustrated in figure 27.1.

REFERENCES

ALLMAN, M and LAWRENCE, D. F. 1972. *Geological Laboratory Techniques*. Blandford, (London). 1-335p.
BECKMANN, H. 1952. The use of acetic acids in micropaleontology. *Micropaleontologist*. **6**, (3), 39.
BENGTSON, S. 1977. Early Cambrian button-shaped phosphatic microfossils from the Siberian Platform. *Palaeontology*. **20**, (4), 751-762.
BELL, W. C. 1948. Acetic acid etching technique applied to Cambrian brachiopods. *Journal of Paleontology*. **22**, (1), 101-102.
BRASIER, M. D. 1980. *Microfossils*. Allen and Unwin (London). xii + 193p.
CHARLTON, D. S. 1969. An improved technique for heavy liquid separation of conodonts. *Journal of Paleontology*. **43**, (2), 590-592.
COLLINSON, C. 1963. Techniques for collecting and processing conodonts. *Illinois State Geological Survey*. Circular 343, 1-16.
COLLINSON, C. 1965. Conodonts, pp 94-102: In; B. Kummel and D. Raup (eds), *Handbook of Paleontological Techniques*. W. H. Freeman and Co., (San Francisco). xiii + 852p.
CONVERSE, H. H. 1989. *Handbook of paleo-preparation techniques* (2nd edition, revised by R. W. McCarty). Florida Paleontological Society. v + 109p.
COOPER, G. A and WHITTINGTON, H. B. 1965. Use of acids in preparation of fossils, pp 294-300: In; B. Kummel and D. Raup (eds), *Handbook of Paleontological Techniques*. W. H. Freeman and Co., (San Francisco). xiii + 852p.
DOW, V. E. 1960. Magnetic separation of conodonts. *Journal of Paleontology*. **34**, (4), 738-743.
DOW, V. E. 1965. Magnetic separation of conodonts, pp 263-267: In; B. Kummel and D. Raup (eds), *Handbook of Paleontological Techniques*. W. H. Freeman and Co., (San Francisco). xiii + 852p.
ETHINGTON, R. L and AUSTIN, R. L. 1993. Note on the use of hydrofluoric acid for the recovery of conodonts from Ordovician cherts in the Southern Uplands of Scotland and the significance of the conodonts. *Journal of Micropalaeontology*. **12**, (2), 194.
EWING, J. F. 1950. A new technique for removing bones from limestone breccia. *Antiquity*. **24**, 102-105.
FORD, P. B and LEE, D. E. 1997. Note on a new method of using hydrofluoric acid for the study of conodonts in cherts in the Torlesse terrane, New Zealand. *Journal of Micropalaeontology*. **16**, (2), 158.
GRIFFITH, J. 1954. A technique for the removal of skeletal remains from bone-bed. *Proceedings of the Geologists' Association*. **65**, (2), 123-124.
GUNNELL, F. H and MOREY, P. S. 1932. The preparation of conodonts for study. *Micropaleontology Bulletin*. **3**, 77-78.
HIGGINS, A. C and SPINNER, E. G. 1969. Techniques for the extraction of selected microfossils. *Geology - Journal of the Association of Teachers in Geology*. **1**, 12-28.
JEPPSSON, L, FREDHOLM, and MATTIASSON, B. 1985. Acetic acid and phosphatic fossils - a warning. *Journal of Paleontology*. **59**, (4), 952-956.
JEPPSSON, L and ANEHUS, R. 1995. A buffered formic acid technique for conodont extraction. *Journal of Paleontology*. **69**, (4), 790-794.
LEIGGI, P and MAY, P. J. (eds). 1995. *Vertebrate Paleontological Techniques*, Volume 1. Cambridge University Press. 368p.
MUIR-WOOD, H. M and OWEN, E. F. 1952. The use of acetic acid in developing brachiopod shells. *Proceedings of the Geologists' Association*. **63**, (4), 313-315.
NORBY, R. D. 1972. Techniques for the study of conodont assemblages. *Geological Society of America, Abstracts with Programs*. **4**, 340-341.

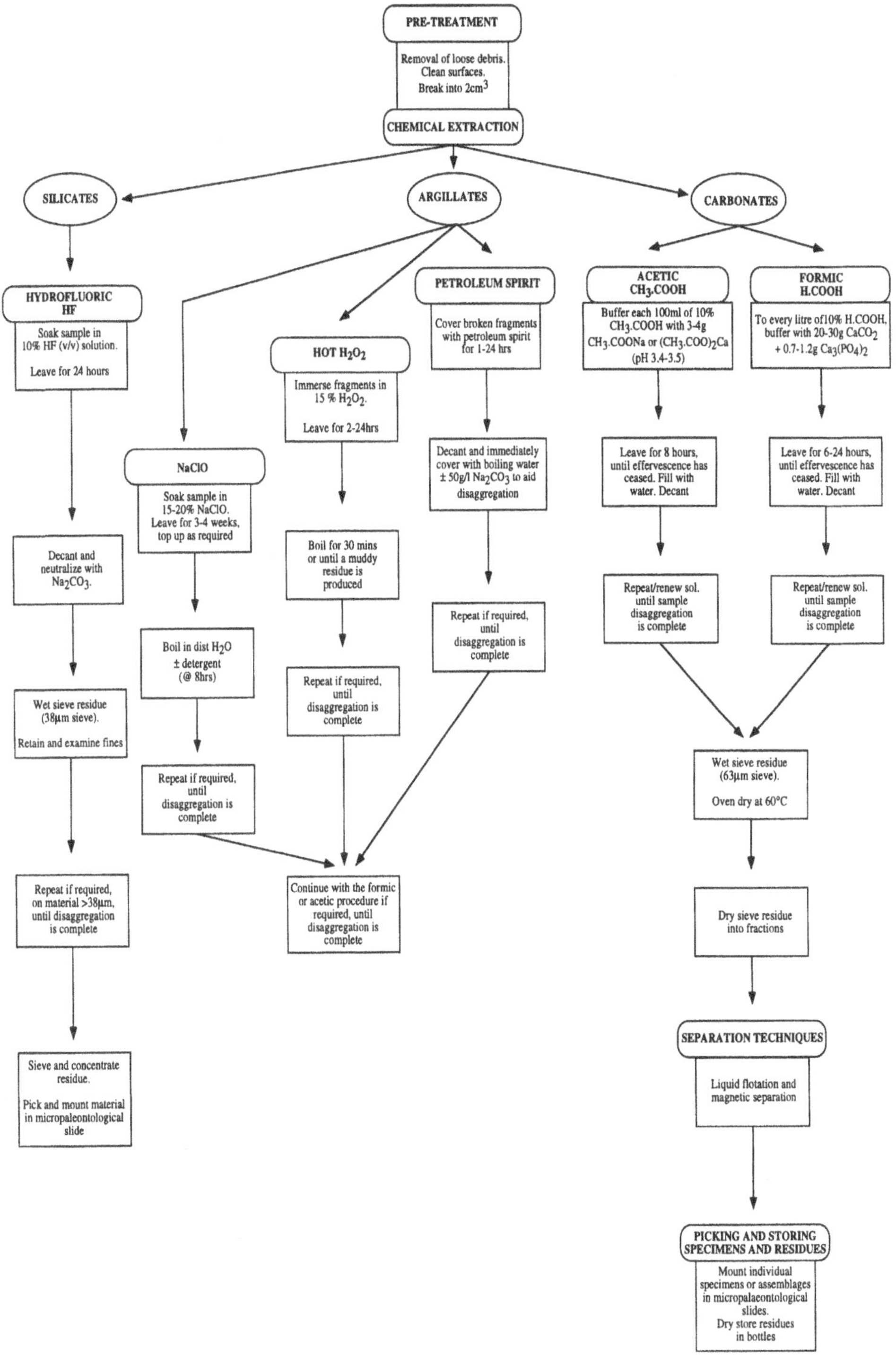

Figure 27.1. Summary flow chart of the main stages in the preparation of phosphatic microfossil residues from calcareous, argillaceous and siliceous sediments.

RIXON, A. E. 1949. The use of acetic and formic acids in the preparation of fossil vertebrates. *Museum Journal*. **49**, 116-117.
RIXON, A. E. 1976. *Fossil animal remains - their preparation and conservation*. Athone Press, (London). vi + 304p.
SASS, D. B. 1963. Controlled acid preparation of fossils. *Journal of Paleontology*. **37**, (5), 1129.
SAVAGE, N. M. 1988. The use of sodium polytungstate for conodont separations. *Journal of Micropalaeontology*. **7**, (1), 39-40.
SIMES, J. E. 1973. Report on conodont techniques and research. *New Zealand Geological Survey Paleontological Reports*. Report **Rp1**. 14p.
SIMES, J. E. 1977. Extraction of insoluble microfossils from limestone. *New Zealand Geological Survey, Department of Scientific and Industrial Research*. Report **NZGS R82**. 18p.
STONE, J. 1987. Review of investigative techniques used in the study of conodonts, pp 17-34: In; R. L. Austin (ed) *Conodonts: Investigative Techniques and Applications*. Ellis Horwood Ltd, (Chichester). 1-422.
TOOMBS, H. A. 1948. The use of acetic acid in the development of vertebrate fossils. *Museum Journal*. **48**, 54-55.
TOOMBS, H. A and RIXON, A. E. 1950. The use of plastics in the transfer method of preparing fossils. *Museum Journal*. **50**, 105-107.
TOOMBS, H. A and RIXON, A. E. 1959. The use of acids in the preparation of vertebrate fossils. *Curator*. **2**, 304-312.
VARKER, W. J. 1967. Conodonts of the genus *Apatognathus* Branson and Mehl from the Yoredale Series of the North of England. *Palaeontology*. **10**, (1), 124-141.
VON BITTER, P. H and MILLAR-CAMPBELL. 1984. The use of "spikes" in monitoring the effectiveness of phosphatic microfossil recovery. *Journal of Paleontology*. **58**, (5), 1193-1195.
ZIEGLER, W, LINDSTRÖM and McTAVISH, R. 1971. Monochloracetic acid and conodonts - a warning. *Nature*. **230**, 584-585.

28. EXTRACTION TECHNIQUES FOR UNCRUSHED GRAPTOLITES

28.1. INTRODUCTION

The common mode of preservation for graptolites is either as carbonised impressions, or whitish films on bedding surfaces of black shales. However, graptolites are not exclusively restricted to these rocks and are occasionally found preserved in carbonates (especially fine-grained argillaceous limestones), sandstones (including coarse grained greywackes), cherts, and low grade metamorphosed slates. It is rare to find uncrushed specimens, but they are occasionally found in limestones, cherts and some mudstones. Consequently, following extraction from the matrix, they are of immense value to palaeontologists studying morphology and taxonomy by scanning electron microscopy (SEM) and light microscopy (LM) techniques (Crowther and Rickards 1977, Crowther 1981).

By using acid digestion procedures, subsequently modified in micropalaeontology for the extraction of conodonts and palynomorphs, it is possible to extract the three-dimensionally preserved graptolites from limestones and cherts. Less carbonised specimens can be prepared by embedding in resin or wax blocks and cutting with a microtome, before examination as serial sections. Maceration, specimen cleaning and mounting techniques for graptolites have been described by a number of workers during the past century (*e.g.* Wiman 1895, pp 253-260; Kraft 1926, pp 212-215; Cox 1933, p 4; Bulman 1944-1947, pp 1-2, Skoglund 1961, pp 389-390; Skevington 1963, pp 3-5; Hutt and Rickards 1967, pp 180-181), and reviewed at length by Berry (1965, pp 103-109; and Strachan *et al.*, 1991, pp 59-68).

Extraction techniques are essentially the same for both limestones or cherts. The acid resistant fossils are removed from the matrix by dissolving in dilute hydrochloric (HCl) or acetic acid (CH_3COOH, for limestones) or hydrofluoric acid (HF, for chert). Impure limestones require a two stage processing procedure; first the removal of any carbonate, and secondly the removal of silica. These procedures are detailed under the sections covering the extraction of conodonts (acetic acid) and palynomorphs (hydrofluoric acid), and are only briefly outlined below.

28.2. EXTRACTION FROM SILICEOUS ROCKS

1. Break the rock, using a hammer or jaw crusher, into fragments between 2-4cm^3.

2. Place fragments in a labelled plastic beaker and immerse in concentrated hydrochloric acid (HCl) until effervescence has ceased.

3. Dilute and decant acid, then wash fragments with distilled water. N.B. The above two stages may require many acid changes over a 2-3 week period to obtain a carbonate free residue.

4. Add concentrated (40% w/v) hydrofluoric acid (HF), and leave for 24 hours.

5 Once a disaggregated residue has formed in the bottom of the beaker, neutralize and decant the acid. Wash residue two or three times until the solution is neutral (test with pH paper).

6 Dry the residue in an oven at 60°C.

7 The residue can now be examined under a binocular microscope and the graptolites picked and mounted. Visual enhancement of the graptolites is outlined below.

28.3. EXTRACTION FROM CARBONATE ROCKS

1 Break the rock, using a hammer or jaw crusher, into fragments between 2-4cm^3.

2 Place fragments in a labelled plastic beaker and immerse in a 15% acetic acid solution. Acetic is preferablc to hydrochloric acid as it less violent to the fragile material. Bulman (1970) suggests adjusting the concentration of the acid so that effervescence is not too strong, and can be maintained by the repeated addition of a few drops of the concentrated acid.

3 Change the acid by decantation every 48 hours. This process may take several weeks to dissolve a large quantity of material because of the production of calcium acetate during the reaction. This salt (which is only slightly soluble in water) coats the sample and retards the reaction.

4 Re-treat the remaining fragments if required or until enough residue has been formed.

5 Thoroughly wash the residue, then dry at 60°C in a laboratory oven before examination.

28.4. ENHANCING OPTICAL APPEARANCE

Individual graptolites can be rendered translucent by the addition of 10% nitric acid (HNO_3) containing 5% potassium chlorate ($KClO_3$). Mix a small amount of each on a watch glass, and then immerse the specimen. Once the technique has been perfected several specimens can be cleaned together, but this should only be attempted by the experienced worker.

The duration of this treatment varies from 10 minutes to 1 hour, and should be observed under a low power binocular microscope. Some specimens may require many days of cleaning to obtain the best results, while others will begin to powder and fall apart once freed from the matrix of the rock (Skoglund 1961). If concentrated acids are used then bleaching must be restricted to under 20 minutes. Longer procedures are possible when using reduced acid concentrations (Bulman 1970). Results are best on low grade metamorphic material (Bulman 1944-1947), while highly carbonised material may not clean at all, but only powder. Only experience will tell if a specimen can withstand the treatment. Practice on one specimen before proceeding with mass preparations. When bleaching has been completed the specimens should be dehydrated by immersing in a series of alcohol's, and finally a microscopical grade xylene ($C_6H_4(CH_3)_2$) low in sulphur.

28.5. MOUNTING AND STORAGE

Individual specimens can be orientated and mounted in Canada Balsam on a glass cavity slide, and covered with a coverslip. Temporary dry mounts between two glasses can also be prepared, but these may result in damage to the specimen. For long term storage the specimens can be placed in a glycerol filled glass vial.

Specimens can also be mounted in wax or epoxy resin, and serially sectioned using a microtome. Careful light microscopy observations must be undertaken to ensure that there are no pyrite grains trapped in the periderm area, as these may damage the microtome blade. The preparation of these ultra-thin sections, and their subsequent transfer to mounts and TEM examination, has been described by Dumican and Rickards (1985).

A two stage method of embedding pyritized graptolites in a clear polyester resin, removing excess matrix and sealing the specimen in with a second layer of resin has been described by Hutt and Rickards (1967). This method enables the specimen to be examined either side, preventing atmospheric deterioration of the unstable pyritized material (see section 10 PRESERVATION, CONSOLIDATION AND REPAIR OF UNSTABLE SPECIMENS).

REFERENCES

BERRY, W. B. N. 1965. Graptolites; pp 103-109: In; B. Kummel and D. Raup (eds), *Handbook of Paleontological Techniques*. W. H. Freeman and Co, (San Francisco). xiii + 852p.

BULMAN, O. M. B. 1944-1947. Monograph of Caradoc (Balclatchie) graptolites from limestones in Laggan Burn, Ayrshire. *Palaeontographical Society (Monograph)*. xi + 78p.

BULMAN, O. M. B. 1970. Graptolithina, with sections on Enteropneusta and Pterobranchia, Part V (revised). *Treatise on Invertebrate Paleontology*. The Geological Society of America and University of Kansas (Boulder Colorado and Lawrence Kansas). xxxii + V163.

COX, I. 1933. On *Climacograptus inuiti* sp. nov. and its development. *Geological Magazine*. **70**, 1-19.

CROWTHER, P. R. 1981. The fine structure of graptolite periderm. *Special Papers in Palaeontology*. **26**, 1-119.

CROWTHER, P. R and RICKARDS, R. B. 1977. Cortical bandages and the graptolite zooid. *Geologica and Palaeontologica*. **11**, 9-46.

DUMICAN, L. W and RICKARDS, R. B. 1985. Optimum preparation, preservation and processing techniques for graptolite electron microscopy. *Palaeontology*. **28**, (4), 757-766.

HUTT, J and RICKARDS, R. B. 1967. An improved transfer technique for the preparation and preservation of pyritised graptolites. *Geological Magazine*. **104**, (2), 180-181.

KRAFT, P. 1926. Ontogenetische Entwicklung und die Biologie von *Diplograptus* und *Monograptus*. *Paläntologische Zeitschrift*. **7**, 207-249.

SKEVINGTON, D. 1963. Graptolites from the Ontikan Limestones (Ordovician) of Öland, Sweden. I. Dendroidea, Tuboidea, Camaroidea and Stolonoidea. *Bulletin of the Geological Institute, University of Uppsala*. **42**, (6), 1-62.

SKOGLUND, R. 1961. *Kinnegraptus*, a new graptolite genus from the Lower *Didymograptus* Shale of Vätergötland, Central Sweden. *Bulletin of the Geological Institute, University of Uppsala*. **40**, (33), 389-400.

STRACHAN, I, WHITE, D, RUSHTON, A and RICKARDS, R. B. 1991. How they are collected and prepared. Chapter 11, pp 59-68: In; D. Palmer and R. B. Rickards (eds), *Graptolites. Writing in the rocks*. Fossils Illustrated. Vol. 1. The Boydell Press (Woodbridge). 166p + 138pl.

WIMAN, C. 1895. Über *Dictyonema cavernosum* n.sp. *Bulletin of the Geological Institute, University of Uppsala*. **3**, 1-13.

29. EXTRACTION TECHNIQUES FOR CALCAREOUS MICROFOSSILS FROM ARGILLACEOUS SEDIMENTS

29.1. INTRODUCTION

Extraction techniques for microfossils (excluding palynomorphs) and micro-mollusca from argillaceous clays and shales can be sub-divided into two groups:

- Techniques suitable for soft partially consolidated rocks, usually samples of Mesozoic and Tertiary ages.
- Techniques suitable for harder indurated rocks, usually of Palaeozoic age.

Higgins and Spinner (1969) suggested that samples from the latter group can be treated using the techniques similar to those employed in conodont extraction. Aldridge (1990) describes boiling under the heading of 'hard clays', while use of petroleum spirit method is outlined under the heading 'soft or partly indurated clays'. There is however, no hard-and-fast rule restricting the use of an extraction technique to sample hardness. The sub-divisions below are used for descriptive convenience, and outline procedures of increasing severity and potential specimen damage.

This section details chemical extraction techniques for both unlithified and lithified sediments, together with sample washing, wet sieving, concentrating the microfossils and picking from the dried residue.

SAFETY NOTE: All procedures should be undertaken in a fume cupboard.

29.2. PRE-TREATMENT

A sample size of between 500-1000 grams is usually sufficient to yield a representative fauna.

1 Scrub the sample with a bristle brush in a sink. This will remove any loose debris.

2 Oven dry at 120°C for 1 hour.

3 Break the sample, using a hammer or jaw-crusher, into fragments approximately 2 cm^3, and place approximately 500g in a labelled 2000ml glass beaker (smaller amounts can probably be processed more quickly (100 grams of sample in 1000ml beaker). Two or three beakers of the same sample can be processed simultaneously).

29.3. CHEMICAL EXTRACTION

29.3.1. Unlithified samples (Allman and Lawrence 1972, Colenutt 1941).

29.3.1a. Boiling (distilled H_2O ± detergent)

See section 30 EXTRACTION TECHNIQUES FOR CALCAREOUS MICROFOSSILS FROM CARBONATE SEDIMENTS for details. The dissolution effects of calcium carbonate in contact with deionised water are accelerated with the addition of a small amount detergent (Pingitore *et al.*, 1993). Avoid using this procedure on delicate calcitic material.

29.3.1b. Boiling (distilled H_2O + Na_2CO_3)

The disintegration of sediments by boiling with a dispersing agent is the most widely used of all techniques in micropalaeontology. Sodium carbonate (Na_2CO_3, washing soda), or 20% sodium hydroxide (NaOH) are commonly used, although a 20% solution of Quaternary O added to the sample and boiled is also widely used.

1 Pre-treatment as outlined above.

2 Boil approximately 500g of the sample in 1 litre of 10% sodium carbonate. **N.B. If the microfossils are fragile do not boil the solution. Heat to 75°C, or use a 5% solution of sodium carbonate and heat to 80°C.**

3 Disaggregation may take many hours. The solution should be renewed every 60 minutes after careful decantation.

4 The disaggregation time may be reduced by placing the beaker, with a small amount of concentrated detergent (*Teepol*, or *Byprox*), in an ultrasonic tank. The possibility of specimen dissolution cannot be ignored (Pingitore *et al.*, 1993, see above), further compounded by the destruction from use of an ultrasonic tank.

5 Pass the residue through a 1mm sieve onto a 75μm sieve.

6 Further concentration can be achieved by flotation, particularly as hollow Mesozoic and Tertiary forms will float.

29.3.2. Lithified samples (Aldridge 1990, Allman and Lawrence 1972, Bolli 1950, 1952, Brasier 1980, Crowley 1952, Duffield and Warshauer 1979, Higgins and Spinner 1969, Layne 1950, Maples and Waters 1990).

Follow the pre-treatment stages' 1-3 outlined above. A variety of techniques can then be employed.

Gross dissolution effects from seven commonly used laboratory solutions (two organic solvents, five aqueous solutions) on powdered calcium carbonate (Iceland spar and aragonite) have been quantified by Pingitore *et al.*, (1993). The results and observations are summarised below in table 29.1. They provide significant evidence concerning the destructive effects of hydrogen peroxide, surfactants, and the time a sample is stored or left standing in deionised water.

SOLUTION	Comments on reactions of carbonates in the grain size ranges 62-125μm and 250-500μm
Ethanol	No calcium carbonate dissolved, even after 10 days of experiment.
Acetone	Dissolved minute amounts after 10 days, which may be related to small amounts of water present in the acetone.
Sodium Hypochlorite	Although high ionic strength increases solubility, none was recorded. The high pH (11.4) strongly inhibits calcium carbonate dissolution.
Sodium Hydroxide	Little dissolution observed. Extensive replacement of calcium carbonate by calcium hydroxide revealed by optical and X-ray diffraction studies.
Deionised water	Dissolution occurs, particularly if material is subjected to long exposure times (in excess of 200 hours).
'Alconox' (Anionic detergent)	Apparent carbonate solubility increased because of the presence of anionic surfactants, while the precipitation of calcite is probably inhibited by the presence of phosphates. Consequently exhibits a similar dissolution pattern to deionised water, but occuring much quicker.
Hydrogen Peroxide	Highly corrosive. Rapid carbonate dissolution in first few hours of contact with specimen, significantly greater than deionised water.

Table 29.1. Summary of results of calcium carbonate dissolution tests in common laboratory solvents undertaken by Pingitore *et al.*, (1993).

29.3.2a. Hot Hydrogen Peroxide (H_2O_2)

1 Pre-treatment as outlined above.

2 Immerse the fragments in a 15% solution of hydrogen peroxide by adding a volume of water to an equal volume of hydrogen peroxide. Leave for several hours. **N.B. Avoid storing hydrogen peroxide at full strength, because of the danger of explosion.**

3 Boil the solution for 30 minutes, or until a mud is produced.

This procedure should only be used if a rapid reconnaissance survey of a sub-sample is required. Following the laboratory experiments of Pingitore *et al.*, (1993) outlined above, use of either the petroleum spirit or sodium hypochlorite extraction procedure will reduce the risk of carbonate particle dissolution.

29.3.2b. Petroleum Spirit (Paraffin, White Spirit) to water ($\pm Na_2CO_3$)

Safety note: All of these solvents are extremely flammable, and irritants to the skin. Ensure that all work is undertaken in a fume-hood or well-ventilated room, and away from direct sources of heat.

1 Pre-treatment as outlined above.

2 Immerse the fragments in petroleum spirit for 1-24 hours. Decant and filter the spirit for reuse.

3 Immediately cover the fragments with water that has just boiled.

4 Many clay rich samples will disintegrate between 5-30 minutes. If not add 50g of sodium carbonate (Na_2CO_3) per litre of water and heat for 30 minutes or until a slurry is produced or the sample has disintegrated, usually about an hour. Renew the solution if required to maintain the solution level. **N.B. If the microfossils are fragile do not boil the solution.**

5 Concentrate and wash the residue as detailed below.

N.B. This technique is not suitable for fine grained sandstones, siliceous clays or black and grey clays (Bolli 1950, Brasier 1980).

29.3.2c. Sodium Hypochlorite (NaClO, domestic bleach)

4 Soak the sample in a 15-20% solution of sodium hypochlorite for 3-4 weeks. Top up as required when the solution evaporates.

5 Boil the fragments in water with detergent. This stage may require up to 8 hours of boiling to thoroughly break down some silty shales.

6 Wash the residues thoroughly to remove all salt crystals. **N.B. If the specimens are of phosphatic composition they may be bleached white.**

29.3.2d. Sodium hydroxide (NaOH)

4 Cover the sample with a solution of 20% NaOH.

5 Boil for between 1-2 hours, stirring occasionally to avoid the sample sticking to the bottom of the container.

6 Replenish the liquid as it evaporates with additional 20% NaOH to maintain the level.

7 Wash the residue through a series of sieves.

8 Repeat the procedure, if required, on the coarse fraction, until all the sample has been disaggregated.

Bolli (1950) did not observe any affects on pyritized or limonitised foraminifera, but bleaching of some agglutinated types did occur. Carbonate dissolution in sodium hydroxide is minimal, and decreases as the temperature rises, (Pingitore *et al.*, 1993), superficially supporting Bolli (1950) in suggesting that calcareous shells are not damaged by the solution. However, optical microscopy and X-ray diffraction studies revealed extensive calcium carbonate replacement by calcium hydroxide ($Ca(OH)_2$) occurs at the surface (Pingitore *et al.*, 1993). Although thorough washing in deionised water can remove the calcium hydroxide, textural alteration is unavoidable (Pingitore *et al.*, 1993). Furthermore, calcium hydroxide contamination may remain which can potentially bias any subsequent geochemical or elemental analysis. Material extracted using this procedure should be carefully labelled.

Ultrasonic disaggregation can be used besides the four techniques outlined above. As Aldridge (1990) states "a combination of techniques may be applied", and this will undoubtedly result in a more successful extraction technique (see also Duffield and Warshauer 1979). Mechanical disaggregation techniques involving the repeated freezing and thawing of specimens in water, or the boiling in sodium thiosulphate ($Na_2S_2O_3.5H_2O$) and crystallization on drying, will also disaggregate most shales.

Combinations of techniques for the disaggregation of argillaceous samples have been used by Bolli (1952), Then and Dougherty (1983) and Maples and Waters (1990). Bolli (1952) recognised a relationship between lithology ($CaCO_3$ content, porosity and bulk density) and the chemicals used. The most successful combination for argillaceous rocks with a high carbonate content was soaking in petrol for between fifteen and twenty minutes, followed by boiling in a 20% solution of sodium carbonate. Argillaceous rocks with a low carbonate content responded better when boiled in sodium hydroxide. In both instances boiling is maintained for sixty minutes, and stages repeated to obtain the maximum amount of residue.

Experimental studies by Then and Dougherty (1983) compared the effectiveness of various reagent combinations in breaking down indurated organic rich shales. The most effective procedure involved soaking the sample in hydrogen peroxide (H_2O_2) and wet sieving, followed by soaking in sodium hypochlorite (NaClO), sodium hydroxide (NaOH), and finally a 5% Quaternary O solution. Samples of 150g were, after treatment, reduced to as little as 3 grams.

Maples and Waters (1990) processed samples of mixed organic/clay matrix. They obtained residues by repeating cycles of soaking the samples in sodium hypochlorite while in an ultrasonic tank, and sodium hydroxide followed by washing.

A flow-chart summary of the disaggregation procedures used in the extraction of microfossils from argillaceous-rich sediments is illustrated in figure 29.1.

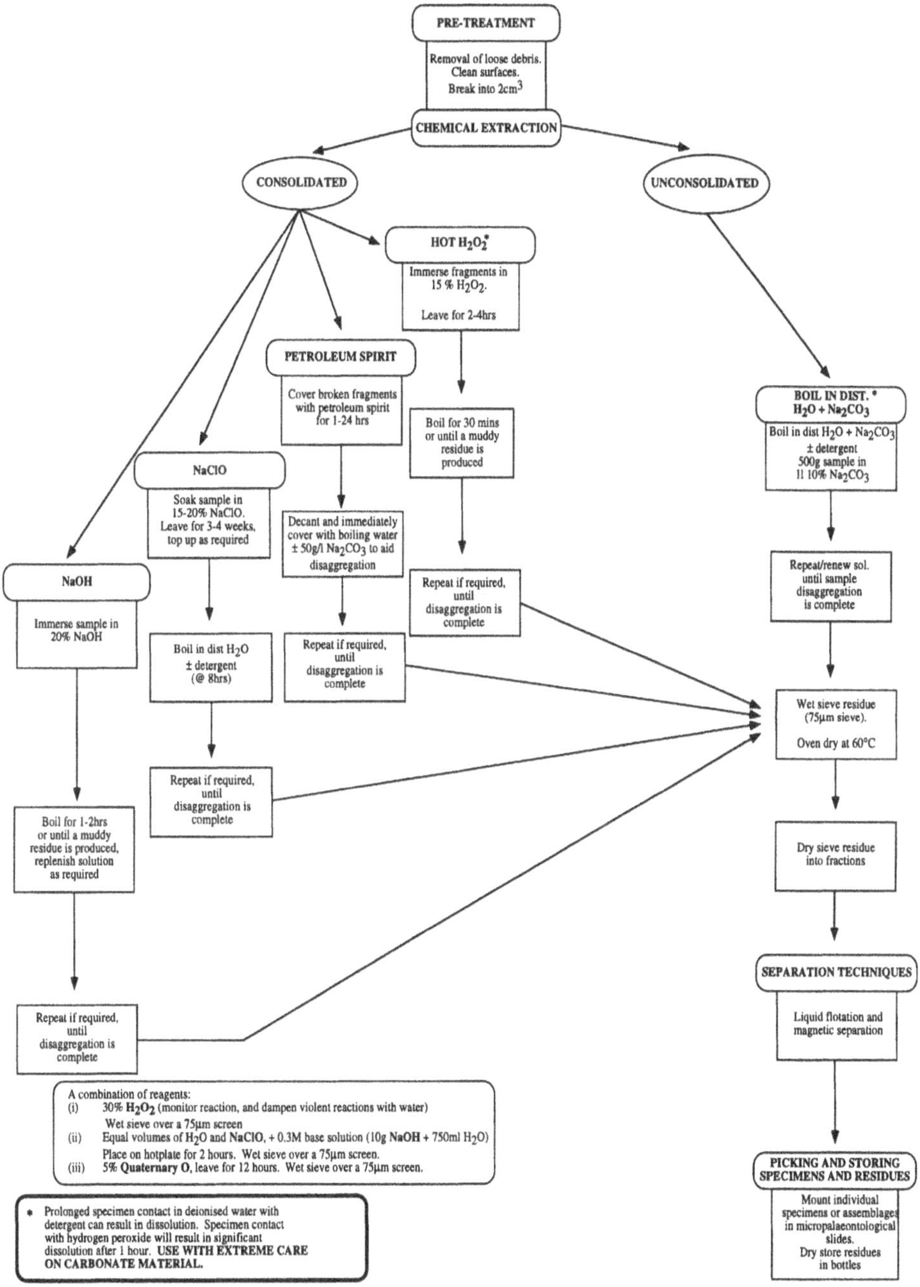

Figure 29.1. Summary flow chart of the main stages in the preparation of calcareous microfossil residues from argillaceous sediments.

29.4. WASHING AND WET SIEVING (Brasier 1980, Hulme 1961, Todd *et al.*, 1965)

Once disaggregation is complete, the residue, consisting of clays, muds and silty sands, must be washed and sieved. The residue is washed over a sieve, using a flexible rubber hose, about 10mm in diameter, to control the force and direction of the water stream. The sieve size has to be chosen with care, as the removal of the clay material may also result in a loss of calcareous nannofossils, acritarchs and diatoms.

1 If required retain a small wet sub-sample (25ml) of the residue for nannofossils and palynological studies.

2 Wash the bulk of the residue on a suitable screen with a gentle jet of water. **N.B. For diatoms, spores and pollen use distilled or filtered water.**

A brief summary of the sieve retention properties of screens is outlined in table 29.2 below, but a more detailed account can be found in section 16 WASHING AND SIEVING TECHNIQUES USED IN MICROPALAEONTOLOGY.

Sieve size (μm)	Main organisms retained
45	Small diatoms and organic walled fossils
63-75	Smaller radiolaria, foraminifera and conodonts
250	Most adult ostracod carapaces
850-2000	Large foraminifera and small shelly fossils

Table 29.2. Sieve screen sizes (μm) and the principal micro-organisms from argillaceous sediments retained. Care must be taken to wash the residue to the centre of the sieve, as excess water pressure towards the side may splash sediment over the sieve edge.

3 Flush the residue into an evaporating dish, and decant off the supernatant liquid.

4 Oven dry the residue at a low temperature (60°C maximum) until dry.

29.5. CONCENTRATING THE SPECIMENS

A number of techniques have been developed for concentrating microfossils in a residue. The technique employed is dependent upon the composition and type of microfossil being studied. Arnold (1965) described a method of foraminiferal concentration with the aid of a modified centrifuge, for use under a binocular microscope. Brasier (1980) describes a method of decanting, suitable for concentrating coccoliths and organic walled microfossils.

Microfossils in washed and dried residues can be further concentrated with the use of heavy liquids, *e.g.* bromoform, tetrabromoethane, zinc bromide. This is a particularly useful technique when the microfossils are phosphatic in composition (Brasier 1980, Swift and Aldridge 1984), see section 27 EXTRACTION TECHNIQUES FOR PHOSPHATIC FOSSILS for details.

29.6. PICKING AND MOUNTING OF DRIED SAMPLES

Dried residues can be sieved before picking. This has the advantage of concentrating size fractions of microfossils, making examination under the binocular microscope easier. Examination of the residues should be carried out using a high quality stereozoom binocular microscope with transmitted light. Spread the sample thinly on a metal or plastic picking tray, preferably with a matt finish to reduce glare and reflection from the light source, and marked with a 1cm^2 grid to aid in methodically scanning the sample. A fine mounted needle aids in the moving of the material around the tray. Use a damp sable hair (OOO or OO) brush to transfer the microfossil to a pre-glued single or multi-celled micropalaeontological slide. Additional water based glue, mixed with a drop of formalin in distilled water to prevent algae and bacteria growth, may be required to adhere the microfossil to the slide. The cardboard slides can be either single or multi-celled, and have black or white bases, the preference being dictated by the type of microfossil being picked.

REFERENCES

ALDRIDGE, R. J. 1990. Extraction of Microfossils, pp 502-504: In; D. E. G. Briggs and P. R. Crowther (eds), *Palaeobiology - A synthesis*. Blackwell Scientific Publications (Oxford). xiii + 583p.

ALLMAN, M and LAWRENCE, D. F. 1972. *Geological Laboratory Techniques*. Blandford, (London). 1-335p.

ARNOLD, Z. M. 1965b. The centrifuge as a tool in foraminiferal preparation, pp 258-261: In; B. Kummel and D. Raup (eds), *Handbook of Paleontological Techniques*. W. H. Freeman and Co., (San Francisco). xiii + 852p.

BOLLI, H. 1950. Disintegration of indurated siliceous rocks. *Micropaleontologist*. **4**, (3), 20-21.

BOLLI, H. 1952. Note on the disintegration of indurated rocks. *Micropaleontologist*. **6**, (1), 46-48.

BRASIER, M. D. 1980. *Microfossils*. Allen and Unwin (London). xii + 193p.

COLENUTT, G. W. 1941. A method of obtaining clean residues from fossiliferous clays and marls. *Proceedings of the Geologists' Association*. **52**, (3), 194-197.

CROWLEY, A. J. 1952. Method of extracting foraminifera from refactory shales. *Bulletin of the American Association of Petroleum Geologists*. **36**, (11), 2185.

DUFFIELD, S. L and WARSHAUER, S. M. 1979. Two-step process for extraction of microfossils from indurated shales. *Journal of Paleontology*. **53**, (3), 746-747.

HIGGINS, A. C and SPINNER, E. G. 1969. Techniques for the extraction of selected microfossils. *Geology - Journal of the Association of Teachers in Geology*. **1**, 12-28.

HULME, S. G. 1961. A mechanized method for breaking down and washing foraminiferal rock samples. *Micropaleontology*. **7**, (1), 107-113.

LAYNE, Jr, N. M. 1950. A procedure for shale disintegration. *Micropaleontologist*. **4**, (1), 21.

MAPLES, C. G and WATERS, J. A. 1990. Use of drain cleaner in processing shale samples. *Journal of Paleontology*. **64**, (3), 484-485.

PINGITORE, Jr, N. E, FRETZDORFF, S. B, SEITZ, B. P, ESTRADA, L. Y, BORREGO, P. M, CRAWFORD, G. M and LOVE, K. M. 1993. Dissolution kinetics of $CaCO_3$ in common laboratory solvents. *Journal of Sedimentary Research*. **63**, (4), 641-645.

SWIFT, A and ALDRIDGE, R. J. 1984. Laboratory techniques in micropalaeontology. *Science Technology*. **7**, 28-30.

THEN, D. R and DOUGHERTY, B. J. 1983. A new procedure for extracting foraminifera from indurated organic shale. *Geological Survey of Canada*, Paper **83-1B**, 413-414.

TODD, R, LOW, D and MELLO, J. F. 1965. Smaller foraminifera, pp 14-20: In; B. Kummel and D. Raup (eds), *Handbook of Paleontological Techniques*. W. H. Freeman and Co., (San Francisco). xiii + 852p.

30. EXTRACTION TECHNIQUES FOR CALCAREOUS MICROFOSSILS FROM CARBONATE SEDIMENTS

30.1. INTRODUCTION

Extraction techniques for calcareous microfossils (foraminifera and ostracoda) and small micro-mollusca from carbonate rich sediments are relatively simple, employing non-destructive disaggregation procedures (Lipps 1993). Conventional acid digestion techniques cannot be applied owing to the similarity in composition between the microfossils and the rock matrix. Mechanical disaggregation techniques causing minimal damage to the microfossils, are used with varying degrees of success. For calcareous rocks acid extraction techniques are restricted to conodonts, scolecodonts, some agglutinated foraminifera and silicified ostracods, all described in detail under separate headings.

Calcareous microfossil extraction is usually restricted to limestones of Cenozoic, and Mesozoic age. These techniques are ineffectual on highly indurated limestones, such as Carboniferous limestones. Foraminifera and ostracoda in these rocks are best examined through the study of thin sections. However, the procedure used by Kirchner (1958), and outlined below, describe a method which in some examples may be suitable for the extraction of fossils from lithified sediments. Microfossil extraction from chalk rocks is one of the earliest micropalaeontological procedures to be documented (Heron-Allen 1894). However, over 50 years elapsed before the procedure was formally described in the 20th century (Williams-Mitchell 1948). Thermal extraction techniques, suitable for lithified Palaeozoic rocks, are dealt with under a separate heading.

30.2. PRE-TREATMENT

A 500-1000 gram sample size is usually sufficient to yield a representative fauna.

1 Scrub the sample with a bristle brush in a sink, removing any loose debris, and reducing the risk of sample contamination. Some samples may readily breakdown at this stage, particularly when they become saturated, or with gentle crushing using a pestle and mortar. In such cases proceed with the soft sample extraction technique described below.

2 Oven dry at 120°C for 1 hour.

3 Break the sample, using a hammer or jaw-crusher, into fragments approximately 2 cm^3, and place approximately 500g in a labelled 2000ml glass beaker (smaller amounts can be processed more quickly, 100 grams of sample in 1000ml beaker, while two or more beakers of the same sample can be processed simultaneously).

30.3. CHEMICAL EXTRACTION

30.3.1. Unlithified samples

30.3.1a. Boiling (distilled H_2O)

4 Fill each beaker with approximately 1 litre of distilled water and place on a hot-plate. Boil the solution. Avoid stirring, even occasionally, as this may damage specimens.

5 Disaggregation may take many hours. The solution should be renewed as it evaporates ensuring that the sample is always completely covered.

6 The disaggregation time may be reduced by placing the beaker, with a small amount of concentrated detergent (*Teepol*, or *Byprox*), in an ultrasonic tank. However, care should be taken as extended ultrasonic treatment may damage delicate microfossils.

7 Pass the residue through a 1mm sieve onto a 75μm sieve, and flush with a directed jet of water. Ensure the residue does not clog finer sieve screens, as part of the washed sample may be lost if overflowing occurs. Sieves must be carefully washed between samples to reduce cross-sample contamination (see section 16 WASHING AND SIEVING TECHNIQUES USED IN MICROPALAEONTOLOGY).

8 The coarse fraction retained on the 1mm sieve can be re-processed if required. The fine residue should be carefully washed into an evaporating dish and allowed to stand. Decant off the supernate and oven dry at 60°C.

9 The dried residues can be further subdivided by dry sieving (see section 16 WASHING AND SIEVING TECHNIQUES USED IN MICROPALAEONTOLOGY), while the microfossils from each fraction can be concentrated by employing flotation techniques (see below), particularly as most Mesozoic and Tertiary forms are hollow, and can easily be separated from the non-fossiliferous fraction.

The effects of dissolution on calcium carbonate from contact with deionised water are accentuated with the addition of a detergent (Pingitore *et al.*, 1993). Avoid this procedure on delicate calcitic shells.

30.3.2. Lithified samples (Bolli 1952, Kirchner 1958, Lethiers and Crasquin-Soleau 1988).

30.3.2a. Petroleum spirit and sodium carbonate (Na_2CO_3)

During a study on disaggregation techniques of indurated rocks, Bolli (1952) treated argillaceous and carbonate rocks to a combination of procedures. The most successful methods were those that used either petroleum spirit or sodium hydroxide (NaOH), (see section 29 EXTRACTION TECHNIQUES FOR CALCAREOUS MICROFOSSILS FROM ARGILLACEOUS SEDIMENTS). Both indurated and porous limestones and argillites with a high percentage of $CaCO_3$ proved to be rather resistant to these methods. On strongly calcareous, highly porous, chalky limestone, Bolli (1952) performed a series of semi-quantitative tests using petroleum spirit and boiling in sodium carbonate (Na_2CO_3). Disaggregating using either petroleum spirit or sodium hydroxide, resulted in the generation of small amounts of residue (particles <105μm). However, a combination of soaking in petroleum spirit and then boiling in sodium carbonate resulted in 75% of the sample passing through the 105μm mesh sieve.

The combined method is as follows:

1 Pre-treatment as outlined above.

2 Immerse each sample in petroleum spirit for 15-20 minutes.

3 Decant and filter the petroleum spirit for re-use.

4 Pour the petroleum saturated sample into a boiling 50% hydrous solution of sodium carbonate (Na_2CO_3, washing soda), and leave until the sample has disintegrated, usually about an hour. Renew the solution as required to maintain the level. **N.B. If the microfossils are fragile do not boil the solution.**

5 Disaggregation may take many hours. The process can be enhanced by immersing the beaker in an ultrasonic tank.

6 Pass the residue through a 1mm sieve onto a 75μm screen. Wash the residue into an evaporating dish and oven dry at 60°C. Repeat from stage 2 on the unprocessed coarse fraction (>1mm).

7 Further concentration can be achieved by flotation, using one of the techniques outlined below (see 30.4).

30.3.2b. Heating in sodium sulphate (Na_2SO_4)

Two longer procedures were described by Kirchner (1958), for the disaggregation of indurated rocks from the Carpathian Flysch, and lithified Palaeozoic and Mesozoic material from the Polish Lowland. The techniques employed are modifications on Wicher's (1942) method, which involved the heating of samples, and subsequent boiling in a concentrated solution of sodium sulphate (Na_2SO_4). This method however, proved highly destructive to the microfauna. The revised procedure is as follows.

1 Pre-treatment as outlined above.

2 Once warmed, pour a concentrated sodium sulphate (Na_2SO_4) solution on the sample.

3 Keep the sample hot enough for the solution to completely penetrate the rock fabric.

4 When completely saturated, decant the excess solution, and allow the sample to cool until crystallization occurs. This may take up to one week, and is complete when the sample looks dry, and is covered with crystals.

5 Cover the sample with hot water, and leave for two days, after which it is usually ready for washing and sieving.

For harder well lithified rocks the following modifications were made.

1 Uncrushed rock is warmed for 1-2 hours at 90°C.

2 Remove the sample from oven and cover with dry sodium sulphate. Return it to the oven.

3 When the crystals have fused, remove from oven, decant off excess liquid from base and use in 'icing' the surface of the sample.

4 The coated sample is returned to the oven for a further 30 minutes.

5 Remove and carefully wash under cold running water for a few minutes.

6 Leave for 12 hours (overnight) until crystallization occurs.

7 Examination the next day should reveal the surface of the sample covered with small crystals, with numerous cracks and fissures in the rock.

8 Stages 1-6 may need to be repeated a number of times (2-5) to achieve a thorough disaggregation.

9 Following the crystallization stages, cover the sample with boiling water and leave for two days, after which it is usually ready for washing and sieving.

These methods are long and time consuming. However, the yield of microfossils may be three times that of standard methods (Kirchner 1958). Williams-Mitchell (1948) succeeded in processing chalk samples using sodium hydroxide (caustic soda, NaOH), (see section 29 EXTRACTION TECHNIQUES FOR CALCAREOUS MICROFOSSILS FROM ARGILLACEOUS SEDIMENTS). For indurated rocks soaking in *di*-sodium hydrogen orthophosphate (Na_2HPO_4), allowing the salt to repeatedly crystallize over a (unspecified) time, successfully disaggregated samples, although some damage was inflicted on the foraminifera (Williams-Mitchell 1948, p 95).

30.3.2d. Acetic acid ($CH_3.COOH$)

Concentrated acetic acid is corrosive and gives off strong fumes. This procedure must be undertaken in a fume cupboard or fume hood with efficient back venting. Use gloves, eye protection and the correct gas/vapour mask when large quantities of the solution are open to the atmosphere.

Indurated sparry and marly limestones can be disaggregated using concentrated acetic acid ($CH_3.COOH$), following the method of Lethiers and Crasquin-Soleau (1988) modified from earlier procedures developed by Bourdon (1957, 1962). Although this procedure was originally applied to the extraction of ostracods, it can be used for the extraction and cleaning of any well-preserved calcitic material. An outline method is as follows.

1 Pre-treatment as outlined above, however for significant amounts of clay matrix, oven dry at a lower temperature (60°C) for an extended period (12 hours +). For this procedure to be successful it is vital that the sample is completely dry, and that all interstitial water has been removed. **Failure to do this will result in the fossils being dissolved by a strong acid!**

2 Ensure the sample is in a well-labelled beaker (*pyrex*) that can be covered with a lid, and a larger bucket or beaker when the fume extraction is turned off. The process is long, and may have to be left unattended overnight or at weekends.

3 Once the sample has cooled to room temperature, carefully cover with pure (99.5%) acetic acid. A laboratory or general purpose grade of reagent is suitable. **No effervescence should be evident.** A reaction may result from (i) acid purity - previously diluted acid not indicated on the bottle; (ii) insufficient sample dehydration.

4 Place the beaker on a laboratory hot-plate (in the fume cupboard), pre-set at a temperature of between 60-80°C. Cover, but do not seal, the beaker with a watch-glass or some other suitable cover.

5 Leave for several hours, but periodically observe to ensure the sample is always immersed in acid. Disintegration is a slow process, but as it proceeds a deposit of mud will form on the bottom of the beaker, larger rock fragments will develop large irregular cracks and fracture, and the acid takes on a distinct dark yellow-brown colour. This stage may last from one day to three weeks (Lethiers and Crasquin-Soleau 1988). If there is no, or little reaction, add a couple of drops of water to initiate a reaction, or continue a slow process (Bourdon 1962).

6 When disaggregation is complete, or is considered to have generated sufficient material for examination, decant the acid through filter paper

(choose a grade with a fine particle retention and fast filtration speed). Although slightly discoloured, the acid can be re-used for this procedure.

7 Rapidly wash the residue through a pre-assembled nest of stainless steel (wet) sieves (2mm-63μm). It is essential to wash the residue under a continuous flow of water, thus reducing the dissolution effects of the acid. **Ensure plastic gloves, sleeve protectors and eye protectors are worn during this stage.** Continue washing until no odour of acetic acid is detectable. Wash each fraction into an evaporating dish.

8 Oven dry the fractions. If required, repeat the procedure from stage 1 on the coarse fraction using the decanted acid.

Lethiers and Crasquin-Soleau (1988) found that two acetic acid immersions where required for each sample. Although the process is long, Lethiers and Crasquin-Soleau (1988), do not recommend the proposal of Bourdon (1962) in boiling the acid, as constant monitoring is essential. Substituting formic for acetic acid may speed up the process, additional experimental work is required to evaluate this (Green *in prep*), and detailed SEM observations are required to confirm that only slight surface corrosion of a few angstroms occurs (Lethiers and Crasquin-Soleau 1988).

As an alternative to the above methods, Allman and Lawrence (1972) suggest using the hydrogen peroxide (H_2O_2) disaggregation procedure on lithified calcareous sediments (see section 29 EXTRACTION TECHNIQUES FOR CALCAREOUS FORAMINIFERA FROM ARGILLACEOUS SEDIMENTS for details). However, boiling in hydrogen peroxide can result in textural and compositional changes in carbonate skeletal material (Gaffey *et al.*, 1991). Its use is also restricted if the extracted calcareous material is required for further geochemical work. Even 30% solutions of hydrogen peroxide buffered with sodium hydroxide (NaOH) to a pH of 7-9 can result in shell chemistry modifications (Gaffey *et al.*, 1991, Gaffey and Bronnimann 1993). The dissolution effects of hydrogen peroxide on carbonates, coupled with its inability to destroy organic tissue, led Pingitore *et al.*, (1993) to suggest it has no role in the preparation of calcium carbonate samples. This is particularly important if high resolution geochemical procedures are to be employed in analyzing samples. Furthermore, it underlies the importance in maintaining a record of all chemical solutions used in sample preparation, particularly if there is a time gap in proceeding with work on material, or a change in personnel analyzing the samples (Green 1995b).

A summary flow-chart of the main procedures used in the extraction of fossils from consolidated and unconsolidated sediments is illustrated in figure 30.1.

30.4. FLOTATION TECHNIQUES

Heavy liquid separation procedures (using bromoform, tetrabromoethane or zinc bromide) are described in detail in section 19 FLOTATION AND LIQUID SEPARATION TECHNIQUES. Care must be taken with these separation techniques, as they can result in selective concentration of certain types of microfossils, distorting the population of a sample. Agglutinated forms present in a sample may be removed by the processes of liquid separation or decanting (Brooks 1954). It is therefore important to retain and

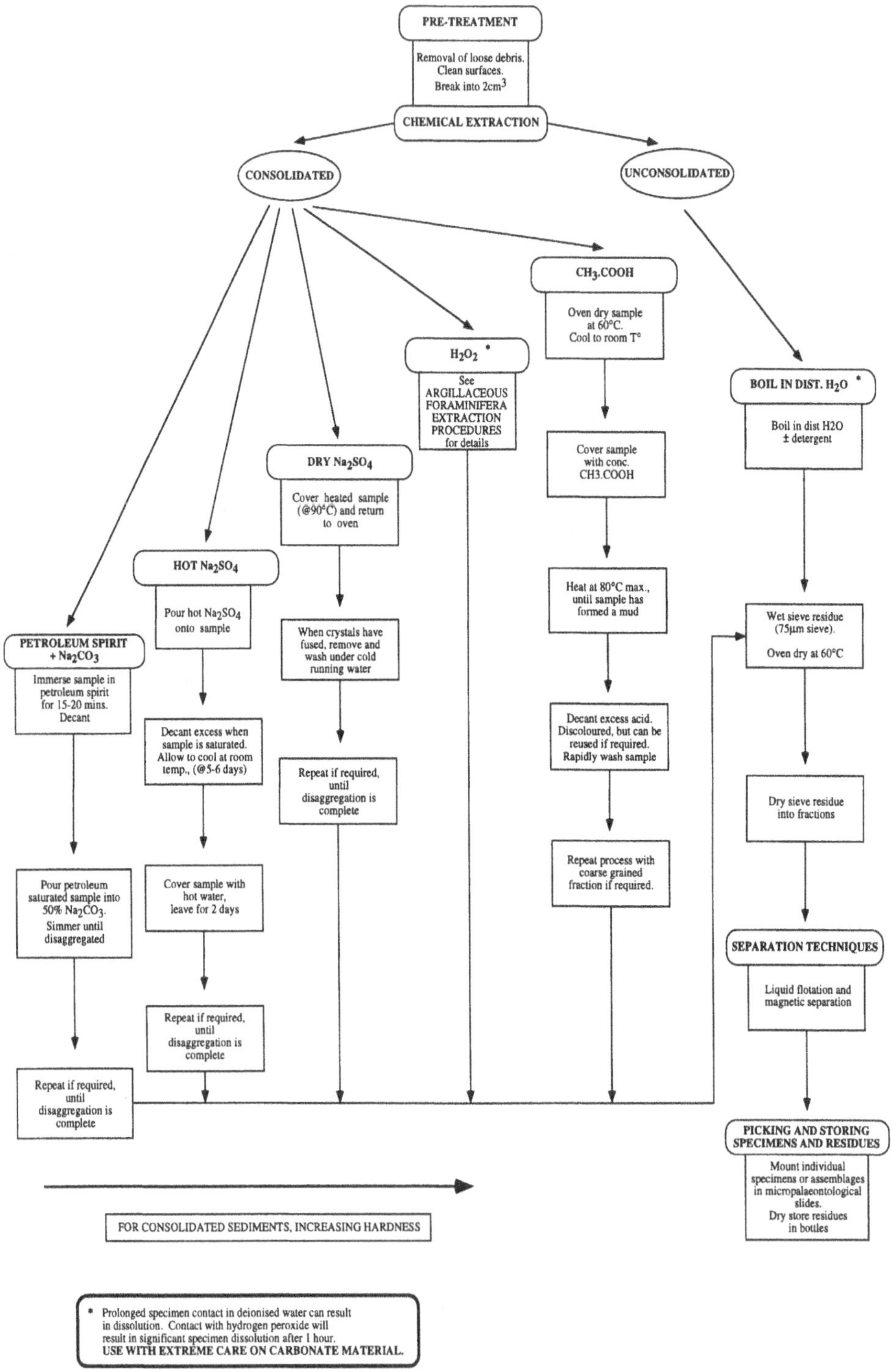

Figure 30.1. Summary flow chart of the main stages in the preparation of calcareous microfossil residues from carbonate sediments.

examine both residue fractions, suspecting one will be richer than the other.

Procedures vary in complexity from using distilled water and detergent to hazardous alcohols and toxic heavy liquids (*e.g.* Brasier 1980, Higgins and Spinner 1969, Kornicker, 1957). All procedures must be undertaken using the appropriate safety equipment (gloves, eye protection, laboratory coat and sleeve protectors), and undertaken in a laboratory fume cupboard.

30.5. MAGNETIC SEPARATION

The procedures for using the Franz Isodynamic magnetic separator are outlined in detail under section 18 MECHANICAL SEPARATION OF MICROFOSSIL RESIDUES. Its use in micropalaeontology in general, and for the separation of foraminifera and ostracoda in particular, has been described by Eckert *et al.*, (1961). Conventional liquid separation methods, as described above, are unsatisfactory in separating glauconite grains from marly residues. However, by first sieving the residue (see section 16 WASHING AND SIEVING TECHNIQUES USED IN MICROPALAEONTOLOGY), to obtain fractions of similar grain sizes, and then using the magnetic separator, microfossils can be concentrated. When undertaking a palaeoecological study, retain all fractions ranging from 2000-63µm as different size ranges may be preferred for particular analytical procedures (*e.g.* Sprechmann 1981).

The picking and mounting of specimens have already been outlined under section 29 EXTRACTION TECHNIQUES FOR CALCAREOUS MICROFOSSILS FROM ARGILLACEOUS SEDIMENTS.

REFERENCES

ALLMAN, M and LAWRENCE, D. F. 1972. *Geological Laboratory Techniques*. Blandford, (London). 1-335p.

BOLLI, H. 1952. Note on the disintegration of indurated rocks. *Micropaleontologist*. **6**, (1), 46-48.

BOURDON, M. 1957. Utilisation de l'acide acétique dans la désagrégation des roches dures. *Revue Institue of French Petroleum*. **12**, (1), 14-15.

BOURDON, M. 1962. Méthode de dégagement des microfossiles par acétolyse à chaud. *Comptes Rendus sommaries de la Societe Géologie France*. 267-268.

BRASIER, M. D. 1980. *Microfossils*. Allen and Unwin (London). xii + 193p.

BROOKS, H. K. 1954. A method for removing platy fragments of fossils and minerals from microfossil concentrates. *Micropaleontologist*. **8**, (4), 39-40.

ECKERT, R, HAY, W. W, LORENZ, G and VOGT, P. 1961. The magnetic separator as a tool in micropalaeontology. *Journal of Paleontology*. **35**, (5), 876-877.

GAFFEY, S. J and BRONNIMANN, C. E. 1993. Effects of bleaching on organic and mineral phases in biogenic carbonates. *Journal of Sedimentary Research*. **63**, (4), 752-754.

GAFFEY, S. J, KOLAK, J. J and BRONNIMANN, C. E. 1991. Effects of drying, heating, annealing and roasting on carbonate skeletal material, with geochemical and diagenetic implications. *Geochimica et Cosmochimica Acta*. **55**, 1627-1640.

GREEN, O. R. 1995b. Pitfalls, problems and procedures in micropalaeontological preparation and conservation. *The Geological Curator*. **6**, (4), 157-166.

HERON-ALLEN, E. 1894. *Prolegomena towards the study of the Chalk Foraminifera*. Nichols, (London). 36p.

HIGGINS, A. C and SPINNER, E. G. 1969. Techniques for the extraction of selected microfossils. *Geology - Journal of the Association of Teachers in Geology*. **1**, 12-28.

KIRCHNER, Z. M. 1958. A new method of hard-rock maceration. *Micropaleontology*. **4**, (3), 327-328.

KORNICKER, L. S. 1957. Concentration of ostracodes by alcohol flotation. *Journal of Paleontology*. **31**, (5), 1030.

LETHIERS, F and CRASQUIN-SOLEAU, S. 1988. Comment extraire les microfossiles a tests calcitiques des roches calcaires dures. *Revue de Micropaléontologie*. **31**, (1), 56-61.

LIPPS, J. H. (ed). 1993. *Fossil Prokaryotes and Protists*. Blackwell Scientific Publications (Oxford). ix + 342p.

PINGITORE, Jr, N. E, FRETZDORFF, S. B, SEITZ, B. P, ESTRADA, L. Y, BORREGO, P. M, CRAWFORD, G. M and LOVE, K. M. 1993. Dissolution kinetics of $CaCO_3$ in common laboratory solvents. *Journal of Sedimentary Research*. **63**, (4), 641-645.

SPRECHMANN, P. 1981. Paleocommunities and paleobathymetry of Maastrichtian sublittoral benthonic foraminifera from western Europe. *Neues Jahrbuch für Geologie und Paläontologie, Abhandlungen*. Band **162**, Heft 2, 188-230.

WICHER, C. A. 1942. *Praktikum der angewandten mikropalntologie*. Berlin: Borntraeger. 143p.

WILLIAMS-MITCHELL, E. 1948. The zonal value of foraminifera in the Chalk of England. *Proceedings of the Geologists' Association*. **59**, (2), 91-112.

31. EXTRACTION TECHNIQUES FOR AGGLUTINATED FORAMINIFERA FROM CALCAREOUS SEDIMENTS

31.1. INTRODUCTION

Agglutinated (arenaceous) foraminifera are of biostratigraphic, palaeoecologic and taxonomic importance, particularly when concentrated in lithified marine sediments of Palaeozoic age. The study of Palaeozoic agglutinated foraminifera is largely by thin section analysis, mainly because of the difficulty in extracting well-preserved specimens from residues. Several attempts were made by Wray (1952) to extract foraminifera from Carboniferous limestones. The few foraminiferal tests that were recovered showed poor preservation, and all taxonomic descriptions were supplemented from observations using thin sections.

Extraction techniques suitable for obtaining calcareous microfossils from calcareous-rich argillaceous and carbonate rich sediments may also yield abundant agglutinated foraminifera if present within the fauna. Consequently these procedures will not be reiterated in this section. Acid preparation techniques used in disaggregating carbonate rich rocks for the recovery of agglutinated foraminifera usually involve matrix dissolution using dilute hydrochloric acid (HCl). Eisenack (1931) used a 5% HCl solution (mainly for the extraction of chitinozoa), while Dunn (1942) used a 4% solution of HCl to dissolve the carbonate. Toomey (1974) modified a technique described by Ireland (1958) used for generating an insoluble residue from mechanically crushed samples suitable for sedimentological analysis. Ireland (1958) digested samples in 50% HCl (a concentration used by Moreman 1930), a solution which Toomey (1974) considered too high for preserving delicate agglutinated tests. Consequently Toomey (1974) processed 100 grams of sediment using formic acid (H.COOH) (see 27 EXTRACTION TECHNIQUES FOR PHOSPHATIC FOSSILS). Formic acid, less violent than hydrochloric, results in a cleaner residue, containing a better preserved and more abundant fauna.

Acid extraction techniques are however, restricted to agglutinated foraminifera constructed of an acid resistant shell cementing media, and non-calcareous detrital grains. Agglutinated foraminifera wall types have been classified into three groups depending on the type of grain cementing media recognized:

- organo-agglutinated - those with an organic cement;
- ferro-agglutinated - those with additional ferric or other non-calcareous microgranular cements;
- calc-agglutinated - those with additional calcareous cement.

Consequently, the acid extraction techniques outlined below are applicable to families and genera of organo- and ferro-agglutinated foraminifera. With the arrival of more sophisticated analytical procedures in determining the microcrystalline nature of the cementing media the taxa in these groups may be reduced in number further (Roberts and Murray 1995).

The use of weak acids for rock digestion was advocated by Cushman (1935), who argued that contact with a weak acid over a long time would result in less damage to the fragile agglutinated tests. Secrist (1934), commenting on the technique used by Moreman (1930), considered the reaction with a 50% HCl solution too violent for many agglutinated tests. Furthermore, he found the 5% HCl solution of Eisenack (1931) was too weak, requiring frequent replacing, resulting in sample agitation sufficient to damage the delicate tests. He concluded that a 20 or 25% HCl solution was suitable to treat most Lower and Middle Palaeozoic limestones. The method described is as follows.

31.2. PRE-TREATMENT

A sample size of 500-1000 grams is usually sufficient to yield a representative fauna.

1 Scrub the sample with a bristle brush in a sink. This will remove any loose debris.

2 Oven dry at 120°C for 1 hour.

3 Break the sample into $2cm^3$ fragments, and place in a 1000ml glass beaker.

31.3. CHEMICAL PROCEDURES

31.3.1. Consolidated sediments

N.B. All work to be undertaken in a fume cupboard, with the operator wearing protective clothing (gloves and goggles).

4 Slowly pour on the acid (20% HCl), if the reaction becomes very violent dilute with the addition of distilled water. Ensure the sample is completely covered.

5 Leave the solution until all effervescence has ceased. Do not stir the solution, as this may damage the delicate foraminifera tests.

6 Remove undissolved fragments from the beaker using a pair of acid resistant tongs.

7 Carefully decant off excess acid into a sink of running water. Try not to disturb the residue at the bottom of the beaker.

8 Refill the beaker with water and repeat decanting process. This stage can be repeated a third time if the sample is clay rich.

9 The undissolved fragments, and residue coarser than 1mm, should undergo a repeat of acid treatment (stages 4 to 6), and the decanting of excess supernate.

The conventional procedure would now involve wet sieving, oven drying, and dry sieving the residue into suitable fractions for microscopical examination. Ireland (1967) described the washing and drying process as "the chief cause for the damage and destruction of very valuable material". To reduce and eliminate this damage, complete the washing by using alcohol or acetone, and allowing it to evaporate after decanting. This procedure can be incorporated into the method described below. Secrist (1934, p246) also found that screening a dry residue, no matter how carefully it was performed, resulted in damage to the foraminifera. Undertake the following procedure.

10 The insoluble residue, still in the beaker, is immersed in alcohol or acetone.

11 Using a dropper pipette with a 2mm modified opening, carefully transfer a small amount of residue to a glass slide for microscopic examination. Examine the residue while it is still wet, as siliceous tests can be detected while damp far more easily than when dry, and test cavities determined as alcohol evaporates first from outside and then inside the test. The hollow nature of agglutinated foraminifera can only be determined when the residue is wet.

12 Once individual foraminifera have been identified, they can be removed from the residue and placed in a normal micropalaeontological faunal cell slide, and allowed to dry. Mounting procedures are described under section 29 EXTRACTION TECHNIQUES FOR CALCAREOUS MICROFOSSILS FROM ARGILLACEOUS SEDIMENTS.

Extraction procedures, detailed above, are summarized in the accompanying flow chart (Fig. 31.1).

31.3.2. Unconsolidated sediments

Recent experimental work (Murray and Alve 1994, Alve and Murray 1995) examines diversity in agglutinated foraminiferal assemblages to determine their origin, palaeoenvironmental position and biostratigraphical significance. Initial experimental work (Murray and Alve 1994) addressed the question of whether fossil agglutinated assemblages are the remnant residues of assemblages originally dominated by calcareous taxa. Subsequent analysis and use of the technique (Alve and Murray 1995) characterized the original dead assemblages (ODA) and resulting acid treated assemblages (ATA), enabling ecological and palaeoenvironmental information to be derived. Working on Holocene material from the North Sea and North East Atlantic, similar procedures were employed in both studies. A composite procedure for the formation of ATA's is outlined below.

By following two preparation techniques for each sample it is possible to generate (1) the total modern assemblage, and (2) the residual acid resistant assemblage.

31.3.2a. Modern assemblage

1 Rose Bengal staining of the assemblage was employed in the recognition of live material (see section 21 STAINING TECHNIQUES USED IN MICROPALAEONTOLOGY). Wash the sediment over a 63μm sieve, and oven dry at 50°C.

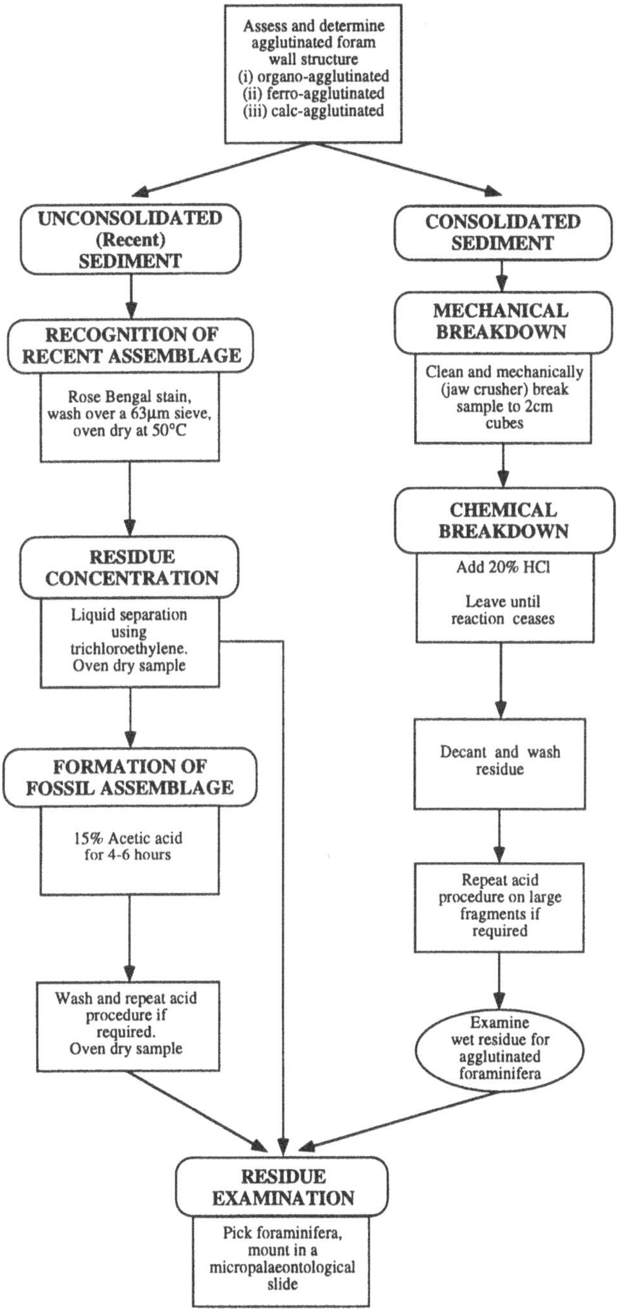

Figure 31.1. Summary flow chart of the main stages in the preparation of fossil and recent agglutinated foraminiferal residues.

2 Modern living foraminifera were separated from ODA's by flotation, using trichloroethylene ($CHCl{:}CCl_2$), and standard micropalaeontological picking procedures.

31.3.2b. Acid resistant assemblage

3 Place a portion of the ODA rich sample in a 15% solution of acetic acid ($CH_3.COOH$) with a pH 2.5-3 for between 4-6 hours. Wash the residue. Repeat the process if any calcareous material still remains.

4 Thoroughly wash (over a 63µm sieve), and oven dry the residue. Concentrating ATA's by flotation methods should be avoided to prevent the loss of agglutinated foraminifera with test components of high specific gravity (Murray and Alve 1994).

Following the experimental work it was concluded that well-preserved ATA's could be derived from primary calcareous dominated ODA's, and that fossil agglutinated assemblages are the result of either a partial or total loss of the calcareous elements of an assemblage by dissolution (Murray and Alve 1994). Furthermore, the experiments showed that ATA's mirror the ODA's relative to species diversity, and that ATA's provided better resolution in defining subenvironments in comparison to ODA's dominated by a single opportunistic calcareous species, thus supplying more ecological (and palaeoecological) information (Alve and Murray 1995). The stages used in preparing these assemblages are incorporated in the flow chart of preparation techniques (Fig. 31.1).

REFERENCES

ALVE, E and MURRAY, J. W. 1995. Experiments to determine the origin and palaeoenvironmental significance of agglutinated foraminiferal assemblages. Proceedings of the Fourth International Workshop on Agglutinated Foraminifera, Kraków Poland, September 12-19, 1993. *Grzybowski Foundation Special Publication*. **3**, 1-11.

CUSHMAN, J. A. 1935. Paleozoic foraminifera, their relationships to modern faunas and their environment. *Journal of Paleontology*. **9**, (3), 284-287.

DUNN, P. H. 1942. Silurian foraminifera of the Mississippi Basin. *Journal of Paleontology*. **16**, (3), 317-342.

EISENACK, A. 1931. Neue mikrofossilien des Baltischen Silurs. *Paläeontologische Zeitschrift*. **13**, (1/2), 74-118.

IRELAND, H. A. 1958. Insoluble residues, pp 74-94: In; J. D. Haun and W. L. LeRoy (eds), *Subsurface geology in petroleum exploration*. Colorado School of Mines, (Golden, Colorado).

IRELAND, H. A. 1967. Preparatory techniques for microfossils and inorganic insoluble residues. *Journal of Paleontology*. **41**, (2), 523.

MOREMAN, W. L. 1930. Arenaceous foraminifera from Ordovician and Silurian limestones of Oklahoma. *Journal of Paleontology*. **4**, (1), 43-59.

MURRAY, J. W and ALVE, E. 1994. High diversity agglutinated foraminiferal assemblages from the NE Atlantic: dissolution experiments. *Cushman Foundation Special Publication*. **32**. 33-51.

ROBERTS, S and MURRAY, J. W. 1995. Characterization of cement mineralogy in agglutinated foraminifera (Protista) by Raman spectroscopy. *Journal of the Geological Society, London*. **152**, (1), 7-9.

SECRIST, M. H. 1934. Technique for the recovery of Paleozoic arenaceous foraminifera. *Journal of Paleontology*. **8**, (2), 245-246.

TOOMEY, D. F. 1974. The biota of the Pennsylvanian (Virgilian) Leavenworth limestone midcontinent region. Part 4: Distribution of agglutinated and silicified foraminifera. *Journal of Paleontology*. **48**, (2), 326-343.

WRAY, J. L. 1952. Endothyrid foraminifera from the Greenbrier Series of northern West Virginia. *Journal of Paleontology*. **26**, (6), 946-952.

32. SPECIALIST TECHNIQUES USED IN THE PREPARATION OF INDIVIDUAL MICROFOSSIL SPECIMENS

32.1. INTRODUCTION

This section deals with specialized techniques used in the preparation of individual microfossils, in particular ostracods and foraminifera. However, techniques outlined are not exclusive to any one particular fossil group. Included in this section are procedures on microfossil replication and peels, and specialist observational methods employed in examining the results. Experimental studies on the biophysical properties of ostracod carapaces are also described.

32.2. PREPARATION OF INDIVIDUAL SPECIMENS

Although general chemical processing and maceration procedures are designed to produce matrix-free material, detailed optical observations of fine structure and any form of specimen replication, require that the surface is free of all secondary material (*e.g.* clay grade particles and organic detritus). Additional surface cleaning of pores, apertures and hinge areas may be necessary, even on some of the better preserved specimens within an assemblage. Techniques for working on individual specimens are very labour intensive (requiring constant monitoring), and require a degree of dexterity not associated with bulk sample maceration or processing. Occasionally it is necessary to work on individual fossils that exhibit fine cracks over their surfaces, and would disintegrate during normal processing. This is often the case with many large highly ornamented Palaeozoic ostracod specimens, or decalcified Mesozoic and Cenozoic molluscs or larger benthic foraminifera. Techniques have been described for the individual preparation of ostracod carapaces by a number of authors (Sohn 1961, Sohn *et al.*, 1965, Martinsson 1965).

Once individual specimens have been isolated, they can be prepared using a combination of mechanical or chemical methods. Stable robust material can be cleaned in an ultrasonic tank, but excessive use can result in damage to material and should be avoided (Sohn 1960, Green 1995b). Cleaning by hand isaided by using fine sable-hair paint brushes, mounted pins and cotton-buds. Sohn (1961) and Sohn *et al.*, (1965) found a "chewed" end of a wooden toothpick formed an ideal stiff bristled brush. Nylon brushes (nail and tooth) may be suitable for some specimens, but in general the bristles do not possess the elasticity of other brushes, and can result in specimen damage. Well-equipped preparation laboratories have a large selection of different brush types and styles.

Preparation work should be observed using either a magnifying light, or for smaller specimens a good quality stereozoom binocular microscope. Detailed preparation work on one side of a small specimen requires it to be held firmly in place. This is essential to avoid structural damage to the organism's shell. The procedure of Martinsson (1965) is suitable for achieving this. The specimen must be firmly attached onto a temporary mount of acetate film using a cellulose adhesive, and on completion of the preparation work, the film and adhesive can be dissolved in acetone. Martinsson (1965) used a small glass embryo dish to hold the temporarily mounted specimen, although having less substantial bases, small evaporating or petri dishes can also be used (Fig. 32.1). The method for attaching and preparing the specimen is as follows:

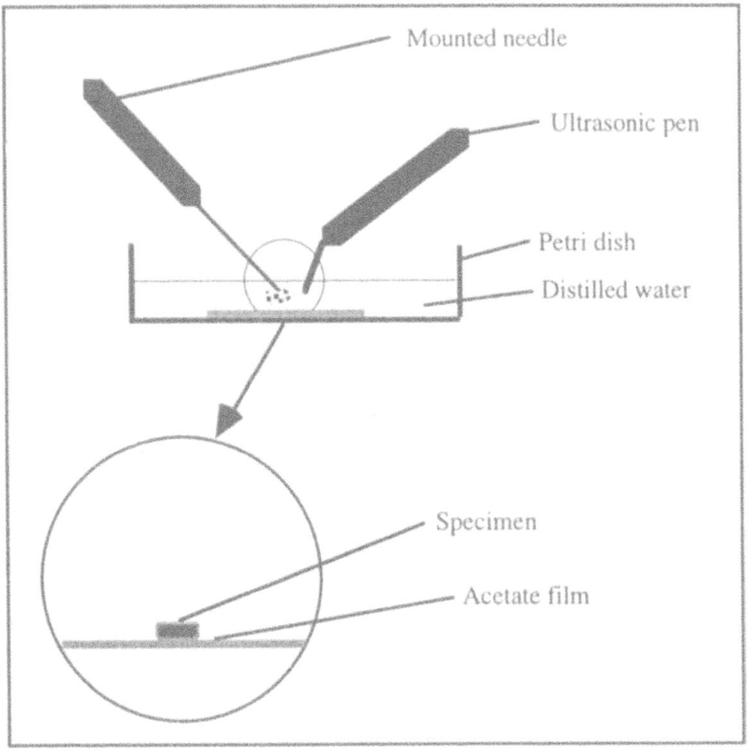

Figure 32.1. The mounting and cleaning of a single specimen in a petri dish using a combination of mounted needle, brush and ultrasonic pen (diagram based on Martinsson 1965).

1 Smear a layer of glue into the bottom of the dish.

2 Place a drop of glue in the centre of the dish. Attach and orientate the specimen, then allow the preparation to dry.

3 Cover the specimen with a few drops of distilled water.

4 By observing the mount under a binocular microscope, and the careful use of mounted pins, brushes and ultrasonic pen, it will be possible to remove any matrix (Fig. 32.1).

5 Loose matrix (forming a slurry with the water) can then be removed by using a small bulb-pipette. Periodically re-flood the dish with clean distilled water to continue with the process.

Martinsson (1965) suggests that the needle should be kept in constant motion, and not allowed to come to rest on the thin shell of the specimen. Preparation of the other side requires the specimen to be remounted, possibly several times. Water soluble glues (*e.g.* gum tragacanth) can be used to temporarily strengthen the surface enabling the specimen to be moved without disintegrating. Once the specimen is clean of matrix, it should be thoroughly washed in acetone to remove all traces of the cellulose based glue. If preparation requires that the specimen remains dry, matrix can be removed by directing a gentle stream of compressed air.

Chemical methods of cleaning are detailed in other sections, including the use of a 15% solution of hydrogen peroxide (H_2O_2) and a 5% solution of sodium hypochlorite (NaClO), carefully observing any reaction under a binocular microscope. Alternatively immersing specimens in a 20% solution of hydrofluoric (HF) acid, for between 2 to 24 hours, removes any siliceous matrix and converts the calcite shells to translucent calcium fluoride (Sohn 1956). Besides ostracod carapaces (Sohn 1956, Sohn *et al.*, 1965), the calcite transformation technique has been successfully applied to foraminifera (Green 1995b) and calcified radiolarian assemblages in argillaceous limestones (Pessagno 1977a).

32.3. REPLICATION OF MICROFOSSILS

32.3.1. Plastic casts

During the examination of microfossil residues, naturally occurring pyritized casts are frequently found. The internal structure seen on casts often provides information not readily observable on original shell surfaces. Consequently artificial casting procedures of foraminifera have been used for well over a century by micropalaeontologists (Carpenter, Parker and Jones 1862). However, the procedure, using Canada Balsam, was not comprehensively described until Hofker (1927), worked on *Nummulites*. The replication technique is not restricted to large benthic foraminifera, where some internal surface detail is lost through grinding (see below), but has been successfully used on complete smaller foraminifera (Hofker 1965). Hageman (1976) used decalcified casts of a recent sessile form, *Miniacina miniacea* from the Mediterranean, in describing the internal morphology. A composite procedure, incorporating Douglass (1965), is detailed below:

1. Soak the test in xylene.

2. Cook in Canada Balsam. Paraffin wax, which melts at 80°C, can also be used, but is more brittle and less transparent than Canada Balsam.

3. Gently grind the specimen to expose a small part of the test wall surface, *i.e.* a thicker than normal thin section. Air-filled specimens will require this procedure done first for Canada Balsam to penetrate all chambers.

4. Continue gentle grinding of the section to remove excess Canada Balsam and expose the test wall.

5. Immerse the slide in a 5% acetic acid solution, and leave for 24 hours. This process will dissolve the test wall. Disodium Ethylenediaminetetra-acetic acid (diNaEDTA) or hydrochloric acid (HCl) provide suitable alternatives, giving similar results.

6 Cover the negative cast with either water or glycerol, then examine slide under high power. Details of pores, stolons and chambers should be clearly visible.

One of the main disadvantages of this technique is getting a satisfactory impregnation of the specimens in which all the chambers and canals are filled with impregnating medium. This difficulty can be overcome by making half-sections (Hansen and Lykke-Anderson 1976):

1 Embed a specimen in a thermo setting plastic and grind down slide to the proloculus.

2 Re-heat and turn over specimen with a hot needle, ensuring that the prepared surface is in contact with the section.

3 Allow the slide to cool down, then while observing under the microscope, remove excess cement with a sharp scalpel.

4 When the shell is exposed immerse slide in a 10% HCl solution to decalcify.

5 Wash slide in water, dry and observe the results using a compound microscope.

Although useful, the preparation of serial sections can be a time consuming enterprise. Hottinger (1978, 1979) and Billman *et al.*, (1980) have exploited the use of artificial foraminiferal casts in the study of shell structure, pore shape and size, and three-dimensional canal system development. The casting media is an epoxy resin (*Araldite N* with hardener *HY 956*). The procedure is as follows:

1 Place the specimen in an aluminium foil lined container.

2 An evacuation system that can independently evacuate both the specimen and resin is preferable (*e.g.* Green 1986). This ensures the complete degassing of the resin.

3 When the resin has ceased to "boil", introduce resin into the specimen container to a depth of 1.5-3mm.

4 Release the vacuum and examine the resin soaking specimen. With the use of mounted needles, and viewing under a binocular microscope, the specimen can be orientated. With practice a number of specimens can be orientated in a single same block. Hottinger *et al.*, (1980) removed the specimen to a second container containing a fresh resin mix, a stage that does not appear to enhance or detract from complete specimen impregnation.

5 Polymerization of the resin occurs at 60°C over 48 hours.

6 When the slab is hard, remove from the container and grind the lower surface on 800 to 1000 grade grit to reveal the test.

7 Remove the test by immersion of the slab in 2% HCl for 15 minutes. Broken slabs may etch more successfully, and provide a better view of the internal structure.

8 Wash and dry the slab.

9 Mount on an SEM stub for examination.

Billman *et al.*, (1980) comment on the importance of correctly orientating rotalid specimens on SEM stubs. This is vital for determining ventral (umbilical) and dorsal (apical) surfaces, and is best undertaken using a stereozoom binocular microscope.

Lacking the internal complexities of test coiling, ostracod valves can be reproduced relatively quickly using the thermoplastic mountant *Lakeside 70*, and dissolving the shells in 5% hydrochloric acid (Kontrovitz 1982). The method is as follows:

1 Thoroughly clean specimen surfaces (*e.g.* if necessary immerse in an ultrasonic tank), and allow to dry.

2 Place the carapace with its convex side down on a clean microscope slide (cavity slides are best). Many specimens can be prepare simultaneously, thus speeding the process.

3 Crush some *Lakeside 70* to a size grade of between 0.1-0.2mm.

4 Using a fine (00) paint brush gently brush the *Lakeside 70* over each specimen, ensuring each of the valves is filled.

5 Place the slides on a hot-plate (pre heated to 150°C), and leave for 10-15 minutes. Observe the flow of *Lakeside 70* into the smallest cavity and the expelling of entrapped air bubbles.

6 Remove the slide, and allow it to cool. Carefully release the solidified *Lakeside 70* enclosed specimen, and apply a couple of drops of 5% HCl to dissolve the carapace. Acid strength is critical, as a stronger solution may damage the mould, while weaker solutions fail to remove all organic carbonate (Kontrovitz 1982).

7 When all effervescence has ceased, use a paint brush to transfer the *Lakeside 70* mould to another slide with a couple of drops of distilled water to wash and neutralize the acid. Once dry specimens can be examined.

The high degree of fine detail reproduced enables moulds to be studied by reflected light microscopy, or if necessary under the scanning electron microscope (Kontrovitz 1982).

32.3.2. Silicone rubber casts

The fine ornament and structure of large Palaeozoic ostracods can be studied in a similar manner using the SEM (Siveter 1982). This is particularly useful considering the identification of these forms relies so much on external valve morphology. As with the procedures outlined above, the preparation of individual carapaces is time consuming when compared with normal processing. However, it allows for the detailed study of decalcified carapaces.

The method uses room temperature vulcanising silicone rubber, compounds well known and used by the macropalaeontological preparator (*e.g.* Allman and Lawrence 1972; Kelly and McLachlan 1980). The use of RTV silicone rubbers, latex and thermo-setting rubbers (*e.g. Vinamold*) in palaeontology in general is beyond the scope of discussion for this section. Methods and procedures used in micropalaeontology (Siveter 1982) will be detailed below.

For micropalaeontological use casting requires accurate reproduction of the original features with little or no shrinkage on curing. Siveter (1982) considers *Silcoset 105* silicone rubber to provide the best quality reproduction, with features as small as 5μm capable of being defined under SEM. The following method was employed:

1. Hand specimens are cleaned of any loose surface material by careful brushing with a soft bristle paint brush, until all the moulds are exposed.

2. Coat the mould with dilute amyl acetate to harden the surface.

3. Build a plasticine wall around the area to be cast, and fill deep undercuts, as this will prevent tearing of the rubber when it is removed.

4. Carefully and thoroughly mix the catalyst and rubber compound together. If reflected light microscopy observation is to be used colouring the rubber with the addition of an opaque filler may be useful.

5. While viewing the specimen under a binocular microscope, place a few drops of catalysed rubber compound on the mould. Carefully using a mounted needle push the rubber into place and remove all air bubbles.

6. Leave for 24 hours and then peel off, photograph or examine under the SEM.

32.3.3. Metal casting

A procedure of casting microfossils using metal alloys instead of plastic resin, was documented by Rasetti (1947). The technique, outlined below, would probably not be used today, as it has been superseded by less hazardous resins following quicker methods.

1 Cut a small disc containing the fossil. The diameter of the rock disc should be small enough to fit into a glass phial (which will be destroyed during the process).

2 Secure the specimen to the base of the glass phial, so it will not float, with the fossil facing upwards.

3 Place a few pieces of low melting alloy in the glass tube. Rasetti (1947) used the following alloy Bismuth 52%, Lead 40%, Cadmium 8%, which has a melting point of 91°C. Close top of tube with a rubber stopper containing a glass tube.

4 Connect the glass tube, via vacuum tubing, to a vacuum pump.

5 Immerse the tube in a hot water bath (set at 100°C). Evacuate the tube for 30 minutes. The alloy will "boil" as air is removed from the specimen.

6 Remove from the hot water bath and allow the tube to cool (cold water immersion can be used).

7 Break the tube to remove the rock disc and solidified alloy.

8 Cut off any excess metal, and dissolve the rock in concentrated HCl (leave for 2 days, the acid will not affect the alloy).

9 Wash and dry the metal cast.

32.3.4. Peels of microfossils

Douglass (1965) briefly mentions a procedure used by Honjo (1960) in which a series of serial sections, less than 10µm apart are prepared using *Bioden* plastic, a type of acetate film. However, unlike the conventional peel technique, a dilute acid cannot be used in the etching process. Because the etching time is critical in the preparation of surfaces, more control over both the depth and time of etching is required. This can be obtained by using an inorganic salt solution, which maintains a constant pH during the reaction.

32.4. EXPERIMENTAL COMPRESSION AND IMPACT LOADING ON MICROFOSSILS

Detailed experimental work described by Whatley *et al.*, (1982) examines compression and impact loading on ostracod carapaces from four families. The sexes, valves and carapaces were all studied separately. Primarily the objective was to account for the over representation of single valves, articulated carapaces or the dominant sex within fossil populations. Furthermore, an indication of any mechanical damage and sample bias of a faunal population following standard processing methods, may also be indicated.

Specialized equipment was designed for the impact tests. This consisted of a glass rod inclined at an angle of 3° to horizontal, in which a steel ball was released down onto the specimen. Before release the steel ball was held in place by magnets, and thus could be set at different distances from the specimen. Testing was on well preserved specimens, set at 1cm distances until the carapace fractured. Consistent results were obtained, and some distinction at family level was detected.

The compression equipment, consisted of a balance, under the pan of which a specimen was placed. A beaker on top was carefully filled with water, the weight of which compressed the ostracod. This was computed to obtain the number of grams loading required to cause fracture. Results, however, were not as consistent as those obtained in the impact tests, and were initially discarded until further tests, using more sophisticated equipment were undertaken. However, results obtained were remarkably consistent with those from the initial tests, and it was concluded that an unknown variable not detected in the impact tests was responsible.

Overall results indicated that females were stronger than males, and that articulated carapaces were stronger than discrete valves. Furthermore, in some species one valve was notably stronger than the other. Of much less importance was shell size, shape, ornamentation and thickness, characters that cannot be used independently in assessing the strength of carapaces. The strength characteristics are a combination of shell chemistry, structure and ultrastructure, features which can be effected by both mechanical and chemical processing techniques.

Similar experimental work has been undertaken on benthic foraminifera (Wetmore 1987, Wetmore and Plotnick 1992). Tests on calcareous and agglutinated benthic foraminifera used a force gauge mechanism from which the electrical resistance from a force beam could be detected by strain gauges and correlated to the force applied to crush the test. The apparatus (Fig. 32.2) worked by slowly raising the specimen, placed on a crushing platform, upwards towards a crushing probe. The upward force gradually bent the aluminium force beam, until sufficient force had been applied to break the test. The maximum force a test bore before breaking was taken as the tests crushing strength.

Mounted on opposing sides of the force beam, two pre-calibrated strain gauges record the deformation imposed on the beam. The force applied generated an electrical resistance detected by the gauges, and amplified by a bridge amplifier before registering on a chart recorder. Care was taken to ensure the crushing platform was raised at the same rate for all measurements, thus keeping the loading rate uniform.

The direction of the compressive force applied to specimens was along the shortest axis of the test. Specimen orientation on the crushing platform was the orientation assumed by the test on a flat surface. Results indicated a complex relationship between crushing strength and morphological characters, so that no single character can be used to predict the strength of a foraminiferan test. However, architectural features such as wall thickness and overall test shape appear to affect shell strength more than coiling morphology or wall composition. Thus single chambered species could resist crushing as effectively as multi-chambered species from the same environment, while calcareous tests are not necessarily stronger or lighter than agglutinated tests from species living in the same habitat.

32.5. X-RAY MICROSCOPY IN MICROPALAEONTOLOGY

Specialized X-radiographic techniques have been adapted for use in the sub-disciplines of palaeontology (*e.g.* Zangerl 1965, Zangerl and Schultze 1989) and petrography (*e.g.* Bouma 1964, 1969; Davis and Walawender 1982, Hamblin 1962, 1971). Standard thin sectioning techniques require the partial destruction of the foraminiferan test during the

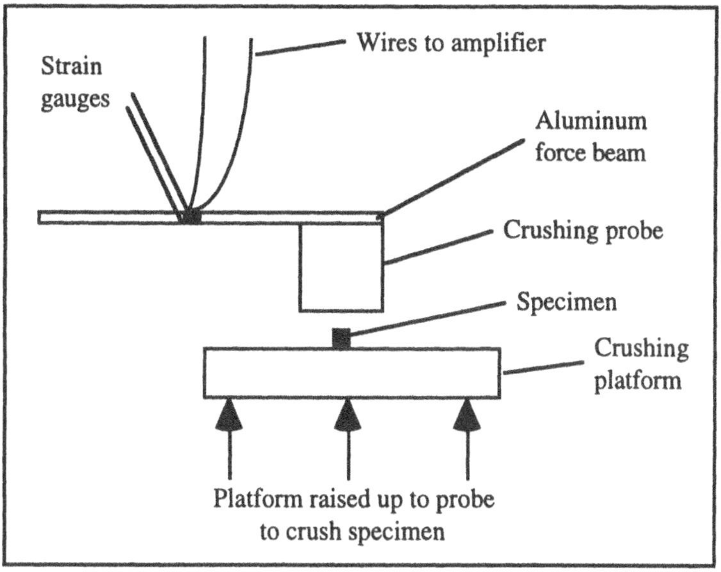

Figure 32.2. Diagrammatic illustration of the force gauge apparatus to measure the crushing strength of foraminiferal tests (re drawn from Wetmore 1987).

preparation process, and consequently the possible loss of much valuable information. A method of revealing the internal structure of foraminifera without shell loss is to use X-ray microradiography. Techniques have much improved since the initial application in studying micropalaeontological specimens (Hedley 1957, Hooper 1959, Bé *et al.*, 1969, Leary and Hart 1988, Hottinger and Mehl 1991). Two methods have been developed and refined:

i Point projection microradiography: where a beam is focused on a metal target that emits x-rays that penetrate the specimen to produce an enlargement on a photographic plate.

ii Contact microradiography: where the specimen is placed in the beam path, very close to a photographic plate. This procedure produces the best resolution.

The x-radiographs are produced by shadows generated by the differential x-ray transparency of an object. The shadow intensity, and ultimately the picture generated, is dependent on two criteria. Firstly the length the ray has to travel through the object, and secondly the differential absorption rates.

Resolution is not achieved by focusing the rays through optical lenses, as in light microscopy, but by the geometric relations between the x-ray source, object and the film recording the shadows produced by the object. In micropalaeontology the objects are small, therefore the absorption will be small, and the resulting contrast in the photograph small. The low energy level produced results in a "soft" x-ray (Hottinger and Mehl 1991).

The technique compliments both light microscopy or scanning electron microscopy studies, avoiding the need to section material. This is important particularly in the study of internal morphology or re-description of type specimens. The procedure used depends on the equipment available and its location within the working area. If the x-ray source is attached to a vacuum column then the exposure time may be considerably shorter than a conventional air filled column. Some types of film described by non U.K. workers may not be available (Leary and Hart 1988). One of the most recent comprehensive descriptions of the technique, methods and equipment is described by Hottinger and Mehl (1991), outlining applications for both quantitative and qualitative biometric research.

32.6. X-RAY DIFFRACTION

The mineralogical constitution of a test can be studied by the x-ray powder diffraction technique (XRD). This procedure is extensively used in petrology and mineralogy, with many references (*e.g.* Lindholm 1987, Hardy and Tucker 1988) providing excellent outlines of the equipment, sample preparation, qualitative and quantitative analytical methods and the interpretation of results. X-ray diffraction provides a fast and reliable means of determining the bulk mineralogy of carbonate material. The first applications in micropalaeontology were described by Emiliani (1955). In this technique the x-ray beam is diffracted from the planes of the atoms within the powdered crystal. The detection of the patterns formed allows for the differentiation of calcitic from aragonitic shells. Furthermore, following the determination of magnesium within the test wall, Blackmon and Todd (1959) recognized the importance of this in the classification of calcareous foraminifera. The non invasive nature of the x-rays led Switzer and Boucot (1955) to examine uncrushed specimens.

32.7. ELECTRON PROBE MICROANALYSIS

The microprobe provides a means of getting chemical analyses of only a few microns in diameter. Equipment is expensive, but the combination of an SEM with an x-ray dispersive spectrometer provides the researcher with an analytical tool of similar power to the electron microprobe (Plotnick and Harris 1989). This technique, first applied in micropalaeontology by Hooper (1964), allows for the elemental analysis of the shell of an organism. Analysis of a larger area of a specimen enables calcium and magnesium determinations, indistinguishable by optical observations, to be mapped and photographed. In some cases similar features and results can be confirmed from either staining or cathodoluminescence. Specimen preparation procedures are similar to that for SEM analysis using EDS (energy dispersive spectroscopy) and WDS (wavelength dispersive spectroscopy) procedures. Specimens must be embedded in epoxy resin, and surfaces carefully prepared as relief-free polished sections (see section 34 ELECTRON MICROSCOPY TECHNIQUES for more details).

REFERENCES

ALLMAN, M and LAWRENCE, D. F. 1972. *Geological laboratory techniques*. Blandford, (London). 335p.
BÉ, A. W. H, JONGEBLOED, W. L and MCINTYRE, A. 1969. X-ray microscopy of Recent Planktonic foraminifera. *Journal of Paleontology*. **43**, (6), 1384-1396.
BILLMAN, H, HOTTINGER, L and OESTERLE, H. 1980. Neogene to Recent Rotaliid foraminifera from the Indopacific Ocean; their canal system, their classification and their stratigraphic use. *Schweizerische Paläontogische Abhandlungen*. **101**, 71-113.

BLACKMON, P and TODD, R. 1959. Mineralogy of some foraminifera as related to their classification and ecology. *Journal of Paleontology.* **33**, (1), 1-15.
BOUMA, A. H. 1964. Notes on x-ray interpretation of marine sediments. *Marine Geology.* **2**, 278-309.
BOUMA, A. H. 1969. *Methods for the Study of Sedimentary Structures.* Wiley-Interscience (New York). xvi + 458p.
CARPENTER, W. B, PARKER, W. K and JONES, T. R. 1862. Introduction to the study of the foraminifera. *Ray Society (London).* 1-319 + 22 plates.
DAVIS, B. L and WALAWENDER, M. J. 1982. Quantitative mineralogical analysis of granitoid rocks: a comparison of x-ray and optical techniques. *American Mineralogist.* **67**, 1135-1143.
DOUGLASS, R. C. 1965. Larger foraminifers, pp 20-25: In; B. Kummel and D. Raup (eds), *Handbook of Paleontological Techniques.* W. H. Freeman and Co., (San Francisco). xiii + 852p.
EMILIANI, C. 1955. Mineralogical and chemical composition of the tests of certain pelagic foraminifers. *Micropaleontology.* **1**, (4), 377-380.
GREEN, O. R. 1986. Vacuum and pressure impregnation of fine grained sediments. Proceedings of Geo-Techniques '85. Part 1. *Science Technology.* **15**, 21-24.
GREEN, O. R. 1995b. Pitfalls, problems and procedures in micropalaeontological preparation and conservation. *The Geological Curator.* **6**, (4), 157-166.
HAGEMAN, J. 1976. The internal morphology of *Miniacina miniacea* (Pallas). *Proceedings Koninklijke Nederlandse Akademie van Wetenschappen* Series B, **79**, (1), 29-35.
HAMBLIN, W. K. 1962. X-ray radiography in the study of structures in homogenous sediments. *Journal of Sedimentary Petrology.* **32**, (2), 201-210.
HAMBLIN, W. K. 1971. X-ray photography, pp 251-284: In; R. E. Carver (ed), *Procedures in Sedimentary Petrology.* Wiley-Interscience (New York). xiii + 653p.
HANSEN, H. J and LYKKE-ANDERSON, A-L. 1976. Wall structure and classification of fossil and recent elphidiid and nonionid foraminifera. *Fossils and Strata.* **10**, 1-37.
HARDY, R. G and TUCKER, M. E. 1988. X-ray powder diffraction of sediments, pp 191-228: In; M. Tucker (ed), *Techniques in Sedimentology.* Blackwell Scientific Publications (Oxford). ix + 394.
HEDLEY, R. H. 1957. Microradiography applied to the study of foraminifera. *Micropaleontology.* **3**, (1), 19-28.
HOFKER, J. 1927. The foraminifera of the Siboga expedition. Part 1 - Tinoporidae, Rotaliidae, Nummulitidae, Amphistiginidae. *Uitk. Zool. Bot. Oceanogr. Geol. Geb.*, monogr. IVa, 1-78.
HOFKER, J. 1965. Methods for study and preparation of foraminifera, pp 251-256: In; B. Kummel and D. Raup (eds), *Handbook of Paleontological Techniques.* W. H. Freeman and Co., (San Francisco). xiii + 852p.
HONJO, S. 1960. A study of some primitive *Neoschwagerina* by a new serial section technique. *Journal of the Faculty of Science, Hokkaido University.* Series 4, Geology and Mineralogy. **10**, (4), 457-470.
HOOPER, K. 1959. X-ray absorption techniques applied to statistical studies of foraminifera populations. *Journal of Paleontology.* **33**, (4), 631-60.
HOOPER, K. 1964. Electron probe microanalysis of foraminifera; an exploratory study. *Journal of Paleontology.* **38**, (6), 1082-1092.
HOTTINGER, L. 1978. Comparative anatomy of elementary shell structures in selected larger foraminifera, pp 203-266: In; R. H. Hedley and C. G. Adams (eds), *Foraminifera.* Volume 3. Academic Press (London). x + 290p.
HOTTINGER, L. 1979. Araldit als Helfer in der Mikropaläontologie. - aspekte *Ciba-Geigy.* 3/1979, 1-10.
HOTTINGER, L and MEHL, J. 1991. X-ray microscopy in micropalaeontology. *Microscopy and Analysis.* No. 21, 21-23.
KELLY, S. R. A and MCLACHLAN, A. 1980. The use of silicone rubbers in the preparation of casts from natural fossil moulds. *Geological Magazine.* **117**, (5), 447-454.
KONTROVITZ, M. 1982. A new method of producing internal molds in ostracod valves. *Journal of Paleontology.* **56**, (3), 818-820.
LEARY, P. N and HART, M. B. 1988. X-raying planktonic foraminifera. *Journal of Micropalaeontology.* **7**, (1), 43-44.
LINDHOLM, R. C. 1987. *A Practical Approach to Sedimentology.* Allen and Unwin (London). xviii + 276.
MARTINSSON, A. 1965. Preparation of individual microfossils on cellulose acetate films, pp 247-251: In; B. Kummel and D. Raup (eds), *Handbook of Paleontological Techniques.* W. H. Freeman and Co., (San Francisco). xiii + 852p.
PESSAGNO, Jr, E. A. 1977a. Upper Jurassic Radiolaria and radiolarian biostratigraphy of the California Coast Ranges. *Micropalaeontology.* **23**, (1), 56-113.
PLOTNICK, R. E and HARRIS, G. 1989. X-ray analysis using Energy Dispersive and Wavelength Dispersive Spectroscopy, pp 179-185: In; R. M. Feldman, R. E. Chapman and J. T. Hannibal (eds), *Paleotechniques.* The Paleontographical Society Special Publication. **4**. iv + 358p.
RASETTI, F. 1947. Notes on techniques in invertebrate paleontology. *Journal of Paleontology.* **21**, (4), 397-399.
SIVETER, D. J. 1982. Casts illustrating fine ornament of a Silurian ostracod, pp 105-122: In; R. H. Bate, E. Robinson and L. M. Sheppard (eds), *Fossil and Recent Ostracods.* Ellis Horwood Ltd, (Chichester). 493p.
SOHN, I. G. 1956. The transformation of opaque calcium carbonate to translucent calcium fluoride in fossil ostracoda. *Journal of Paleontology.* **30**, (1), 113-114.
SOHN, I. G. 1960. Cleaning ostracode valves with ultrasonic vibrations. (Abstract only). *Bulletin of the Geological Society of America.* **71**, 1982.
SOHN, I. G. 1961. Techniques for preparation and study of fossil ostracodes, pp Q64-Q70: In; R. C. Moore (eds), *Treatise on Invertebrate Paleontology. Part Q; Arthropods 3.* Geological Society of America and the University of Kansas (Lawrence, Kansas). xxiii + 442p.
SOHN, I. G, BERDAN, J. M and PECK, R. E. 1965. Ostracods, pp 75-89: In; B. Kummel and D. Raup (eds), *Handbook of Paleontological Techniques.* W. H. Freeman and Co., (San Francisco). xiii + 852p.
SWITZER, G and BOUCOT, A. J. 1955. The mineral composition of some microfossils. *Journal of Paleontology.* **29**, (3), 525-533.
WHATLEY, R. C, TRIER, K and DINGWALL, P. M. 1982. Some preliminary observations on certain mechanical and biophysical properties of the ostracod carapace, pp 76-104: In; R. H. Bate, E. Robinson and L. M. Sheppard (eds), *Fossil and Recent Ostracods.* Ellis Horwood Ltd, (Chichester). 493p.
ZANGERL, R. 1965. Radiographic techniques, pp 305-320: In; B. Kummel and D. Raup (eds), *Handbook of Paleontological Techniques.* W. H. Freeman and Co., (San Francisco). xiii + 852p.
ZANGERL, R and SCHULTZE, H-P. 1989. X-radiographic techniques and applications, pp 165-178: In; R. M. Feldman, R. E. Chapman and J. T. Hannibal (eds), *Paleotechniques.* The Paleontographical Society Special Publication. **4**. iv + 358p.

33. EXTRACTION TECHNIQUES FOR CALCAREOUS NANNOFOSSILS

33.1. INTRODUCTION

This group comprises a heterogeneous collection of organisms, or their disaggregated parts ranging up to 20μm in size. The most common examples encountered in residues are coccoliths; the basic elements of chrysomonad flagellates. Coccoliths are button shaped rings arranged in a complex manner forming hollow spheres (coccolithospheres). The individual coccoliths of a coccolithosphere may be alike or dimorphic. During a complex life history coccolithospheres may produce many different kinds of coccolith.

Processing is essentially a mechanical procedure, where chemically cleaned elements are concentrated through the use of a centrifuge. Sample collecting, processing, slide preparation and electron microscopy techniques are outlined in this section.

33.2. SAMPLE COLLECTING PROCEDURES

Because of the small nature of the elements, field recognition of coccoliths is virtually impossible, unless the aid of a field microscope is employed (Hay 1965). Rarely do they form the primary reason for collecting many of the specimens in which they are found. Confirmation of their presence is obtained after laboratory processing. However, they are frequently occur in association with planktonic foraminifera in fine-grained pelagic sediments. Concentrations in deep sea sediments such as clays and oozes often exceeds a million per cubic centimetre. Consequently, processing rarely requires samples exceeding 25 grams in weight. Barss and Williams (1973) provide a useful guide to lithology type and the amount required:

Lithology	Weight (grams)
Chalk	0.5
Shale, limestone	0.5 - 2.0
Siltstone	2.0 - 3.0
Sandstone	3.0 - 5.0

The advantage gained in collecting smaller samples is frequently lost in the additional number collected. This has the disadvantage of increasing the likelihood of cross sample contamination. To reduce this factor, ensure all collecting equipment is thoroughly cleaned between each sample, and that hands, and particularly finger nails are clean. Only clean samples should be collected, avoiding areas where surface run-off can result in natural contamination. Hay (1977) considers contamination, both from field collecting and laboratory processing, to be the singularly most important factor in an incorrect diagnosis of nannofossil rich and poor material (see section 15 MICROFOSSIL SAMPLE CONTAMINATION AND RELIABILITY PROBLEMS for a full discussion of both stratigraphical and laboratory contamination).

33.3. SAMPLE PREPARATION

In the laboratory begin with a preliminary examination of the sample to determine if full processing is required. Start by crushing a few grams of sediment in distilled water using a mortar and pestle. On a research quality polarizing microscope using high powered objectives, optically examine temporarily prepared strew slides (Hay 1961). The presence of commonly occurring heliolithid calcareous nannofossils will be confirmed by the swastika-like negative uniaxial pseudo-figure. The preparation of quick strew slides during processing is a useful guide in assessing cleaning and elemental concentration within a residue. It also provides some indication of the physical and chemical effects of processing, many stages of which, if incorrectly performed, may result in etched specimens (Hay 1977).

Once the presence of nannofossils has been established, preparation of a cleaner more concentrated residue is undertaken. Techniques for sample preparation have been discussed by Hay (1965, 1977), Barss and Williams (1973), and Taylor and Hamilton (1982). These form the basis of the procedure outlined below, with reference made to other authors who employed specialized equipment during extraction and concentration. If multiple processing of samples is undertaken, ensure that all centrifuge tubes are clearly labelled with the appropriate sample number.

To avoid etching and damaging the coccoliths surface, avoid the use of acidic solutions. Hay (1977) recommends using solutions with a pH of between 8 to 8.5, although higher pH's will enhance the deflocculation properties of a solution. Saturated solutions of calcium carbonate ($CaCO_3$) can be used, and a pH of 9.4 achieved by dissolving 4 grams of sodium bicarbonate ($NaHCO_3$) in 20 litres of distilled water, and adjusting to the correct pH with the addition of sodium carbonate (Na_2CO_3) (approximately 3 grams). This solution usually has a shelf life of 2 to 3 months before the pH deteriorates. Although successfully used by Finch (1972), sodium carbonate solutions alone are unstable losing their pH in a few days. Stored stock solutions must be checked before use, and discarded if the pH is unreliable. Both tap and distilled water should be avoided for processing, particularly if they have pH's below 7.

1 Break down about 5 grams of the sample by soaking in buffered distilled water, and either crush or use a platform sample shaker. The sediment should be fine enough to pass through a 45μm sieve.

2 To disperse the clay fraction it may be necessary to add a small amount of alconox solution (7 grams *Alconox* in 1 litre of buffered distilled water). Finch (1972) uses *Quebracho* solution but does not state the concentration. The use of sodium hexametaphosphate [$(NaPO_3)_6$, *Calgon*] should be avoided as its pH is below 8, and can etch the specimens. The calcium phosphate ($Ca(PO_3)_2$) precipitate coating the nannofossils prevents high quality TEM replication (see below). Thoroughly shake or stir the sample using a polypropylene stirring rod. Avoid using an ultrasonic tank during processing as this will destroy the elements (Finch 1972).

3 Organic matter can be removed by using a 30% hydrogen peroxide (H_2O_2) solution buffered to a pH of 9.4 with the addition of $NaHCO_3$. This eliminates the effects of the production of formic acid that could etch or dissolve the nannofossils. Gently heat the sample for about 1 hour, until the dark organic-rich mixture turns a pale grey, indicating oxidation is complete.

4 Pour the suspension into a 50ml centrifuge tube, and top up with buffered sodium carbonate, and stir. It is essential that the mixing is done quickly to ensure the minimum amount of settling within the tubes.

5 Following the method of Edwards (1963), centrifuge the sample at 300 rpm (Barss and Williams 1973 use 250 rpm) for 15 seconds. Decant the suspension and leave for a few minutes. Repeat the process between three to eight times, until the supernate is clear.

6 Finally short centrifuge at 850 rpm for 30 seconds. This will concentrate a nannofossil residue within the particle size range of 3-25µm, those usually studied using LM techniques. A longer centrifuge time will settle a finer fraction that can be studied under the SEM. Barss and Williams (1973) centrifuged at 1200 rpm for 3 minutes.

7 Residues should be stored in glass or plastic phials in buffered distilled water until required for slide preparation. The pH of long term stored residues must be constantly monitored. Avoid storing in phenol as this dissolves the nannofossils. The addition of a few drops of 10% ethanol to each phial will act as a short-term fixative (see below, Taylor and Hamilton 1982).

Through SEM and particle size analysis of non-lithified non-pelagic marine sediments, Katz (1978) modified the procedure of Edwards (1963) by extending the "short centrifuging" time. Approximately 90ml of slurried sample suspension is passed through a 63µm sieve and poured into 100ml centrifuge tubes, with any shortfall made up by the addition of double distilled water. The sample is then centrifuged at 500 rpm for 20.6 seconds, and the supernate containing particles less than 25µm diameter retained, while material ranging up to 63µm in size is discarded. Repeating the procedure is usually sufficient to remove all fine sand and course to medium silt, leaving a suspension containing particles finer than 25µm diameter. The retained supernate is transferred to clean 100ml tubes and topped to the 90ml mark before centrifuging for 90 seconds at 2000 rpm, time and speeds also preferred by Finch (1972). At this stage it is the supernate, containing material finer than 3µm, that is discarded. Double distilled water is added to the precipitate, and the particles returned to suspension and centrifuging repeated until the supernate is clear. This procedure may require repeating up to 15 times particularly if the sediment is clay rich. Katz (1978) confirmed, through SEM and particle size analysis, an exceptionally clean precipitate of particles ranging in size from 3-25µm that contained less than 10% non-nannofossil material. The procedure is also suitable for concentrating diatoms and silicoflagellates (see section 26 EXTRACTION TECHNIQUES FOR ACID INSOLUBLE MICROFOSSILS).

Following the removal of organic material careful consideration should be given to the long term storage of processed residues, even with the addition of a fixative. Taylor and Hamilton (1982) have documented their experimental study of residue storage, and have recorded samples exhibiting a marked deterioration over a 12 week period. This was attributed to the decomposition of residual organic matter, and the production of a weak acid and H_2S gas, in concentrations sufficient to cause nannofossil dissolution. To avoid this complication permanent slides of residues must be prepared, and individual specimens examined under an SEM within a few days of the completion of sample processing.

Note: The centrifuge times given by Edwards (1963) and Katz (1978) to obtain residues of a particular size relate to THEIR centrifuges only. Centrifuge times must be calculated for individual preparators models (see section CENTRIFUGING TECHNIQUES IN MICROPALAEONTOLOGY for calculating the correct time).

Hay (1977) describes the application of the elutriator in separating a residue. This specialized equipment is not always available, and is very difficult to construct and maintain. During use it requires a constant head of water, which in nannofossil preparation, where a controlled pH is required, will be difficult to maintain. However, the resulting fractionation may be so effective that strew slides will not represent the residue as a whole.

33.4. SLIDE PREPARATION

Light microscopy studies can be undertaken on two types of slide preparation: strew slides and viscous mounts. Strew slides are prepared from centrifuge suspensions in the following way:

1. Place a drop of the suspension on a cover slip using a disposable pipette. The use of these in the preparation of each sample will help reduce the possibility of contamination during slide preparation (Edwards 1963). Allow the water to evaporate, or dry on a hot plate.

2. Place a drop of mounting medium (*Caedax* used by Hay 1965; or phenol free *Cellosize* and *Elvacite* used by Barss and Williams 1973), on a glass slide. As with solutions used in residue preparation, the mounting media must be non-acidic. The optical resolution is further enhanced if the mountant has an RI between the extremes of calcite (n = 1.4864 - 1.6585), and is viewed under phase contrast microscopy (Hay 1977). High relief can be obtained by conversion of the nannofossils to fluorite (n=1.43) using HF treatment of Bramlette and Sullivan (1961), see section 20 THIN SECTION AND SLIDE PREPARATION TECHNIQUES OF MACRO- AND MICROFOSSIL SPECIMENS.

3. Invert the cover slip and place on the slide. Ensure that the mounting media extends to the edges of the glass slide. Return the slide to the hot plate and allow the mounting media to cure before examination.

Distinct differences between the proximal side enclosing the cellular body, and the distal side, facing the water, is used by taxonomists in classifying the organisms. Material can be viewed from either side if permanently mounted between two cover slips (Stradner and Papp 1961). However, these mounts are particularly fragile, although it is possible to use thicker cover slips (size 1 instead of the customary size 0), without forming adverse optical properties that result from the inversion of a standard slide. By placing a small drop of non-drying oil on the slide, mixing it with the residue, and encasing it in mountant with a cover slip, it is possible to produce a slide that enables the examination of nannofossils in any orientation (Bramlette and Sullivan 1961).

To obtain the maximum amount of information from optical observations, nannofossil examination must be carried out at a magnification of x800, or using x1200 to x1600 oil immersion objectives. The use of a polarizing microscope is invaluable in studying heliolithid nannofossil structure, and can be complemented by dark field illumination and phase contrast options present on compound biological microscopes.

The method of Clark (1973), in coating the nannofossils with a 500-1000Å thick aluminium layer, and viewing under x100 oil immersion objectives succeeded in enhancing specimen contrast and photographic properties. Material treated in this way can be viewed under the SEM, or permanently protected by applying a cover slip.

33.5. CARBON REPLICATION FOR THE TEM

Observations with the transmission electron microscope, using carbon replica preparations (Black and Barnes 1959; Hay and Towe 1962), have long been the standard method for studying nannofossils, although unlike scanning electron microscopy it results in the destruction of original material. The successful construction of replicas requires an organic matter free residue (if not already been performed).

The sample must be mixed with a weak solution of *Clorox* and allowed to stand for 24 hours. Once the organic matter has been removed, further disaggregation can be achieved by immersing the solution in an ultrasonic tank, and then short centrifuging. This procedure thoroughly cleans the nannofossils, with the effectiveness checked by preparing temporary mounts.

1. On completion of sample cleaning, spread a few drops of the suspension on a freshly cleaved surface of a piece of mica, or a very clean glass slide, and allow it to dry.

2. Place a mica sheet in a vacuum evaporator positioning the carbon rods approximately 12cm above. A drop of glycerol on the tip of each carbon rod will aid in the removal of the replica from the mica sheet.

3. Evacuate the chamber to a vacuum of 5×10^{-5} mm Hg, and evaporate carbon onto the specimens in four 15 second bursts.

4. Remove the mica from the chamber, and score the carbon film, and float the carbon off in water.

5. Add a few drops of HCl, sufficient to produce a 1% solution, and leave for a few hours (until the calcium carbonate has been removed). If $(NaPO_3)_6$ has been used in the dispersal of clays, then an insoluble coating of $Ca(PO_3)_2$ can form on the specimen. This in turn reacts with HF of the following stage, to form the insoluble calcium fluoride (CaF_2) coating, and spoil the replica.

6 Transfer the replica to a polythene dish containing distilled water, and slowly add HF to produce a 25-40% solution. Leave the replica in this solution for about 8 hours. This dissolves all siliceous material. The addition of a small amount of hydrogen sulphide (H_2SO_4) to the HF will aid in producing a cleaner replica.

7 Decant and neutralize the solution, and transfer the replica to a dish containing distilled water for a final washing. Pick and place the replica on a copper electron microscope grid. A thick carbon film will support itself. If not, a *Formvar* backing may be applied.

33.6. SCANNING ELECTRON MICROSCOPY

The long and complicated techniques used in TEM work have been superseded by SEM studies, where preparations are quick and easy to produce, and results far superior. Both applications are outlined by Hay (1977).

Processed residues and rock fragment can be examined with the aid of the SEM. A brief outline of the procedure is as follows:

1 Pipette a small amount of the suspension on to a stub, and dry on a hot plate.

2 Vacuum coat the stub with a 20nm (200Å) Gold/Palladium thick coat. Rock specimens require a 30-40nm (300-400Å) thick coat.

For suspensions, the most important factor to consider is the residue thickness on the stub. Every effort must be taken to ensure that this is kept to a single layer of material, allowing an unrestricted study of any single specimen. The procedure of Mai (1988) enables the examination of both sides of the same specimen using the SEM.

1 Place a few drops of cleaned nannofossil rich residue on a clean 26x48mm glass slide. Allow residue to air dry (oven dry at 70°C maximum temperature).

2 Carbon coat the slide with a 15-20nm (150-200Å) thick layer.

3 Cut the film with a sharp mounted razor into 3x3mm squares.

4 Slowly immerse the slide in 10% ammonia solution so that the squares float free, *i.e.* thin carbon film with coccoliths adhering to it).

5 Remove the squares using a TEM Finder Grid held by fine forceps. Ensure each quadrate is labelled so that the exact location of each specimen can be achieved.

6 Dry the film under an infrared light.

7 Gold coat both sides of the specimen.

8 Examine under the SEM. Use an enlarged template of the grid to mark the position of the photographed object.

9 On finishing one side of the grid, turn it over and repeat the procedure for the opposite side of the specimen.

Electron microscopy procedures complement the light microscopy viewing techniques of Stradner and Papp (1961) and Bramlette and Sullivan (1961) detailed above. Furthermore, they confirm the SEM as an essential tool in calcareous nannofossil taxonomic diagnosis. Identifying and comparing features under light microscopy with transmission and scanning electron microscopy has resulted in the continued evolution of procedures over recent years (Fig. 33.1). Perch-Nielsen (1967) describes a method of observing the same specimen under the light microscope and then TEM. The method's of Thierstein, Franz and Roth (1971) and Moshkowitz (1974) enables examination of the same specimen in the SEM followed by light microscopy. Smith (1975) observed the same material in both the scanning and transmission electron microscopes. Further developments and modifications of these techniques have resulted in the examination of single specimens by light microscopy, transmission and scanning electron microscope (Hansen, Schmidt and Mikkelsen 1975, Mai 1983), with the additional benefit of viewing proximal and distal sides (Mai 1988).

Quantitative studies, for determining the absolute abundance's of coccoliths can be achieved following the filtration technique of Andruleit (1996). This method requires the splitting the suspension and filtering by means of a vacuum pump. After drying a small amount of the filter is cut, mounted on an SEM stub, sputter coated and examined. Numerical calculations of the number of coccoliths per gram of dry sediment can then be determined.

A flow-chart summary of the main processing stages in generating nannofossil residues is presented in figure 33.2.

REFERENCES

ANDRULEIT, H. 1996. A filtration technique for quantitative studies of coccoliths. *Micropaleontology*. **42**, (4), 403-406.

BARSS, M. S and WILLIAMS, G. L. 1973. Palynology and nannofossil processing techniques. *Geological Survey of Canada*, Paper **73-26**. 1-25.

BLACK, M and BARNES, B. 1959. The structure of coccoliths from the English Chalk. *Geological Magazine*. **96**, (5), 321-328.

BRAMLETTE, M. N and SULLIVAN, F. R. 1961. Coccolithophorids and related nannoplankton of the early Tertiary in California. *Micropaleontology*. **7**, (2), 129-188.

CLARK, D. F. 1973. Improved light microscopy techniques for viewing nannofossils. *Micropaleontology*. **19**, (4), 481-484.

EDWARDS, A. R. 1963. A preparation technique for calcareous nannoplankton. *Micropaleontology*. **9**, (1), 103-104.

FINCH, E. M. 1972. The calcareous nannofossils and their usefulness as biostratigraphic indicators. *BP Exploration and Production Research Division, (Sunbury-on-Thames, Middlesex), Technical Note EPR/TN3107* ii + 21p.

HANSEN, H. J, SCHMIDT, R. R and MIKKELSEN, N. 1975. Convertible techniques for the study of the same nannoplankton specimen. *Proceedings of the Koninklijke Nederlandse Akademie van Wetenschappen*. **B 78**, 225-230.

HAY, W. W. 1961. Note on the preparation of samples for discoasterids. *Journal of Paleontology*. **35**, (4), 873.

HAY, W. W. 1965. Calcareous nannofossils, pp 3-7: In; B. Kummel and D. Raup (Eds), *Handbook of Paleontological Techniques*. W. H. Freeman and Co., (San Francisco). xiii + 852p.

HAY, W. W. 1977. Calcareous nannofossils, pp 1055-1200 (techniques 1056-1069), In: A. T. S. Ramsay (Ed), *Oceanic Micropalaeontology* (2 Volumes). Academic Press (London). xii + 1453p.

HAY, W. W and TOWE, K. M. 1962. Electron-microscopic examination of some coccoliths from Donzacq (France). *Eclogae geologicae Helvetiae*. 55, (2), 497-519.

KATZ, B. J. 1978. Preparation of calcareous nannofossil assemblages for chemical examination. *Journal of Paleontology*. **52**, (2), 497-500.

MAI, H. 1983. Korrelation cretazischer Coccolithen im Lichtmikroskop, Rasterelektronenmikroskop und Transmissionselektronenmikroskop. *Paleontographica Abteilung B*. 189, 139-161.

MAI, H. 1988. A new method for SEM analysis of both proximal and distal sides of the same Coccolith. *Journal of Paleontology*. **62**, (1), 151-152.

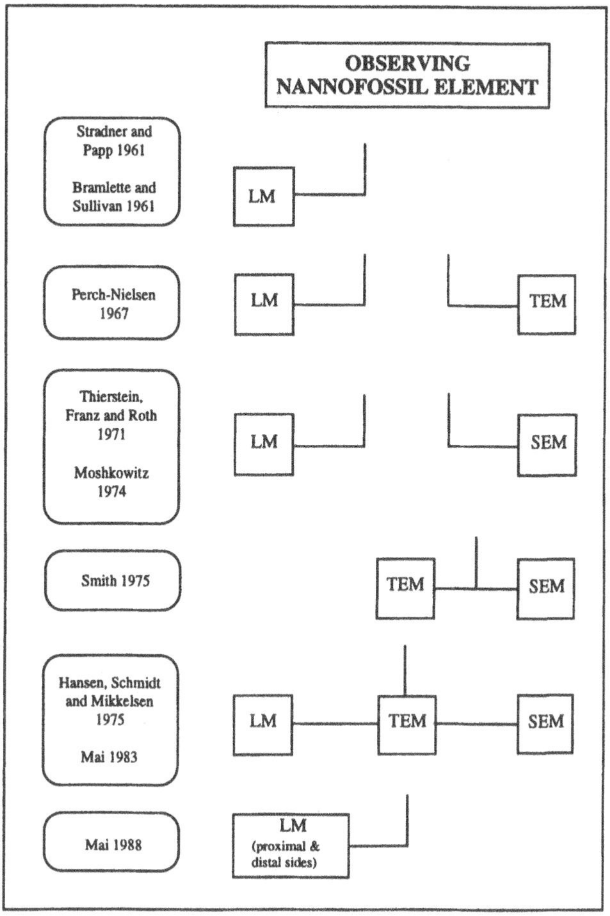

Figure 33.1. Diagrammatic evolution of methods used in examining the same nannofossil specimen by light microscopy (LM), transmission electron microscopy (TEM), and scanning electron microscopy (SEM) procedures. Despite electron microscopy providing superior quality micrographs, light microscopy is still favoured as a quick and inexpensive means of observing material.

MOSHKOWITZ, S. 1974. A new method for observing the same nannofossil specimens both by light microscope and scanning electron microscope and preservation of types. *Israel Journal of Earth Science*. **23**, 145-147.

PERCH-NIELSEN, K. 1967. Eine Präparationstechnik zur Untersuchung von Nannoplankton im Lichtmikroskop und im Elektronenmikroskop. *Dansk Geologisk Forening Meddelelser*. **17**, 129-130.

SMITH, C. C. 1975. A new method for studying calcareous nannofossils using scanning and transmitted light optics. *Revista Española de Micropaleontologia*. **7**, 43-48.

STRADNER, H and PAPP, A. 1961. Tertiaere Discoasteriden aus Oesterreich und deren stratigraphische Bedeutung. *Jahrbuch der Geologischen Bundesanstalt* Sonderband (Wien). **7**, 1-160.

TAYLOR, R. J and HAMILTON, G. B. 1982. Techniques, pp 11-15: In; A. R. Lord (ed), *A Stratigraphical Index of Calcareous Nannofossils*. Ellis Horwood Ltd, (Chichester). 1-192.

THIERSTEIN, H. R, FRANZ, H. E and ROTH, P. H. 1971. Scanning electron microscopy of the same small objects. *Micropaleontology*. 17, 501-502.

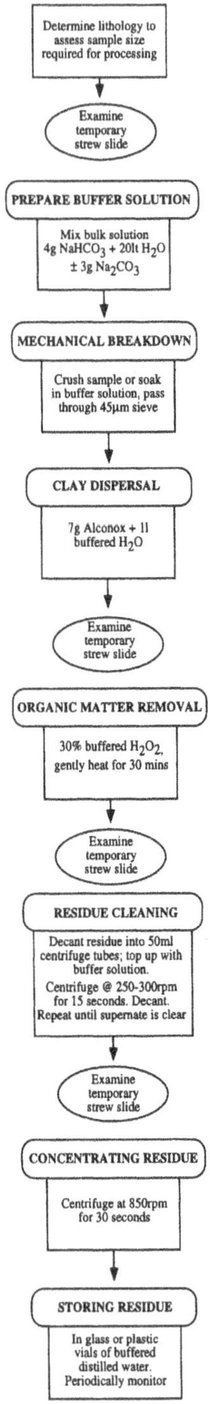

Figure 33.2. Summary flow chart of the main stages in the preparation of calcareous nannofossil residues.

PART III

LABORATORY TECHNIQUES

D. ANALYTICAL PROCEDURES

34. Electron microscopy techniques.

35. X-radiography techniques.

34. ELECTRON MICROSCOPY TECHNIQUES

34.1. INTRODUCTION

The human eye is capable of separating two objects not closer than 0.1mm at a normal viewing distance of 25cm. Finer detail is resolved by enlarging the image. This is conventionally achieved by placing the object under a simple magnifying lens, or in the laboratory, under a compound microscope (Fig 34. 1).

Scientists have used optical microscopes for over three hundred years, and within every palaeontological laboratory it remains one of the most important (and relatively cheapest) tools for examining specimens. Fine structural detail is made distinct when images are viewed in a beam of light intensified by passing through condensing lenses, magnified by an objective lens and finally focused at the eyepiece (Fig 34. 2a). The ability of the objective lens to perform its task is expressed in terms of its numerical aperture (N.A.), a value usually marked on the side of a high quality lens. The limits of an objective lens are such that a numerical aperture of 1.4 is the maximum that can be attained with a white light source in the visible spectrum. Detail can be further improved by viewing specimens in different types of filtered light (*e.g.* differential interference contrast (DIC), phase contrast, polarized light), or where appropriate in reflected rather than transmitted light (see section 20 THIN SECTION AND SLIDE PREPARATION TECHNIQUES OF MACRO- AND MICROFOSSIL SPECIMENS for details).

Resolution can be significantly improved by shortening the wavelength of the light source, *e.g.* fluorescence, where particles of 130nm (1300Å) can be resolved using ultraviolet (UV) radiation, or with cathodoluminescence where longer wave infra-red (IR) emissions become visible (Marshall 1988). By using the electron, with wavelengths many times shorter than light particles, instruments can resolve objects to only a few nanometres (Fig 34.1). Fine skeletal detail of opaque and poorly preserved material provides maximum data when viewed using the electron microscope.

This brief review of electron microscopy does not have the scope to cover operating and maintenance of transmission and electron microscopes, these aspects are covered in greater detail in other texts (*e.g.* Chescoe and Goodhew 1990, Glauert (ed) 1974-1994, Goldstein *et al.*, 1981, Hayat 1989). However, many of these specialized texts are directed towards "biological" users of electron microscopes, and of only limited use (particularly in specimen preparation) to palaeontologists. Consequently greater discussion is given to the preparation of palaeontological specimens, and obtaining high quality images and photographs.

34.2. ELECTRON MICROSCOPY

The increased resolving power provided by electron microscopes, in many instances to 0.1nm, is achieved using specialized equipment and working environments. Electron microscopes focus electron particles through a series of electro-magnetic lenses. For the electrons to reach their acceleration speed, they must operate in a high vacuum (approximately 10^{-4} torr). A vacuum controlled environment is established and maintained within the column of the microscope (Figs. 34.2b, 34.3), the purpose of which is to prevent electrical discharge, and stabilize the beam by eliminating air molecule interference. Two variations of the electron microscope are outlined below: the

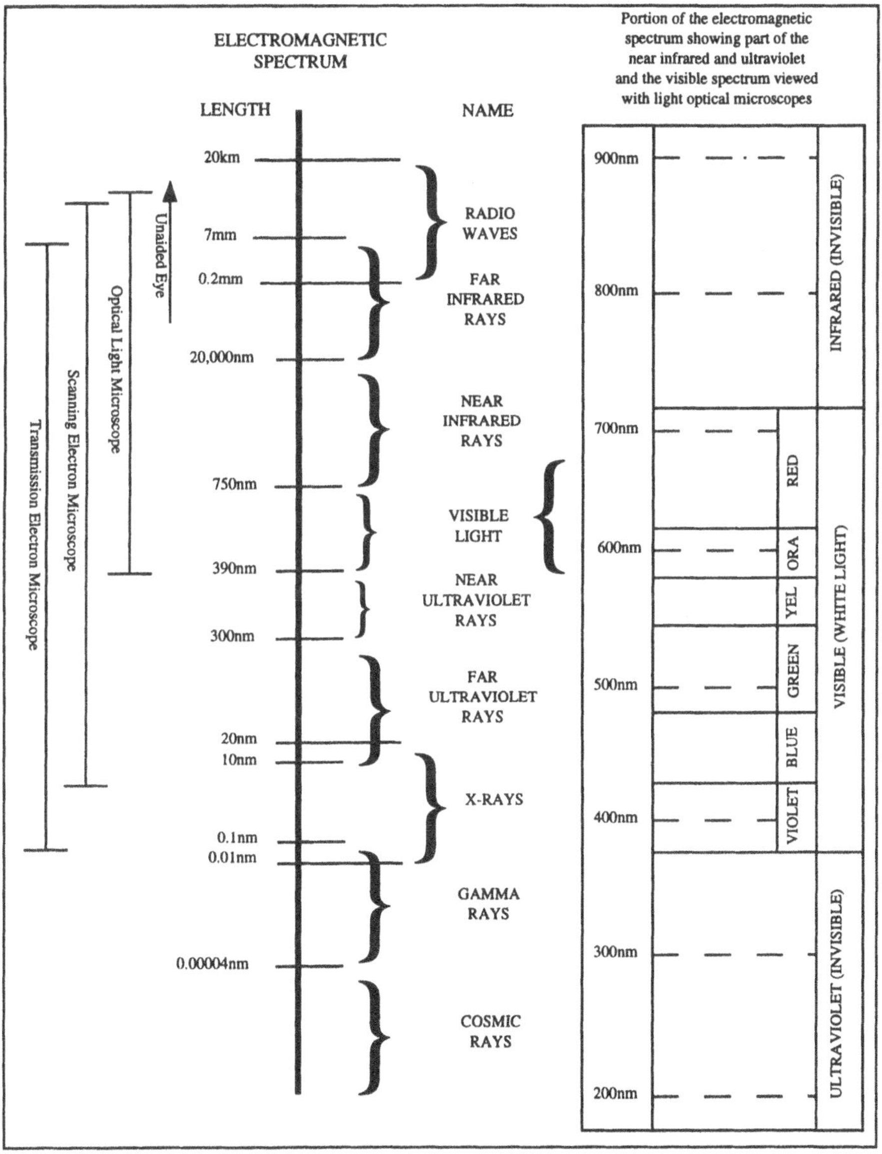

Figure 34.1. The electromagnetic spectrum, illustrating the optimum ranges of the optical and electron microscopes discussed in this section.

transmission electron microscope (TEM) and the scanning electron microscope (SEM). Combinations of the two (TSEM and STEM) have for many years provided palynologists with high resolution images (*e.g.* Muir 1970, Skvarla *et al.*, 1987/1988a, 1988b), and as a consequence extended the operating range of each basic machine. The biological applications of the Scanning Tunneling Microscope (STM), with a resolving power of 0.1nm (1Å), are still under development (Besenbacher *et al.*, 1989).

34.2.1. Transmission electron microscope (TEM)

The TEM was the first generation of microscope to replace light rays with an electron beam, and thus heralded a quantum leap in the optical examination of objects. Viewing specimens follows a similar principle to that of the compound optical microscope, with correcting lenses situated above and below the specimen (Fig 34.2b), although in the TEM's case the lenses are magnetic coils that affect the beam. The specimen is in the form of an ultra-thin section, as the electron beam passes through the specimen and is focused, by an electro-magnetic objective and projector lens, to form an enlarged image on a fluorescent screen. This is viewed through an eyepiece, but unfortunately, and in common with light microscopes, the image bears little resemblance to the specimens overall appearance as the TEM analyses the internal structure by identifying phases and defects. Many modern TEM's can operate in a scanning mode. This may occur in one of two forms, either in transmission (STEM) or using secondary electron emission (SEM) (Chescoe and Goodhew 1990). This facility will, however, reduce the resolution seen on the screen. The high voltage electron beam is generated from a filament operating at an accelerating voltage of between 40-100kv.

The TEM's major application has been in the field of biological sciences, and in particular soft tissue studies. This is in part do to the fact that specimen preparation is particularly well suited to soft tissue. Specimen preparation for the TEM can, at its most basic, be summarized as a three stage process: (1) specimens placed on an agar block or grid with additional agar drops added to cover the specimen, (2) block dehydrated by passing it through a series ethyl alcohol concentrations and impregnated with an epoxy resin, (3) from cured resin block an ultra-thin section is cut using a microtome. Specimens are sputter coated (see below) before examination. Details of histological preparations and biological applications are described in Goodhew (1984) and Hayat (1989), while additional information on the general operating and monitoring of TEM's can be found in Chapman (1986) and Chescoe and Goodhew (1990).

Two x-ray detecting analytical instruments are frequently available on TEM or STEM machines. Energy dispersive analysis (EDS) detects characteristic x-rays excited from small areas of the specimen irradiated by the beam (see below). Electron energy loss analysis (EELS) detects the energy lost by electrons that have interacted with the specimen. In the normal configuration the EDS detector is fitted to the side of the microscope close to the specimen, whereas the EELS spectrometer is situated beneath the fluorescent screen (Fig 34.2b).

34.2.2. Scanning electron microscope (SEM)

With the development of the SEM during the late 1960's, it became possible to view specimen surfaces' three dimensionally, and directly observe fine surface detail. Instead of the beam passing through the specimen, the electro-magnetic lenses of the SEM form the electrons into a very fine beam. The beam scans across the specimen surface by varying voltages of two sets of scan coils acting on it (Fig 34.3, double deflection coils). Excitement of the specimen surface causes it to release a shower of electrons. These low energy secondary electrons are attracted to a positively charged electron detector (*e.g.* Everhart-Thornly scintillator-photomultiplier). The amplified signal from the secondary electrons is used to modulate the electron beam brightness of a cathode ray image display tube, which, via a visual raster, is scanned in synchronism with the specimen, and generates a picture visible on the cathode ray tube (CRT, Fig 34.3). In this sense the SEM works in a similar principle to a closed-circuit television system, as information about the specimen surface is examined. For a more detailed explanation of the SEM's operating system consult Chescoe and Goodhew (1990).

SEM development during the late 1970's expanded the instrument's application from the limited imagery of gold surfaces of specimens (*e.g.* Hay and Sandberg 1967, Honjo and Berggren 1967, Sandberg and Hay 1967). With modern sophisticated and highly stable instruments it is now possible to produce many other image types, and to interface the microscope with computer controlled image acquisition, enhancing, archival and analytical systems. A well-equipped SEM will be able to generate an image in one of four modes:

- **secondary electron imagery** - the normal scanned image generated from metallic gold coated specimens. Images result from a transfer of energy from the beam to the specimen, as low-energy secondary electrons are emitted from a shallow zone up to the specimen surface. Images with a resolution of 5nm may be possible, while at low magnifications large depths of field remain in focus.
- **backscattered electron imagery** - high-energy electrons are backscattered from larger, deeper areas of the specimen. Although considerably more backscattered electrons are emitted than secondary electrons, they are subject to various energy losses. No energy transfer occurs between the electron beam and the specimen. This type of emission is normally used for observing high resolution compositional contrast differences of polished, carbon-coated sections. Imaging can be effective for resolutions better than 10nm, and used in differentiating between minerals with mean atomic number differences of approximately 0.1. Mineral boundaries, cement zonations and fine grained rock fabrics can be mapped using this facility. Martill *et al.*, (1992) experimented with the use of this technique in examining exceptionally well preserved phosphatized soft tissue. Unfortunately the contrast between the phosphatized areas and infilled late diagenetic calcite was insufficient for fine detail to be resolved. However, an elemental mapping procedure (see below) was successful.
- **backscattered electron-channelling imagery** - the high energy electrons are channelled by the crystal lattice structure, and are preferentially scattered in certain directions. Consequently, crystal orientations and lattice structures can be determined. This specialized facility is more commonly used by mineralogists and crystallographers.
- **x-ray emission imagery (EDS or EDX and WDS or WDX)** - as the high energy backscattered electrons cause excitation of the specimen, the component atoms emit x-rays of characteristic energies. Two features of the x-ray spectrum are measured: (i) energy, (ii) wavelength. Energy dispersive x-ray spectrometers (EDS) are an

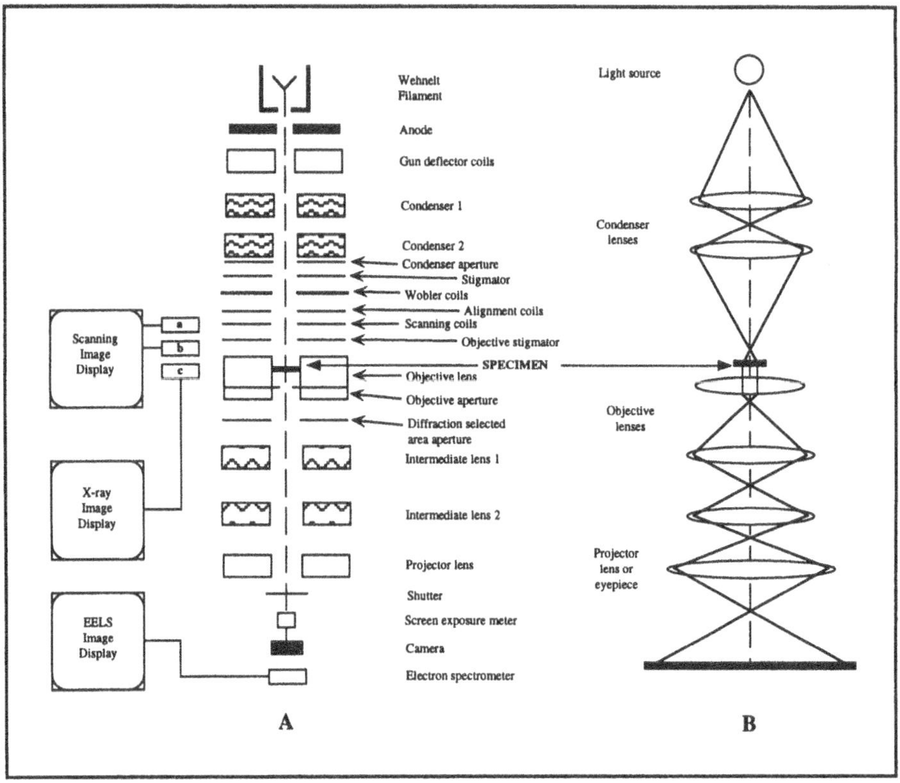

Figure 34.2. A schematic comparison of the TEM (A) and optical microscope (B). The usual position of the detectors and their relevant displays is indicated on the TEM diagram. (a) backscattered electron detector; (b) secondary electron detector; (c) x-ray detector.

essential facility for the integrated identification and imagery of minerals. This sophisticated electronic device has a detection limit of about 0.2wt%, and provides a quick method for determining elements ranging from sodium (Na) to uranium (U). The resulting spectrum provides a qualitative analysis of energy distribution, showing characteristic peaks of the chemical elements present. The wavelength dispersive spectrometer (WDS) is a mechanical device that detects individual elements using diffraction from a crystal of a known lattice spacing. Crystal spectrometers enable a better resolution to be obtained, with a detection limit of about 0.002wt%, and can detect mounted thin section, although three-dimensional specimens can be analysed using the EDS system. Martill *et al.*, (1992) discovered that sharper elemental mapped images could be obtained from sections of

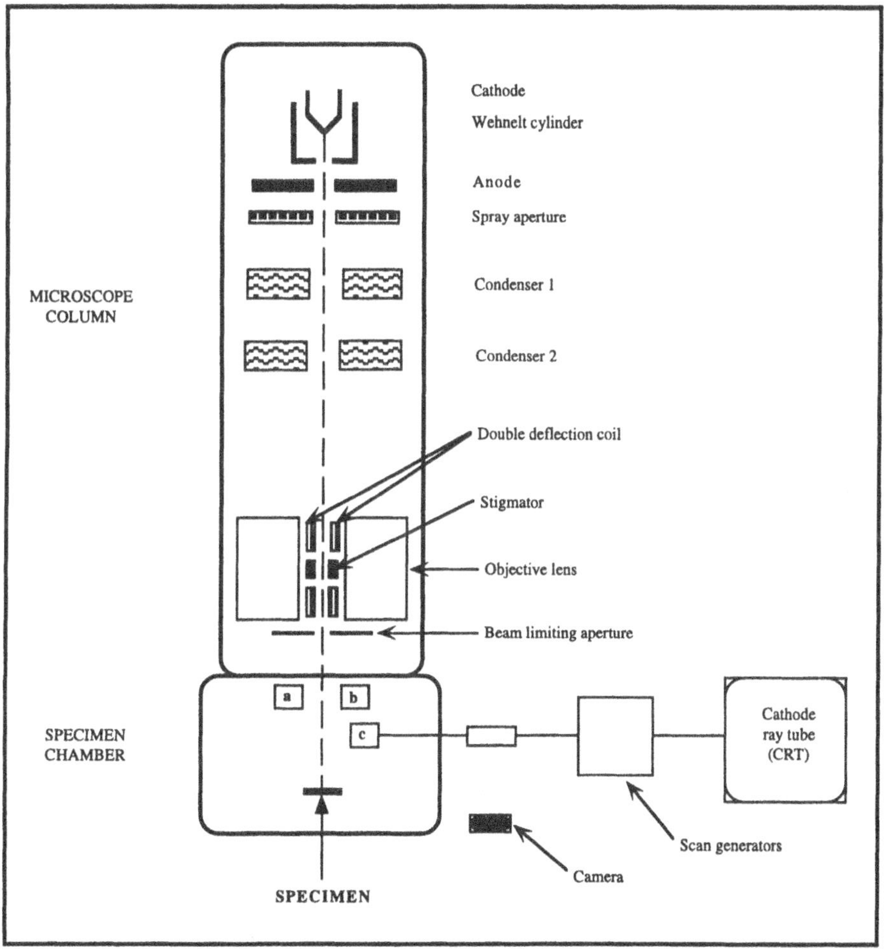

Figure 34.3. A schematic diagram of the SEM illustrating the relevant positions of electro-magnetic lenses, specimen and electron detectors. (a) Backscattered electron detector; (b) x-ray detector (WDS and/or EDS); (c) secondary electron detector. A similar vacuum exists in the microscope column and specimen chamber of the SEM, while differential vacuums are required in the case of the "Environmental Chamber" (see section 34.5 below).

50μm thickness instead of the standard 30μm. This was attributed to the lack of interference from the specimen mounting media. It was also suggested that this procedure may be suitable for the examination of pyritized and silicified soft tissues. Analytical applications in palaeobotany, and to a greater extent palaeontology, are limited to studies of biominerals and mineral alteration products involved in various modes of preservation (Hill 1990). For the majority of major and minor element analyses EDS is adequate. Studies requiring greater precision, or where elemental concentrations are low, can be analysed using WDS or X-ray fluorescence (Plotnick and Harris 1989). Initial mineral identifications can be obtained from X-ray diffraction (see section 35 X-RADIOGRAPHY TECHNIQUES).

In addition to the conventional method of viewing the image on a high resolution CRT screen, black and white photography using role or Polaroid film and video printing can be employed for recording a hard copy image. As electrons cannot convey colour, the images are black and white. However, false colour images can be produced by intensity coding of integrated images or using an array of detectors set at different angles to register shadowing. This new technique assigns a colour to each detector and tints the shadows to a natural colour. The integration of powerful computer controlled image analysis software equipment has helped produce a versatile powerful tool with multi-disciplinary adaptations within Earth Sciences.

The high resolution image analysis and photographic facilities so frequently used by palaeontologists requires well prepared (cleaned and orientated) samples, carefully mounted on aluminium stubs and sputter coated.

34.3. SAMPLE PREPARATION

Any problems relating to the image quality, barring those induced by photographic processing and printing, are usually attributable to one or two aspects of sample preparation. Firstly, the physical act of matrix removal and specimen cleaning before stub mounting. Secondly the process of sputter coating material. The removal of matrix and cleaning of specimens (physical and chemical) forms the bulk of this text (see Part III, LABORATORY TECHNIQUES, A. PHYSICAL PROCEDURES, B. CHEMICAL PROCEDURES), and no further reference will be made to it within this section. The procedure for mounting material on SEM stubs, and their subsequent coating will be detailed.

Good image contrast is essential for the interpretation of structures illustrated in electron microscope photomicrographs. Specimen contrast is related to differences in mass density (mass per unit area) generated from the specimen following electron bombardment, and subsequent electron scattering and detection. A combination of these factors is seen on the CRT screen as differences in intensity (*i.e.* contrast).

Furthermore some techniques used during stub preparation may modify morphological features and chemical analyses. Muir *et al.*, (1974) have documented changes in these for stub material prepared by air drying, freeze drying and critical point drying (see section 11 PREPARATION OF RECENT MATERIAL FOR COMPARATIVE STUDIES). Good retention of morphological structures and replication of elemental analyses in soft bodied micro-organisms is successfully achieved when freeze-drying is employed (Muir *et al.*, 1974).

34.3.1. Stub preparation

Aluminium stubs (pin or cylinder) must be labelled and cleaned before material is attached. The preference of an electron microscopist for using either pin or cylinder stubs (Fig. 34.4) is dependent on the make of machine, the stub adapters available or a predisposition for using a particular type of stub. Cylinder stubs have the advantage that they can free-stand beneath a stereozoom microscope when specimens require orientating. Pin stubs have to be secured in either a stub holder or pin base. Rimmed pin stubs (although cylinder stubs can be adapted or modified) have been used for viewing loose samples (*e.g.* sand grade size material), suitable for unattached foraminiferal assemblages and loose grains (Hüttemann and Hemleben 1985).

1 Using a diamond marker, number each stub. Cylinder stubs can be inscribed on both the side and the base.

2 Clean the surface of each stub using acetone and tissue paper. Avoid direct handling of cleaned stubs by using stub forceps and wearing disposable latex gloves. Cleaned stubs must be kept in a stub box to prevent airborne particles from settling on the surface.

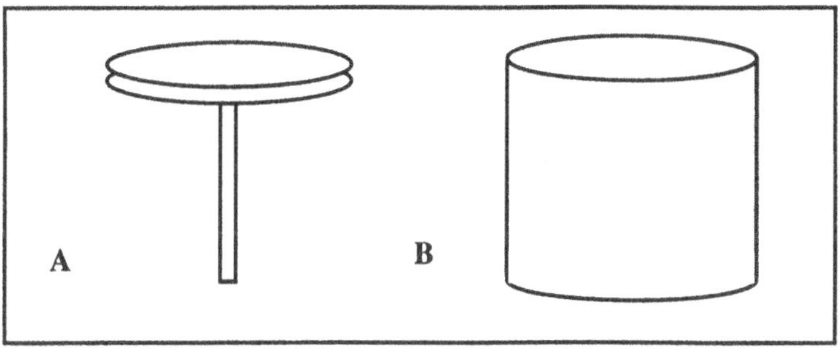

Figure 34.4. Diagrammatic illustrations of the common stub types used in electron microscopy. Usually constructed from aluminium and less commonly brass or copper. Pure carbon stubs can be used if background radiation causes a problem. A: pin stubs are available in 12.5, 25 and 32mm diameters with pins 3.2mm long. B: cylinder stubs are usually available in 10, 12.5, and 32mm diameters, and are 10mm high. *Hitachi* instruments require stubs with an internal M4 thread drilled in the base. Specialized stubs of unusual dimensions can usually be provided by suppliers.

Extensive testing and discussion have been given to the subject of adhesives suitable for SEM work (*e.g.* Hansen 1968, Leffingwell and Hodgkin 1971, Whitcomb 1981, 1985), while the use of small pieces of film or glass coverslips as an image enhancing backing for specimens has also been tested and reviewed (Haslett and Kersley 1997, Whittaker and Hodgkinson 1991).

34.3.1a. Sticky tabs and double sided tape

The conventional procedure of mounting specimens on stubs using peel-off double sided sticky tabs is suitable for most cases. The sticky tabs are carefully pressed onto the stub, and the protective cover peeled off before mounting the specimen. This must be undertaken in a clean dust-free environment. A disadvantage in using sticky tabs, is that it is difficult to remove and reposition specimens without damaging them. Image contrast between the specimens and the background is good. However, the sputter coated surface is prone to cracking and blistering (Haslett and Kersley 1997). Similar problems have also been encountered using standard double-sided tape (cut into small squares to fit within the stub diameter), with the disadvantage that specimens are liable to sink into the slightly thicker tape (Haslett and Kersley 1997). Image contrast can be increased and background quality enhanced by using carbon tabs.

34.3.1b. Epoxy resins and direct mounting

Specimens can also be directly mounted onto stubs using an epoxy resin adhesive. Ensure that the resin is not applied too thickly, as until cured, specimens are liable to sink into it. Epoxy resins are liable to crack, although this can be partially eliminated by polishing the stub (1000 grade grit) before applying the resin (Haslett and Kersley 1997). Resin applications should be thinly spread, then carefully scraped off leaving only a thin veneer for the specimen to adhere to (Hill 1990). In evaluating this procedure, Haslett and Kersley (1997) found that despite a good specimen:stub contact, images generated usually had low grey contrast backgrounds.

34.3.1c. Photographic film and glass coverslips

The use of glass coverslips and photographic film as specimen backgrounds have been advocated by both Haslett and Kersley (1997), and Whittaker and Hodgkinson (1991), although small pieces of mica sheet and silicon mounts are also effective. The principle in using a backing to small specimens is essentially the same.

1. Glass coverslips (10mm circles on 12.5mm stubs) or photographic film (emulsion side up) is stuck (using double sided sticky tabs) directly onto the aluminium stub.

2. Coat the surface with a thin layer of gum tragacanth or similar water soluble glue. Ensure that only a thin covering of glue is applied as a thick layer can result in fine surface details of the specimen becoming obscured.

3. Carefully attach specimens to the surface of the stub using a damp 000 paint brush. Haslett and Kersley (1997) report water soluble glues give variable contrast backgrounds to images, and sometimes it is difficult to attach specimens firmly to the stub.

If the contact between the specimen and the stub is poor, charging (flaring bright areas of parts or the whole specimen) and thermal damage may be evident. Whittaker and Hodgkinson (1991) describe similar problems, even when using adhesive metal tape. Slightly better results are obtainable using epoxy resins in preference to water soluble glues, but in using irreversible adhesives, specimens cannot be repositioned or removed from the stub without causing damage (Haslett and Kersley 1997). If water soluble glues (or other reversible glues) or gelatine saturated suspensions must be used, it is important to allow sufficient time for the glue to thoroughly dry before continuing with sputter coating. The degassing of glues and adhesives can reduce the operational effectiveness of the vacuum within the column, and infringe upon filament performance (Whitcomb 1981, 1985).

As a rule, small specimens (*e.g.* calcareous nannofossils, spores, diatoms) mounted on film do not require additional glue to make them adhere to the surface. The emulsion covering the film becomes sticky when wet, and is usually sufficient to position and maintain most micropalaeontological organisms, even those which require tilting within the electron microscope column. Spherical specimens, which are liable to move, can be secured by making a small indentation within the film (using either a mounted needle or razor blade). This aspect has been successfully used by both Haslett and Kersley (1997), and Whittaker and Hodgkinson (1991) in positioning planktonic foraminifera, ostracods and conodonts. Specimens can be orientated accordingly (*e.g.* Baumgartner *et al.*, 1981, De Wever 1980, Pessagno 1976).

Although both Haslett and Kersley (1997) and Whittaker and Hodgkinson (1991) conclude the preparation procedure using photographic film is the most time consuming, the final images produced were aesthetically more pleasing, particularly when preparing plates for publication.

34.3.2. Sputter coating

Stub mounted material requires an electron conducting surface in order for images to be generated. Specimens can be coated in a sputter coater under a gold (Au), gold-palladium (Au/Pd), and less commonly nickel (Ni), platinum (Pt) or silver (Ag) target. Coating is conducted between a cathode (the target) and anode (the specimen stage) in a small argon gas filled chamber, over a set time. When a current is initiated a purple omni-directional discharge occurs between the two. The bombardment of the target with gas ions erodes it, and sputtered atoms form an even coating on the specimen surface. A variety of machines are available for this operation. The principal function is essentially similar, but points critical to successful coating are reinforced below. Firstly, the distance between the specimen and the target must be set at between 3-4cm. Secondly, the actual process of coating (controlled by the timer) should take no longer than 2-3 minutes. During the ionising process (when a purple discharge is generated within the chamber) approximately 28nm of gold will be deposited per minute when a current of 16-18mA is maintained (by adjusting the argon leak valve). In brief, the procedure is as follows:

1. Turn on the gas supply from the argon cylinder, and set the pressure to between 3-5psi.

2. Switch on the electrical power to the sputter coater, and set the timer for 3 minutes.

3. Admit air into the chamber through the air leak valve. Open the lid and remove the stub holder, usually held in place with a large hexagonal bolt.

4 Place the stubs in the holder, securing each with the small retaining (Allen key) nut. Replace the holder in the chamber, and secure it by tightening the larger Allen key bolt.

5 Replace the chamber lid, ensuring it is centrally situated and the flange is clean. Set the height of the target 3-4cms above the specimen.

6 Turn on the pump to evacuate the chamber to 0.1 torr or better. The rotary vacuum pump must have an operating capacity of no lower than 47 litres per minute.

7 Fully open the argon leak valve, and leave for 10 seconds (ensures the chamber is filled with argon). Close the valve, and allow the pumping to continue to a vacuum of 0.8 torr or better. **N.B. never over tighten the needle valve** - these are the most frequently damaged and replaced parts of these instruments.

8 Repeat the purging procedure until a vacuum of 0.07 torr or greater is attained.

9 Select "COAT" or "HT" and gradually increase the voltage to 1.4kV. This will initiate a purple discharge clearly visible within the chamber, and a reading of 5-6mA on the control monitor.

10 As the argon leak valve is gradually opened, the discharge will increase. Adjust the leak valve to get a steady current of 16-18mA.

11 Select and activate the "TIMER" or "AUTO" setting for coating to begin.

12 By gently controlling the argon leak valve during coating, the current can be maintained at 16-18mA.

13 On completion of coating, close the argon leak valve, turn the current to zero, and turn the selector off setting.

14 Admit air into the chamber (using the air leak valve), and remove the specimens. Wipe the internal glass of the chamber clean and replace the specimen holder, and close the lid. Evacuate the chamber to 0.01 torr and switch the machine off (*i.e.* store the chamber under vacuum).

15 Switch off the electrical power, and the argon supply.

16 Store coated specimens in a desiccator until they are to be viewed under the SEM.

Hüttemann and Hemleben (1985) describe an illustrate a battery operated planetary rotating device that can be used within a sputter coater. A cylindrical container with fluted interior walls allows specimens to turn continuously and uniformly during the sputtering process. Electron conductivity (and therefore image contrast) is substantially improved when the entire specimen is coated. Fractions of similar grain size ensure even coating of individual material, as finer components have a lower turnover rate than the coarse component. As the system holds three of the specially constructed sample containers, separated fractions can be combined after sputtering (Hüttemann and Hemleben 1985).

34.4. REMOVAL OF GOLD COATING

Sputter coating is usually considered permanent, even for micropalaeontological material removed from stubs and stored (more securely) in micropalaeontological slides. Restudy of metallic coated specimens in reflected light using a stereozoom binocular microscope is difficult because of the high degree of reflectance from the specimen surface (Taylor 1986). However, Hansen (1968) describes a procedure for the removal of a pure gold coating, which is time consuming, requires a great deal of patience, and does not guarantee that the specimen will remain undamaged. The procedure has also been described by Sela and Boyde (1977).

The procedure, which generates cyanide fumes, must be undertaken within a fume cupboard, and is as follows:

1. Carefully remove the specimen from the stub, and place in a small conical flask containing a 0.1-1% aqueous solution of sodium cyanide (NaCN) and one or two drops of a 1 molar sodium hydroxide (NaOH) solution.

2. In a rubber stopper, perforated by two 5mm diameter holes, insert two lengths of glass tubing. One length should be sufficiently long to reach the bottom of the flask. Insert the stopper in the flask.

3. Connect a small water pump to the shorter length of glass tubing, and bubble air through the solution for 15 minutes or until fossils are clean. According to Hansen (1968) the following process occurs:
$$2Au + H_2O + {}^1/_2O_2 + 4NaCN \rightarrow 2NaAu(CN)_2 + 2NaOH$$

4. When clean, filter the solution and wash specimens in distilled water. Following drying, specimens can be picked in the conventional way and returned to micropalaeontological slides. Hansen (1968) reports that calcareous fossils show no ill effects following this cleaning procedure.

34.5. SCANNING UNCOATED SPECIMENS

Damage inflicted upon specimens from which removal of the gold coating has been attempted is a principal reason for the non-examination by SEM of some valuable specimens. Examining uncoated fragile, large specimens, *in-situ* specimens and type material has considerable advantages (Barnes 1991, Taylor 1986, Taylor and Jones 1996). By using low beam accelerating voltages (*e.g.* 5KV) it is possible to obtain images from some uncoated specimens (*i e.* pyritized fossils), although contrast is low and resolution generally poor (Howden and Ling 1974). Non-conducting specimens (*i.e.* calcareous fossils) must be coated, as outlined above, with a metallic layer. This prevents the specimen surface from charging (a build-up of negative charge seen as enhanced emission image distorting bright spots), by providing the means for the charge to earth via the specimen stub.

Conventional scanning procedures require the maintenance of a high vacuum in both microscope column and specimen chamber. Viewing uncoated specimens requires slightly different conditions. While the column is maintained under high vacuum (*e.g.* 10^{-4} torr), the specimen chamber is maintained at a lower operating vacuum (*e.g.* 10^{-1} torr). The electron beam passes through a 200µm aperture, sufficiently small enough to allow for the differential vacuum to be maintained. The strong positive charge of the secondary electron detectors prohibits its use in the environmental chamber for detecting electrons and viewing images. However, an uncharged scintillator back scattered electron detector (*e.g.* Robinsons' back scattered electron detector) situated directly above the specimen, provides a means for detecting and viewing high resolution images. The positioning of the detectors within the column (Fig 34.3), ensuring the majority of backscattered electrons emitted at high angles are detected, generates flat images in comparison with secondary electron images (Taylor 1986). These images have many characteristics in common with those observed in standard light microscopy. This is because image surfaces are angled perpendicular to the electron beam, and backscattered electrons detected from sub-horizontal surfaces (*i.e.* a similar configuration to viewing down a stereozoom binocular microscope with illumination from a ring light). Unlike secondary electron imaging, optimum results cannot be achieved from specimens with high relief, or those tilted within the column (Taylor 1986). Taylor (1986) compares and contrasts the significant differences between secondary electron images and backscattered electron images, and these are tabulated below (Table 34.1). An appreciation, and recognition of these differences is required by the EM operator when interpreting morphological features.

The palaeobiological applications of this procedure are numerous. From a practical point of view, specimen preparation time is reduced, and specialized storage conditions (beyond those required to maintain specimen longevity) are not required. Furthermore, the potential for damaging the specimen during preparation is also reduced. Not only can fragile specimens be examined using this procedure, but also valuable type material, observations of which are usually confined to the light microscope, and large specimens that would be difficult to coat and probably charge when viewed in a conventional scanning electron microscope (Taylor 1986). The most thorough and complete observations of fossil samples, providing the maximum amount of data, will utilise a combination of procedures. This approach is already used to much effect in palynology (*e.g.* Leffingwell and Hodgkin 1971), and in particular where key diagnostic facilities (*e.g.* phase and differential contrast, fluorescence) are present on the compound microscope. A similar approach has also been used in the study of calcareous nannofossils (see section 33 EXTRACTION TECHNIQUES FOR CALCAREOUS NANNOFOSSILS), where both TEM and SEM observation's complement (although they will never replace) standard light microscopical procedures. Hendey (1971) recognized that electron microscope studies of diatom frustules provided new ultrastructural data. However, if this were used within a taxonomic classification system, it would separate taxa that, from conventional light microscopy studies, were recognized to have strong structural, ecological and phylogenetic relationships. Hendey (1971) argued that electron microscopical data alone was unlikely to provide the basis for a new classification system.

IMAGE FEATURE		SECONDARY ELECTRON (Coated specimens)	BACKSCATTERED ELECTRONS (Uncoated specimens)
Resolution	Mag. <500	Very good	Very good
	>500	Very good	Inferior
Depth of field		Very good	Appears flat
Differential contrast from emission a. "Edge effect"		High	Low - similar to optical appearance
b. "Cavity effect"		High	Low - similar to optical appearance
c. "Brightness"		Relief dependent	Compositionally dependent

Table 34.1. A comparison of secondary electron and backscattered electron image characteristics. (Data from Taylor (1986), with observations based on constant machine settings for working distance, brightness and contrast at an accelerating voltage of 15KV).

REFERENCES

BARNES, S. H. 1991. Electron microscopy and analysis at the Natural History Museum. *Microscopy and Analysis*. No. 25, 29-31.

BAUMGARTNER, P. O, BJØRKLUND, K. R, CAULET, J-P, DE WEVER, P, KELLOGG, D, LABRACHERIE, M, NAKASEKO, K, NISHIMURA, A, SCHAAF, A, SCHMIDT-EFFING, R and YAO, A. 1981. EURORAD II, 1980 - Second European meeting of radiolarian paleontologists: current research on Cenozoic and Mesozoic radiolarians. *Eclogae geologicae Helvetiae*. **74**, (3), 1027-1061.

BESENBACHER, F, LAEGSGAARD, E and STENSGAARD, I. 1989. The scanning tunneling microscope. *Microscopy and Analysis*. No. 12, 17-20.

CHAPMAN, S. K. 1986. *Maintaining and monitoring the transmission electron microscope*. Royal Microscopical Society, Handbook No. 8. Oxford Science Publications, Oxford University Press.

CHESCOE, D and GOODHEW, P. J. 1990. *The operation of Transmission and Scanning Electron Microscopes*. Royal Microscopical Society, Handbook No. 20. Oxford Science Publications, Oxford University Press. vi + 86p.

DE WEVER, P. 1980. A new technique for picking and mounting radiolarians for scanning electron microscopy. *Micropaleontology*. **26**, (1), 81-83.

GLAUERT, A. M. (ed). 1974 to 1994. *Practical methods in electron microscopy* (Volumes 1-15). North Holland Publishing Company, Elsevier Science Publishers, (Hollond) and Portland (London).
GOLDSTEIN, J. I, NEWBURY, D. E, ECHLIN, P, JOY, D. C, FIORI, C and LIFSHIN, E. 1981. *Scanning electron microscopy and X-ray microanalysis*. Plenum Press (New York). xiii + 673p.
GOODHEW, P. J. 1984. *Specimen preparation for transmission electron microscopy of materials*. Royal Microscopical Society, Handbook No. 3. Oxford Science Publications, Oxford University Press. iv + 41p.
HANSEN, H. J. 1968. A technique for removing gold from plated calcareous microfossils. *Micropaleontology*. **14**, (4), 499-500.
HASLETT, S. K and KERSLEY, C. L. 1997. Mounting planktonic foraminifera for scanning electron microscopy. *Microscopy & Analysis*. May No. 59, 33-34.
HAY, W. W and SANDBERG, P. A. 1967. The scanning electron microscope, a major break-through for micropalaeontology. *Micropaleontology*. **13**, (4), 407-418.
HAYAT, M. A. 1989. *Principles and techniques of electron microscopy, Biological applications* (3rd Edition). Macmillan Press (Basingstoke). xv + 469.
HENDEY, N. I. 1971. Electron microscope studies and the classification of diatoms, pp 625-631: In; B. Funnel and W. Riedel (eds), *Micropalaeontology of the Oceans*. Cambridge University Press. (Cambridge). x + 828p.
HILL, C. R. 1990. Scanning electron microscopy in palaeobotany, In; D. Claugher (ed), *Scanning Electron Microscopy in Taxonomy and Functional Morphology*. Clarendon Press, (Oxford). Systematics Association Special Volume No. 41. 193-204.
HONJO, S and BERGGREN, W. A. 1967. Scanning electron microscope studies of planktonic foraminifera. *Micropaleontology*. **13**, (4), 393-406.
HOWDEN, H. F and LING, L. E. C. 1974. Low-magnification study of uncoated specimens, In; M. A. Hayat (ed), *Principles and techniques of scanning electron microscopy. Volume 1. Biological Applications*. Van Nostrand Reinhold Co. (New York). 273p.
HÜTTEMANN, H and HEMLEBEN, CH. 1985. An improved rotating device for sputter preparation of SEM particulate specimens. *Journal of Foraminiferal Research*. **15**, (2), 108-110.
LEFFINGWELL, H. A and HODGKIN, N. 1971. Techniques for preparing fossil palynomorphs for study with the scanning and transmission electron microscopes. *Review of Palaeobotany and Palynology*. **11**, 177-199.
MARSHALL, D. J. 1988. *Cathodoluminescence of geological materials*. Unwin Hyman (Boston). xiv + 146p.
MARTILL, D. M, WILBY, P. R and WILLIAMS, N. 1992. Elemental mapping: a technique for investigating delicate phosphatized soft tissues. *Palaeontology*. **35**, (4), 869-874.
MUIR, M. D. 1970. Scanning electron microscopy in palynology. *Review of Palaeobotany and Palynology*. **10**, 85-97.
MUIR, M. D, SPICER, R. A, GRANT, P. R and GIDDENS, R. 1974. X-ray microanalysis in the SEM for the determination of elements in modern and fossil micro-organisms. *Proceedings of the Eighth International Congress of Electron Microscopy, Canberra*. Volume 2, 104-105.
PESSAGNO, E. A. 1976. Radiolarian zonation and stratigraphy of the Upper Cretaceous portion of the Great Valley Sequence, California Coast Ranges. *Micropaleontology, Special Publication*. **2**. 1-95.
PLOTNICK, R. E and HARRIS, G. 1989. X-ray analysis using Energy Dispersive and Wavelength Dispersive Spectroscopy, pp 179-185: In; R. M. Feldman, R. E. Chapman and J. T. Hannibal (eds), *Paleotechniques*. The Paleontographical Society Special Publication. **4**. iv + 358p.
SANDBERG, P and HAY, W. W. 1967. Study of microfossils by means of the scanning electron microscope. *Journal of Paleontology*. **41**, (4), 999-1001.
SELA, J and BOYDE, A. 1977. Cyanide removal of gold from SEM specimens. *Journal of Microscopy*. **111**, 229-231.
SKVARLA, J. J, FARABEE, M. J and CHISSOE, W. F. 1987/1988a. Fossil pollen ultrastructure by transmission imaging in a scanning electron microscope (TSEM). *Journal of Palynology*. **23/24**, 205-211.
SKVARLA, J. J, ROWLEY, J. R and CHISSOE, W. F. 1988b. Adaptability of scanning electron microscopy to studies of pollen morphology. *Aliso, Journal of the Rancho Santa Ana Botanic Garden, Claremont, California*. **12**, 119-208.
TAYLOR, P. D. 1986. Scanning electron microscopy of uncoated fossils. *Palaeontology*. **29**, (4), 685-690.
TAYLOR, P. D and JONES, C. G. 1996. Use of the environmental chamber in uncoated SEM of Recent and fossil bryozoans. *Microscopy and Analysis*. No. 52, 27-29.
WHITCOMB, M. J. 1981. The suitability of various adhesives as mounting media for scanning electron microscopy. *Journal of Microscopy*. **121**, (3), 289-308.
WHITCOMB, M. J. 1985. The suitability of various adhesives as mounting media for scanning electron microscopy. II. General purpose glues. *Journal of Microscopy*. **139**, (1), 75-114.
WHITTAKER, J. E and HODGKINSON, R. L. 1991. On the preparation of specimens for scanning electron microscopy and a simple technique for plate making, using a black background. *Journal of Micropalaeontology*. **9**, (2), 219-220.

35. X-RADIOGRAPHY TECHNIQUES

35.1. INTRODUCTION

SAFETY NOTE: Ionizing radiation can result in cancer if inadequate protection is not used. Equipment must be used in rooms protected by a lead shield, and preferably inoperable when the door is open. Machine controls should be located outside the room, while trained operators must wear monitoring badges that are periodically checked.

This section will examine how x-ray procedures have been adapted for use in palaeobiology, and specifically techniques of photomicroscopy, applicable to microfossils, contact x-radiography for use on macrofossils, and a brief outline of computer tomography and its application in palaeontology. X-rays occur in that part of the electromagnetic spectrum with wavelengths between 20-0.01nm (see figure 34.1, section 34 ELECTRON MICROSCOPY TECHNIQUES). These very short wavelengths have the ability for greater penetration and higher resolution than visible light observations (Hooper 1965). This factor, coupled with the ability of x-rays to penetrate matter, make it an ideal medium for the non-destructive observation of the internal features of an object. Examples readily known and encountered by most people include medical and dental applications.

Within Earth Sciences specialized X-radiographic techniques have been adapted for use in the disciplines of palaeontology (*e.g.* Zangerl 1965, Zangerl and Schultze 1989) and petrography (*e.g.* Bouma 1964, 1969, Davis and Walawender 1982, Hamblin 1962, 1971, Hill *et al.*, 1979). The development of these procedures owes much to the techniques and applications used within the medical industry. However, because the desired objectives within geology differ from those within medicine, techniques have required some modification. Medical applications are constrained by the potentially injurious effects of ionizing radiation on patients. Consequently, exposure times must be short and fast films and intensifying screens used. These limitations are not as critical in geological applications, where slow film speeds, low ray intensities and long exposure times can be employed (Zangerl and Schultze 1989). In some cases this may require careful examination of the equipment and facilities on offer. The absence of X-ray equipment as standard in most Palaeobiological Laboratories, results in the use of machines situated in medical laboratories for palaeontological studies. If these machines operate with air or oil-cooled anodes (the x-ray generating target), interruptions during long exposure times are required to allow for the cooling of the tube. Water-cooled machines do not present this problem, and are preferable for palaeontological work (Zangerl and Schultze 1989).

However, before embarking on this expensive procedure, ensure all conventional optical facilities and procedures have been exhausted. Savrda *et al.*, (1985), compared x-ray and oil immersion procedures in the biogenic structural enhancement of textural elements within diatomaceous sediments. The resolution and structures observed on 3-5mm prepared slabs under oil immersion were comparable or better than that achieved with x-radiography, although some remnant primary laminae and structures were more visible in the radiograph. For biogenic structures in slabbed specimens Savrda *et al.*,

(1985) suggest that oil immersion examination should be used before embarking on more expensive and time consuming x-ray procedures.

The specialized procedures used in x-ray examination and photography cannot be undertaken without the specific help and assistance of an institution with the necessary x-ray equipment and qualified radiographer to operate the machine. Expert help is also required in film processing and image production and interpretation.

35.2. PALAEONTOLOGICAL CONSIDERATIONS

Two key objectives are pursed by preparators in the development of fossil material. At there simplest these are:

(i) observing obscured internal and external structures;
(ii) observing the relationship of life or burial position within the sediment.

These objectives are principally achieved by preliminary exploratory mechanical and chemical development. Complex interpretations may require conventional thin and serial sectioning techniques, involving the partial or complete destruction of fossil material, limiting procedures to "non-valuable" material. Furthermore, any preparation of partially exposed fossil material involves an element of guess-work in determining the orientation of the unexposed fossil. X-ray analysis by either photomicroscopy or computer tomography (the two principal techniques employed) provides a method of examining internal structures. Following this examination, fossils can then be safely developed (*e.g.* Whybrow 1982), and three-dimensional reconstruction's made. However, x-ray microscopy of prepared bone sections can provide a valuable tool in ultra-structural examination (Cook *et al.*, 1962), while the successful use of computer tomography procedures used during the past few years (*e.g.* Conroy and Vannier 1984, 1985, Haubitz *et al.*, 1988, McGowan 1989a, 1989b, 1990, 1991) in vertebrate palaeontology have reinforced it as virtually a standard procedure (Clark and Morrison 1995).

The penetrating ability of x-rays is a function of the electromotive force or tube energy (kilovoltage or kv) generated from the target within the x-ray tube, and the absorption ability of the sample material under analysis, related to its atomic weight and thickness. In essence, absorption increases proportionally with sample density and thickness, and tube energy used. Objects composed of substances with high atomic numbers, or thickly prepared materials both require high tube voltages. X-ray images and radiographs are a compromise between x-ray penetration (requiring high voltages) and contrast (requiring lower voltages) (Harbersetzer 1994). Most palaeontological work is undertaken between 10-130kv, with low voltages used for small or delicate structures. The x-rays are generated in a high vacuum tube, where a heated cathode produces electrons which are directed towards an anode (the target). Most of this energy is converted to heat, but a small amount of the spectrum is given off as "soft" long wave x-radiation, characterized by low penetration and high contrast.

35.3. X-RAY PHOTOGRAPHY

The most commonly used and oldest, x-ray technique within palaeontology is contact radiography (CR). At its simplest, the procedure requires that specimens are situated close to the photographic plate, in the path of the x-ray beam. Small specimens are best examined using point-projection or mammography procedures (described below), as these generate better contrast photographs (Harbersetzer 1994). The best pictures are generated from specimens in which the enclosing matrix consists of material with a lower atomic weight than the fossilized fragments. A clear distinction between the matrix and fossil results in the preferential absorption of x-rays, and assists in producing a high contrast image. Additionally, the specimen thickness must be thin enough to permit x-rays to penetrate to the underlying film layer.

35.3.1. Specimen preparation

The factors of matrix and fossil composition, coupled with sample thickness, must be independently considered. Both aspects effect the quality of the x-ray image produced.

35.3.1a. Composition

Optimum results are obtained following the removal of the maximum amount of matrix, particularly in calcareous examples. Compositional differences are an important factor in image quality production. Fossils preserved within an organic rich matrix (*e.g.* shales, resins) produce better pictures than calcareous fossils contained within a matrix of similar composition. Prepared material, of any composition, will produce superior quality images in comparison to unprepared specimens.

35.3.1b. Thickness

An evenly thick specimen assists in reducing the amount of distracting noise generated both above and below the plane of the fossil. Thinning can be achieved by the conventional procedures of cutting, splitting, grinding or chemical preparation, and prepared specimens resin embedded, while uneven, matrix free specimens must be immersed in water before analysis. The ability of water to absorb x-rays makes it an ideal medium to simulate soft tissue (Clark and Morrison 1995, Zangerl and Schultze 1989), forming part of the procedure routinely used in ichthyology (Bartlett and Haedrich 1966). However, before undertaking work on fossil material, Zangerl and Schultze (1989) advise experimentation on small fragments.

35.3.2. Procedure

1 Centre the specimen beneath the x-ray tube, on the cassette holder containing the x-ray film. This must be situated above a lead sheet that absorbs the x-rays scattered beneath the film.

2 Ensure the fossil is placed close to the film. This will assist in obtaining the sharpest definition, and give a one to one shadow picture of the fossil.

3 Machine settings depend upon specimen thickness, target distance and film speed, while the distance between target and the film depends upon the object size. Long distances show less image distortion, and must be compensated for by increasing the exposure time (Zangerl 1965). Choose the lowest voltage (kv) setting at which the x-rays will

penetrate the matrix and fossil, but not obscure fine detail (Zangerl and Schultze 1989).

The density of radiation, expressed in milliampres (ma), is inversely proportional to the exposure time (seconds), consequently machine settings of 5ma at 30 seconds and 10ma at 15 seconds produce similar exposures of 150 ma/sec (Zangerl 1965). However, to the experienced eye differences in picture quality are evident, resulting from variations in film emulsion reaction times to the exposure settings.

35.3.3. Image generated

The radiogram produced is essentially a shadow picture of homogeneity within the specimen. It is comparable with a standard black and white (monochrome) photographic negative, with the image represented in various shades of grey. Black areas represent the background and thin areas of the object. Considerable experience is required in the recognition and interpretation of grey tone values and contrast differences (Harbersetzer 1994, Zangerl 1965, Zangerl and Schultze 1989). Fossil material will appear lighter than the background, however, if the fossil is penetrated by the x-rays more readily than the surrounding matrix, a reverse shadow configuration is seen, as in a conventional monochromatic photographic positive print. A balance of tube current intensity and exposure time achieves optimum picture quality.

Structural interpretation of the image can be considerably enhanced by taking stereo-radiographs, and viewing results with the aid of a stereoscope to produce a three-dimensional image (Hamblin 1971, MacIntyre 1976).

35.3.4. Stereo-radiograph production

Pictures are produced by (a) moving the tube, or (b) moving the specimen equidistant around the mid-point of the beam. In summary the procedure is as follows:

1 Centre the specimen and film beneath the tube target.

2 Move the tube to one side of centre, and expose the film.

3 Move the tube an equal distance to the other side of the centre line of the tube target. Replace the film holder with new unexposed film. MacIntyre (1976) exposes alternate sides of a single sheet of film, covering each half with a sheet of lead.

4 Expose the film at similar settings to the first.

5 Process both films together to ensure equal emulsion density, and examine with a stereoscope.

35.3.5. X-ray films

Unlike conventional films, x-ray films are typically emulsion coated on both sides, and loaded in cassettes with intensifying screens situated at the front and back. These films are characterized by fast speeds, and are coarse grained in comparison with those used in industry, which are less sensitive but finer grained. A wide variety of films are available (Harbersetzer 1994, Zangerl 1965), varying in contrast, speed and emulsion grain. For palaeontological applications, the best results are obtained using a fine grained, high contrast, slow speed film. Both the contrast and definition of small fossil specimens can be adversely effected by the physical properties of the intensifying screens. The addition of aluminium and copper filters can assist in hardening the x-ray spectrum, and reducing the effects of fogging. In general, however, exposure times are longer than those required with medical films.

Zangerl (1965) suggests experimenting with medical films, as these are readily available, and the radiographer will have considerable experience in processing and obtaining the best results. Experimentation with intensifying screens and filters may assist in enhancing results. Generally however, medical films give poor results for delicate material, obscuring fine detail. Harbersetzer (1994) discusses the availability of film types for palaeontological x-ray work, and concludes that special industrial Non Destructive Testing (NDT) films (with thick emulsion coatings and high silver contents) are the most suitable for high definition work.

35.3.6. Hard copy photographs

Exposed films require special developer and fixing solutions, and must be processed in tanks (not trays) by either hand or machine. Processing usually requires some sort of deviation from standard procedures (Zangerl and Schultze 1989).

Printing is also a specialized procedure, best undertaken by a skilled expert, as many x-ray photographs usually require some sort of manipulation (*e.g.* contrast correction, enlarging or reducing). Specialized equipment is available to assist in reducing the overall negative contrast, while increasing the fine detail (Zangerl 1965). Photographs can be printed as either (a) negatives (black background, specimen structure appears in various shades of grey), or (b) positive (reversal of all shadow values, specimen structure appears dark against a white background).

35.3.7. Radiograph and photograph interpretation

Conventional medical examination and interpretation are usually done from the original (negative), as the characteristics observed may not be evident in a standard print (Harbersetzer 1994). Because x-ray images and analysis are not standard palaeobiological procedures, image interpretation may present some difficulties to the non-radiographer. At first sight many images appear as nothing more than a meaningless pattern of grey emulsion densities in amorphous shapes. However, greater understanding will be achieved by becoming familiar with the morphology, anatomy and ultra-structure of the organism. The use of stereoscopic pictures (described above), will also assist in interpretation (Zangerl 1965).

Clear areas seen on the image are produced by the most dense or thickest materials, or those absorbing and blocking the x-rays, and usually represent fossilized material. When present, fossil components may be instantly recognizable, and complete skeletons determined. In other specimens, details may be obscured by diagenesis or the presence of ferric salts embedded within the fossil, preventing x-ray penetration and concealing detail (Harbersetzer 1994). Areas of equal shadow intensity within a single image usually indicate material of similar composition. Direct comparison with other images is not usually possible, as exposure conditions (kv, exposure time, machine settings and film type) may be different (Zangerl and Schultze 1989). Special techniques, now used in vertebrate palaeontology, for retaining detail and contrast, are outlined by Harbersetzer (1994).

35.4. X-RAY PHOTOMICROSCOPY

Hooper (1965) describes three x-radiography photographic procedures:

(i) **Point projection microradiography**: where an electron gun generates a beam that is focused by an electro-magnetic lens system to a point upon a metal target (usually copper, tungsten, gold or silver). This emits x-rays that penetrate the specimen to produce an enlarged image on a photographic plate. The procedure allows for great depth of focus and penetrating power, with resolutions of 0.1μm obtainable. Many instruments used are essentially modified electron microscopes, where x-ray exposure times in vacuum columns may be considerably shorter than conventional air filled columns.

(ii) **Contact microradiography**: where specimens are placed in the beam path, very close to a photographic plate. The procedure produces good resolution, but not of the quality of that achieved in point projection microradiography. This method has been outlined above under x-ray photography, and will not be described further.

(iii) **Reflection x-ray microscopy**: where x-rays are totally reflected from a mirror at a small angle, and focused by a curved mirror to give an image. The equipment and method are difficult to master, and no palaeobiological application is yet developed.

The advantages of using x-ray photomicroscopy to complement conventional light and scanning electron microscopy in micropalaeontology has changed little since their initial definition over 30 years ago (Hooper 1965). X-ray microradigraphy enables:

- detailed internal morphological studies
- highly accurate measurements of internal structures
- non-destructive sample preparation (*i.e.* non-thin sectioning)
- information is gained very quickly.

Recent advances in x-ray equipment resolution, photographic recording, and improvements in micropalaeontological applications and procedures are well documented (*e.g.* Schmidt 1952, Hedley 1957, Hooper 1959, 1965, Bé *et al.*, 1969, Leary and Hart 1988, Hottinger and Mehl 1991). The small nature of micropalaeontological specimens makes them ideal for this type of x-ray analysis. Sample preparation time is reduced, as specimens do not require cutting or thinning. Consequently, low x-ray energy levels producing "soft" x-ray images suffice. However, in using low energy levels, x-ray absorption is also small, generating low contrast images (Hottinger and Mehl 1991). Further disadvantages, related to small specimen size, can be encountered during image interpretation where the complexity of overlapping internal structures may inhibit the measurement of fine detail (Hooper 1965). A comprehensive summary of methods and equipment used in micropalaeontological quantitative and qualitative biometric research can be found in Hottinger and Mehl (1991).

35.5. COMPUTERIZED AXIAL TOMOGRAPHY (CT OR CAT-SCANNING)

This procedure is much safer than x-ray photography, as the beam is concentrated to a limited area, eliminating secondary radiation and radiation around the machine. As with x-ray photography, most equipment used by palaeontologists is situated in hospitals or medical centres, and primarily has a medical application. It is thus important to ensure that the establishment in which work is undertaken has the necessary facilities to transfer digital images (cross sections) directly into three-dimensional pictures (Clark and Morrison 1995, Zangerl and Schultze 1989).

In similarity with x-ray photography, specimens can be examined dry or immersed in water, to simulate soft tissue, with the advantage that slab thinning is not required. Once initiated, data acquisition proceeds automatically, moving the x-ray tube around the specimen a pre-determined distance each time. A series of accurate cross sections, recording density differences through the specimen to depths of 1-2mm is obtained. A computer interprets the digital information, and adds cross lines to complete the section, processing it to generate a video picture. Accumulated data is stored on either magnetic tape or optical diskette. Computer graphic modelling is employed to convert the two-dimensional image into a three-dimensional reconstruction. Ideally, specimens should be free of heavy metallic minerals (*e.g.* pyrite) as these can cause image distortion. However, unlike conventional x-ray photography, the CAT-scan resolution can be enhanced by adjusting the beam. Images are accurate enough for angular, linear, areal and volumetric measurements to be calculated (Zangerl and Schultze 1989).

REFERENCES

BARTLETT, M. R and HAEDRICH, R. L. 1966. Techniques in the radiography of fishes. *Transactions of the American Fisheries Society*. **95**, 99-101.

BÉ, A. W. H, JONGEBLOED, W. L and MCINTYRE, A. 1969. X-ray microscopy of Recent Planktonic foraminifera. *Journal of Paleontology*. **43**, (6), 1384-1396.

BOUMA, A. H. 1964. Notes on x-ray interpretation of marine sediments. *Marine Geology*. **2**, 278-309.

BOUMA, A. H. 1969. *Methods for the Study of Sedimentary Structures*. Wiley-Interscience (New York). xvi + 458p.

CLARK, S and MORRISON, I. 1995. CT scan of fossils, pp 323-329 in Chapter 12, Methods and use of CT scan and X-ray: In; P. Leiggi and P. May (eds), *Vertebrate Paleontological Techniques*, Volume 1. Cambridge University Press. 368p.

CONROY, G. C and VANNIER, M. W. 1984. Noninvasive three-dimensional computer imaging of matrix-filled fossil skulls by high resolution computer tomography. *Science*. **226**, 456-458.

CONROY, G. C and VANNIER, M. W. 1985. Endocranial volume determination of matrix-filled fossil skulls using high resolution computer tomography, pp 419-426: In; P. V. Tobias (ed), *Homonid evolution: Past, present and future*. Alan R. Liss (New York).

COOK, S. F, BROOKS, S. T and EZRA-COHN, H. E. 1962. Histological studies on fossil bone. *Journal of Paleontology*. **36**, (3), 483-494.

DAVIS, B. L and WALAWENDER, M. J. 1982. Quantitative mineralogical analysis of granitoid rocks: a comparison of x-ray and optical techniques. *American Mineralogist*. **67**, 1135-1143.

HAMBLIN, W, K. 1962. X-ray radiography in the study of structures in homogenous sediments. *Journal of Sedimentary Petrology*. **32**, (2), 201-210.

HAMBLIN, W, K. 1971. X-ray photography, pp 251-284: In; R. E. Carver (ed), *Procedures in Sedimentary Petrology*. Wiley-Interscience (New York). xiii + 653p.

HARBERSETZER, J. 1994. Radiography of fossils, pp 329-339 in Chapter 12, Methods and use of CT scan and X-ray: In; P. Leiggi and P. May (eds), *Vertebrate Paleontological Techniques*, Volume 1. Cambridge University Press. 368p.

HAUBITZ, B, PROKOP, M, DOHRING, W, OSTROM, J. H and WELLENHOFER, P. 1988. Computer tomography of Archaeopteryx. *Paleobiology*. **14**, 206-213.

HEDLEY, R. H. 1957. Microradiography applied to the study of foraminifera. *Micropaleontology*. **3**, (1), 19-28.

HILL, G, W, DORSEY, M. E, WOODS, J. C and MILLER, R. J. 1979. A radiographic scanning technique for cores. *Marine Geology*. **29**, 93-106.

HOOPER, K. 1959. X-ray absorption techniques applied to statistical studies of foraminifera populations. *Journal of Paleontology*. **33**, (4), 631-60.

HOOPER, K. 1965. X-ray microscopy in morphological studies of microfossils, pp 320-326: In; B. Kummel and D. Raup (eds), *Handbook of Paleontological Techniques*. W. H. Freeman and Co., (San Francisco). xiii + 852p.

HOTTINGER, L and MEHL, J. 1991. X-ray microscopy in micropalaeontology. *Microscopy and Analysis*. No. 21, 21-23.

LEARY, P. N and HART, M. B. 1988. X-raying planktonic foraminifera. *Journal of Micropalaeontology*. **7**, (1), 43-44.
MacINTYRE, G. T. 1976. A simplified method for x-ray stereographs of small specimens. *Journal of Paleontology*. **50**, (2), 357-358.
McGOWAN, C. 1989a. Computer tomography reveals further details of *Excalibosaurus*, A putative ancestor for the swordfish-like ichthyosaur *Eurhinosaurus*. *Journal of Vertebrate Paleontology*. **9**, 269-281.
McGOWAN, C. 1989b. The ichthyosaurian tailbend: a verification problem facilitated by computer tomography. *Paleobiology*. **15**, 429-436.
McGOWAN, C. 1990. Computer tomography confirms that Eurhinosaurus (Reptilia: Ichthyosauria) does have a tailbend. *Canadian Journal of Earth Science*. **27**, (11), 1541-1545.
McGOWAN, C. 1991. An ichthosaur forefin from the Triassic of British Columbia exemplifying Jurassic features. *Canadian Journal of Earth Science*. **28**, (10), 1553-1560.
SAVRDA, C. E, BOTTJER, D. J and GORSLINE, D. S. 1985. An image-enhancing oil technique for friable, diatomaceous rocks. *Journal of Sedimentary Petrology*. **55**, (4), 604-605.
SCHMIDT, R. A. M. 1952. Microradiography of microfossils. *Science*. **115**, 91.
WHYBROW, P. J. 1982. Preparation of the cranium of the holotype of *Archaeopteryx lithographica* from the collections of the British Museum (Natural History). *Neues Jahrbuch für Mineralogie, Geologie und Paläontologie, Monatshefte*. H3, 184-192.
ZANGERL, R. 1965. Radiographic techniques, pp 305-320: In; B. Kummel and D. Raup (eds), *Handbook of Paleontological Techniques*. W. H. Freeman and Co., (San Francisco). xiii + 852p.
ZANGERL, R and SCHULTZE, H-P. 1989. X-radiographic techniques and applications, pp 165-178: In; R. M. Feldman, R. E. Chapman and J. T. Hannibal (eds), *Paleotechniques*. The Paleontographical Society Special Publication. **4**. iv + 358p.

PART III

LABORATORY TECHNIQUES

E. EXHIBITION OF FOSSIL MATERIAL

36. Fossil replication.

37. Macro and microphotographic techniques.

38. Fossils and illustrations for display and publication.

36. FOSSIL REPLICATION TECHNIQUES

36.1. INTRODUCTION

Replication is an important tool to the palaeontologist, enabling high quality positive and negative impressions to be obtained (Linsley 1965). The accurate replication of specimens requires a number of specialized moulding and casting procedures, with the replicates complexity and working conditions influencing the technique's degree of sophistication. Field applications require simplified and rapid procedures, while in controlled laboratory environments more elaborate techniques are employed. Many of the procedures and techniques commonly used by geotechnologists have their origins in artistic modelling - in particular wax and clay sculpting, metal and resin casting.

Although most commonly used by palaeobiologists in the study of fossils, other applications within earth sciences are documented. Garner (1953) prepared rubber latex casts of animal burrows, while Allen (1971) used Plaster of Paris, polyvinychloride and latex in the replication and preservation of sedimentary structures that had been formed experimentally in a flume tank. Frost and Pettifer (1975) simulated tectonic strain in fossils by deforming a flexible mould of an undeformed fossil, casting in plaster, and measuring the amount of deformation. A modified procedure, with the inclusion of a strain ellipsoid in the mould, deforming fossils three-dimensionally, models the effects of homogenous and inhomogenous strain (Frost and Pettifer 1975).

A range of replication techniques has been described, discussed and compared by palaeobiologists following the introduction of the procedure a little over sixty years ago (*e.g.* Fischer 1939, Fuehrer 1938, 1939, Quinn 1940, Wittard and Sisson 1940). Recent comprehensive reviews of procedures and product availability (*e.g.* Rigby and Clark 1965, Allman and Lawrence 1972, Rixon 1976, Kelly and McLachlan 1980, Chaney 1989b, Converse 1989, Goodwin and Chaney 1994) testify to the fact that, until recently, products specifically designed for palaeontological use were not available. Consequently, many of the techniques are adapted for macro- and micro-palaeontological use (for macrofossils see part II, FIELD TECHNIQUES: section 5 CONSOLIDATION, STABILIZATION AND REPLICATION TECHNIQUES, and for microfossils see section 32 SPECIALIST TECHNIQUES USED IN THE PREPARATION OF INDIVIDUAL MICROFOSSIL SPECIMENS). This has usually required some form of modification to the products viscosity, cure time and active pot life (the time the mixed compounds can used before the onset of curing), critical factors when the procedure is adapted for field use.

Within this section the main products used in moulding and casting will be described, although greater emphasis will be placed on the use of flexible Room Temperature Vulcanising (RTV) silicone rubbers, the most commonly used product in moulding, and Glassfibre Reinforced Plastic (GRP) resins used in casting. The technique's success, applicable in field and laboratory use, usually follow a three stage approach:

(i) Preparing the surface of the original by cleaning and stabilizing.

(ii) Selecting the appropriate moulding and casting materials. The following criteria should be considered:

- single layer application for ease of use;
- rapid cure time;

- low shrinkage, and good undercutting properties;
- production of a high quality cured product.

(iii) Finishing procedures (*e.g.* painting) for photography and display.

As field procedures have been detailed earlier (see part II, section 5 FIELD TECHNIQUES: CONSOLIDATION, STABILIZATION AND REPLICATION TECHNIQUES), this section will centre on laboratory applications.

36.2. MOULDING AND CASTING MATERIALS

A review of moulding and casting literature reveals a wide variety of materials have been used in the duplication of specimens. Many of these materials (*e.g. Vinagel* putty, gelatine, beeswax and paraffin wax, modelling clays, dental plasters, etc.) are no longer manufactured or used in palaeontological preparation work, having been superseded by superior quality products, in particular silicone rubber (although some RTV brands have been discontinued by manufacturers). Details of these products, materials and techniques can be found in Rigby and Clark (1965), Allman and Lawrence (1972), Rixon (1976) and Chaney (1989).

Stanley (1975) compares the merits of the most commonly used materials and techniques with *Alginate* impression compounds, commonly used for oral impressions by dentists. Although this product is ideal for rapid "one-off" and field reproductions as it sets quickly ("slow setting" type: pot-life 3-4 minutes, set 8-10 minutes, "normal setting" type: set 2-3 minutes), moulds are susceptible to tearing during removal, and fungal attack if stored for longer than a week. Furthermore, mould quality deteriorates when multiple casts are produced, particularly if plaster of Paris is used for making the replica. The exothermic reaction associated with the plaster setting contributes to the deterioration of the *Alginate* mould, by accelerating the process of desiccation and shrinkage (Conrad 1989). This technique must be carefully assessed before use, and the mixing of compounds to the correct constituencies practised before commencing. Highly viscous mixtures usually set very quickly, however if insufficient powder is added to the water curing is inhibited, and the mould lacks strength (Conrad 1989). Water is an excellent release agent for the specimen from the rubber compound. Before commencing the procedure, spray the specimen surface with a fine mist. This is particularly important when undertaking the procedure on a hot day in the field (Conrad 1989). Sharpe (1990) describes additional applications and tips on preparing the specimen and compound.

36.2.1. RTV silicone rubber

Safety note: RTV rubber compounds and catalysts must be stored in a cool dry environment. Storing below 25°C increases shelf life beyond six months. Catalysts are particularly harmful if swallowed, and can be irritants when in contact with skin or eyes. Mixing of moulding materials must be undertaken in well-ventilated areas, avoiding inhalation of alcohol rich vapours during silicone rubber curing.

RTV's are considered by many to be the most versatile, and easiest to use moulding materials available. They are pourable compounds that solidify upon the addition and mixing of a catalyst. The fluid catalysed rubber compound is poured or sprayed over the specimen and left to cure. Once solidified, the replicate mould can be carefully removed.

The strength, quality and time taken to produce the mould is controlled by the ratio of the rubber:catalyst system, and the addition of any additives (*e.g.* colouring pigment, thixotropic paste, Anon 1990a) to the system. Materials are relatively easy to use, producing high quality reproductions that do not shrink or deteriorate with age (and can tolerate a wide range of environmental storage conditions), and remain unaffected by water or dilute acids (Kelly and McLachlan 1980). *Rhodorsil* silicone rubbers, manufactured by Rhone-Poulenc (see supplier's list) have been specifically tested for archaeological and palaeontological. Techniques are detailed below.

36.2.2. Latex rubber

Safety note: No toxic fumes are produced from latex, although the presence of ammonia in the compound produces an unpleasant odour. Avoid getting coloured latex compounds on clothing, as it may permanently stain. Clean brushes and mixing utensils immediately after use in running water. Dried and cured rubber can be removed easily by peeling off. An opened latex container has a shelf life of 12-18 months, although is best used within 3 months. Store the compound in a cupboard at 7-21°C, and never allow it to freeze.

Prior to the introduction of silicone rubber systems, natural latex moulds were the most commonly used, and the technique has changed little over the years (compare Fischer 1939 with Parsley 1989). Moulds are built up by the addition of successive latex coats, beginning with a thin coat and, once dried (approximately 2 hours), the subsequent addition of thicker coats (allowing at least 1 hour between each coat). The layers can be reinforced by the addition of a gauze mesh or open weave backing. The production time for high quality moulds is long in comparison with silicone rubber, as each layer must dry before the next is applied. However, despite the use of high intensity lamps in hastening the drying process, the entire procedure is very time consuming (Parsley 1989). Care is required in ensuring a uniform thickness of each application over the entire specimen (otherwise mould distortion may occur on removal).

Further disadvantages are, unfortunately many. These include the excessive shrinkage of moulds during curing and storage (Chaloner and Gay 1973), eventually resulting in the rubber perishing (Stanley 1975). Moreover, latex tends to react with certain other oil based materials (Gunther *et al.*, 1979, Kelly and McLachlan 1980). Care must be taken to avoid a latex to latex contact in recently cured moulds, as they immediately adhere unless the surface is wet or dusted with talcum powder (Stanley 1975). Moulds deteriorate following the production of replicas, and are not suitable for polyester resin casting, although they are adequate for experimental enlargement (see 36.7 MOULD ENLARGEMENT below).

Heaton (1980) maintains that suitable types of latex (containing 72% solids), correctly coloured (red, orange, green or grey) to ensure the best contrast, "are more amenable to changes in, or development of, techniques to offset the disadvantages". He also describes a method of latex casting in natural moulds in claystone nodules. The latex is applied using a *Paasche H-1* external-mix airbrush, with the advantages of producing higher quality casts, avoiding surface brush marks, the introduction of air bubbles or detached bristles into the un-cured rubber, and the destruction of the brush after one application. However, ensure that the airbrush is of a type that can be dismantled and cleaned after use. Furthermore, Heaton (1980) suggests this technique may prove useful in casting low-relief palaeobotanical compression fossils. For conventional applications and moulding methods see part II, FIELD TECHNIQUES: section 5 CONSOLIDATION, STABILIZATION AND REPLICATION TECHNIQUES.

36.2.3. Thermo-setting vinyl rubber

Safety note: Cured thermo-rubbers are generally harmless, although people with sensitive skin should avoid any prolonged contact. Application of a barrier cream and gloves is advisable. During melting some fumes are generated, which although unpleasant, are harmless. To avoid any irritation, undertake melting in a fume cupboard. Overheating the rubber compound will result in decomposition fumes, from which a health hazard cannot be excluded. During its molten state thermo-rubbers are a fire risk. Any fire should be smothered with dry sand. Skin contact with molten rubber can cause severe burns. Affected areas should be held under running water until medical attention can be sought.

Thermal polyvinyl chloride compounds, termed "plastic rubber" by Rigby and Clark 1965, are ideally suited for the production of flexible moulds. They melt on heating, are pourable, and cool to a tough flexible rubber. Moulds are produced relatively quickly, and cheaply with an advantage over RTV silicone rubber that materials can be recycled. However, in comparison with silicone rubber, mould quality is poor (Vernon 1957), while in common with latex moulds thermo-setting vinyl rubber tends to decay over time, and with use following multiple castings.

The procedure can be used on non-porous specimens requiring the minimum preparation, and with slight modification, porous specimens (Rowland 1961). Light coloured porous specimens may be subject to slight discoloration following contact with oils within the rubber (Stanley 1975). The technique has been widely described, using compounds with a variety of brand names (*e.g. Green Mold Plastiflex*, Vernon 1957, *Vinamold*, Allman and Lawrence 1972, Frost and Pettifer 1975, Keyes 1959, Rowland 1961). The current vinyl compound available is *Gelflex* (which possesses similar properties to *Vinamold*, enabling the two to be mixed), suitable for the production of a range of flexible moulds. *Gelflex* is available in either a soft or hard form in comparison with seven forms of *Vinamold* (Allman and Lawrence 1972). Additional support in the form of a mother mould (a rigid plaster or polyester support) is necessary to maintain the flexible mould in a proper configuration. Application is as follows:

1. Cleaned specimens with at least one flat surface should be placed on a smooth glass or metal plate, or in a flat bottomed container with sides twice the height of the specimen. The most important surface should be uppermost. Uneven undersides can be embedded in plasticine, ensuring the plasticine does not protrude around the edges. In irregular shaped examples, where all surfaces of the specimen are important, technical ingenuity may be required in designing a support. Most

commonly used are pins, positioned so as not to leave any impression on the surface of the completed cast. Alternatively, specimens can be suspended by a light thread attached to a retort stand, and secured to the base by a weight embedded in the plasticine.

Keyes (1959) avoided using this technique on porous specimens in which the rubber might penetrate too deeply for removal without damaging the specimen or tearing the mould. However, to assist in moulding non-porous light specimens, Rowland (1961) employs a small vacuum desiccator attached to a simple water pump. Specimens are secured to the base by embedding in plasticine, or suspended by thread enclosed within glass tubing. This ensures that the specimen will not float within the liquid rubber.

2 Around the supported specimen, a corral or dam must be constructed to retain the liquid rubber. A strip of plasticine forming a ring around the specimen, leaving a minimum gap of 6cm between the specimen and the plasticine, provides an ideal bed for strong cardboard or a large diameter plastic bottle with the base cut off to be embedded into. Mould the plasticine to ensure there are no gaps for liquid rubber to leak between the two. If cardboard is used seal the upright edge with sellotape. The height of the retaining corral must be sufficient to provide a minimum of 6cm clearance (effectively providing the inverted mould with a strong 4cm supporting base).

3 The rubber is prepared for use by cutting it into small $1cm^3$ cubes, and placing it in a heating vessel. Small quantities can be melted in a saucepan over a bunsen flame or on a laboratory hot-plate, however, this is not to be recommended as the rubber compound easily burns (at 180°C), destroying its properties and rendering it unusable. Properly constructed, thermostatically controlled heating vessels are commercially available (by far the largest expense of the technique), and ensure heat is evenly distributed throughout the liquid rubber, prolonging its life. The rubber melts at temperatures of between 150-160°C, and is pourable at 140-150°C. By following the manufacturer's recommendations for use, thermo-rubbers can be reused between seven to ten times (Anon 1980).

4 Gently pour the liquid rubber into the mould, allowing it to rise up the sides and flow over the top of the specimen to a depth of about 4cm. **Do not** pour the rubber directly onto the specimen, as this may cause hot rubber to splash onto hands, and trap air pockets around the base and sides of the specimen. Gently tap the side of the mould to assist in the settling of the rubber and dislodge small air bubbles. These will be seen to rise to the surface of the rubber. Allow the mould to cool and set (probably best left overnight in a dust free area). Moulds prepared in a vacuum desiccator should be half-filled with liquid rubber, evacuated for 3-4 minutes, degassing the rubber and causing it to rise to about 1.5cm from the top of the corral. Following the release of the vacuum, top up the mould with additional liquid rubber.

5 Following solidification, the outer supporting corral can be removed, and the contents removed from the base plate. Gently ease the specimen out of the mould. The elastic properties of these rubber compounds allow for considerable stretching for the removal of irregularly shaped objects. Take care with sharp projections on the original specimen as these may tear the rubber.

6 Before taking a cast, the mould may require a light wash with detergent to remove any small particles. **Do not** use organic solvents to clean the moulds, as they may cause swelling and distortion (Anon 1980).

36.2.4. Glassfibre Reinforced Plastic (GRP)

Safety note: Most of the compounds, catalysts, hardeners and additives used in polyester and epoxy resin work contain volatile solvents and generate nauseous fumes that can induce dizziness, headache, and irritation of mucous membranes. All mixing procedures must be undertaken in a well-ventilated area, away from naked flames. Avoid ingestion, as this can cause severe irritation of the alimentary tract, and the preparation of food or beverages in the work area. Contact with skin or eyes can cause severe irritation. Appropriate PPE must be worn, and medical attention sought in case of accident.

Resin compounds (epoxy and polyester) are commonly used in casting (see below). However, when used for constructing contact moulds the preparation techniques are similar to those described in making rubber latex moulds, and differ in only two respects. Firstly, the laminating material is a two-part polyester resin (compound and catalyst, ± additives; *e.g.* colouring paste, thixotropic paste, wax styrene). Secondly, the final high quality product is a rigid replica. The best reproductions are obtained from large specimens with large, smooth surfaces (Anon 1995b), and lacking of any areas of undercut (areas of overhang). Very large moulds may require multiple stage construction, initially applying a thixotropically enhanced gel coat before laminating with fibre matting and resin layers (Anon 1983).

36.3. MOULDING: FIELD APPLICATIONS

The main requirements for a field method are:

- Single layer application, requiring a minimum of equipment and a simplified procedure.
- Rapid cure time, for use at temporary exposures or coastal sections.
- Low shrinkage, so that little distortion of measurable data occurs.
- Good undercutting abilities, leading to the production of a high quality mould.
- High strength and low tear (flexibility), allowing easy removal of moulds, whilst retaining a high quality finish.
- Stability in wet environments.

Two part RTV silicone rubbers are best suited for this purpose, although dental impression rubber compounds (*e.g. Alginate*) with hardening times of 8-10 minutes are useful to experiment with for moulding at temporary, unstable of water logged localities. For a more detail discussion see part II, section 5 FIELD TECHNIQUES: CONSOLIDATION, STABILIZATION AND REPLICATION TECHNIQUES.

Climatic extremes encountered during the preparation of compounds can have adverse effects on the pot-life (the maximum time mixed components of base and hardener retain their lowest viscosity, before the onset of solidification). Temperature fluctuations and gusts of wind can disrupt the mixing and pouring, by introducing airborne contaminants and shortening the cure time (Obata *et al.*, 1989, Maceo and Riskind 1989).

36.4. LABORATORY TECHNIQUES WITH RTV SILICONE RUBBER

A large number of silicone rubber products have been tried and tested in palaeontological applications (*e.g.* Benton and Walker 1981, Chaney 1989b, Chaney and Goodwin 1989, Converse 1983, 1989, Kelly and McLachlan 1980, Obata *et al.*, 1989, Watson and Alvin 1976). Unfortunately, many of the products described in these publications are quickly outdated as manufacture's discontinue and modify product ranges. However, regardless of the make of product used or preferred by a preparator, specimen preparation and rubber compound application techniques are similar. The method outlined below is of a general nature, applicable to most two-part rubber compounds.

Silicone rubber moulds are usually constructed as either single or more complex, multiple moulds. The catalysed rubber is usually poured onto the original, although application by brush (as with the latex procedure described above), and compressed air have also been used (Chaney 1989b).

36.4.1. Specimen preparation

1 Carefully clean the surface of the specimen, using a directed jet of air from a compressed air duster (as used in a photographic darkroom). This will remove any dust and loose debris.

2 Any label information (*e.g.* locality data and number) must be recorded separately, and kept in the specimen tray during the moulding process, as it is susceptible to removal and loss (Chaney 1989b).

3 Ensure all specimen surfaces, but particularly those which will come into contact with the moulding compound, are stabilized and consolidated. Chemicals used in these processes must not react with the moulding compound. Poly(vinyl acetate) (PVA), dissolved in either acetone or alcohol, can be used prior to the application of the RTV rubber compound (Chaney 1989b).

4 Remove excessive overhangs, reducing areas of undercut, and temporarily infill holes and grooves which do not interfere with the fossil. This will prevent deep penetration of the rubber, forming a strong bond with the specimen, which might result in damage to either the specimen or mould during the latter's removal. Clay can be used for this purpose, although it may prove difficult to remove on completion of moulding. Chaney (1989b) suggests a plug of cotton

and a thin layer of melted poly(ethylene glycol) (PEG) over the top, while Baird (1955) used a latex and plaster mix to achieve the same results. Both of these methods can be reversed by immersion and washing the original in water. Allow the specimen to dry before proceeding.

5 Small specimens can be placed in a plastic (semi-rigid) plastic container, offering at least 2cms of clearance between the upper most surface of the specimen and the top of the container. For large specimens construct a corral, retaining wall or dike around the area of interest. Methods, using plastic bottles with the base removed, or cardboard embedded in plasticine, as described under thermo-setting vinyl rubbers above are usually adequate to retain the liquid rubber.

6 When all surface preparation work is complete, apply an aerosol coating of release agent (*e.g. Ambersil DP 100*), or brush on (with a soft brush to avoid leaving any marks) a warmed mixture of 5% Petroleum Jelly in White Spirit. If kept in a stock bottle the jelly may solidify, but it can be re-dissolved by standing the container in hot water and gently mixing (Anon 1990). Because silicone rubber bonds well with plasticine it must be initially coated with a layer of shellac before treating with a release agent (Anon 1990a).

36.4.2. One piece moulds

These are the simplest types of mould to construct. Two variations are employed:

(i) Open moulds: are more commonly used in palaeontology, and form the basis of the description below. Specimens are enclosed in a frame or box, and encased in a rubber. Matting reinforcement can be added to strengthen the mould, and additional layers of rubber applied. Plaster or resin backings can also be added.

(ii) Strip-off skin moulds: are occasionally used in palaeontological preparations of very large areas (*e.g.* Obata *et al.*, 1989). A combination of high strength and thixotropic rubbers are used in the construction of the mould. A plaster or resin mother mould is usually required to support the rubber mould before casting can commence. In specimens where there is considerable relief, the finished mould is removed from the original by gently pulling and rolling it off inside out (Anon 1990a).

The method is as follows:

1 Stir the silicone rubber before the addition of catalyst. This reduces the base compound thixotropy, which generally increases during the shelf life of an unopened can, and will ensure the best flow during application.

2 Pour the base compound into a dry clean container and add the catalyst, up to a maximum ratio of 10:1, and mix thoroughly. Ensure all

material adhering to the sides is thoroughly folded in. Mix by hand using either a flat-bladed metal, or wooden spatula, or if preferred an electric mixer can be used, but not at high speeds or over prolonged time periods as this may cause the compound to heat up, resulting in a shortened pot-life.

3 Allow the mixed compound to stand for a few minutes to de-aerate, ensuring no air becomes entrapped at the mould:specimen interface. Trapped bubbles can effect the replica quality resulting in mould weakness, which may lead to tearing upon removal from the specimen surface.

De-aeration can be accelerated by evacuating the mixed rubber compound in a vacuum dessicator for approximately three minutes. If this option is employed it is advisable to put the catalysed rubber mix in a container that will accommodate a 4-5 fold volumetric expansion (Chaney 1989b).

4 Pour and evenly spread the activated silicone rubber, ensuring there is sufficient depth of coverage, over the entire specimen surface. Any bubbles reaching the top surface of the rubber, should be broken with the mixing spatula or a mounted needle.

5 Curing of the rubber will begin immediately the catalyst is in contact with the base rubber compound. The cure cycle can be shortened by gently heating the rubber to 65°C. The volume of mix will effect the cure time, with larger volumes taking longer to cure. Once the silicone rubber mould has cured (is solid), gently ease it off the specimen, so as not to cause any tearing of the skin and maintaining undercuts.

6 Leave the mould open to the air for at least 48 hours before attempting a cast from it.

7 Clean the specimen, removing all remnants of cured rubber, fillers and corral supporting plasticine (see below).

36.4.3. Multiple piece moulds

Occasionally, for the reproduction of internal and external surfaces (*e.g.* skulls), and complex specimens, two or multiple piece moulds are required. These are constructed in the following way:

36.4.3a. First Half

1 Depending on the shape of the original, at least half of it, or to a suitable seam line, should be embedded in a clay platform, the surface of which should have at least four cone-shaped clay plugs or wooden dowels inserted (these provide toggle holes for locating bosses produced in the second half of the mould). Place the original in a frame or

moulding box. Frame supports are preferable, as the entire structure requires inverting for the second part of the procedure. Metal cake tins with removable bases are ideal containers for this method. Ensure the sides of the frame provide adequate clearance for a minimum 1cm thick base of rubber. If this cannot be achieved, a supporting plaster base must be constructed **before** proceeding with the second half of the mould construction.

2 Apply a coating of release agent to the sides of the frame and the exposed section of the original and the surrounding clay base (*i.e.* all areas that come into contact with the liquid rubber). Allow the surfaces to dry.

3 Mix the silicone rubber (as described in one piece moulds above).

4 Gently pour the rubber into one corner of the box until the model is half covered. Gently tap the side to assist the rubber in settling down. Allow the rubber to thoroughly cure.

36.4.3b. Second Half

5 Without removing the mould carefully invert the frame box, exposing the supporting clay. Remove all the embedding clay, clay plugs and dowels and carefully clean exposed areas of the original. It is important to avoid moving the original (although this is nearly always impossible) to reduce the possibility of forming unsightly seams (these are always difficult to remove from casts). Thin seam lines obscure less detail, and where possible they should be situated to coincide with less important surfaces and features.

6 Re-apply a coating of release agent to the frame sides, the exposed original and the cured silicone rubber (now forming the supporting base of the original). Allow the surfaces to thoroughly dry.

7 Mix a fresh batch of silicone rubber, and following de-aeration, carefully pour the rubber into the corner as described above. When the specimen is covered (to a minimum depth of 1cm), gently tap the sides of the frame to assist in settling the rubber. Allow the rubber to cure.

8 Remove and separate the two halves of the mould from the frame. Thoroughly clean the specimen, ensuring all remnants of cured rubber and release agents are removed.

If the original specimen was not in contact with the frame support (as would be the case in the construction of very complex multi-piece moulds), vent holes, slits or sprue (openings in the mould in which the casting material was poured) are required. These are necessary if casting from the mould is required. As an alternative to cutting the mould, Chaney (1989b) suggests inserting a clay plug into one side of the mould during construction.

When undertaking moulding procedures, always consider the possibility of achieving the desired results with a simple single piece mould. Constructing multiple-piece moulds is always a time consuming process. A great deal of care and thought must be applied during preparation, deciding where seams should be placed and the position of the pouring hole for casting. The complexity of the original specimen may facilitate the need for a mould composed of more than two elements (Smith and Latimer 1989a). This, of course, adds stages in mould preparation, compounding the possibility of introducing erroneous unsightly seams and surface blemishes to the finished cast. Besides the problem of getting the completed mould components mould to fit together, it is also very difficult to construct a supporting jacket. This may require as much thought and ingenuity as the construction of the mould itself. For a detailed account of production problems associated with two and three or more part moulds, refer to Smith and Latimer (1989a).

36.4.3c. Post moulding cleaning of specimen

A time consuming procedure of moulding is the cleaning of the original specimen to its pre-mould condition. This involves the tedious, but necessary procedure of removing small infills of silicone rubber, plasticine and fillers. Poly(ethylene glycol) plugs are easily removed by gently heating the specimen, and finally washing the surface in warm water. Clay fillers can be loosened and removed by the careful directing of a high pressure stream of water or using an ultrasonic descaling pen (see section 12 MECHANICAL METHODS FOR PREPARING FOSSIL SPECIMENS). Any broken pieces of the specimen must be carefully re-attached.

36.5. CASTING: FIELD APPLICATIONS

The detail required in casts prepared from moulds, and the time for casting compounds to cure, plus unfavourable environmental conditions, usually prohibit casting of specimens under field conditions. However, large natural moulds, such as dinosaur footprints, which cannot be removed to the laboratory, may have to be cast in the field (Heintz 1963).

Most objects are cast in a solid compound of either plaster or a polyester resin. These are ideal for simple specimens with little or no undercut. More complex moulds, particularly those which cannot be extracted and brought back to the laboratory, must be cast using a flexible compound such as silicone rubber (*e.g.* Johnson and McKerrow 1995). Many of the compounds used in mould preparation are employed in casting.

36.6. LABORATORY TECHNIQUES

36.6.1. Plaster casts

Gypsum based plasters, generally referred to as Plaster of Paris, have been used in fossil reproduction considerably longer than moulding compounds (*e.g.* Green 1832, Ward 1866). Plaster compounds are easy to use and simple to mix. A variety of compounds are now available, including acrylic plaster polymers, and used instead of water to mix up the plaster. High alumina cements (*e.g. Ciment Fondu*) are fast setting and extremely hard. They too can be mixed with a waterproof acrylic plaster polymer. Fast setting and hard casting plasters are also widely used in palaeontological preparations. Additional hardeners, colouring powders and pigments, can be introduced while preparing the mix. Hardened casts are painted on completion. Most of the quality sculpting plasters will accurately reproduce the fine detail of the original mould, leaving the type of plaster chosen governed by the use to which the cast is intended. For example, casts of fossils prepared for teaching purposes must be extremely hard and durable, as should exhibition specimens in which there is a possibility of frequent handling. Fine surface detail may be more important for specimens cast for research purposes.

Plasters are commercially available in a variety of grades and formulations (Babcock 1989). The three most common types available are summarised below:

Moulding plaster: composed of pulverised, calcined (roasted) gypsum in which the water has been removed. Rehydration of the powder recovers water lost during heating. Following thorough mixing with water, the plaster sets to a uniform, inert, solid mass, with essentially a similar composition as the original gypsum.

Dental plaster: similar in composition to moulding plasters, with the addition of additives that accelerate the setting time.

Gypsum cements: differ from moulding plasters in that the gypsum is more finely ground, and therefore more water is removed during the roasting stage. The addition of alum (Aluminium potassium sulphate, $AlK(SO_4)_2.12H_2O$) or borax (*di*-Sodium tetraborate *10-hydrate*, $Na_2B_4O_7.10H_2O$) saturated salt solutions (1 part to 20 parts water) adds strength to the finished product. On setting, gypsum cements are finer grained, harder and denser (less porous) than either moulding or dental plasters. They are ideal for palaeontological reproductions where fine detail is important.

Procedures used in geological applications have been described by Babcock (1989) and Spence (1960). A composite procedure is as follows:

1 Wash the mould to remove fine dust and dirt using soap and water.

2 Use the mould to estimate the volume of water required, *i.e.* fill it with water and pour it into the mixing bowl. This avoids mixing too smaller or excessive and wasteful amounts of plaster. When two or more batches of plaster are used the cast will develop weak seams (Babcock 1989).

3 Use only water at room temperature to mix the plaster, as warm water will cause the plaster to set quicker. Cold or chilled water slows the setting procedure. Add the plaster to the water by gently sprinkling or sifting small quantities on to the water. Do not mix immediately, but allow it to stand for two minutes, or until the plaster becomes thoroughly saturated. Trapped air in the plaster is released, reducing the risk of imperfections within the finished cast. Gently stir the mixture. Water added to plaster results in a lumpy mixture that is very difficult to work smooth. Ratios of water to plaster vary according to the compound formulation and type (Babcock 1989). This directly affects the density, hardness and strength of the resulting cast.

4 Continue adding plaster until a crust is formed, and smoothly stir until the mixture has the consistency of cream. Rest the mould on a flat surface. When poured the mixture should readily flow, running into grooves and undercuts. Avoid excessive mixing, as although this increases the strength of the cast, it reduces the setting time. Also, avoid removing the mixing spatula from the plaster, as this introduces air bubbles into the mixture, weakening the cast. Once the mould is full, pour the plaster back into the mixing basin. This first pouring ensures a wet plaster coating of the entire inner surface of the mould, and is particularly important when casting into small moulds (Rowland 1960).

5 Carefully re-pour the plaster back into the mould, and gently tap the sides to force entrapped air and bubbles to the surface.

6 Cover the pouring vent with a piece of clean glass, and wipe off any excess plaster before the cast hardens.

7 Once set, plaster casts are easily removed from flexible moulds by carefully distorting the mould to free the cast. Removal involves gently pulling the cast away from the mould. If using rigid moulds, first coat the surface with a release agent

8 Any slight surface imperfections and small air bubbles can be repaired by (a) wetting the surface, and (b) filling with freshly mixed plaster (Babcock 1989).

Complete drying of the cast depends on its size, and whether it is air or oven dried. Babcock (1989) recommends painting with water colours, although colour intensity may become diffuse as the paint soaks into the plaster, alternatively apply a thin coat of acrylic or oil based paint, as too thick a coat will obscure fine detail on the cast surface. Additional cast painting procedures are described by Tiffany and Iwama (1994).

36.6.2. Resin casts

Resin casts are considerably more durable than plaster casts, making them excellent compounds for reproduction techniques. Glass-fibre reinforced plastics (GRP's) have been used in palaeontological reproductions for over thirty years (Rixon and Meade 1960). Care must be taken in mixing the base and catalyst, with the operator wearing appropriate PPE, mixing and using components in a well-ventilated area. It is also wise to test the compatibility between the moulding compound and the casting resin before undertaking the mass production of casts (Smith and Latimer 1989b), and to determine if a release agent must be applied to the mould. The mixing of all compounds with catalysts should be undertaken using disposable mixing pots with disposable spatulas.

Large vertical casts may require the construction of a frame to support the mould in a position reflecting the natural attitude and curvature (Obata *et al.*, 1989). The "hand lay-up" procedure for constructing two part replicas (2x6m, 2x4m) adopted by Obata *et al.*, (1989) is summarised as follows:

1. Erect the mould with a curvature reflecting that of the natural outcrop.

2. Prepare a plastic resin mix (140kg resin), with the addition of gypsum or carbonate of lime (60kg), and colouring agents. For the best results and reproduction of detail, always use the manufacturers recommended ratios for mixtures of resin, catalyst, colouring paste and any additional accelerator.

3. Apply a thin resin coat by brush onto the mould, and a layer of fibreglass matting. Repeat the process to strengthen the cast.

4. Coat the back with more additional resin and matting until the cast is rigid. Insert additional supports to the frame if required.

5. Allow the resin to cure. Remove the mould and frame support.

6. Apply finishing touches to the surface, and any additional colouring to areas as required.

Good quality casts and mounts are used in museum displays, particularly for the construction of large skeletons. Solid casts require a supporting framework, and on completion can produce heavy cumbersome structures. Hollow steel and aluminium tubing provides an effective alternative, which can be concealed in casts by drilling holes through the bones. Croucher and Howie (1976) describe a light weight structure of "resinated glass-fabric tubing" supporting hollow GRP casts of the bones. The plastic tubing was calculated to be approximately 60 % lighter than equivalent steel tubing, and 30% lighter than aluminium. This type of tubing used in combination with hollow GRP casts, provides the principal advantage in favour of casting with resins as opposed to plaster. Other advantages and disadvantages of casting with plaster and resin compounds of are summarised in table 36.1.

Although the drilling of solid resin casts is a relatively straight forward procedure, specialized rotary equipment is required in the production of single piece hollow casts (Lanooy 1984, Smith and Latimer 1989c). Plaster (Quinn 1940), epoxy and polyester resins (Schrimper 1973), polyacrylic resins (Burke *et al.*, 1983), and polyurethane's (Smith and Latimer 1985) have all been successfully used in the production of hollow casts. The advantage of this procedure is that it uses considerably less casting material, and is thus more economical in producing a strong, light weight product. Procedures are not simple, requiring considerable patience and practice to produce acceptable results.

Examples of other casting and modelling compounds used in museum displays of palaeontological specimens (*e.g.* asbestos mâché, paper mâché, foam with resin laminates, modelling clays, beeswax, poly(vinyl chloride)) are described, along with the more conventional compounds (*e.g.* plaster, latex and silicone rubbers, polyester and epoxy resins), in detail by Chase (1979).

36.7. MOULD ENLARGEMENT

Natural latex, polybutadiene rubber and silicone silastomer moulds and casts can be enlarged, in some cases permanently, following solvent immersion. This feature was

CASTING MEDIUM	ADVANTAGES	DISADVANTAGES
PLASTER	No deterioration over time. Easy to mix. Quick setting. Universally available. Low expansion of coefficient (<0.4%). Inexpensive	Not suitable for light weight casts, or in moulds with numerous undercuts. Breakable. Relatively soft, surfaces easily scratched or marked, particularly if not sealed.
RESIN	Permanence and durability of casts (particularly useful if casts are required for teaching, where they undergo extensive handling). Cured compounds are very stable. Large, light weight casts can be produced.	Degrades the mould quickly, therefore fewer casts can be made from one mould. Products are expensive. Safety hazards in mixing, cutting and grinding resin compounds.

Table 36.1. Advantages and disadvantages in using plaster and resin casting mediums. (Data from Babcock 1989, and Smith and Latimer 1989b).

first described for latex soaked in kerosene (paraffin oil) by Fuehrer (1939), and subsequently reported in Rigby and Clark (1965) and Wilson (1989). Rixon (1972) reported similar findings for silicone silastomer casts. Gunther *et al.*, (1979) describe the enlargement of moulds and casts made of poly(butadiene) rubber by soaking in toluene, noting that the most satisfactory results were achieved when working with natural moulds or negative specimens, suggesting it may be necessary to make a reverse mould from a positive part before attempting any enlargement. Furthermore, low relief specimens will produce better results, enlarging more uniformly than moulds and casts with strong relief and angular surfaces (Wilson 1989). Internal reinforcing of the rubber (*e.g.* cloth or fabric matting) must be avoided, as this will result in uneven distortion of the rubber during enlargement. The procedure is particularly useful for enlarging detail of small specimens for both museum displays and teaching purposes, enabling ontogenetic stages and differences in shape to be compared (Wilson 1989).

The method for enlarging latex rubber moulds and casts is as follows:

Safety note: solvents used in this procedure are flammable and potentially explosive, while fumes are respiratory irritants. This procedure should only be undertaken in a well-ventilated area with fume extraction facilities.

1 Place the "original" cast or mould in a solvent resistant container, sufficiently large enough to accommodate the enlarged specimen, with the detailed surface uppermost.

2 Immerse the specimen in a sufficient quantity of kerosene so that it remains covered even when enlarged. Cover the container to avoid excess loss of the solvent through evaporation. Periodically monitor the process of enlargement. The time required for obtaining the desired size is related to the thickness of the original rubber mould or cast. A doubling in size may occur in less than two hours, or require up to 24 hours (Wilson 1989). Enlargements over 50% should be avoided, as the mould structurally deforms and weakens (Wilson 1989).

3 When the desired enlargement is obtained, remove the specimen from the kerosene, and thoroughly wash and dry. Latex enlargements of less than 50%, if stored correctly, have a shelf life equivalent to that of standard latex moulds (Wilson 1989).

4 Standard plaster or resin replicas can be made from the enlargement (these should be undertaken immediately if silicone silastomers and poly(butadiene) rubbers are used, as both are liable to shrink and distort on removal from the solvent, or if excessive amounts of solvent evaporate from the surface before casting (Gunther *et al.*, 1979)), and second or third generation latex moulds produced for further enlargement. The procedure can be repeated indefinitely.

REFERENCES

ALLEN, J. R. L. 1971. Some techniques in experimental sedimentology. *Journal of Sedimentary Petrology* . **41**, (3), 695-702.

ALLMAN, M and LAWRENCE, D. F. 1972. *Geological Laboratory Techniques*. Blandford Press, (London). 1-335.
ANONYMOUS. 1980. *Vinamold: flexible mould making*. Alec Tiranti Ltd. (Theale) Technical Booklet. 17p.
ANONYMOUS. 1983. *Crystic polyester handbook*. Scott Bader Co. Ltd. (Wellingborough). x + 111p.
ANONYMOUS. 1990a. *Cold cure silicone rubber for mould making*. Alec Tiranti Ltd. (Theale) Technical Booklet. 24p
ANONYMOUS. 1995b. *Polyester resin, glass fibre, cold cast resin metals, embedding resin*. Alec Tiranti Ltd. (Theale) Technical Booklet. 30p.
BABCOCK, L. E. 1989. Casting with plaster of Paris, pp 320-325: In; R. M. Feldman, R. E. Chapman and J. T. Hannibal (eds), *Paleotechniques*. The Paleontological Society Special Publication. 4. iv + 358.
BAIRD, D. 1955. Latex micro-moulding and latex plaster moulding. *Science*. 122, 202.
BENTON, M. J and WALKER, A. D. 1981. The use of flexible synthetic rubbers for casts of complex fossils from natural moulds. *Geological Magazine*. **118**, (5), 551-556.
BURKE, A. C, ANDERSON, M, WELD, A and GAFFNEY, E. S. 1983. The reconstruction and casting of a large extinct turtle, *Meiolania*. *Curator*. **26**, (1), 5-26.
CHALONER, W. G. and GAY, M. M. 1973. Scanning electron microscopy of latex casts of fossil plant impressions. *Palaeontology*. **16**, (3), 645-649.
CHANEY, D. S. 1989b. Mold making with room temperature vulcanizing silicone rubber, pp 284-304: In; R. M. Feldman, R. E. Chapman and J. T. Hannibal (eds), *Paleotechniques*. The Paleontological Society Special Publication. **4**. iv + 358.
CHANEY, D. S and GOODWIN, M. B. 1989. RTV silicone rubber compounds used for molding fossil vertebrate specimens: a comparison. *Journal of Vertebrate Paleontology*. **9**, (4), 471-473.
CHASE, T. L. 1979. Methods for the preparation of palaeontological models, pp 225-267; In: M. G. Bassett (ed), *Curation of palaeontological collections*. Special Papers in Palaeontology. **22**, 1-279.
CONRAD, K. 1989. The use of Alginate for footprint replication, pp 417-418: In; D. D. Gillette and M. G. Lockley (eds), *Dinosaur Tracks and Traces*. Cambridge University Press, (New York). xvii + 454p.
CONVERSE, H. H. 1983. A new procedure in silicone rubber moldmaking. *Bulletin of the Society of Vertebrate Paleontologists*. **128**, 65-66.
CONVERSE, H. H. 1989. *Handbook of paleo-preparation techniques* (2nd edition, revised by R. W. McCarty). Florida Paleontological Society. v + 109p.
CROUCHER, R and HOWIE, F. M. P. 1976. A G.R.P cast and mount of the Tertiary mammal *Arsinoitherium*. *Newsletter of the Geological Curators Group*. **1**, (7), 348-368.
FISCHER, A. 1939. Rubber casts and moulds of fossils. *Journal of Paleontology*. **13**, (6), 621.
FROST, M. J and PETTIFER, G. S. 1975. Simulation of deformed fossils. *Geology Teaching*. **7**, 153.
FUEHRER, O. F. 1938. Liquid rubber, a new casting medium. *Museum News*. **15**, (16), 7.
FUEHRER, O. F. 1939. Liquid rubber as an enlarging medium. *Museum News*. **16**, (14), 8.
GARNER, M. R. 1953. The preparation of latex casts of soil cavities for the study of tunnelling activities of animals. *Science*. **118**, 380-381.
GOODWIN, M. B and CHANEY, D. S. 1994. Molding and casting: techniques and materials, pp 235-271 in Chapter 10, Molding, casting, and painting: In; P. Leiggi and P. May (eds), *Vertebrate Paleontological Techniques*, Volume 1. Cambridge University Press. 368p.
GREEN, J. 1832. *A monograph of the Trilobites of North America: with coloured models of the species*. Published by the author. (Philadelphia). 94p.
GUNTHER, V. G, GUILLOT, D. G and RIGBY, J. K. 1979. Enlargement of casts and molds of polybutadiene rubber. *Journal of Paleontology*. **53**, (3), 742-744.
HEATON, M. J. 1980. New advances in latex casting techniques. *Curator*. **23**, (2), 95-100.
HEINTZ, N. 1963. Casting dinosaur footprints at Spitsbergen. *Curator*. **6**, 217-225.
JOHNSON, M. E and McKERROW, W. S. 1995. The Sutton Stone: an early Jurassic rocky shore deposit in South Wales. *Palaeontology*. **38**, (3), 529-541.
KELLY, S. R. A and McLACHLAN, A. 1980. The use of silicone rubbers in the preparation of casts from natural fossil moulds. *Geological Magazine*. **117**, (5), 447-454.
KEYES, I. W. 1959. Paleontological casting and moulding techniques. *New Zealand Journal of Geology and Geophysics*. **2**, (1), 56-65.
LANOOY, R. 1984. A new casting method with a modified mold rotator. *Der Präparator*. **30**, (4), 381-388.
LINSLEY, R. M. 1965. Collecting molds of fossils, pp 189-191: In; B. Kummel and D. Raup (eds), *Handbook of Palaeontological Techniques*. W. H. Freeman and Co., (San Francisco). xiii + 852p.
MACEO, P. J and RISKIND, D. H. 1989. Field and laboratory moldmaking and casting of dinosaur tracks, pp 419-420: In; D. D. Gillette and M. G. Lockley (eds), *Dinosaur Tracks and Traces*. Cambridge University Press, (New York). xvii + 454p.
OBATA, I, MARUO, H, TERAKADO, H, MURAKAMI, T, TANAKA, T and MATSUKAWA, M. 1989. Replicas of dinosaur tracks using silicone rubber and fibreglass reinforced plastics, pp 413-415: In; D. D. Gillette and M. G. Lockley (eds), *Dinosaur Tracks and Traces*. Cambridge University Press, (New York). xvii + 454p.
PARSLEY, R. L. 1989. Latex casting of macroinvertebrate fossils, pp 275-281: In; R. M. Feldman, R. E. Chapman and J. T. Hannibal (eds), *Paleotechniques*. The Paleontological Society Special Publication. **4**. iv + 358.
QUINN, J. H. 1940. Rubber moulds and plaster casts in the paleontological laboratory. *Fieldiana; Field Museum of Natural History, (Chicago), Field Technique Series*. **6**, 3-21.
RIGBY, J. K and CLARK, D. L. 1965. Casting and molding, pp 389-413: In; B. Kummel & D. Raup (eds), *Handbook of Palaeontological Techniques*. W. H. Freeman and Co., (San Francisco). xiii + 852p.
RIXON, A. E. 1976. *Fossil Animal Remains: their preparation and conservation*. Athlone Press, (University of London). vi + 304p.
RIXON, A. E and MEADE, M. J. 1960. Glass fibre resin casts of fossils. *Palaeontology*. **3**, (1), 124-126.
ROWLAND, E. O. 1961. A rapid method for making plaster casts for geological work. *Proceedings of the Geologists' Association*. **72**, (4), 461-466.
SCHRIMPER, G. D. 1973. Hollow casting of fossil skulls in epoxy plastic. *Curator*. **16**, (4), 286-305.
SHARPE, M. 1990. *Plaster waste-moulding, casting and life casting*. Alec Tiranti Ltd. (Theale) Technical Booklet. 52p.
SMITH, J. A and LATIMER, B. 1985. The joy of casting. *Explorer*. **27**, (4), 4-5 & 22-24.
SMITH, J and LATIMER, B. 1989a. Making a multiple-piece mold, pp 305-319: In; R. M. Feldman, R. E. Chapman and J. T. Hannibal (eds), *Paleotechniques*. The Paleontological Society Special Publication. **4**. iv + 358.

SMITH, J and LATIMER, B. 1989b. Casting with resins, pp 326-330: In; R. M. Feldman, R. E. Chapman and J. T. Hannibal (eds), *Paleotechniques*. The Paleontological Society Special Publication. **4**. iv + 358.
SMITH, J. and LATIMER, B. 1989c. Preparation of hollow casts, pp 331-335: In; R. M. Feldman, R. E. Chapman and J. T. Hannibal (eds), *Paleotechniques*. The Paleontological Society Special Publication. **4**. iv + 358.
SPENCE, T. F. 1960. A technique of plaster-casting for museum specimens. *Science Technology*. **6**.
STANLEY, J. W. 1975. Alginate impression compound:- A comparative assessment of a cheap moulding material with potential applications in palaeontology. *Newsletter of the Geological Curators Group*. **1**, (4), 191-193.
TIFFANY, M and IWAMA, B. 1994. Cast painting, pp 271-284 in Chapter 10, Molding, casting, and painting: In; P. Leiggi and P. May (eds), *Vertebrate Paleontological Techniques*, Volume 1. Cambridge University Press. 368p.
VERNON, R. O. 1957. New technique for casting fossils and forming molds. *Journal of Paleontology*. **31**, (2), 461-463.
WARD, H. A. 1866. *Catalogue of casts of fossils from the principal museums of Europe and America, with short descriptions and illustrations*. Benton and Andrews (Rochester). 228p.
WATSON, J. and ALVIN, K. L. 1976. Silicone rubber casts of silicified plants from the Cretaceous of the Sudan. *Palaeontology*. **19**, (4), 641-650.
WILSON, M. A. 1989. Enlarging latex molds and casts, pp 282-283: In; R. M. Feldman, R. E. Chapman and J. T. Hannibal (eds), *Paleotechniques*. The Paleontological Society Special Publication. **4**. iv + 358.
WITTARD, W. F and SISSON, J. E. 1940. Phenolformaldehyde resin as a casting material. *Geological Magazine*. **77**, (6), 478-481.

37. PHOTOMACROGRAPHY AND PHOTOMICROGRAPHY TECHNIQUES

37.1. INTRODUCTION

A critical array of techniques, employed by virtually all palaeobiologists at one point during a research project, are those concerned with the photography and visual display of specimens and data. High resolution image capturing through digital cameras and computer image manipulation techniques are of increasing importance within laboratories undertaking palaeobiological interpretation. Applications within micropalaeontology are well established (Athersuch and Jones 1991). Time-lapse cinematography has also been used in recording reproductive stages and burrowing activities of living benthic foraminifera (Röttger 1984, Wetmore 1988). The increasing use of these media has revolutionised image capturing, manipulation and analysis, and may result in the demise of the "classical" photographer specializing in colour slide (transparency), print and monochrome formats. However, until that time the recording of palaeobiological material through video or conventional photographic slide and negative film techniques will continue.

There are some concerns with the use of digitally enhanced images of palaeobiological material. A pre-requisite of photographing fossil material is that a very accurate rendition of the subject is achieved, introducing no ambiguity in form, texture or detail (Scovil 1996). By virtue of using computer enhancing techniques, authenticity and originality may, inadvertently, be compromised. To avoid this every attempt must be made to ensure the hard copy reflects the form of the original, with any indication of the areas of image enhancement provided in the text or figure caption.

This section describes some of the procedures and techniques used in macro and microphotography, with particular emphasis placed on photographing difficult specimens with low relief and low matrix to fossil contrast. Following a brief outline of the equipment required for photomacrography, contrast enhancement procedures through lighting alterations and specimen coating (dusting or smoking) are described. Plate preparation techniques are also discussed. Electron microscopy photographic techniques are discussed elsewhere (see section 34 ELECTRON MICROSCOPY TECHNIQUES).

Procedures discussed within this section are confined to laboratory based problems. Photography in the field is not covered. A wide variety of well-illustrated references (*e.g.* Bracegirdle 1995, Calder and Garrett 1990, Hicks and Schultz 1997, Jacobson *et al.*, 1988) direct workers to suitable camera and lens systems, film types, and techniques specific to the photography of subjects within the studio or outside. Similarly, darkroom procedures (film processing and printing) are not discussed. Detailing these aspects is beyond the scope of this manual.

Specific photographic techniques are defined by specialist equipment and size range produced (magnification or reproduction ratio). Photomacrography covers magnifications of objects from x1 to x20 (reproduction ratios of 1:1 - 1:20). Magnifications ranging from x0.1 to x1 (10:1 - 1:1) are achieved on camera systems fitted with close-up lenses or by attaching extension tubes. This is termed close-up photography. When photographs are taken with a camera attached to a microscope (stereozoom or compound), the procedure is known as photomicrography. Magnifications of between x20 to x2000 (20:1 - 2000:1) are possible. Retaining resolution at higher magnifications requires the use of an electron microscope. However, distinctions between these photographic techniques are arbitrary, and considerable overlap is used to ensure the optimum results are obtained from any procedure employed. Consequently, high resolution, low magnification views of surface features are best achieved using the SEM, providing material can be appropriately prepared and viewed in the machine.

In comparison to photographing minerals and lapidary materials, the photographing of fossils is a much simpler process (Scovil 1996). Most fossils lack the highly lustrous, planar surfaces, commonly present on mineralogical specimens. Furthermore, besides Holocene and Recent material and exceptionally preserved rarities, the variety of natural colours exhibited by minerals is rarely seen in fossil specimens. The most notable colour differences seen with fossil specimens are those contrasting the fossil with the surrounding and enclosing matrix. The successful exploitation of this contrast difference, coupled with differences in relief, is essential in the production of high contrast images.

A similar philosophy is employed in photomicrography, even with specimens prepared as peels and thin sections. In many optical procedures, colour differences are exploited to assist observations by enhancing contrast differences between grain boundaries, the supporting matrix and the mounting media. Colour differences may result from natural or artificial staining, optical induction through birefringence when viewed in cross polarized light, or fluoresce when illuminated using a non standard light source.

When undertaking either photomacrography or photomicrography it is essential that cameras and lighting settings are correct, and reproducible if required later. It is also essential to use a suitable film compatible with the lighting conditions and subject material. To reduce costs, and familiarize one's self with the camera and lens system, it is prudent to run a test film. The majority of laboratory work is undertaken using a fine grained (50 ASA or less) film (*e.g. Ilford "Pan-F'* and *Kodak 'Panatomic X*), ensuring that when the image is enlarged, the minimum loss of definition occurs. However, McNair (1941) used positive film when photographing low negative relief impressions of bryozoa under even low-angled lighting to assert contrast. Although the printed image reversed the apparent position of the fossil, the procedure proved more successful than the conventional means of photographing a silicone rubber cast (see section 36 FOSSIL REPLICATION TECHNIQUES).

As with most techniques employed in palaeobiology, the highest quality results are obtained from carefully prepared specimens. Photographic enhancement of inadequately prepared material, is very difficult, if not impossible to achieve.

37.2. PHOTOMACROGRAPHY

Specialized equipment used in photomacrography of three dimensional and compression specimens have been extensively described (*e.g.* Gutschick 1960, Sander 1955, Siveter 1990, Whittington 1956, 1960). Although these systems are still extensively used today, much of the equipment described is no longer manufactured. Theoretical considerations in relation to the properties of lenses fall beyond the scope of this manual, although descriptions can be found in other palaeontological publications (*e.g.* Fournier 1956, Rasetti 1946, 1965), and in more detail in dedicated photographic references (*e.g.* Bracegirdle 1995, Jacobson *et al.*, 1988).

37.2.1. Camera and copy stand set up

Key considerations in setting the optical conditions for photography include correctly focusing the lens and setting the correct aperture, influencing light wavelength diffraction (Rasetti 1946). Thin objects (with little relief), are in sharpest focus, and diffraction at its minimum when viewed through a wide aperture (small *F* numbers). Three-dimensional objects, in which focusing is usually on a plane through the middle of the depth of field, require a small aperture setting. This assists in decreasing the effects of the lack of focusing within the field of view. There will, however, come a point when reducing the aperture size is ineffectual, as diffraction impairs the image more than greater depth of focus improves it. Irrespective of the print's final magnification, low relief objects are best photographed larger than natural size, and three dimensional objects smaller than natural size to produce the best defined images (Rasetti 1946).

A wide variety of camera and lens systems are suitable for scientific work. The preferred system used is very much at the discretion and preference of the operator and its availability within a laboratory. It should, however, be remembered that the quality of the finished print is affected by the quality of the camera and lens system. Basic 35mm single reflex camera system requirements have been defined by Feldman (1989a), enabling the photographer to:

- focus on the specimen through the lens,
- have the capability to manually operate settings (exposure and aperture); if a semi-automatic camera is used, setting the aperture-priority exposure mode ensures the camera's microcomputer automatically selects the correct shutter speed to match the aperture selected,
- vary the focal length of the camera by physically changing the lens and the addition of accessory extension tubes.

Exposure times may be long, up to several seconds. It is therefore imperative to secure the camera and specimen to eliminate the effects of vibration during exposure. Carefully attach the camera to either a tripod or the movable arm of a copy stand. If a copy stand is used (camera vertically mounted), a right-angle eyepiece attached to the viewfinder facilitates focusing in a more convenient manner. A copy stand provides a solid base table on which to position the specimen in a stable manner, reducing the effects of vibration. Where horizontal mounting of the camera is preferred, the secure mounting of specimens on the vertical background can be achieved by attaching one or more bar magnets to the back of the specimen using a quick drying polyester cement (Sass 1962). Specimens can be positioned on a metal mounting board, with fabric covering the surface. The magnets can be removed by dissolving the cement in acetone. A cable release to activate the camera shutter is also essential in eliminating camera vibration.

37.2.2. Lighting

Correct lighting is an essential aspect required for the successful production of high quality scientific photographs (Fitzpatrick 1991, Howe 1969, see also table 37.1). With improper or inadequate lighting a variety of problems may manifest. These problems may be compounded by specimen shape, or the specific features that require illustration. The importance or correct illumination, and care taken in setting-up the correct lighting levels cannot be over-emphasised. Howe (1969) considers this not only "the most important but most commonly abused steps in photographic reproduction".

Incident tungsten illumination is preferable to softer electronic flash lighting. The intense conditions created by the tungsten lights are more conducive to photographing specimens of poor contrast. Photomacrography using a copy stand set-up, usually involves illuminating the specimen with a minimum of four photoflood lamps, two from each side. Angle and position each lamp individually, with the others turned off, in order to obtain the desired illumination. Additional free-standing photoflood lamps may be required, and can be used in conjunction with the copy stand lights, and should be situated opposite and further away from the specimen than the main light. The object of this light is to reduce contrast by lighting the shadows cast by the main light. Where light is directed from one main source, placement is critical in determining the form of the subject (Howe 1969). Positioning lights sub-horizontally to the specimen avoids strong reflections from the surface (Jeffords and Miller 1960). The angle can be slightly raised for flat specimens, and higher still for specimens with strong relief, however, the best results are achieved when the beam just skims the upper surface (Jeffords and Miller 1960).

The completed lighting arrangement should result in all shadows toned down or removed. Secondary shadows, or shadows within shadows, forming a criss-cross pattern, must also be eliminated with the careful placing of secondary illumination. Failure to remove strong shadows' results in inferior quality prints, with well-defined black areas (Howell 1977). Optimum lighting conditions are obtained with a directed main light source illuminating the top or upper left quadrant of the subject, a convention in photographing fossils for publication (Lund 1980). Furthermore, the correct conditions ensure detail is retained, and do not impart a washed out effect (Howe 1969, Jeffords and Miller 1960). The rendition of texture on the specimen, usually an important factor in palaeontological specimens, is best achieved by coating the specimen (see below) and positioning the lights at a very low angle to the specimen surface (Howe 1969). Experimentation with infrared and ultraviolet light sources (Lund 1980), coloured and polarizing filters, parabolic diffusers and reflectors (Howe 1969), and frosted bulbs (Rasetti 1965) provide additional lighting options for increasing the contrast between the fossil and matrix. More sophisticated lighting procedures used in the studio are outlined by Scovil (1996).

Black velvet is frequently used as a specimen back-drop, providing an even matt surface to both monochrome and colour photographs. However, different coloured backgrounds can be used if black is unsuitable. The angle of the incident light is adjusted to eliminate background shadows around the specimen edge. For a more detailed account of background materials see Scovil (1996).

The immersion of specimens in suitable liquids (*e.g.* water n = 1.332, heavy distillate, glycerol n = 1.473) can, in some instances be advantageous to photographing their features. Rasetti (1965) recognizes the following attributes when undertaking immersion methods:

- increased contrast between the fossil and matrix,
- eliminates unwanted reflections from shiny surfaces,
- renders the fossil surface uniformly dark when covered with irregular microcrystalline calcite and iron oxide deposits,
- renders the fossil transparent aiding the detailed observation of internal structures.

The enhanced observation of the above features are dependent upon the transparency and refractive index of the liquid, which must be close to that of the fossils (Rasetti 1965). Liquid immersed specimens reflect less light than those dry. Consequently it may be necessary to increase the length of exposure time, by a maximum four-fold, depending upon the specimen's colour and degree of transparency (Fournier 1956). However, considerable care must be taken when using liquid immersion, ensuring there is no adverse reaction with either the fossil or matrix, and that the liquid **is not** volatile (heat generated from the lights can cause convection currents and evaporation). Glycerol's higher refractive index produces greater contrast. Furthermore, its viscosity inhibits the formation of convection currents, allowing micro-slide specimens to remain in a fixed position during photography (Fournier 1956). It is advisable to test a small sample before proceeding with liquid immersion, particularly as there is rarely any advantage in this procedure for opaque specimens exhibiting strong relief (Rasetti 1965). The immersion of microfossil specimens in glycerol is outlined below in the section on photomicrography (37.5.1c).

Specimens stored in bottles or embedded in a resin, which cannot be removed for photography, should be carefully examined to ensure that unwanted reflections and glare are reduced to a minimum. The addition of coloured glass and polarising filters may assist in this. Table 37.1 contains a summary of the optimum lighting conditions most usefully employed in the photography of common macrofossil types or preparations.

37.2.3. Dark-field illumination

Photographing large peeled sections for publication is achieved by using off-axis uni- or bi-directional low angle transmitted light from below the specimen stage (Thomson and Bradbury 1987). Procedures used in the photography of fossil Cnidaria have been outlined by Easton (1942) and Sorauf and Tuttle (1988). Both methods used a dark-field set-up (Fig. 37.1), controlling the angle of incident light upon the section surface, and eliminating any illumination on a dark background situated behind the specimen.

Lighting is positioned at a low, oblique angle (approximately 30° or less) to the section. However, for optimum result's experimentation with the lighting angle, uni- or bi-directional source, and the insertion of crossed or plane polarized filters should be undertaken (Sorauf and Tuttle 1988). Specimen contrast is a factor related to the amount of light refracted as it passes through the peel. Areas where the light ray passing through the section is reflected off crystal surfaces of cleavage planes, appear brighter than those where only refraction occurs (Fig. 1b *in* Easton 1942). Contrast within the photograph is further enhanced during printing, by the use of graded paper and the insertion of filters (*e.g.* No. 3 polycontrast filter, Sorauf and Tuttle 1988).

This procedure does not work well with fine grained specimens, where detail is obscured by crystal face and cleavage surface interference. A rapid procedure for obtaining photographs of peels is to insert the section in a photographic enlarger. However, this results in a low-contrast **negative** print, which is not suitable for publication (Sorauf and Tuttle 1988).

37.2.4. Specimen orientation and data recording

Specimen orientation is a factor requiring some consideration in the preparation of palaeontological plates for publication. Formal taxonomic descriptions usually require

FOSSIL TYPE	OPTIMUM LIGHTING CONDITIONS
STONY FOSSILS **Three-dimensional, high relief (*e.g.* shells, bones)**	Undiffused (unless the object has a high lustre), directed from the "upper left". Fill in lighting to provide the detail.
RELIEF IMPRESSIONS **(*e.g.* bryozoa, fish, leaf imprints, trace fossils)**	Very low angle light "grazing" the surface. The angle depends on the amount of relief. Low relief = low relief = low fill in illumination. Make silicone rubber positive of the original. Photograph under even lighting using positive film. Printing reverses the image relief (McNair 1941).
FOSSILS AS SURFACE "STAINS" **(*e.g.* trace fossils)**	Low contrast, non-directional. Poor relief can be enhanced by inserting a colour filter of similar colour to the fossil (Lund 1980). Surface wetting and liquid immersion can assist in enhancing contrast.
CARBONACEOUS OR HIGHLY REFLECTIVE LOW RELIEF FOSSILS **(*e.g.* fossil leaves)**	High angle diffuse lighting produces high contrast images. Axial or near-axial lighting produces detailed light objects against a dark background.
EMBEDDED AND ENCLOSED FOSSILS **(*e.g.* resin encased and amber specimens)**	Specimen surfaces must be clean (see section 23 PREPARING FOSSIL IN AMBER). Dirty surfaces can result in distortions and reflections.

Table 37.1. The optimum lighting conditions used in the photomacrography of selected fossils. (Data from Scovil 1996 and other sources used in the text).

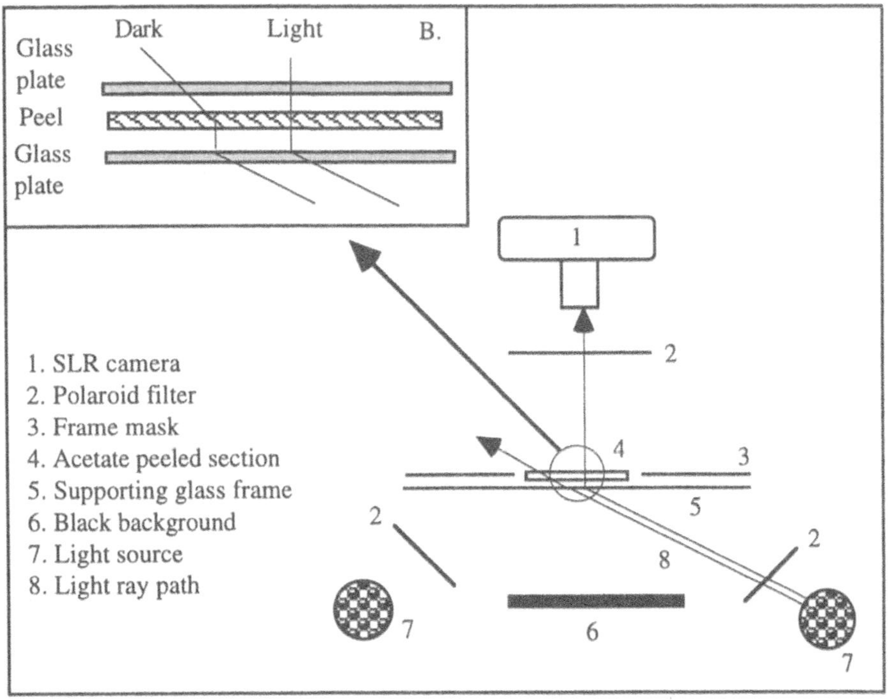

Figure 37.1. The arrangement of camera, lighting, filters and specimen for dark-field transmitted illumination of acetate peeled sections. (b) the insert diagrammatically shows the extremes of light rays refracted and reflected on passing through the section in generating a light and dark image.

illustrating, and it is often necessary to view specimens from several different angles. The number and views of orientation vary from one taxonomic group to another. Preferred views of a specimen may limit the choice of photographic angle and viewpoint, requiring additional mechanical or chemical preparation of the specimen.

It is impractical to describe all the views and orientations encountered, and reference is best made to the *Treatise* and monographs of the particular taxonomic group of interest. However, scientific photographers should attempt to standardize procedures of data recording, as this will be particularly beneficial during research work, avoiding confusion to subsequent workers should they use the photographs or negatives later. Feldman (1989a) and Scovil (1996) suggest a simple sequence to include all the conventional views required: (1) dorsal, (2) ventral or lateral, (3) distal, (4) proximal, (5) close-ups of details, and (6) oblique views to focus on a particular characteristic not evident in any conventional view, or in applied studies emphasizing occurrence and mode of life. Before photographing specimens (in any orientation) ensure a suitable (and legible) scale is inserted within the frame, and that it is in the same plane of focus as the specimen.

37.3. SPECIMEN COATING

Where possible it is preferable to photograph specimens in their natural state, as images of untreated fossil surfaces are usually flat, lacking contrast and depth, and generally considered as uninformative (Feldman 1989a). For these reasons palaeontological specimens are subjected to artificial whitening (also known as smoking) before photography. This specialized procedure can be applied to obliterate colour banding (Jeffords and Miller 1960) or enhance and define surface contours and sculpture (Siveter 1990), focusing observations towards detail of morphological significance. The procedure has been extensively described in the geological literature in the coating of invertebrate fossils (*e.g.* Feldman 1989b, Kier *et al.*, 1965, Siveter 1990), modified and adapted for use on coating microfossils (Benson 1965) and much larger artefacts (Weide and Webster 1967).

37.3.1. Ammonium chloride

The origins of this procedure, applicable to palaeontological specimens, can be traced to Grabau and Shimer (1910), and has been successfully used and illustrated by Ulrich and Bassler (1923), Bassler and Kellett (1934) and Bassler (1953). The original apparatus consisted of two Pyrex glass flasks, one containing a strong solution of hydrochloric acid (HCl) and the other ammonia (NH_4OH). Rubber bungs sealed the top of each tube, and by means of glass and rubber tubing it was possible for the operator to pass air through each solution. The fumes of each unite at combined nozzles, forming the sublimate. The sublimate stream is then directed towards the specimen, and coating controlled to give a uniform thin film over the surface. Numerous modifications recorded by palaeontological preparators, hint that this part of the procedure was easier described than performed. Modifications to the procedure refine the technique in relation to the heating source (Cooper 1935), substituting the solutions for solid ammonium chloride lumps or powder (Bassler 1953, Sass 1962, Teichert 1948), employing a carrier gas to ensure the even distribution of the sublimate (Marsh and Marsh 1975), and stabilizing the humidity of the working environment whilst undertaking this procedure (Bassler 1953, Poulsen 1957). These modifications will now be discussed, and a composite procedure outlined.

37.3.1a. Specimen preparation

1 For specimens with high relief, clean all infilled areas of remaining matrix (an ultrasonic bath is suitable for this if the specimen is non-porous). Where specimens are still embedded in matrix, direct efforts towards cleaning the margins. This will provide a sharp boundary between the specimen and matrix, eliminating the need to retouch or cut around the fossil outline when preparing the final print (Siveter 1990).

2 Handling and touching the specimen surface must be kept to a minimum to avoid finger prints. All grease and oil deposits must be removed with an organic solvent (alcohol or acetone). This will also aid in the removal of residual moisture from within the specimen.

3 The specimen must be completely free of all moisture. If in doubt oven dry at a low temperature (40-60°C) overnight.

4 When the specimen is thought to be clean and dry, carefully examine the surface under a binocular microscope. Use a directed jet of compressed air from an air duster to remove any loose dust.

37.3.1b. Coating

The danger of using strong acidic solutions (95% HCl) has been removed by the use of ammonium chloride (NH_4Cl) powder. Heating the powder results in subliming the ammonium chloride, producing ammonium and hydrogen chloride gas. Although the sublimate fumes are not known to be hazardous to health (Feldman 1989b), it is advisable to perform this procedure in a fume hood, as this will be less offensive to other laboratory users.

1 Using a soft paint brush, apply a uniformly even coat of photographic opaque or black graphite. Indian ink can also be used, but this can impart a slight sheen to the specimen surface. Some washable inks may be suitable for enhancing details of pores and faint suture lines (Fournier 1956). Ideally, the specimen will now possess a matt black or grey appearance. Position the specimen close to the apparatus within the fume cupboard. Artificially darkening specimens can result in problems if the surface coating leaves a residue following removal (see comments below). Considerable thought should be given to the importance of the specimen, mode and condition of preservation, in evaluating the long term effects the coating procedure will have.

2 The sublimate applicator most commonly used in this procedure is illustrated in figure 37.2. Fill the bulbous portion of the tube with powdered ammonium chloride, and re-insert the rubber stopper connected to the rubber bulb or air supply. Take care not to overfill the glass bulb, as the heated sublimate quickly cools and deposits at the tip of the drying tube, blocking the end. The deposition of the sublimate over the specimen is only effective if the airflow through the drying tube remains unobstructed.

3 When the assembly is complete, ignite the bunsen burner, and adjust the flame to a cone shape. Hold the tube horizontally, and gently warm the bulb by passing it through the flame. Rotating the bulb ensures uniform heating of the ammonium chloride powder. As the tube is heated to nearly red hot, driving off all the moisture, gently angle it towards the specimen, and watch as the ammonium chloride sublimate flows towards the tapered small diameter orifice. Keep the tapered end of the drying tube heated to prevent sublimate deposition clogging the tube vent.

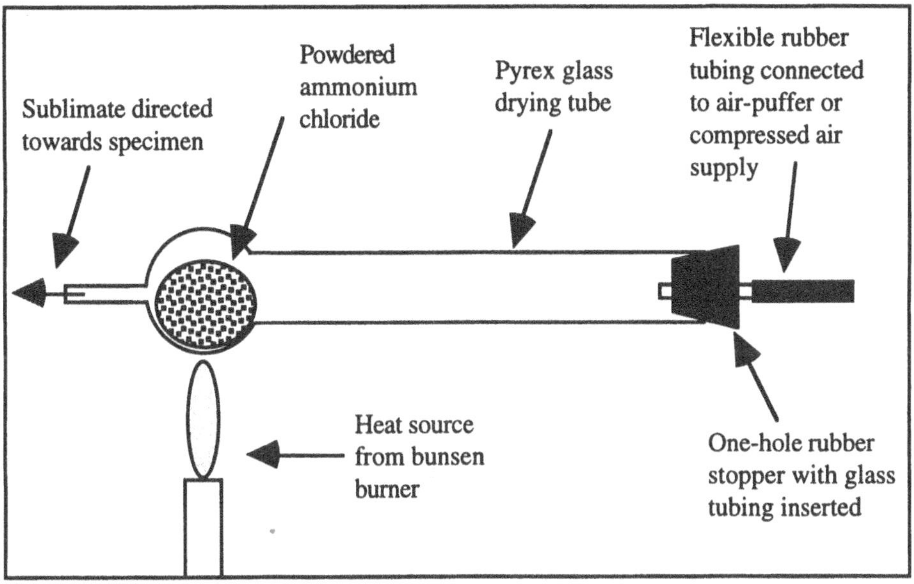

Figure 37.2. The specialized glass bulb drying tube used in specimen whitening, first illustrated by Teichert (1948) and successfully used by Hessland (1949) and countless other Palaeontologists since.

4 Gently pumping a small amount of air through the tube reduces the chance of the sublimate accumulating and blocking the tube (Feldman 1989b). By keeping the tube hot there is less chance of the sublimate rehydrating. It will become readily apparent that the hotter the tube is, the thicker the sublimate. This results in the production of a very white coat on the specimen, a feature not always desired.

5 When the tube is full of sublimate, direct it towards the specimen, and gently introduce air into the tube expelling a cloud of sublimate towards the surface of the specimen. Avoid the excessive application of sublimate from a single direction, as it produces an uneven, high contrast surface coating on the specimen. Feldman (1989b) advises proceeding slowly, periodically rotating the specimen to assist in applying a uniform coating.

6 Once coated, specimens should be photographed immediately. If photography has to be delayed, the coated specimens must be kept in a desiccator with silica gel to prevent deliquescence. If storing is unavoidable, check the quality of the coating by examining the specimen under a stereozoom binocular microscope before photographing.

7 Following the completion of photography, washing the specimen removes the ammonium chloride. Failure to do this can result in

permanent damage to specimens of carbonate composition, as the sublimate combines with water vapour in the air forming hydrochloric acid capable of etching the surface. By monitoring the relative humidity of the working environment, it is possible to avoid using this technique in areas, or on days, of high relative humidity, when the sublimate rapidly rehydates, becoming coarse grained and obscuring fine specimen detail (Siveter 1990). Warm soapy water will remove the photographic opaque (although use of an ultrasonic tank may be required for specimens with a porous matrix). Black graphite paint can be removed with a dilute solution of a photographic wetting agent. Inks are successfully removed with a mixture of ammonia and hydrogen peroxide.

Although quite simple, the technique requires great dexterity and considerable patience coupled with experience to achieve acceptable results. Modifications to the technique have revolved around the heating source, diameter of the glass drying tube, and gas carrier for the sublimate. These are aimed at providing the preparator with greater versatility in observing and controlling the amount and degree of sublimate applied to the specimen. Cooper (1935) used a nickel-chrome coil heating source enclosing (and obscuring) the drying tube, while Teichert (1948) and Marsh and Marsh (1975) remained faithful to the standard laboratory bunsen burner, as it is provides a means of directly observing the ammonium chloride heating.

Obstructions within the drying tubes tapered orifice through which the sublimate is expelled, have always been a problem with this technique. The heated ammonium chloride is rapidly deposited as it cools along the glass tube, quickly blocking the narrow orifice. Cooper (1935) opted for a large (5mm) opening, while Teichert (1948) recommended an opening of no less than 2.5mm. A 1-2mm diameter hole allows for more control in directing the sublimate, while reducing the length of the glass tube from the opening to the ammonium chloride reservoir bulb makes for easier heating of this critical end of the apparatus. However, by having a number of glass tubes of varying orifice diameters prepared with ammonium chloride before starting, this enables the operator to proceed in coating large specimens, having only to pause to change glass tubes, and without continually having to clean and dry blocked tubes. Furthermore, after experimenting, a tube with a suitable nozzle can be selected to suit the coating of a specific specimen size (Marsh and Marsh 1975). Tubes must be cleaned directly after use by washing in warm soapy water, and oven dried before re-use.

Further suggestions from Marsh and Marsh (1975) include the use of a dry carrier gas (*e.g.* nitrogen), supplied at a pressure of 2-4psi, to carry the sublimate to the specimen surface. This enables a constant flow to be maintained, facilitating an even coating on the specimen as the operator can concentrate on directing the flow. If a conventional air-puffer bulb is used, ensure that it is a bulb with a non-return valve, avoiding the possibility of drawing the sublimate back into the drying tube where it condenses as it cools (Teichert 1948).

Solid lumps of ammonium chloride, opposed to the more commonly used powder, were favoured by Marsh and Marsh (1975), who suggested it was less likely to block the drying tube orifice. However, experience gained with gently heating ammonium chloride powder and its careful application nullifies the need for this change.

Because of the problems associated with deliquescence described above, primarily the production of large sublimate grains in conditions of high humidity, ammonium chloride coating may not be suitable for microfossil specimens. Under the correct conditions, however, and no doubt with many years experience, it is possible to retain specimen detail (*e.g.* Benson 1965, p. 440, fig 5).

37.3.2. Magnesium oxide coating

Safety note: The burning of magnesium generates an extremely bright light source. Prolonged direct viewing may cause permanent injury to eyesight. Wear dark glasses and avoid looking directly at the light source (Jago 1973).

Rasetti (1947) described an alternative coating method using magnesium oxide (MgO), thus avoiding the irritating, offensive and corrosive HCl fumes generated by the ammonium chloride (NH_4Cl) procedure. Specimens' surfaces are coated with an extremely fine layer of powdered crystalline magnesium oxide following their movement through fumes of a burning magnesium ribbon. The principal advantages concerning the ammonium chloride procedure are listed below in table 37.2.

Advantages	Disadvantages
• Less offensive and corrosive fumes generated • Finer grained sublimate • Sublimate insoluble in water • Removable by lightly brushing • Selectively deposited on areas of higher relief (Kier et al., 1965) • Second or subsequent coats can be applied when previous is dry	• Difficult to cover large areas uniformly • Use of a direct heat source near the specimen precludes its use on heat sensitive materials • Difficulty in holding and manoeuvering the specimen and magnesium ribbon during sublimate deposition

Table 37.2. Summary of the advantages and disadvantages of coating specimens with magnesium oxide in comparison with ammonium chloride.

The method of application is as follows:

1. In a bunsen flame ignite a 2.5cm long piece of magnesium ribbon.

2. While holding the specimen approximately 5cm above the burning magnesium, rapidly move it back and forth through the fumes generated. The motion ensures an even coating. Initially, this may be difficult to achieve, so practice on inferior, unimportant specimens. Small fossils can be held using tweezers, and require between 2-10 seconds in contact with the fumes, depending on the thickness required. Some difficulty will be encountered if the area to be coated is greater than 2cm^2. Larger specimens require the process to be repeated until the entire surface to be photographed is covered.

A modified method of application, using an artists air-brush, has been outlined by Jeffords and Miller (1960). A mixture of 1 part magnesium oxide powder mixed with 20 part's alcohol is applied from a distance of approximately 5cm from the specimen surface. The mixture must dry rapidly (within a minute) to prevent the formation of flow structures. Initially, the coating may not be evident until it has dried, but subsequent coats can be applied as required.

37.3.3. Antimony oxide coating

Safety note: Poisonous fumes are generated from the heating of antimony. This work must be undertaken within a fume cupboard. Safety concerns limit this method to only the most experienced preparators.

The method of application is as follows:

1 Heat small pieces of antimony oxide (Sb_2O_3) in a drying tube until glowing. A white sublimate of tetra-oxide is generated.

2 By passing a gentle air current through the tube (by the process as detailed in 37.3.1), the sublimate can be directed and deposited in a similar manner to that of the ammonium chloride technique. The specimen surface takes on a light blue to ivory white appearance, depending on the thickness of the deposit.

The use of any of the above three methods require considerable practice before embarking on research and display quality specimens. Surface darkening materials can be difficult to remove, particularly from porous surfaces. Siveter (1990, p 505) provides a cautionary note, "the implications for future conservation of the specimen should be considered before employing these techniques", of added importance when examining and photographing borrowed, type and loan material.

37.4. STEREOPHOTOMICROGRAPHY

The specialist technique of stereophotography is a valuable tool in recording palaeontological specimens exhibiting high relief, with stereographs (the images) supplementing, and not replacing, conventional photomicrographs. Stereographs are produced by photographing specimens in two slightly different perspectives, differing by a horizontal angle of rotation between 8-10° (Siveter 1990). Printed images are viewed with the aid of a stereoscope. This makes the three-dimensional nature of the object readily appreciated, as this style of viewing exaggerates relief. The amount of vertical exaggeration is a function of the horizontal displacement about the axis or rotation between the two images. The greater the displacement, the greater the exaggeration (Feldman 1989c).

The procedure has been used in palaeontology for over sixty years (Branson *et al.*, 1933 p 17, Gott 1945). Equipment has been adapted and described for photographing both macrofossils (Feldman 1989c) and microfossils (Fournier 1956, Lehmann 1956), although the principle of operation remains similar. The procedure is as follows:

1 Mount the specimen directly beneath the centre of the camera lens on a flat table that rotates around a horizontal axis. A crystallographic four-axis universal stage or goniometer (with three axes fixed) can be adapted for photomicroscopy of microfossil specimens (Lehmann 1956).

2 Tilt the platform to one side (by an angle of 4-5° from the vertical) and photograph the specimen. Then, rotating the table an equal angle in the opposite direction, re-photograph the specimen.

Variations of this procedure include a sliding platform, where the specimen is moved in a horizontal plane an equal distance to the left and right of the centre line of the camera (Feldman 1989c, Rasetti 1965). Alternatively, the table can be fixed, and the camera moved in a horizontal or rotational axis around the specimen. During the procedure, care must be taken to ensure the table and specimen is not moved in any way other than the required rotation - particularly important between a pair of exposures. Similarly, lighting conditions must also remain constant between exposures, as even small differences in shadow can produce a poor stereogram (Rasetti 1965). Consideration must also be given to the distance of separation on the final plate (Evitt 1949). This is usually set at about 75mm, to correspond to the average interpupillary distance, when using a pocket stereoscope (Feldman 1989c). Alignment and positioning of **one** of the stereopair **must** be done while viewing the images stereographically. The other image must be permanently positioned. During plate construction the space between the stereo images can be infilled with a separate image, maximising all available space on the plate (Evitt 1949, Siveter 1990).

37.5. PHOTOMICROGRAPHY

The techniques required in obtaining high quality photomicrographs are just as involved as those undertaken for photomacrography. The increased magnifications obtained by photography through the microscope reinforce the need for careful specimen preparation, and the positioning of filters and high intensity lights. Any surface blemishes on the specimen are considerably enlarged during this photographic process. This is applicable for both three dimensional material (*e.g.* small fossils and microfossils) and prepared thin and cellulose peel sections. Hastily prepared sections may retain artefacts from the cutting and grinding stages of preparation. The coarse sublimate grain size produced by ammonium chloride coating may be evident on small three dimensional objects. Both of these factors contribute to the loss of fine detail and picture quality.

Before undertaking photography using a microscope, ensure all lenses, objectives (particularly oil immersion) and sub-stage condensers are dust free. Sections should also be cleaned before examining. After use replace all dust covers over equipment. Eliminate potential sources of dust from around the microscope, and set-up the equipment on a vibration free bench. Avoid contact with the microscope and bench, while the camera shutter is open.

37.5.1. Three dimensional objects

Small, three dimensional objects, such as microfossils, provide the greatest challenge, and it is not surprising that considerable debate has been given over to their successful photography (*e.g.* Cummings 1956, Fournier 1950, 1954, 1956, 1957, Gutschick 1960, Kugler 1957, Sander 1955). Two factors influence the success of photographing this type of material. Firstly, the object's depth of field and the many curved or angular planes, may be so great as to prevent focusing over the entire specimen surface. Secondly, the relief of the object may be so slight as to result in insufficient contrast for good photographic reproduction.

37.5.1a. Silver nitrate coating

Some techniques used in enhancing the contrast of individual microfossil specimens have been detailed in previous sections (see sections 21 STAINING TECHNIQUES USED IN MICROPALAEONTOLOGY and 20 THIN SECTION AND SLIDE PREPARATION TECHNIQUES OF MACRO- AND MICROFOSSIL SPECIMENS AND RESIDUES). Considerable success has been achieved using either water soluble food dyes, or coating the specimen in silver nitrate ($AgNO_3$) (Triebel 1947). Cummings (1956) variation of the silver nitrate procedure is particularly useful in reducing the natural shine of microfossil shells and enhancing morphological details. A significant advantage to the micropalaeontologist of this procedure is that it can be performed while observing the effects under the stereozoom binocular microscope. The method is as follows:

1. Ensure the specimen is thoroughly clean, by removing all loose matrix by washing with a fine sable-hair brush dipped in distilled water.

2. Transfer the specimen to a circle of filter paper, and using a disposable pipette apply a few drops of a 10% silver nitrate solution. Ensure the specimen remains saturated for five minutes by repeating the procedure. Allow the specimen to dry.

3. Brush-wash and soak the specimen with ordinary photographic developer for two minutes. Allow the specimen to dry. Shell darkening can be intensified by additional applications of developer.

4. Mount the specimen on a clean, white background to observe and photograph. The dark coating can be removed by washing and vigorously scrubbing with a brush in warm water.

This procedure has a distinct advantage over that described by Triebel (1947), in that it is applicable to heat sensitive specimens (compare the above procedure with that outlined in section 21 STAINING TECHNIQUES USED IN MICROPALAEONTOLOGY). However, the vigorous action of washing required to remove the coating may provide a cautionary note to the preparator before proceeding with delicate and or type specimens.

37.5.1b. Carbon coating

Other contrast enhancing alternatives include carbon coating, in a manner similar to gold coating for electron microscopy (see section 34 ELECTRON MICROSCOPY TECHNIQUES). This technique has been adapted for the study of acetate peel replicas, which can then be viewed using a conventional light microscope (*e.g.* Krinsley *et al.*, 1964, Krinsley and Bé 1965). The procedure is as follows:

1 Follow the preparation of acetate peels (see section 22 PREPARATION OF CARBONATE STAINED ACETATE PEELS AND THIN SECTIONS), discarding early peels which may contain loose material from the specimen surface.

2 Transfer the peel to a vacuum evaporator, and vaporize a thin layer from a heavy metal (*e.g.* chromium), followed by a thin carbon film. This process results in the deposition of a metal film, which is thickest on surface's perpendicular to the direction of shadowing, and lightest on surfaces parallel to that direction (Krinsley *et al.*, 1964).

3 Transfer the peel to a glass embryo dish or small petri dish, and flood with acetone to dissolve the acetate. This will leave a thin (<200Å) transparent metal carbon film.

4 The carbon film can be transferred to an electron microscope grid (for viewing in a TEM), or onto a conventional glass microscope slide, covered with a cover slip and viewed using a light microscope. Grain size, shape and surface relief, as well as texture can be studied using this procedure.

37.5.1c. Glycerol immersion

The technique of immersing microfossils in glycerol before photographing, has been described by Fournier (1956) and, with particular reference to agglutinated forms, Gutschick (1960). A composite procedure of application is as follows:

1 Place the specimen on a clean microscope slide (glass cavity slides are suitable).

2 Using a disposable pipette, apply a drop of glycerol to the specimen. This will evenly disperse over and around the specimen.

3 Air bubbles are frequently trapped in the viscous fluid, and are removed by carefully dabbing with a fine paint brush. Air trapped within the foraminiferal chambers can be removed by the addition of a drop of water (Fournier 1956) or alcohol (Gutschick 1960). Excess water is removed with a dry brush, and an additional drop of glycerol applied if required. The placing of a cover glass over the specimen improves optical resolution (Fournier 1956).

37.5.2. Petrographic thin sections

Although considerable emphasis is placed on the quality of the optical system (and by virtue cost) used in obtaining images, an expensive microscope may not always be required. Rusbridge (1989) illustrates the use of a zoom photographic slide duplicator in obtaining publication quality micrographs with reproduction ratios of up to 35:1. Douglass (1965) considered the use of an adequate light source and a satisfactory method of focusing the image at a desired magnification as the most critical factors in obtaining high quality images.

Because of the rapidity and low cost by which they can be produced, and the range of textures illustrated, acetate peel photography has received considerable attention over the past 45 years (*e.g.* Abineri 1989, Easton 1942, Sorauf and Tuttle 1988). From well-prepared surfaces a considerable amount of textural detail can be observed from the peel, with the recognition of biotic examples from the invertebrates, particularly those with a calcareous skeletal component, and microfossil groups including foraminifera, ostracods, calcareous nannofossils and plant debris clearly illustrated (Abineri 1989).

The lighting intensity is critical, and must be set to evenly illuminate the entire field of view, at a level sufficient to allow details to be photographed. Once the correct interval has been attained, it **should not** be altered during the process of photography. Any subsequent alterations in intensity must be achieved by the insertion of filters, condenser lenses, and the opening and closing of diaphragms. Closed diaphragms increase the depth of field. Research quality microscopes possess these as standard fittings, while some high quality objectives on compound microscopes may also have this feature. High magnification work requires the insertion of an additional sub-stage plano-convex condenser lens (*i.e.* between the light source and the section). The larger the diameter of the lens, the larger the area of even illumination (Douglass 1965).

In general, the depth of focus decreases with an increase in magnification, and at high magnifications, may be less than the section thickness (Douglass 1965). The magnification can be determined by inserting a stage micrometer in place of the section, and with reference to the ocular (eyepiece) micrometer, measuring part of the field of view. Alternatively, the stage micrometer can be photographed (remembering to re-photograph when switching between objective lenses). The micrometer image must be printed at a similar scale to the other images. The final magnification can then be checked by direct measurement from the enlarged image.

37.5.3. Filter selection and use

In macrophotography filters are commonly used to create "special" or artistic effects. The unique requirements of scientific photomicroscopy mean greater care must be employed during filter use to ensure image degradation is minimal. Four main types of filter are used in colour photography:

- **Colour conversion filters**: the Wratten 80 series of filters are used to match a daylight type film with an artificial light source (*e.g.* Kodachrome 25 (daylight film) used with 3200°K tungsten lights must be used with a 80A filter (increasing the colour temperature from 3200°K to 5500°K); the 85 series are used to match tungsten films for daylight use (*e.g.* *Ektachrome 160* (a 3200°K tungsten film) can be converted to daylight use with the use of an 85B filter (decreasing the temperature from 5500°K to 3200°K).
- **Light balancing filters**: are used when the mismatch between film and light source is small, the 82 series are used to increase the colour temperature slightly for a cooler (bluer) tone. The 81 series is used to decrease the colour temperature slightly for a warmer (redder) tone.

- **Colour correction** or **compensating filters**: come in the three primary colours (green, red, blue) and three secondary colours (magenta, cyan, yellow). These filters (usually prefixed CC, followed by the density value and the colour) are used to correct differences in hue caused by reflections, reciprocity failure (colour shifts from either long or short exposure times), and films coming from different batches. Although colour correction is best done at time of exposure, with print films it can be done during printing.
- **Special light modifying filters**: include ultraviolet, polarizing, and neutral density filters. Ultraviolet (UV) filters absorb ultraviolet rays, and in conventional outdoor photography make images sharp and clear by eliminating fogginess. Modern coated lenses mean they have virtually no effect, although many photographers use them as a lens protector. A polarizing filter restricts the light wave motion to one direction, resulting in plane polarized light. Neutral density filters are used to control the amount of light entering the camera lens, and in photomicroscopy their insertion assists in reducing the depth of field.

Individually, or in combination these filter systems can be used in black and white photography, where enhancing the contrast between the grey tones is a primary objective. The insertion of a green or yellow-green filter is highly effective, particularly when using a tungsten light source. A similar effect can be obtained when using a polarizing filter. The use of green and other coloured filters as controls of contrast in black and white photography have been described by Tapp and Prezbindowski (1990), and is summarised in table 37.3. The foremost principle to understand is that filters lighten their own colour (and those in that area of the spectrum), and darken complementing (opposite) colours. However, it must be remembered that the insertion of a filter reduces the light passing through a section, and consequently the exposure time must be increased to compensate for light absorbed. The number by which an exposure must be increased, for a particular filter and film type, is known as the filter factor. The precise factor, however, is determined with consideration to the film speed and light source, and determined by running a film speed test (Tapp and Prezbindowski 1990).

Like optical lenses, filters are available in a range of qualities (Scovil 1996). Gelatin plastic coloured filters, offer the best optical quality. However, they easily scratch, mark with fingerprint impressions, and wrinkle or fade through poor storage. When sandwiched between glass the gelatin is protected and given rigidity. Generally, these filters are of good quality. Coloured glass filters are not produced to the quality of gelatin filters, and are subject to fading with time. Coated glass filters (*e.g.* neutral density filters) have a thin layer of a light altering material applied to their surface. Store filters at room temperature, out of direct sunlight.

37.5.4. Record keeping

As with all the processes described in this manual, a record of camera and microscope settings should be maintained. This enables the duplication (or correction) of equipment settings that may be required later. Records should include exposure times, light intensity, as well as any filters used. Notes should also include the objective magnification, lens and diaphragm settings plus the film type and speed. A general description and thumbnail sketch of the object and field of view will also be useful.

KODAK WRATTEN NUMBER	FILTER COLOUR	PHOTOMICROSCOPY FILTER CHARACTERISTICS
11	Light green/ Yellow green	Also commonly known as the tungsten light correction filter. This filter will enhance edge and boundary detail, effectively increasing the sharpness of crystal and grain cleavage planes, boundaries and contacts. Printed correctly, images exhibit a contrast range similar to that perceived by the human eye with tungsten light illumination.
58	Dark green/ Green	A tricolour filter that removes red and blue light. This filter enhances grain boundaries, and slightly darkens blue stained epoxy resins and any grains that transmit red light.
25	Bright red	This filter removes red and blue light. Blue stained epoxy appears dark in the image. Red grains appear lighter in the final image, making this filter useful when examining stained sections.
29	Dark red	A tricolour filter that removes blue and green light. In final printed images blue stained epoxy appears nearly black. This filter gives the most dramatic darkening of blue epoxy and also renders calcite stain lighter, giving detail in stained sections.
44	Cyan	This filter removes (darkens) red light and enhances blue and green light. The filter is best used to enhance stained calcite relative to other grains and cements.

Table 37.3. The colour compensating filters most commonly used in black and white photomicrography. (Data from Tapp and Prezbindowski 1990).

37.6. PREPARATION OF PLATES FOR PUBLICATION

The culmination of successful photography is frequently seen in the published plate. Many of the specifications required, such as style, tone, arrangement and size, can be obtained from examining previously published examples, journals "notes to authors", or directly from the editor. Some journals go as far as illustrating examples of plate defects, or provide notes on "how not to make a plate" (Anon 1990b), particularly useful to potential authors as it graphically illustrates the diversity of problems encountered. The most common reasons given for the rejection of plates can usually be assigned to one of three areas of preparation (Table 37.4). These include poor techniques employed in (a) photographing the original specimen, (b) darkroom procedures, and (c) inadequate care during plate preparation, with contributing factors that influence print quality and ultimately plate preparation listed in table 37.4. The completed plate will, however, only be as good as the photographs used in composing it (Palmer 1965).

In general, the best published results are obtained from plates prepared with slightly more contrast than that required in the finished product. Modern printing techniques enable plates to be prepared to fill part or all of a single column or full page width. Formatting to a specific journal's page size in this manner may influence an author in the choice of publication.

Techniques of plate preparation have been outlined, in increasing complexity, by Whittaker and Hodgkinson (1991), Palmer (1965), Feldman (1989d) and Bengtson (1986). At its most simplest, the procedure must include stages of cutting and cropping individual images. These stages are usually performed on a computer following scanning or digitally importing images. Re-touching areas and morphing (working at pixel level if required), has replaced the felt and fibre-tip pen in removing the marginal white line effect, for so-long the threat to high quality plates (Whittaker and Hodgkinson 1991). However, if the well-established technique of preparing plates from photographs is required, the following steps will assist in producing a high quality plate.

1. Carefully blacken around the edge of each image with a felt or fibre tip pen. Cut around the image using a sharp pair of scissors, leaving a margin of a few millimetres.

2. Blacken the cut edge of the photograph. If left, this will show as a white line in the final plate production (Whittaker and Hodgkinson 1991).

3. After positioning, mount the individual photographs on matte board using either a dry mounting process or a rubber adhesive. Feldman (1989d) advocates dry mounting, as this process leaves no excess adhesive around the edge of the print. The time consuming process using rubber adhesives requires a weight (usually books) to ensure the edges of the photographs are firmly attached to the board. A sheet of tracing paper or grease-proof paper placed over the plate before pressing will avoid spoiling (Whittaker and Hodgkinson 1991). Following the removal of the weight, allow the adhesive to harden. Excess around the edges of the print can be removed by gently rubbing with a cloth or paper tissue.

AREAS OF PHOTOGRAPHIC PLATE PREPARATION	CAUSES AND RESULTS OF INFERIOR IMAGE AND PLATE
1. POOR PHOTOGRAPHIC TECHNIQUE	(a) bad illumination of specimen - poor contrast. (b) poor focusing. (c) incorrect aperture setting - lacks depth of field, produces poor negative. (d) poor maintenance of photographic equipment - incapable of obtaining quality desired.
2. POOR FILM PROCESSING & NEGATIVE/PRINT STORAGE	(a) scratched negatives. (b) fogged film processing - incorrect chemicals, concentrations and time for developing - produces inferior quality negative. (c) poor prints - printing on incorrect grade of paper - variaton in tone quality.
3. POOR PLATE PREPARATION	(a) individual prints inappropriately placed, wasteful of space. Asthetically lacks conformity in layout of the grouping of species, print sizes and types. (b) inconsistent numbering of images, lack of scale. (c) poor cropping of images and badly touched up edges of prepared prints.

Table 37.4. The three principal areas of plate preparation, and the possible causes and reasons of problems resulting in inferior plate quality.

4 Check the edges of the print, and touch up with the felt or fibre tip pen if required.

Feldman (1989d) suggests that fossil illustrations should be printed at unit scales (*e.g.* x0.5, x1, x5, x10). If this can be achieved while making the plate a minimum of twice its normal size, final reduction by the printer will reduce any small surface blemishes, such as the marginal white line effect (Whittaker and Hodgkinson 1991).

REFERENCES

ABINERI, K. W. 1989. Photomicrographs of cellulose peels from the Mesozoic rocks of Dorset. *Proceedings of the Geologists' Association*. **100**, (2), 161-174.

ANON. 1990b. The Palaeontological Association - notes for authors submitting papers for publication in Palaeontology and Special Papers in Palaeontology. *Palaeontology*. **33**, (4), 993-1000.

ATHERSUCH, J and JONES, C. 1991. Applications of video image capturing to palaeontology. *Microscopy and Analysis*. No. 24, 7.

BASSLER, R. S. 1953. Techniques, pp G16-G21: In; R. C. Moore (eds), *Treatise on Invertebrate Paleontology, Part G, Bryozoa*. Geological Society of America and the University of Kansas. 253p.

BASSLER, R. S and KELLETT, B. 1934. Bibliographic Index of Paleozoic Ostracoda. *Geological Society of America*. Special Paper. **1**. 1-500.

BENGTSON, S. 1986. Preparing clean backgrounds in published photographic illustrations. *Lethaia*. **19**, (4), 361-362.

BENSON, R. H. 1965. Photography of microfossils, pp 433-446: In; B. Kummel and D. Raup (eds), *Handbook of Paleontological Techniques*. W. H. Freeman and Co., (San Francisco). xiii + 852p.

BRACEGIRDLE, B. 1995. *Scientific photoMACROgraphy*. Microscopy Handbooks 31. Royal Microscopical Society, Bios Scientific Publishers (Oxford). 105p.

BRANSON, E. B, MEHL, M. G and BRANSON, E. R. 1933. Conodont studies. *The University of Missouri Studies*. **8**, (1-4), 1-343.

CALDER, J and GARRETT, J. 1990. *The new 35mm photographer's handbook* (Revised edition). Crown Publishers (New York). 239p.

COOPER, C. L. 1935. Ammonium chloride sublimate apparatus. *Journal of Paleontology*. **9**, (4), 357-359.

CUMMINGS, R. H. 1956. Preparation of microfossils for photography. *Micropaleontology*. **2**, (4), 402.

DOUGLASS, R. C. 1965. Photomicrography of thin sections, pp 446-453: In; B. Kummel and D. Raup (eds), *Handbook of Paleontological Techniques*. W. H. Freeman and Co., (San Francisco). xiii + 852p.

EASTON, W. H. 1942. An improved technique for photographing peel sections of corals. *Journal of Paleontology*. **16**, (2), 261-263.

EVITT, W. R. 1949. Stereophotography as a tool of the palaeontologist. *Journal of Paleontology*. **23**, (5), 566-570.

FELDMAN, R. M. 1989a. Photographic procedures, pp 336-341; In: R. M. Feldman, R. E. Chapman and J. T. Hannibal (eds), *Paleotechniques*. The Paleontological Society Special Publication. **4**. iv + 358p.

FELDMAN, R. M. 1989b. Whitening fossils for photographic purposes, pp 342-346; In: R. M. Feldman, R. E. Chapman and J. T. Hannibal (eds), *Paleotechniques*. The Paleontological Society Special Publication. **4**. iv + 358p.

FELDMAN, R. M. 1989c. Preparation of stereoscopic photographs, pp 347-350; In: R. M. Feldman, R. E. Chapman and J. T. Hannibal (eds), *Paleotechniques*. The Paleontological Society Special Publication. **4**. iv + 358p.

FELDMAN, R. M. 1989d. Preparation of paleontological plates, pp 356-358; In: R. M. Feldman, R. E. Chapman and J. T. Hannibal (eds), *Paleotechniques*. The Paleontological Society Special Publication. **4**. iv + 358p.

FITZPATRICK, K. 1991. Anatomical photography of the human skeleton. *Science Technology*. **24**, 12-17.

FOURNIER, G. 1950. Photographing small foraminifera. *Micropaleontologist*. **4**, (1), 19-21.

FOURNIER, G. 1954. Use of the pinhole diaphragm in photomicrography. *Micropaleontologist*. **8**, (3), 58-62.

FOURNIER, G. 1956. New methods and techniques in the photography of microfossils. *Micropaleontology*. **2**, (1), 37-56.

FOURNIER, G. 1957. Construction of pinhole diaphragms for use in photomicrography. *Micropaleontology*. **3**, (1), 85-87.

GOTT, P. F. 1945. Procedure of simplified stereophotography of fossils. *Journal of Paleontology*. **19**, (4), 390-395.

GRABAU, A. W and SHIMER, H. W. 1910. *North American Index Fossils*. Vol. 2. 818-819.

GUTSCHICK, R. C. 1960. Photography of Paleozoic arenaceous foraminifera. *Journal of Paleontology*. **34**, (4), 756-762.

HESSLAND, I. 1949. Investigations of the Lower Ordovician of the Siljan district, Sweden: 1. Lower Ordovician ostracods of the Siljan District, Sweden. *Bulletin of the Geological Institute of Uppsala*. **33**, 97-408.

HICKS, R and SCHULTZ, F. 1997. *The Black and White handbook*. David and Charles, (Newton Abbot, Devon). 224p.

HOWE, H. J. 1969. Lighting techniques in photomacrography. *Journal of Paleontology*. **43**, (2), 589-592.

HOWELL, J. 1977. Photography of low-contrast fossils. *Journal of Paleontology*. **51**, (2), 419-420.

JACOBSON, R. E, RAY, S. F and ATTRIDGE, G. G. 1988. *The manual of photography* (8th edition). Focal Press, (Guildford). xi + 394p.

JAGO, J. B. 1973. A hazard in the magnesium oxide method of whitening fossils. *Journal of Paleontology*. **47**, (3), 591-592.

JEFFORDS, R. M and MILLER, T. H. 1960. Air brush whitening fossils, and notes on photography. *Journal of Paleontology*. **34**, (2), 275-276.

KIER, P. M, GRANT, R. E and YOCHELSON, E. L. 1965. Whitening fossils, pp 453-456: In; B. Kummel and D. Raup (eds), *Handbook of Paleontological Techniques*. W. H. Freeman and Co., (San Francisco). xiii + 852p.

KRINSLEY, D and BÉ, A. W. H. 1965. Electron microscopy of internal structures of foraminifera, pp 335-344: In; B. Kummel and D. Raup (eds), *Handbook of Paleontological Techniques*. W. H. Freeman and Co., (San Francisco). xiii + 852p.
KRINSLEY, D, MORTON, N, and CUTBILL, J. L. 1964. The application of vacuum shadowing to the examination and photography of small objects. *Geological Magazine*. **101**, (4), 357-359.
KUGLER, H. G. 1957. Preparation of microfossils for photography. *Micropaleontology*. **3**, (2), 190.
LEHMANN, E. P. 1956. A technique of stereophotomicrography for illustrations in micropaleontology. *Journal of Paleontology*. **30**, (3), 757-759.
LUND, R. 1980. Photomacrography of fossils for publication. *Journal of Paleontology*. **54**, (1), 264-266.
MARSH, R. C and MARSH, L. F. 1975. New techniques for coating paleontological specimens prior to photography. *Journal of Paleontology*. **49**, (3), 565-566.
McNAIR, A. H. 1941. A method of photographing impressions of fossils. *Journal of Paleontology*. **15**, (1), 91.
PALMER, A. R. 1965. Preparation of plates for paleontological publication, pp 456-459: In; B. Kummel and D. Raup (eds), *Handbook of Paleontological Techniques*. W. H. Freeman and Co., (San Francisco). xiii + 852p.
POULSEN, C. 1957. Improved method for whitening fossils for study. *Journal of Paleontology*. **31**, (5), 1029.
RASETTI, F. 1946. Optimum conditions for the photography of fossils. *Journal of Paleontology*. **20**, (5), 514-516.
RASETTI, F. 1947. Notes on techniques in invertebrate paleontology. *Journal of Paleontology*. **21**, (4), 397-399.
RASETTI, F. 1965. Photography of fossils, pp 423-430: In; B. Kummel and D. Raup (eds), *Handbook of Paleontological Techniques*. W. H. Freeman and Co., (San Francisco). xiii + 852p.
RÖTTGER, R. 1984. Summaries of three scientific films on larger foraminifera. *Benthos '83; 2nd International Symposium of Benthic Foraminifera*. Pau and Bordeaux. 535.
RUSBRIDGE, M. M. 1989. Photomicrography without a microscope. *Proceedings of the Geologists' Association*. **100**, (2), 209-213.
SANDER, N. J. 1955. An apparatus for photographing foraminifera and other small objects. *Micropaleontology*. **1**, (3), 251-256.
SASS, D. B. 1962. Improved techniques for the photographing of fossils. *Journal of Paleontology*. **36**, (1), 171-172.
SCOVIL, J. A. 1996. *Photographing minerals, fossils and lapidary materials*. Geoscience Press, Inc. Tucson, Arizona. xvi + 224p.
SIVETER, D. J. 1990. Photography, pp 505-508: In; D. E. G. Briggs and P. Crowther (eds), *Palaeobiology - A synthesis*. Blackwell Scientific Publications, (Oxford). xiii + 583p.
SORAUF, J. E and TUTTLE, D. W. 1988. Dark-field illumination in photography of acetate peels. *Journal of Paleontology*. **62**, (1), 153-156.
TAPP, J. B and PREZBINDOWSKI, D. R. 1990. Application of contrast control filters in black and white photomicrography. *Journal of Sedimentary Petrology*. **60**, (4), 617-619.
TEICHERT, C. 1948. A simple device for coating fossils with ammonium chloride. *Journal of Paleontology*. **22**, (1), 102-104.
THOMSON, D. J and BRADBURY, S. 1987. *An introduction to photomicrography*. Microscopy Handbooks 13. Royal Microscopical Society, Oxford University Press, (Oxford). 74p.
TRIEBEL, E. 1947. Methodische und Technische Fragen der mikropalaontologie. *Senckenberg Buch, 19 W Kramer, (Frankfurt)*. 47p.
ULRICH, E. O and BASSLER, R. S. 1923. Paleozoic Ostracoda: their morphology, classification and occurrence. *Maryland Geological Survey*, Silurian. 281-283.
WEIDE, D. L and WEBSTER, G. D. 1967. Ammonium chloride powder used in photography of artefacts. *American Antiquity*. **32**, 104-105.
WETMORE, K. L. 1988. Burrowing and sediment movement by benthic foraminifera, as shown by time-lapse cinematography. *Revue de Paléobiologie. Benthos '86*. 921-927.
WHITTAKER, J. E and HODGKINSON, R. L. 1991. On the preparation of specimens for scanning electron microscopy and a simple technique for plate making, using a black background. *Journal of Micropalaeontology*. **9**, (2), 219-220.
WHITTINGTON, H. B. 1956. Photographing small fossils. *Journal Paleontology*. **30**, (3), 756-757.
WHITTINGTON, H. B. 1965. Photography of small fossils, pp 430-433: In; B. Kummel and D. Raup (eds), *Handbook of Paleontological Techniques*. W. H. Freeman and Co., (San Francisco). xiii + 852p.

38. ILLUSTRATING AND EXHIBITING FOR DISPLAY AND PUBLICATION

38.1. INTRODUCTION

In many respects the illustrating and exhibiting of palaeobiological data and specimens is the culmination of the study and analysis of fossils and fossil bearing samples. The form and medium of illustrating or exhibiting decided upon are becoming ever wider with the increasing use of electronic media and the advances in Information Technology. Detailing the use and applications of specialized computer software programmes is beyond the scope of this publication, as are the specialized conventional pen and ink drafting techniques. Within this section attention will revolve around aspects in recognizing the potential of a specimen through illustrating and exhibition.

Although illustrations (computer generated, pen and ink, and photographic images) form a key part in any palaeontological study (Chase 1979a), the preparation and display of original specimens or high quality reproductions and scaled displays, form the focal points of many exhibitions. Specialized specimen preparation procedures (fossil extraction, mechanical and chemical preparation and replication) described in other chapters within this manual can, however, provide the starting point for the exhibition. Critical to a visually stunning display are high quality photographic images (see section 37 PHOTOMACROGRAPHY AND PHOTOMICROGRAPHY TECHNIQUES), expertly prepared specimens, and occasionally high quality replicates as a backdrop or focal point (*e.g.* Obata *et al.*, 1989, see section 36 FOSSIL REPLICATION TECHNIQUES).

In this section consideration will be given to the preparation of specimens for exhibition and display, an area in which geological curators have given considerable thought, reflected in the 281 references cited by Sharpe (1983, from the keywords: exhibition, illustrating and models). With such a volume of literature available on this subject, planning must commence with the collector in the field, and should subsequently be considered by the specimen preparator.

38.2. PLANNING

It is perhaps the multiplicity of skills required to produce an exhibition that effectively communicates data and information to a wide ranging, and probably unknown audience, that presents problems to both the professional curator and researcher alike. In essence, similar problems are addressed in the planning and construction of a poster display outlining new research, or exhibitions illustrating broad based concepts with well-known examples. In both cases, *distance learning* provides the means by which information is communicated. Although the author of a poster might be available to discuss scientific data and any other aspect, the display must also stand-alone. Despite there being no consensus on what constitutes a "good" palaeontological exhibition (Brunton *et al.*, 1985), success in understanding the principles behind the data provided is a direct reflection of how successful the display has been thought out and assembled (Miles 1990). Ultimately, the success in communicating information involves the "author" knowing his audience, and attending to the content (what is said) and the form (how it is said), ultimately inspiring and increasing understanding.

FACTOR	JUSTIFICATION
1. Background	Where and how the idea originated.
2. Objectives	All stages of planning and development should relate to points raised in the objectives. Assumes greater importance if project is aimed at general public. Predict the effect it will have on target audience. Can then be used as a means of testing (measuring the success, see below) and promoting concepts, enhancing knowledge, creating awareness of the topic, and relating it to the environment. All objectives should be compatible.
3. Audience	Target groups, ages and comprehension levels. Market research forms an important factor in determining the audience for public exhibitions (e.g. Loomis 1987, Miles et al., 1988). Knowledge of the audience assists the author in targeting the message, and physically setting key concepts and data at eye-level. Avoid creating a false image through (i) self-interest, (ii) not knowing the audience.
4. Description	How the objectives are to be fulfilled. Will also describe (i) size, (ii) duration, (iii) importance of specimens, (iv) use of graphics and photographs, (v) amount of text, (vi) supporting publications, (vii) possible expansion.
5. Budget	Of greater importance in large scale public museum exhibitions. Estimate costs of staging and running. Allow a contingency fund. Detailed planning may be constrained by budget.
6. Timetable	Detail construction, including time and co-operation with other people. Provides an effective and efficient way of planning the whole operation. Allow extra time to accommodate any unforseen circumstances.

Table 38.1. Areas of consideration during the initial planning stage of a display or exhibition. The necessity of marshalling and justifying ideas may increase with importance if (a) financial controls are a key factor, or (b) outside designers or constructors must interpret ideas. (Based on data from Brunton *et al*., 1985 and Miles 1990).

Planning is the all important aspect, applicable to large or small, permanent or temporary, and displays or exhibitions. An effective check list (Table 38.1), uses the approach offered by Brunton (*et al*., 1985) and Miles (1990), augmented by additional references from examples specific to palaeontology. Planning a project in this manner assists in formulating and refining ideas, particularly useful when presenting new concepts or data in a research poster, and provides a method for planning an approach for multiple viewing at different audience levels. For example, poster displays can be designed and modified for different conferences and seminars, where audiences might range from a select group of specialists to those with only a passing interest. Influencing the latter through selling-ideas and applications, is just as important as convincing one's peers in discussion of the data.

Planning and detailing a synopsis of a poster display for a symposium and conference might appear a tedious and a somewhat time consuming exercise, particularly if data acquisition remains a prime concern. However, many of the factors outlined above are undertaken sub-consciously. For example a graphic plan for a poster display may be drawn (Fig. 38.1), illustrating the relative size and positions of text, diagrams and images. Similarly, a timetable for constructing the display is followed, as this invariably involves assistance from specialists in photography and graphics. Larger scale museum-type exhibitions require a well defined and informative plan. The justification can, however, take on considerable significance if budgetary controls form a factor in decision making. If a designer is contracted, a brief conceptual framework of thoughts, ideas and practical knowledge (including examples of text and figure captions) of the exhibition, will be required to fulfil the plan (Brunton *et al.*, 1985).

Miles (1990) presents many useful strategies and pointers that can be employed in the planning of a large exhibition. Many of these are just as applicable to smaller exhibits and poster displays. Ideas and concepts must be presented in a logical well-ordered manner, with one concept understood before proceeding to the next. Displays are much more effective if a central theme flows through the work, supported by a framework that unifies the facts. The introduction can be used to outline how the work and display are organized. In large exhibitions it may be necessary to repeat information at different places, and provide signposts and maps to enable people to orientate themselves and select areas of interest. By telling only one story at a time the reader will quickly appreciate the direction of the work, and the clear nature and status of each message as either a conclusion or a supporting argument (Miles 1990).

38.3. SELECTING AND CONSTRUCTING

Once the planning of the exhibition or display is complete, the selection of materials (specimens, illustrations) and construction can continue. If planning has been thorough, this might be a relatively quick process. It must, however, be followed by a period of testing, enabling any faults and uncertainties to be corrected.

38.3.1. Specimens

Specimens provide two essential elements to a display: (i) beauty and self-importance in their own right, and (ii) as illustrations of examples (Brunton *et al.*, 1985). Open displays must be used with care, particularly if material is on loan. The security of all specimens must be adequate, although material is probably at greater risk through abrasion and lack of environmental control. By considering the use of good quality replicas, this problem can be overcome, although real specimens have a higher attracting and holding ability (Johnston and Sharpe 1997). Rare and financially valuable specimens may require special security measures, and should not be placed on open display. Specimens should not be highlighted for value's sake, particularly if displayed in association with less valuable or more common material. Examples should be

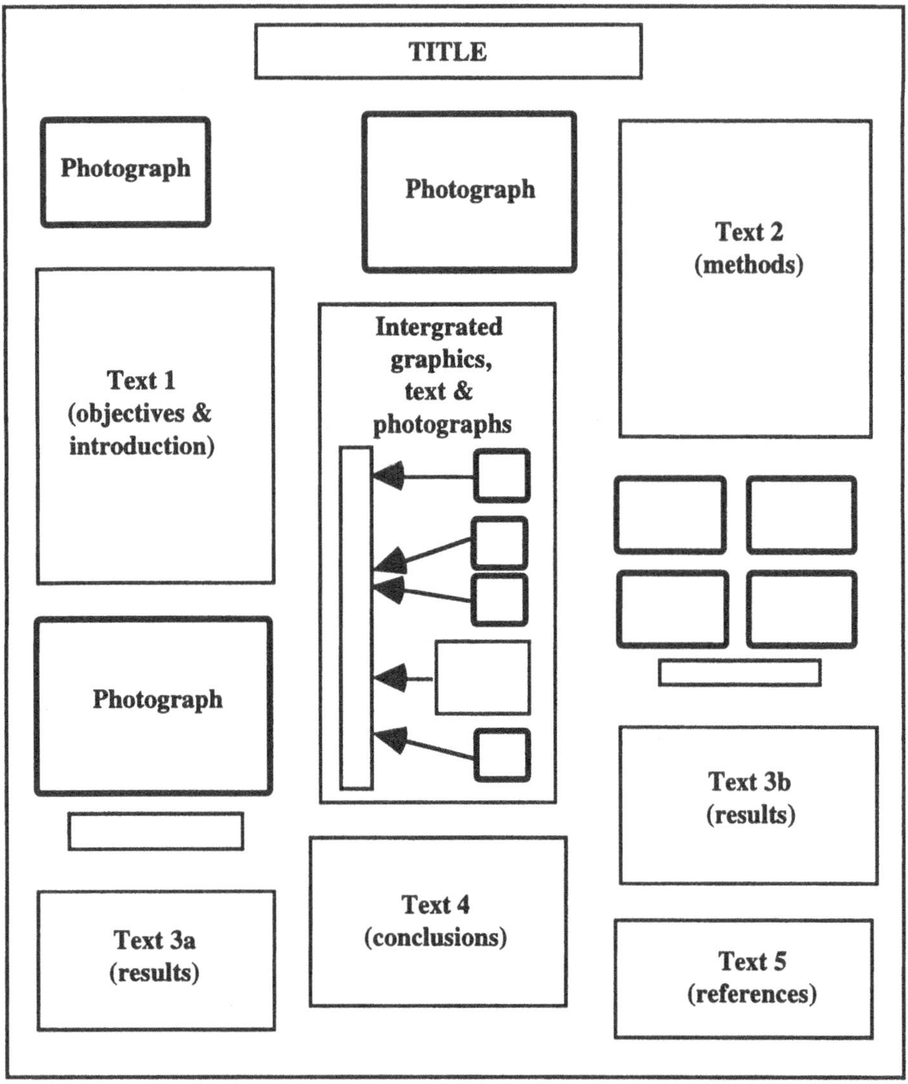

Figure 38.1. Diagrammatic illustration of a poster plan. Text, graphics and images can be individually mounted for ease of transportation and re-use (good photographs are worth keeping!). Positioning of data (particularly new findings) and photogenic illustrations should be carefully considered to provide maximum impact and interest.

aesthetically displayed with similar forms or type, and not for value's sake (Brunton *et al.*, 1985).

The complexity involved in preparing fossil material for exhibition varies considerably within a display as well as between displays. It may involve nothing more than securing a specimen to a horizontal or vertical display board using pins or a contoured base to prevent movement (Rixon 1976). However, considerably more thought is required in preparing three-dimensional dioramas, incorporating a painted background with specimens and models in the foreground (*e.g.* Chase 1979b).

Complexity is increased further when an element of movement is attempted. This is particularly so when displaying the articulated skeletons of vertebrates, and the structural pose of a mount must be considered (Converse 1989, Carpenter *et al.*, 1995). Action clearly adds life to an exhibition, and may be desirable to illustrate the interaction between specimens. A simple walking pose is most commonly employed, although there is increasing pressure on museums to display dinosaurs in more active poses of running or fighting. The pose adopted governs the anatomical relationship of limbs and joints, and even the relationship of the skull to the vertebrae and spinal column. Many of the anatomical errors and physiological inaccuracies seen in museum displays arise from an inadequate knowledge of basic vertebrate anatomy (Carpenter *et al.*, 1995). To a large extent, reference to one of many specialized anatomical guides on mounting skeletons can avoid this, as can seeking advice from a vertebrate specialist or selecting a less adventurous and more realistic pose for the display. Consideration should also be given to the palaeoecological interpretation of an organism, as significant aspects of the life history may be revealed from associated tracks and traces.

Unless casts are used, it is unlikely that any two fossil mounts are preserved or prepared in exactly the same way. Missing parts of the skeleton may have to be cast before mounting can be undertaken, while the continuation of scientific research may require casting of original material to facilitate studies. This can also influence the type of display mount employed, which in recent years has seen a change from a support frame of welded iron rods (*e.g.* Bessom 1963), to the more superior and widely used internal mounting techniques (*e.g.* Bathel 1966) which provides a greater flexibility in dismantling an exhibit. The specialized procedures used in vertebrate palaeontology are outlined by Carpenter *et al.*, (1995), Converse (1989) and Rixon (1976).

38.3.2. Media selection

Assuming the data and subject covered are of interest, the single most influential factor governing the effect of a display will be the physical means used in transporting the message from the sender to the receiver. The most common method is through a combination of graphics and text, nowadays interspersed with audio-visual and interactive modes, and real or replica specimens and models. Although there are few rules governing media use, it is important to select the correct form for a particular example (Miles 1990). It is unlikely though, that any one form will ever supersede or replace any other, but maintenance costs against educational value must always be considered.

Although many user surveys show the general public spends less than 30 seconds at each display, and less than 15 minutes within an exhibition (Brunton *et al.*, 1985), the retention and uses of key phrases and words depicted in text provides some evidence of the usefulness of conventional copy writing (Johnston and Sharpe 1997). Specialists and "experts" spend considerably longer at displays, supporting the notion that text still has a duty to inform the occasional serious visitor as well as entertaining the casual majority (Brunton *et al.*, 1985).

The preparation of line diagrams (*e.g.* Chase 1979a, Isham 1965) has, in many cases, been replaced by high quality photographs, although considerable detail can be indicated from carefully drawn camera lucida (drawing tube) diagrams (*e.g.* Briggs *et al.*, 1983, p5). The specialized preparation of photographic plates (*e.g.* Feldman 1989d,

Palmer 1965) is detailed in section 37 PHOTOMACROGRAPHY AND PHOTOMICROGRAPHY TECHNIQUES. A wide variety of pen and ink techniques and board types can be used in preparing two-dimensional diagrams of three dimensional specimens. Procedures applicable to palaeontological illustrations are described in some detail by Chase (1979a) and Isham (1965). Table 38.2 differentiates the principle types of diagrams, tools and material used in their construction.

The use of dynamic (movable) audio-visual media provides an excellent and important method of imparting information, particularly if a change over time is being portrayed. Multi-media (motion and sound) provide key aspects to improving both the attracting and holding power of the display (Johnston and Sharpe 1997), although the more complex the modes of communication, the more specialist and expert the help required to set up, operate and work the systems. The ambient lighting and environmental conditions created around a display are crucial in realistically portraying many biological processes, and recreating the palaeobiological setting. A successful display and exhibition will achieve this by actively engaging the audience in learning through a "hands-on" approach. However, to achieve this goal a great deal of skill and care is required in assembling a combination of objects (specimens) and graphics in the correct order (Miles 1990).

38.3.3. Structure of content

Within a research institution specialist designers might be employed in the construction of murals and models (Chase 1979a, b). The careful design of a display within an exhibition can be the key to how the work is understood by the public and accepted amongst specialists (Brunton *et al.*, 1985). Communicating ideas formulated during the planning stages is an extremely difficult task, requiring the interpretation of detailed statements for individual exhibits, and the collective skills of many individuals. Invariably, there is more to say about a subject than space and resources available (Miles 1990). Thus the basis selection of material forms a key aim of the exhibition, or what Miles (1990) defines as a *statement of purpose*. The more detailed statements provided for individual exhibits should include a list of the salient teaching points (*e.g.* facts, concepts, relationships, procedures). Miles (1990) suggests that these should be divided into key concepts and ancillary points, some of which will be included to define other concepts, or remove misconceptions, while others ensure the transfer of knowledge. These statements will also help to promote clear communication between the author and designer responsible for the exhibit, and provide a basis for judging its success as a piece of communication.

38.3.4. Testing

Testing the success of a display and exhibition is a continuous process, initiated during planning and defining objectives, and culminates with developmental testing (Jarrett 1988). Developmental testing or formative evaluation of a new exhibition is usually undertaken by museums and institutions mounting large public displays (Miles 1990). Employing a procedure called cued testing, rough mock ups of the exhibition are assembled and, along with a simple questionnaire, are tested on a small number of people. The qualitative feed-back provided from such an exercise generates valuable information on areas that might be ambiguous, and lessen the chance of visitor interpretation being different from that intended (Miles 1990). As the financial

TYPE OF DRAWING	TOOLS AND MATERIALS
Stipple drawing	Ink diagram stippled with individual dots to represent areas in shade. Areas of darkest shadow are most densely stippled.
Line shading or hatching	Takes less time than stippling. (a) stiff mapping pen - for lines of uniform width. (b) flexible pen - changes the width of the line. Cross hatching - produces a shaded effect by criss-crossing lines.
Scratchboard	Uses a drawing paper which has a smooth chalk covered surface. The surface is inked, allowed to dry, and then using a sharp knife or blade, white lines are carefully scratched. The board is very fragile, and the technique requies a great deal of artistic skill.
Stipple board	A textured board (of raised dots or lines). Outline and important details are drawn with Indian ink, and then lithographic crayons of china-graph pencils are used. The finished product resembles a stipple or line drawing.

Table 38.2. The principle procedures and materials for constructing line diagrams.

implication of incorrectly judging audience reaction can be considerable, it is not surprising that an extensive literature on the subject has arisen (see Miles 1990 for review).

Besides the evaluation of completed exhibitions (*e.g.* Loomis 1987, Miles and Tout 1979, Miles *et al.*, 1988), studies monitoring human nature (how people react to the exhibits) have been documented (*e.g.* Johnston and Sharpe 1997). This specialized survey used behavioural mapping, conducted by tracking and observing selected visitors as they moved through the exhibition and semi-structured interviews. The survey followed on from conventional visitor surveys, similar to those conducted by major museums (*e.g.* Alt 1980), and supplemented by the evaluation of visitor comment cards. On the basis of their findings, Johnston and Sharpe (1997) suggest factors produce displays that both attract and hold visitor attention, and conclude that multi-media (motion and sound) is a key to improving the attracting and holding power of a display, along with its novelty value and positioning within an exhibition. Miles and Tout (1979) attempt to incorporate a similar approach with the scientific knowledge of a subject, and in a philosophical discussion outline a technology (the best learning process) and design strategy of exhibits to achieve this.

38.4. MAINTAINING AND UPDATING

A completed display forms a bench-mark for the information, knowledge and resources available at the time of its completion. Maintaining and updating a display is an essential part of ensuring it remains topical, scientifically correct and modern. This may involve nothing more than the replacement of labels to make them more legible (Sorsby and Horne 1980), or replacing photographs that have deteriorated in the presence of ultraviolet light (Brunton *et al.*, 1985). Alternatively it may involve alteration of the infrastructure of the displays or building, increasing the services and facilities to enable a wider audience to view the exhibition (Keen 1984).

Occasionally extensive restoration and specialized conservation of valuable antique specimens are required. Procedures employed in the restoration of mounted ichthyosaur specimens suffering pyrite decay have recently been documented (Lindsay and Comerford 1996). Damage to the wall mounted specimens were accelerated by their exposure and contact with a damp gallery wall. Preliminary examination revealed that the mounted construction was more complex than a simple prepared slab supported within a frame. The slab was attached to a frame with steel screws around the edge, countersunk in infilled holes. A joint evident around the perimeter of the fossil and supporting matrix also suggested a more complex construction. Damage in the form of cracks, radiating from behind the skull, with evidence of active pyrite decay (confirmed by x-ray diffraction) occurring within the matrix. Careful dissection of the underside of the mount could only continue following the application of a glassfibre jacket to the poly(vinyl acetyl) coated exposed surface. The painstaking removal work uncovered wood panelling that concealed a bituminous felt layer underlain by chicken wire embedded in a patchy covering of bituminous cement. Beneath this was a layer of yellow elemental sulphur (not associated or to be confused with pyrite decay products). Having clarified the order of construction, conservation work (applications of poly(vinyl acetyl) and filler) could then begin. The original layers were replaced with poly(ethylene) sheets supported by a glassfibre backing, reducing the variety of materials used with a more simple and reversible structure (Lindsay and Comerford 1996).

This story illustrates the careful approach required for the conservation of material prepared during Victorian times. It further highlights the need for preparators and conservators to record details of all chemicals and materials employed during the preparation and construction of specimens for permanent display, which ideally, should be maintenance free.

REFERENCES

ALT, M. B. 1980. Four years of visitor surveys at the British Museum (Natural History) 1976-79. *Museums Journal*. **80**, (1), 10-19.
BATHEL, K. W. 1966. Mounting a skeleton of *Smilodon californicus* Bovard. *Curator*. **9**, (2), 119-124.
BESSOM, L. C. 1963. A techniques for mounting skeletons with fibreglass. *Curator*. **6**, (3), 231-239.
BRIGGS, D. E. G, CLARKSON, E. N. K and ALDRIDGE, R. J. 1983. The conodont animal. *Lethaia*. **16**, (1), 1-14.
BRUNTON, C. H. C, BESTERMAN, T. P and COOPER, J. A. 1985. Guidelines for the Curation of Geological Materials. *Miscellaneous Paper of the Geological Society (London)*. No. 17.
CARPENTER, K, MADSEN, J. H and LEWIS, A. 1995. Mounting of fossil vertebrate skeletons, pp 285-322; In: P. Leiggi, and P. J. May (eds) *Vertebrate Paleontological Techniques*, Volume 1. Cambridge University Press. 368p.
CHASE, T. L. 1979a. Illustrating techniques for palaeontological exhibits, pp 189-208; In: M. G. Bassett (ed) *Curation of palaeontological collections*. Special Papers in Palaeontology. **22**, 1-279.
CHASE, T. L. 1979b. Methods for the preparation of palaeontological models, pp 225-267; In: M. G. Bassett (ed) *Curation of palaeontological collections*. Special Papers in Palaeontology. **22**, 1-279.
CONVERSE, H. H. 1989. *Handbook of paleo-preparation techniques* (2nd edition, revised by R. W. McCarty). Florida Paleontological Society. v + 109p.
FELDMAN, R. M. 1989d. Preparation of paleontological plates, pp 356-358; In: R. M. Feldman, R. E. Chapman and J. T. Hannibal (eds), *Paleotechniques*. The Paleontological Society Special Publication. **4**. iv + 358.
ISHAM, L. B. 1965. Preparation of drawings for paleontological publication, pp 459-468: In; B. Kummel and D. Raup (eds), *Handbook of Paleontological Techniques*. W. H. Freeman and Co., (San Francisco). xiii + 852p.
JARRETT, J. E. 1988. Learning from developmental testing of exhibits. *Curator*. **29**, 295-306.
JOHNSTON, D and SHARPE, T. 1997. Visitor behaviour at *The Evolution of Wales* exhibition, National Museum and Gallery, Cardiff, Wales. *The Geological Curator*. **6**, (7), 255-266.
KEEN, C. 1984. Visitor services and people with a disability. *Museums Journal*. **84**, (1), 33-38.
LINDSAY, W and COMERFORD, G. M. 1996. An unusual method of mounting an ichthyosaur. *The Geological Curator*. **6**, (6), 221-225.

LOOMIS, R. J. 1987. *Museum visitor evaluation: new tool for management*. American Association of State and Local History. Nashville, Tennesee.
MILES, R. S. 1990. Exhibit strategies, pp 519-522: In; D. E. G. Briggs and P. R. Crowther (eds), *Palaeobiology - A synthesis*. Blackwell Scientific Publications (Oxford). xiii + 583p.
MILES, R. S, ALT, M. B, GOSLING, D. C, LEWIS, B. N and TOUT, A. F. 1988. *The design educational exhibits* (2nd edition). Allen & Unwin (London).
MILES, R. S and TOUT, A. F. 1979. Outline of a technology for effective science exhibits, pp 209-224; In: M. G. Bassett (ed), *Curation of palaeontological collections*. Special Papers in Palaeontology. **22**, 1-279.
OBATA, I, MARUO, H, TERAKADO, H, MURAKAMI, T, TANAKA, T and MATSUKAWA, M. 1989. Replicas of dinosaur tracks, using silicone rubber and fibreglass-reinforced plastics, pp 413-415: In; D. G. Gillette and M. G. Lockley (eds) *Dinosaur tracks and traces*. Cambridge University Press, (Cambridge). xvii + 454p.
PALMER, A. R. 1965. Preparation of plates for paleontological publication, pp 456-459: In; B. Kummel and D. Raup (eds), *Handbook of Paleontological Techniques*. W. H. Freeman and Co., (San Francisco). xiii + 852p.
RIXON, A. E. 1976. *Fossil animal remains: their preparation and conservation*. Athlone Press, (London). viii + 304p.
SHARPE, T. 1983. *Geology in museums: a bibliography and index*. National Museum of Wales (Cardiff), Geological Series No. 6. 128p.
SORSBY, B. D and HORNE, S. D. 1980. The readability of museum labels. *Museums Journal*. **80**, (3), 157-159.

REFERENCES

ABBOTT, M, BAREFOOT, A. C and HANKINS, F. W. 1982. Appendix 2, Specimen preparation, pp 165-172: In; A. C. Barefoot and F. W. Hankins, *Identification of modern and Tertiary Woods.* Clarendon Press, Oxford. vii + 189.

ABELMANN, A. 1988. Freeze-drying simplifies the preparation of microfossils. *Micropaleontology.* **34**, (4), 361.

ABINERI, K. W. 1989. Photomicrographs of cellulose peels from the Mesozoic rocks of Dorset. *Proceedings of the Geologists' Association.* **100**, (2), 161-174.

ADAMCZAK, F. 1961. Eridostraca - a new suborder of ostracods and its phylogenetic significance. *Acta Palaeontologica Polonica.* **6**, (1), 29-104.

AGNEW, N, GRIFFIN, H, WADE, M, TEBBLE, T and OXNAM, W. 1989. Strategies and techniques for the preservation of fossil tracksites: an Australian example, pp 397-407: In; D. G. Gillette and M. G. Lockley (eds), *Dinosaur tracks and traces.* Cambridge University Press, (Cambridge). xvii + 454p.

AKPATI, B. N. 1969. New technique for serial sectioning of foraminifera. *Journal of Paleontology.* **43**, (4), 1046-1048.

ALDRIDGE, R. J. 1990. Extraction of Microfossils, pp 502-504: In; D. E. G. Briggs and P. R. Crowther (eds), *Palaeobiology - A synthesis.* Blackwell Scientific Publications (Oxford). xiii + 583p.

ALLEN, J. R. L. 1971. Some techniques in experimental sedimentology. *Journal of Sedimentary Petrology.* **41**, (3), 695-702.

ALLISON, P. A. 1990. Variations in rates of decay and disarticulation of echinodermata: implications for the application of actualistic data. *Palaios.* **5**, (5), 432-440.

ALLISON, P. A, and BRIGGS, D. E. G. (eds). 1991. *Taphonomy: releasing the data locked in the fossil record.* Plenum Press (New York). 560p.

ALLISON, P. A, and BRIGGS, D. E. G. 1993. Exceptional fossil record: distribution of soft-tissue preservation through the Phanerozoic. *Geology.* **21**, (6), 527-530.

ALLISON, P. A, and PYE, K. 1995. Early diagenetic mineralization and fossil preservation in modern carbonate concretions. *Palaios.* **9**, (6), 561-575.

ALLISON, P. A, SMITH, C. R, KUKERT, H, DEMING, J. W and BENNETT, B. A. 1991. Deep-water taphonomy of vertebrate carcasses: a whale skeleton in the bathyal Santa Catalina Basin. *Paleobiology.* **17**, (1), 78-89.

ALLMAN, M and LAWRENCE, D. F. 1972. *Geological Laboratory Techniques.* Blandford, (London). 1-335p.

ALT, M. B. 1980. Four years of visitor surveys at the British Museum (Natural History) 1976-79. *Museums Journal.* **80**, (1), 10-19.

ALVE, E and MURRAY, J. W. 1995. Experiments to determine the origin and palaeoenvironmental significance of agglutinated foraminiferal assemblages. Proceedings of the Fourth International Workshop on Agglutinated Foraminifera, Kraków Poland, September 12-19, 1993. *Grzybowski Foundation Special Publication.* No. 3, 1-11.

ANDERSON, L and HOFFMAN, E. 1963. Concentration of microfaunal samples by use of a *Syntron. Micropaleontology.* **9** (1), 105-106.

ANDERSON, T. F. 1950a. A new technique for the study of biological structures in the electron microscope (Abstract only). *Biological Bulletin.* **99**, 315-316.

ANDERSON, T. F. 1950b. The use of critical point phenomena in preparing specimens for the electron microscope. *Journal of Applied Physics.* **21**, (7), 724.

ANDERSON, T. F. 1951. Techniques for the preservation of three-dimensional structure in preparing specimens for the electron microscope. *Transactions of the New York Academy of Sciences Series II.* **13**, (4), 130-134.

ANDREWS, C. W and COOPER, C. F. 1928. *On a specimen of Elephas antiquus from Upnor.* British Museum (Natural History), London.
ANDREWS, H. N. Jr. 1961. *Studies in Paleobotany.* John Wiley & Sons Inc., (New York). 487p.
ANDRULEIT, H. 1996. A filtration technique for quantitative studies of coccoliths. *Micropaleontology.* **42**, (4), 403-406.
ANIMA, R. J. 1981. A diver operated reverse corer to collect samples of unconsolidated coarse sand. *Journal of Sedimentary Petrology.* **51**, (2), 653-654.
ANONYMOUS. 1935. Recovery of foraminifera by means of flotation, Laboratory of N. V. Bataafsche Petroleum Maatschappij. *Journal of Paleontology.* **9**, (8), 745-746.
ANONYMOUS. 1977. Proposals for the documentation of conservation in museums. *Museum Documentation Association. Occasional Paper.* **1**, 1-40.
ANONYMOUS. 1980. *Vinamold: flexible mould making.* Alec Tiranti Ltd. (Theale) Technical Booklet. 17p.
ANONYMOUS. 1983. *Crystic polyester handbook.* Scott Bader Co. Ltd. (Wellingborough). x + 111p.
ANONYMOUS. 1986. *Bransonic* ultrasonic cleaner, operator's manual. Branson Europa, Netherlands. 80p.
ANONYMOUS. 1990a. *Cold cure silicone rubber for mould making.* Alec Tiranti Ltd. (Theale) Technical Booklet. 24p
ANONYMOUS. 1990b. The Palaeontological Association - notes for authors submitting papers for publication in Palaeontology and Special Papers in Palaeontology. *Palaeontology.* **33**, (4), 993-1000.
ANONYMOUS. 1992. 2. *Standards in the Museum Care of Biological Collections 1992.* Museum and Galleries Commission (London). 55p.
ANONYMOUS. 1995a. HF fatality. *The American Association of Stratigraphic Palynologists Newsletter.* **28**, (1), 14-15.
ANONYMOUS. 1995b. *Polyester resin, glass fibre, cold cast resin metals, embedding resin.* Alec Tiranti Ltd. (Theale) Technical Booklet. 30p.
ANONYMOUS. 1996. *BDH Spillage Chart* (10th Edition). BDH Ltd., (Poole).
ARNOLD, Z. M. 1958. A precision sectioning instrument for microfossils. *Micropaleontology.* **4**, (1), 103-112.
ARNOLD, Z. M. 1965a. An improved instrument for sectioning microfossils, pp 232-238: In; B. Kummel and D. Raup (eds), *Handbook of Paleontological Techniques.* W. H. Freeman and Co., (San Francisco). xiii + 852p.
ARNOLD, Z. M. 1965b. The centrifuge as a tool in foraminiferal preparation, pp 258-261: In; B. Kummel and D. Raup (eds), *Handbook of Paleontological Techniques.* W. H. Freeman and Co., (San Francisco). xiii + 852p.
ARNOLD, Z. M. 1965c. The vacuum oven for impregnating microfossils, pp 262-263: In; B. Kummel and D. Raup (eds), *Handbook of Paleontological Techniques.* W. H. Freeman and Co., (San Francisco). xiii + 852p.
ARNOLD, Z. M. 1974. Field and laboratory techniques for the study of living foraminifera, pp 153-206: In; R. H. Hedley and C. G. Adams (eds), *Foraminifera*, Volume 1. Academic Press (London). iv + 276p.
ARTUSY, R. L and ARTUSY, J. C. 1956. The use of food colouring as a new technique for staining microfossils. *Journal of Paleontology.* **30**, (4), 969-970.
ASHLEY SMITH, J. 1987. Environmental considerations, pp 403-405: In; P. R. Crowther and C. J. Collins (eds), The Conservation of Geological Material. *The Geological Curator.* **4**, (7), 375-474.
ASSARSSON, G and GRANLUND, E. 1924. En metod för pollenanalys av minerogena jordarter. *Geologiska Föreingen Förhandlinger.* Band **46**, 76-82.
ATHERSUCH, J and JONES, C. 1991. Applications of video image capturing to palaeontology. *Microscopy and Analysis.* No. 24. 7.
AUFFENBERG, 1967. What is a plaster jacket? *The Plaster Jacket.* **6**, 1-7.

AUSTIN, W. E. N. 1994. Disturbed foraminiferal stratigraphies - a cautionary tail. *Cushman Foundation Special Publication*. **32**. 155-159.

AWADALLAH, S. A. 1991. A simple technique for vacuum impregnation of unconsolidated, fine-grained sediments. *Journal of Sedimentary Petrology*. **61**, (4), 632-633.

AZAR, D. 1997. A new method for extracting plant and insect fossils from Lebanese amber. *Palaeontology*. **40**, (4), 1027-1029.

BAAS, M, BRIGGS, D. E. G, van HEEMST, J. D. H, KEAR, A. J and de LEEUW, J. W. 1995. Selective preservation of chitin during the decay of shrimp. *Geochimica et Cosmochimica Acta*. **59**, (5), 945-951.

BABCOCK, L. E. 1989. Casting with plaster of Paris, pp 320-325: In; R. M. Feldman, R. E. Chapman and J. T. Hannibal (eds), *Paleotechniques*. The Paleontological Society Special Publication. **4**. iv + 358.

BAIRD, D. 1955. Latex micro-moulding and latex plaster moulding. *Science*. **122**, 202.

BAIRD, D. 1980. The burnt dope technique and other intertidal ploys from America. *Newsletter of the Geological Curators Group*. **2**, (8), 519-520.

BAIRD, G. C, SHABICA, C. W, ANDERSON, J. L and RICHARDSON, Jr., E. S. 1985. Biota of a Pennsylvanian muddy coast: habitats within the Mazonian delta complex, Northeast Illinois. *Journal of Paleontology*. **59**, (2), 253-281.

BALDAUF, J. G. 1985. Cenozoic diatom biostratigraphy and palaeoecology of the Rockall Plateau region, North Atlantic, DSDP leg 81. *Initial Reports of the Deep Sea Drilling Project*. 86. 439-478.

BARNES, S. H. 1991. Electron microscopy and analysis at the Natural History Museum. *Microscopy and Analysis*. No. 25, 29-31.

BARRY, J and JEPSON, T. 1988. *Safety on Mountains*. British Mountaineering Council, (Manchester). 37p.

BARSS, M. S and CRILLEY, B. 1976. A mounting media for palynological residues. *Geological Survey of Canada*, Paper **76-1B**. 131-132.

BARSS, M. S and WILLIAMS, G. L. 1973. Palynology and nannofossil processing techniques. *Geological Survey of Canada*, Paper **73-26**. 1-25.

BARTLETT, M. R and HAEDRICH, R. L. 1966. Techniques in the radiography of fishes. *Transactions of the American Fisheries Society*. **95**, 99-101.

BASSETT, M. G. (ed) 1979. Curation of palaeontological collections. *Special Papers in Palaeontology*. **22**, 1-279.

BASSLER, R. S. 1953. Techniques, pp G16-G21: In; R. C. Moore (ed),.*Bryozoa, Part G of Treatise on Invertebrate Paleontology*. Geological Society of America and the University of Kansas.

BASSLER, R. S and KELLETT, B. 1934. Bibliographic Index of Paleozoic Ostracoda. *Geological Society of America*. Special Paper No. 1. 1-500.

BATES, C. D, COXON, P and GIBBARD, P. L. 1978. A new method for the preparation of clay-rich sediment samples for palynological investigation. *New Phytologist*. **81**, 459-463.

BATHEL, K. W. 1966. Mounting a skeleton of *Smilodon californicus* Bovard. *Curator*. **9**, (2), 119-124.

BATTARBEE, R. W. 1973. A new method for the estimation of absolute microfossil numbers, with reference especially to diatoms. *Limnology and Oceanography*. **18**, (4), 647-653.

BATTEN, D. J and MORRISON, L. 1983. Methods of palynological preparation for palaeoenvironmental, source potential and organic maturation studies. *Norwegian Petroleum Directorate-Bulletin 2 (Oljedirektoratet)*. 35-53.

BAUMGARTNER, P. O, BJØRKLUND, K. R, CAULET, J-P, DE WEVER, P, KELLOGG, D, LABRACHERIE, M, NAKASEKO, K, NISHIMURA, A, SCHAAF, A, SCHMIDT-EFFING, R and YAO, A. 1981. EURORAD II, 1980 - Second European meeting of radiolarian paleontologists: current research on Cenozoic and Mesozoic radiolarians. *Eclogae geologicae Helvetiae*. **74**, (3), 1027-1061.

BAUMGARTNER-MORA, C and BAUMGARTNER, P. O. 1994. Shell structure of fossil foraminifera studied by cathodoluminescence. *Microscopy and Analysis*. No. 40, 35-38.

BÉ, A. W. H. 1959. A method for rapid sorting of foraminifera from marine plankton samples. *Journal of Paleontology*. **33**, (5), 846-848.

BE, A. W. H, JONGEBLOED, W. L and MCINTYRE, A. 1969. X-ray microscopy of Recent Planktonic foraminifera. *Journal of Paleontology*. **43**, (6), 1384-1396.

BECKMANN, H. 1952. The use of acetic acids in micropaleontology. *Micropaleontologist*. **6**, (3), 39.

BELKA, Z, SKOMPSKI, S and WAKSMUNDZKI, B. 1989. A new interfacial method for conodont separation. *Journal of Paleontology*. **63**, (2), 253-255.

BELL, W. C. 1948. Acetic acid etching technique applied to Cambrian brachiopods. *Journal of Paleontology*. **22**, (1), 101-105.

BELLEMO, S. 1974. Ultrastructures in Recent radial and granular calcareous foraminifera. *Bulletin of the geological Insitute, University of Uppsala*. New Series. **4**, 117-122.

BENGTSON, S. 1977. Early Cambrian button-shaped phosphatic microfossils from the Siberian Platform. *Palaeontology*. **20**, (4), 751-762.

BENGTSON, S. 1986. Preparing clean backgrounds in published photographic illustrations. *Lethaia*. **19**, (4), 361-362.

BENJAMIN, R. E. K. 1971. Recovery of heavy liquids from dilute solutions. *American Mineralogist*. **56**, (3/4), 613-619.

BENSON, R. H. 1965. Photography of microfossils, pp 433-446: In; B. Kummel and D. Raup (eds), *Handbook of Paleontological Techniques*. W. H. Freeman and Co., (San Francisco). xiii + 852p.

BENSON, R. H and KAESLER, R. L. 1965. The Benson-Kaesler bolapipe dredge. *Micropaleontology*. **11**, (3), 396-372.

BENTON, M. J. 1988. British fossil reptile sites, pp 73-84: In; P. R. Crowther and W. A. Wimbledon (eds), *The use and conservation of palaeontological sites*. Special Papers in Palaeontology. **40**, 1-200.

BENTON, M. J and WALKER, A. D. 1981. The use of flexible synthetic rubbers for casts of complex fossils from natural moulds. *Geological Magazine*. **118**, (5), 551-556.

BENTON, M. J and WIMBLEDON, W. A. 1985. The conservation and use of fossil vertebrate sites: British fossil reptile sites. *Proceedings of the Geologists Association*. **96**, (1), 1-6.

BERDAN, J. M. 1989. Mechanical extraction of microfossils, pp 99-100: In; R. M. Feldman, R. E. Chapman and J. T. Hannibal (eds), *Paleotechniques*. The Paleontological Society Special Publication. **4**. iv + 358.

BERNHARD, J. M. 1988. Postmortem vital staining in benthic foraminifera: duration and importance in population and distributional studies. *Journal of Foraminiferal Research*. **18**, (2), 143-146.

BERRY, W. B. N. 1965. Graptolites; pp 103-109: In; B. Kummel and D. Raup (eds), *Handbook of Paleontological Techniques*. W. H. Freeman and Co, (San Francisco). xiii + 852p.

BESENBACHER, F, LAEGSGAARD, E and STENSGAARD, I. 1989. The scanning tunneling microscope. *Microscopy and Analysis*. No. 12, 17-20.

BESSOM, L. C. 1963. A techniques for mounting skeletons with fibreglass. *Curator*. **6**, (3), 231-239.

BESTERMAN, T. P. 1988. The meaning and purpose of palaeontological site conservation, pp 9-19: In; P. R. Crowther and W. A. Wimbledon (eds), *The use and conservation of palaeontological sites*. Special Papers in Palaeontology. **40**, 1-200.

BILLMAN, H, HOTTINGER, L and OESTERLE, H. 1980. Neogene to Recent Rotaliid foraminifera from the Indopacific Ocean; their canal syste, their classification and their stratigraphic use. *Schweizerische Paläontogische Abhandlungen*. **101**, 71-113.

BISHOP, M. J and STANLEY, M. F. 1987, (for 1986). Bruynzeel systems in use: 2, Buxton Museum. *The Geological Curator*. **4**, (9), 548-550.

BLACK, G. P. 1988. Geological conservation: a review of past problems and future promise, 105-111: In; P. R. Crowther and W. A. Wimbledon (eds), *The use and conservation of palaeontological sites*. Special Papers in Palaeontology. **40**, 1-200.

BLACK, M and BARNES, B. 1959. The structure of coccoliths from the English Chalk. *Geological Magazine*. **96**, (5), 321-328.

BLACKIE, R. A. S. 1988. Three dimensional microscopy using holography. *Microscopy and Analysis*. No. 3. 15-17.

BLACKMON, P and TODD, R. 1959. Mineralogy of some foraminifera as related to their classification and ecology. *Journal of Paleontology*. **33**, (1), 1-15.

BODÉN, P. 1991. Reproducibility in the random settling method for quantitative diatom analysis. *Micropaleontology*. **37**, (3), 313-319.

BOLLI, H. 1950. Disintegration of indurated siliceous rocks. *Micropaleontologist*. **4**, (3), 20-21.

BOLLI, H. 1952. Note on the disintegration of indurated rocks. *Micropaleontologist*. **6**, (1), 46-48.

BOLTOVSKOY, E. 1966. Methods for sorting of foraminifera from plankton samples. *Journal of Paleontology*. **40**, (5), 1244-1246.

BONEM, R. M and PERSHOUSE, J. R. 1981. Inexpensive, portable underwater coring device. *Journal of Sedimentary Petrology*. **51**, (2), 654-656.

BOOTH, G. H and SEFTON, G. V. 1970. Vapour phases inhibition of *thiobacilli* and *ferrobacilli*. A potential preservative for pyritic museum specimens. *Nature*. **226**, 185-186.

BOTHNER, M. H and VALENTINE, P. C. 1982. A new instrument for sampling flocculent material at the sediment/water interface. *Journal of Sedimentary Petrology*. **52**, (2), 639-640.

BOUMA, A. H. 1964. Notes on x-ray interpretation of marine sediments. *Marine Geology*. **2**, 278-309.

BOUMA, A. H. 1969. *Methods for the Study of Sedimentary Structures*. Wiley-Interscience, (New York). xvi + 458p.

BOURDON, M. 1957. Utilisation de l'acide acétique dans la désagrégation des roches dures. *Revue Institue of French Petroleum*. **12**, (1), 14-15.

BOURDON, M. 1962. Méthode de dégagement des microfossiles par acétolyse à chaud. *Comptes rendus sommaire des seances de la Societe géologique de France*. 267-268.

BRACEGIRDLE, B. 1995. *Scientific photoMACROgraphy*. Microscopy Handbooks 31. Royal Microscopical Society, Bios Scientific Publishers (Oxford). 105p.

BRADBURY, S. 1989. *An introduction to the optical microscope* (2nd ed). Microscopy Handbooks 1. Royal Microscopical Society, Oxford University Press (Oxford). 86p.

BRADLEY, S. M. 1987. An introduction to the use of silanes in stone conservation, pp 427-432: In; P. R. Crowther and C. J. Collins (eds), The Conservation of Geological Material. *The Geological Curator*. **4**, (7), 375-474.

BRAMLETTE, M. N and SULLIVAN, F. R. 1961. Coccolithophorids and related nannoplankton of the early Tertiary in California. *Micropaleontology*. **7**, (2), 129-188.

BRANSON, E. B, MEHL, M. G and BRANSON, E. R. 1933. Conodont studies. *The University of Missouri Studies*. **8**, (1-4), 1-343.

BRASIER, M. D. 1980. *Microfossils*. Allen and Unwin (London). xii + 193p.

BRASIER, M. D. 1981. Microfossil transport in the tidal Humber basin, pp 314-322: In; J. W. Neale and M. D. Brasier (eds), *Microfossils from Recent and Fossil Shelf Seas*. Ellis Horwood, (Chichester). 380p.

BRETHERICK, L. 1986. Hazards in the Chemical Laboratory (4th Edition). *Royal Society of Chemistry (London)*. xiii + 604p.

BRETT, C. E and BAIRD, G. C. 1986. Comparative taphonomy: a key to paleoenvironmental interpretation based on fossil preservation. *Palaios*. **1**, (3), 207-227.

BRIGGS, A. 1985. *An introduction to scanning acoustic microscopy*. Microscopy Handbooks 12. Royal Microscopical Society, Oxford University Press (Oxford). 70p.

BRIGGS, D. E. G. 1995. Experimental taphonomy. *Palaios*. **10**, (4), 539-550.

BRIGGS, D. E. G, CLARKSON, E. N. K and ALDRIDGE, R. J. 1983. The conodont animal. *Lethaia*. **16**, (1), 1-14.

BRIGGS, D. E. G and CROWTHER, P. R. (eds). 1990. *Palaeobiology - A synthesis*. Blackwell Scientific Publications (Oxford). xiii + 583p.

BRIGGS, D. E. G and KEAR, A. J. 1993a. Decay and preservation of polychaetes: taphonomic thresholds in soft-bodied organisms. *Paleobiology*. **19**, (1), 107-135.

BRIGGS, D. E. G and KEAR, A. J. 1993b. Fossilization of soft tissue in the laboratory. *Science*. **259**, 1439-1442.

BRIGGS, D. E. G and KEAR, A. J. 1993c. Decay of *Branchiostoma*: implications for soft-tissue preservation in conodonts and other primitive chordates. *Lethaia*. **26**, (3), 275-287.

BRIGGS, D. E. G and KEAR, A. J. 1994. Decay and mineralization of shrimps. *Palaios*. **9**, (5), 431-456

BRIGGS, D. E. G and KEAR, A. J, BASS, M, de LEEUW, J. W and RIGBY, S. 1995. Decay and composition of the hemichordate *Rhabdopleura*: implications for the taphonomy of graptolites. *Lethaia*. **28**, (1), 15-23.

BRIGGS, D. E. G, KEAR, A. J, MARTILL, D. M and WILBY, P. R. 1993. Phospatization of soft-tissue in experiments and fossils. *Journal of the Geological Society, London*. **150**, (6), 1035-1038.

BRIGGS, D. E. G, , WILBY, P. R, PÉREZ-MORENO, B. P, SANZ, J. L, and FREGENAL-MARTÍNEZ, M. 1997. The mineralization of dinosaur soft tissue in the Lower Cretaceous of Las Hoyas, Spain. *Journal of the Geological Society, London*. **154**, (4), 587-588.

BRITISH STANDARD 410:1986. Metric units specification for test sieves. *British Standard Institute*.

BRITISH STANDARD 1769:1952. Methods for the use of B.S. fine mesh test sieves. *British Standard Institute*.

BROMLEY, R. G. 1980. Enhancement of visibility of structures in marly chalk: modification of the Bushinsky oil technique. *Bulletin of the Geological Society of Denmark*. **29**, 111-118.

BROOKS, H. K. 1954. A method for removing platy fragments of fossils and minerals from microfossil concentrates. *Micropaleontologist*. **8**, (4), 39-40.

BROTZEN, F. 1936. Foraminiferen aus dem schwedischen untersten Seon von Erikskal in Schonen Sweden. *Sveriges geologiska undersokning. Afhandlingar*. Serie C, **30**, (396), 1-206.

BROWNE, T. D. 1982. The treatment of hydrofluoric acid burns. *Journal of the Society of Occupational Medicine*. 24, 80-88.

BRUTON, T. 1974. Pneumatic hand tools. *Newsletter of the Geological Curators Group*. **1**, (2), 70-72.

BRUNTON, C. H. C. 1979. The development of a computer-based curitorial system for palaeontology at the British Museum (Natural History), pp 159-173: In; M. G. Bassett (ed), *Curation of palaeontological collections*. Special Papers in Palaeontology. **22**, 1-279.

BRUNTON, C. H. C, BESTERMAN, T. P and COOPER, J. A. 1985. Guidelines for the Curation of Geological Materials. *Miscellaneous Paper of the Geological Society (London)*. No. 17.

BULMAN, O. M. B. 1944-1947. Monograph of Caradoc (Balclatchie) graptolites from limestones in Laggan Burn, Ayrshire. *Palaeontographical Society (Monograph)*. xi + 78p.

BULMAN, O. M. B. 1970. *Graptolithina, with sections on Enteropneusta and Pterobranchia,* Part V (revised). Treatise on Invertebrate Paleontology. The Geological Society of America and University of Kansas (Boulder Colorado and Lawrence Kansas). xxxii + V163.

BURCKLE, L. H. 1978. Marine Diatoms, pp 245-266: In; B. U. Haq and A. Boersma (eds), *Introduction to Marine Micropaleontology.* Elsevier Biomedical (New York). viii + 376.

BURKE, A. C, ANDERSON, M, WELD, A and GAFFNEY, E. S. 1983. The reconstruction and casting of a large extinct turtle, *Meiolania. Curator.* **26**, (1), 5-26.

BURMA, B. H. 1965. Radiolarians, pp 7-14: In; B. Kummel and D. Raup (eds), *Handbook of Paleontological Techniques.* W. H. Freeman and Co., (San Francisco). xiii + 852p.

BURMANN, G. 1968. Diacrodien aus dem Unteren Ordovizium. *Paläontolgische Abhandlungen Band 2.* **4**, 635-793.

BURNHAM, R. J. 1988. A large-diameter coring device for use in shallow water. *Journal of Paleontology.* **24**, (4), 479-484.

BUSHINSKY, G. I. 1947. Textures and structures of Cretaceous chalk and flints (Russian with English summary). *Bulletin of Society of Naturalists, Moscow.* New Series 52, (Geology). **22**, 37-43.

BUZAS, M. A. 1965. Foraminifera from late Pleistocene clay near Waterville, Maine. *Smithsonian Miscellaneous Collection.* **145**, (8), 1-30.

CALDER, J and GARRETT, J. 1990. *The new 35mm phtographer's handbook* (Revised edition). Crown Publishers (New York). 239p.

CALLAHAN, J. 1987. A nontoxic heavy liquid and inexpensive filters for separation of mineral grains. *Journal of Sedimentology.* **57**, (4), 765-766.

CAMP, C. L and HANNA, G. D. 1937. *Methods in Paleontology.* University of California Press, (Berkeley, California). xxiii + 153p.

CANNO, R, POINAR, H, ROUBIK, D. W and POINAR, Jr., G. O. 1992a. Enzymatic amplification and nucleotide sequencing of portions of the 18s rRNA gene of the bee *Proplebeia dominicana* (Apidae: Hymenoptera) in 25-40 million year old Dominican amber. *Medical Science Research.* **20**, (7), 249-251.

CANNO, R, POINAR, H. N and POINAR, Jr , G. O. 1992b. Isolation and partial characterization of DNA from the bee *Proplebeia dominicana* (Apidae: Hymenoptera) in 25-40 million year old amber. *Medical Science Research.* **20**, (17), 619-623.

CARATINI, C. 1980. Ultrasonic sieving to improve palynological processing of sediments: a new device. *International Commission for Palynology Newsletter.* **3**, (1), 4.

CARPENTER, K, MADSEN, J. H and LEWIS, A. 1995. Mounting of fossil vertebrate skeletons, pp 285-322; In: P. Leiggi, and P. J. May (eds) *Vertebrate Paleontological Techniques,* Volume 1. Cambridge University Press. 368p.

CARPENTER, W. B, PARKER, W. K and JONES, T. R. 1862. *Introduction to the study of the foraminifera.* Ray Society (London). 1-319 + 22 plates.

CARRECK, J. N and ADAMS, S. J. 1969. Field extraction and laboratory preparation of fossil bones and teeth, using expanded polyurethane. *Proceedings of the Geologists' Association.* **80**, (1), 81-89.

CARSON, C. M. 1933. A method of concentrating foraminifera. *Journal of Paleontology.* **7**, (4), 439.

CARSON, C. M. 1953. Heavy liquid concentration of foraminifera. *Journal of Paleontology.* **27**, (6), 880-881.

CARVER, R. E. 1971. *Procedures in sedimentary petrology.* Wiley-Interscience, (New York). xiii + 653p.

CHALONER, W. G. and GAY, M. M. 1973. Scanning electron microscopy of latex casts of fossil plant impressions. *Palaeontology.* **16**, (3), 645-649.

CHANEY, D. S. 1989a. Hand-held mechanical preparation tools, pp 186-203: In; R. M. Feldman, R. E. Chapman and J. T. Hannibal (eds), *Paleotechniques.* The Paleontological Society Special Publication. **4**. iv + 358.

CHANEY, D. S. 1989b. Mold making with room temperature vulcanizing silicone rubber, pp 284-304: In; R. M. Feldman, R. E. Chapman and J. T. Hannibal (eds), *Paleotechniques*. The Paleontological Society Special Publication. **4**. iv + 358.

CHANEY, D. S and GOODWIN, M. B. 1989. RTV silicone rubber compounds used for molding fossil vertebrate specimens: a comparison. *Journal of Vertebrate Paleontology*. **9**, (4), 471-473.

CHAPMAN, S. K. 1986. *Maintaining and monitoring the transmission electron microscope*. Royal Microscopical Society, Handbook No. 8. Oxford Science Publications, Oxford University Press.

CHARLTON, D. S. 1969. An improved technique for heavy liquid separation of conodonts. *Journal of Paleontology*. **43**, (2), 590-592.

CHASE, T. L. 1979a. Illustrating techniques for palaeontological exhibits, pp 189-208; In: M. G. Bassett (ed) *Curation of palaeontological collections. Special Papers in Palaeontology*. **22**, 1-279.

CHASE, T. L. 1979b. Methods for the preparation of palaeontological models, pp 225-267; In: M. G. Bassett (ed) *Curation of palaeontological collections. Special Papers in Palaeontology*. **22**, 1-279.

CHESCOE, D and GOODHEW, P. J. 1990. *The operation of Transmission and Scanning Electron Microscopes*. Royal Microscopical Society, Handbook No. 20. Oxford Science Publications, Oxford University Press. vi + 86p.

CHIOVETTO, R, McGUFFE, L. J, LITTLE, S. A, WHEELER-CLARK, E and BRASS-DALE, J. 1987. Combined quick freezing, freeze-drying, and embedding tissue at low temperature and in low viscosity resins. *Journal of Electron Microscopical Techniques*. **5**, (1), 1-15.

CHOQUETTE, P. W and TRUSELL, F. C. 1978. A procedure for making the Titan-yellow stain for Mg-calcite permanent. *Journal of Sedimentary Petrology*. **48**, (2), 639-641.

CIESIELSKI, P. F. 1983. The Neogene and Quaternary diatom biostratigraphy of subantarctic sediments, deep sea drilling project Leg 71. *Initial Reports of the Deep Sea Drilling Project, 71. Washington, U.S. Government Printing Office*. 635-665.

CLARK, D. F. 1973. Improved light microscopy techniques for viewing nannofossils. *Micropaleontology*. **19**, (4), 481-484.

CLARK, S and MORRISON, I. 1995. CT scan of fossils, pp 323-329 in Chapter 12, Methods and use of CT scan and X-ray: In; P. Leiggi and P. May (eds), *Vertebrate Paleontological Techniques*, Volume 1. Cambridge University Press. 368p.

CLAUGHER, D and TAYLOR, P. D. 1990. Electron Microscopy, pp 508-511: In; D. E. G. Briggs and P. R. Crowther (eds), *Palaeobiology - A synthesis*. Blackwell Scientific Publications (Oxford). xiii + 583p.

CLAY-POOLE, S. T. 1990. Calcium dissolution and clay flocculant removal in pollen extractions from sediments of the arid southwest, USA. *Journal of Paleontology*. **64**, (3), 483-484.

CLEAL, C. J. 1988. British palaeobotanical site, pp 57-71: In; P. R. Crowther and W. A. Wimbledon (eds), *The use and conservation of palaeontological sites*. Special Papers in Palaeontology. **40**, 1-200.

CLEMENS, W. A. 1988. Challenges of management of palaeontological site resources in the United States, pp 173-180: In; P. R. Crowther and W. A. Wimbledon (eds), *The use and conservation of palaeontological sites*. Special Papers in Palaeontology. **40**, 1-200.

COHEE, G. V. 1937. Inexpensive equipment for reclaiming heavy liquids. *Journal of Sedimentary Petrology*. **7**, (1), 34-35.

COHEN, A. D and GUBER, A. L. 1968. Production of pollen-sized "microforaminifera" from "normal" foraminifera. *Micropaleontology*. **14**, (3), 361-362.

COLBATH, G. C. 1985. A comparison of palynological extraction techniques using samples from the Silurian Bainbridge Formation, Missouri, USA. *Review of Palaeobotany and Palynology*. **44**, 153-164.

COLENUTT, G. W. 1941. A method of obtaining clean residues from fossiliferous clays and marls. *Proceedings of the Geologists' Association.* **52**, (3), 194-197.

COLLINS, C. J. (ed) 1995. *The Care and Conservation of Palaeontological Material.* Butterworth-Heinemann, (Oxford). xi + 139p.

COLLINSON, C. 1963. Techniques for collecting and processing conodonts. *Illinois State Geological Survey.* Circular 343, 1-16.

COLLINSON, C. 1965. Conodonts, pp 94-102: In; B. Kummel and D. Raup (eds), *Handbook of Paleontological Techniques.* W. H. Freeman and Co., (San Francisco). xiii + 852p.

COLLINSON, M. E. 1987. Special problems in the conservation of palaeobotanical material, pp 439-445: In; P. R. Crowther and C. J. Collins (eds), The Conservation of Geological Material. *The Geological Curator.* **4**, (7), 375-474.

COLTHURST, J. R. J and SMITH, D. G. 1977. Palaeontological evidence for the age of the Lower Palaeozoic rocks of the Slievenamon Inlier, County Tipperary. *Proceedings of the Royal Irish Academy.* **77B**, 143-148.

COMMITTEE OF HEADS OF UNIVERSITY GEOSCIENCE DEPARTMENTS and THE GEOLOGICAL SOCIETY, LONDON. 1995. *Safety in geoscience fieldwork: precautions, procedures and documentation.* (London). 18p.

COMPTON, R. R. 1962. *Manual of field geology.* John Wiley and Sons, Inc. (New York). xi + 378p.

CONRAD, K. 1989. The use of Alginate for footprint replication, pp 417-418: In; D. D. Gillette and M. G. Lockley (eds), *Dinosaur Tracks and Traces.* Cambridge University Press, (New York). xvii + 454p.

CONROY, G. C and VANNIER, M. W. 1984. Noninvasive three-dimensional computer imaging of matrix-filled fossil skulls by high resolution computer tomography. *Science.* **226**, 456-458.

CONROY, G. C and VANNIER, M. W. 1985. Endocranial volume determination of matrix-filled fossil skulls using high resolution computer tomography, pp 419-426: In; P. V. Tobias (ed), *Homonid evolution: Past, present and future.* Alan R. Liss (New York).

CONVERSE, H. H. 1983. A new procedure in silicone rubber moldmaking. *Bulletin of the Society of Vertebrate Paleontologists.* **128**, 65-66.

CONVERSE, H. H. 1989. *Handbook of paleo-preparation techniques* (2nd edition, revised by R. W. McCarty). Florida Paleontological Society. v + 109p.

COOK, S. F, BROOKS, S. T and EZRA-COHN, H. E. 1962. Histological studies on fossil bone. *Journal of Paleontology.* **36**, (3), 483-494.

COOPER, C. L. 1935. Ammonium chloride sublimate apparatus. *Journal of Paleontology.* **9**, (4), 357-359.

COOPER, G. A and WHITTINGTON, H. B. 1965. Use of acids in preparation of fossils, pp 294-300: In; B. Kummel and D. Raup (eds), *Handbook of Paleontological Techniques.* W. H. Freeman and Co., (San Francisco). xiii + 852p.

COOPER, J. A. 1987. Conservation documentation, pp 407-412: In; P. R. Crowther and C. J. Collins (eds), The Conservation of Geological Material. *The Geological Curator.* **4**, (7), 375-474.

COPE, J. C. W. 1988. The conservation of biostratigraphical sites, pp 95-104: In; P. R. Crowther and W. A. Wimbledon (eds), *The use and conservation of palaeontological sites.* Special Papers in Palaeontology. **40**, 1-200.

CORLISS, B. H and EMERSON, S. 1990. Distribution of rose bengal stained deep sea benthic foraminifera from the Nova Scotia continental margin and Gulf of Maine. *Deep Sea Research.* **37**, (3), 381-400.

CORNISH, L. 1987. The treatment of decaying pyritiferous fossil material using ethanolamine thioglycollate, pp 451-454: In; P. R. Crowther and C. J. Collins (eds), The Conservation of Geological Material. *The Geological Curator.* **4**, (7), 375-474.

CORNISH, L and DOYLE, A. M. 1983. Ethanolamine Thioglycollate as a chemical agent for the neutralization and removal of oxidised pyrite. *The Geological Curator.* **3**, (8), 512-513.

CORNISH, L and DOYLE, A. M. 1984. Use of Ethanolamine Thioglycollate in the conservation of pyritized fossils. *Palaeontology.* **27**, (2), 269-280.

COSTA, L. I. 1983. Safety in palynology laboratories. *Norwegian Petroleum Directorate-Bulletin 2 (Oljedirektoratet).* 75-84.

COX, I. 1933. On *Climacograptus inuiti* sp. nov. and its development. *Geological Magazine.* **70**, 1-19.

CRELLING, J. C and DUTCHER, R. R. 1980. *Principles and applications of coal petrology.* Society of Economic Paleontologists and Mineralogists - Geological Society of America meeting, Indiana. Short Course No. 8 (Tulsa, Oklahoma). 1-127.

CREVELLO, P. D, RINE, J. M and LANESKEY, D. M. 1981. A method for impregnating unconsolidated cores and slabs of calcareous and terrigenous muds. *Journal of Sedimentary Petrology.* **51**, (2), 658-660.

CROUCHER, R and HOWIE, F. M. P. 1976. A G.R.P cast and mount of the Tertiary mammal *Arsinoitherium. Newsletter of the Geological Curators Group.* **1**, (7), 348-368.

CROWLEY, A. J. 1952. Method of extracting foraminifera from refactory shales. *Bulletin of the American Association of Petroleum Geologists.* **36**, (11), 2185.

CROWTHER, P. R. 1981. The fine structure of graptolite periderm. *Special Papers in Palaeontology.* **26**, 1-119.

CROWTHER, P. R. 1990a. Collections Care and Status Material, pp 515-517: In; D. E. G. Briggs and P. R. Crowther (eds), *Palaeobiology - A synthesis.* Blackwell Scientific Publications (Oxford). xiii + 583p.

CROWTHER, P. R. 1990. Collection Management and Documentation Systems, pp 517-519: In; D. E. G. Briggs and P. R. Crowther (eds), *Palaeobiology - A synthesis.* Blackwell Scientific Publications (Oxford). xiii + 583p.

CROWTHER, P. R and COLLINS, C. J. (eds). 1987. The Conservation of Geological Material. *The Geological Curator.* **4**, (7), 375-474.

CROWTHER, P. R and RICKARDS, R. B. 1977. Cortical bandages and the graptolite zooid. *Geologica and Palaeontologica.* **11**, 9-46.

CROWTHER, P. R and WIMBLEDON, W. A. (eds) 1988. The use and conservation of palaeontological sites. *Special Papers in Palaeontology.* **40**, 1-200.

CULVER, S. J and BANNER, F. T. 1978. Foraminiferal assemblages as Flandrian environmental indicators. *Palaeogeography, Palaeoclimatology, Palaeoecology.* **24**, 53-72.

CUMMINGS, R. H. 1950. Orientated thin-sectioning of microfossils. *Contributions to the Cushman Laboratory for Foraminiferal Research.* **1**, (3-4), 66-67.

CUMMINGS, R. H. 1956. Preparation of microfossils for photography. *Micropaleontology.* **2**, (4), 402.

CUSHMAN, J. A. 1935. Paleozoic foraminifera, their relationships to modern faunas and their environment. *Journal of Paleontology.* **9**, (3), 284-287.

CUSHMAN, J. A. 1948. Chapter 3, Collecting and preparing material, pp 25-28: In; *Foraminifera, their classification and economic use,* (4th Edition), revised and enlarged, with an illustrated key to the genera. Harvard University Press (Cambridge, Massachusetts). 605p.

CWYNAR, L. C, BURDEN, E, McANDREWS, J. H. 1979. An inexpensive sieving method for concentrating pollen and spores from fine-grained sediments. *Canadian Journal of Earth Sciences.* **16**, (5), 1115-1120.

DAVIS, B. L and WALAWENDER, M. J. 1982. Quantitative mineralogical analysis of granitoid rocks: a comparison of x-ray and optical techniques. *American Mineralogist.* **67**, 1135-1143.

DAVIS, L and WEBSTER, G. D. 1985. A modified funnel for heavy mineral separations. *Journal of Paleontology.* **59**, (6), 1505-1506.

DAVIES, W. 1865. On the preservation of fossil mammalian remains found in Tertiary deposits. *Geological Magazine*. **2**, 239-240.

DEAN, A. 1989. Palynomorphs from deformed low-grade metamorphic rocks: an SEM-based technique. *Journal of the Geological Society, London*. **146**, (4), 597-599.

DEAN, B. 1902. The preservation of muscle fibres in sharks of the Cleveland Shale. *American Geologist*. **30**, 273-278.

DESALLE, R, GATESY, J, WHEELER, W and GRIMALDI, D. 1992. DNA sequences from a fossil termite in Oligo-Miocene amber and phylogenetic implications. *Science*. **257**, 1933-1936.

DE WEVER, P. 1980. A new technique for picking and mounting radiolarians for scanning electron microscopy. *Micropaleontology*. **26**, (1), 81-83.

DE WEVER, P, RIEDEL, W. R, BAUMGARTNER, P. O, DUMITRICA, P, BJØRKLUND, K. R, CAULET, J-P, DROBNE, K, GRANLUND, A, KOCHER, R and SCHAAF, A. 1979. Recherches actuelles sur les Radiolaires en Europe. *Annales de la Societé geologique du Nord, Lille*. **98**, 205-222.

DICKSON, J. A. D. 1965. A modified staining technique for carbonates in thin section. *Nature*. **205**, 587.

DICKSON, J. A. D. 1966. Carbonate identification and genesis as revealed by staining. *Journal of Sedimentary Petrology*. **36**, (2), 491-505.

DILLE, J. E. 1983. Foot-operated fine-focusing device for the light microscope. *Stain Technology*. **58**, (1), 62.

DIMICHELE, W. A and WING, S. L. (eds). 1988. Methods and Applications of Plant Paleoecology. *Paleontological Society Special Publication*. **3**, iv + 1-171.

DOHER, I. L. 1980. Palynomorph preparation procedures currently used in the paleontology and stratigraphy laboratories, U.S. Geological Survey. *U. S. Geological Survey Circular*. 830. 1-29.

DONOVAN, S. K. (ed). 1991. *The Processes of Fossilization*. Belhaven Press (London). ix + 303p.

DORAN, J. B. 1982. A modification of the bulk maceration technique for paleobotany. *Journal of Paleontology*. **56**, (3), 823-825.

DOUGHTY, P. S. 1981. The state and status of Geology in U. K. Museums. *Miscellaneous Paper of the Geological Society*, (London). No. 13, 118p.

DOUGLASS, R. C. 1965a. Larger foraminifers, pp 20-25: In; B. Kummel and D. Raup (eds), *Handbook of Paleontological Techniques*. W. H. Freeman and Co., (San Francisco). xiii + 852p.

DOUGLASS, R. C. 1965b. Photomicrography of thin sections, pp 446-453: In; B. Kummel and D. Raup (eds), *Handbook of Paleontological Techniques*. W. H. Freeman and Co., (San Francisco). xiii + 852p.

DOW, V. E. 1960. Magnetic separation of conodonts. *Journal of Paleontology*. **34**, (4), 738-743.

DOW, V. E. 1965. Magnetic separation of conodonts, pp 263-267: In; B. Kummel and D. Raup (eds), *Handbook of Paleontological Techniques*. W. H. Freeman and Co., (San Francisco). xiii + 852p.

DOYLE, A. M. 1983. The conservation of sub-fossil bird bone. *The Geological Curator*. **3**, (7), 447-451.

DOYLE, A. M. 1987. The conservation of sub-fossil bone, pp 463-465: In; P. R. Crowther and C. J. Collins (eds), The Conservation of Geological Material. *The Geological Curator*. **4**, (7), 375-474.

DRIVER, H. L. 1928. Aid in disintegrating samples for micro-organic study. *Journal of Paleontology*. **1**, (4), 253-254.

DUFFIELD, S. L and WARSHAUER, S. M. 1979. Two-step process for extraction of microfossils from indurated shales. *Journal of Paleontology*. **53**, (3), 746-747.

DUMICAN, L. W and RICKARDS, R. B. 1985. Optimum preparation, preservation and processing techniques for graptolite electron microscopy. *Palaeontology*. **28**, (4), 757-766.

DUMITRICA, P. 1970. Cryptocephalic and crytothoracic Nassellaria in some Mesozoic deposits of Romania. *Revue roumaine de geologie, geophysique et geographie, Serie de Geologie*. **14**, (1), 1-124.

DUNN, P. H. 1942. Silurian foraminifera of the Mississippi Basin. *Journal of Paleontology*. **16**, (3), 317-342.

DYER, R and COPESTAKE, P. 1990. A review of Late Jurassic to earliest Cretaceous radiolaria and their biostratigraphic potential to petroleum exploration in the North Sea, pp 215-235: In; D. J. Batten and M. C. Keen (eds), *Northwest European micropalaeontology and palynology*. Ellis Horwood.

EAGAR, S. H and SARJEANT, W. A. S. 1963. Fossil hystrichospheres concentrated by sieving techniques. *Nature*. **198**, 81.

EASTON, W. H. 1942. An improved technique for photographing peel sections of corals. *Journal of Paleontology*. **16**, (2), 261-263.

EATON, J. G, KIRKLAND, J. I and DOI, K. 1989. Evidence of reworked Cretaceous fossils and their bearing on the existence of Teriary dinosaurs. *Palaios*. **4**, 281-286.

ECKERT, R. 1960. Reinigungs- und Anreicherungsversuche an Kleinforaminiferen. *Eclogae Geologicae Helvetiae*. **53**, (2), 645-648.

ECKERT, R, HAY, W. W, LORENZ, G and VOGT, P. 1961. The magnetic separator as a tool in micropaleontology. *Journal of Paleontology*. **35**, (5), 876-877.

EDIGER, V. S. 1986. Sieving techniques in palynological sample processing with special reference to the MRA system. *Micropaleontology*. **32**, (3), 256-270.

EDWARDS, A. R. 1963. A preparation technique for calcareous nannoplankton. *Micropaleontology*. **9**, (1), 103-104.

EDWARDS, L. E. 1993. Dinoflagellates, pp 105-129: In; J. H. Lipps (ed), *Fossil Prokaryotes and Protists*. Blackwell Scientific Publications (Oxford). ix + 342p.

EISENACK, A. 1931. Neue mikrofossilien des Baltischen Silurs. *Paläeontologische Zeitschrift*. **13**, (1/2), 74-118.

ELDER, A, MADSEN, S, BROWN, G, HERBEL, C, COLLINS, C, WHELAN, S, WENZ, C, ALDERSON, S and KRONTHAL, L. 1997. Adhesives and consolidants in geological and paleontological conservation: a wall chart. *Society for the Preservation of Natural History Collections Leaflets*. **1**, (2), 1-4.

EMERY, G. R and BROUSSARD, D. E. 1954. A modified Kullenberg piston corer. *Journal of Sedimentary Petrology*. **24**, (3), 207-211.

EMILIANI, C. 1951. Notes on thin sectioning of smaller foraminifera. *Journal of Paleontology*. **25**, (4), 531-532.

EMILIANI, C. 1955. Minerological and chemical composition of the tests of certain pelagic foraminifers. *Micropaleontology*. **1**, (4), 377-380.

EMO, G. T and SMITH, D. G. 1978. Palynological evidence for the age of the Lower Palaeozoic rocks of Slieve Aughty, Counties Clare and Galway. *Proceedings of the Royal Irish Academy*. **78B**, 281-292.

EMPSON, T. R. 1982. Holographic microscopy of fossil ostracods. Chapter 7, pp 123-146: In; R. H. Bate, E. Robinson and L. M. Sheppard (eds), *Fossil and Recent Ostracods*. Ellis Horwood Ltd., (Chichester). 493p.

ERDTMAN, G. 1943. *An introduction to pollen analysis*. Plant Science Books, vol. 12, and Chronica Botanica. (Waltham, Massachusetts and New York). 239p.

ETHINGTON, R. L and AUSTIN, R. L. 1993. Note on the use of hydrofluoric acid for the recovery of conodonts from Ordovician cherts in the Southern Uplands of Scotland and the significance of the conodonts. *Journal of Micropalaeontology*. **12**, (2), 194

EVERETT, K. 1992. Safety planning and laboratory design, pp 18-29: In; S. G. Luxon (ed), *Hazards in the Chemical Laboratory* (5th Edition). Royal Society of Chemistry (London). xix + 675p.

EVITT, W. R. 1949. Stereophotography as a tool of the palaeontologist. *Journal of Paleontology*. **23**, (5), 566-570.

EVITT, W. R. 1961. Observations of fossil dinoflagellates. *Micropaleontology*. **7**, (4), 385-420.

EVITT, W. R. 1963. A discussion and proposals concerning fossil dinoflagellates, hystrichospheres and acritarchs, 1. *Proceedings of the National Academy of Science (USA)*. **49**, (2), 158-164.

EVITT, W. R. 1984. Some techniques for preparing, manipulating and mounting dinoflagellates. *Journal of Micropalaeontology*. **3**, (2), 11-18.

EWING, J. F. 1950. A new technique for removing bones from limestone breccia. *Antiquity*. **24**, 102-105.

FAEGRI, K and IVERSEN, J. 1989. *Textbook of pollen analysis* (4th Edition). Revised by K. Faegri, P. E. Kaland and K. Krzywinski. John Wiley and Sons (Chichester).

FARLEY, M. B. 1988. Environmental variation, palynofloras, and paleoecological interpretation, pp 126-146: In; W. A. DiMichele and S. L. WING (eds), *Methods and Applications of Plant Paleoecology*. Paleontological Society Special Publication. **3**, iv + 1-171.

FARUQI, S. A and COPLEY, A. J. 1966. A technique for separating pollen and spores from tar. *Micropaleontology*. **12**, (3), 370.

FEIGL, G. 1937. *Qualitative analysis by spot tests*. Nordemann Publishing Co., (New York). 400p.

FELDMAN, R. M. 1989a. Photographic procedures, pp 336-341; In: R. M. Feldman, R. E. Chapman and J. T. Hannibal (eds), *Paleotechniques*. The Paleontological Society Special Publication **4**. iv + 358p.

FELDMAN, R. M. 1989b. Whitening fossils for photographic purposes, pp 342-346; In: R. M. Feldman, R. E. Chapman and J. T. Hannibal (eds), *Paleotechniques*. The Paleontological Society Special Publication **4**. iv + 358p.

FELDMAN, R. M. 1989c. Preparation of stereoscopic photographs, pp 347-350; In: R. M. Feldman, R. E. Chapman and J. T. Hannibal (eds), *Paleotechniques*. The Paleontological Society Special Publication **4**. iv + 358p.

FELDMAN, R. M. 1989d. Preparation of paleontological plates, pp 356-358; In: R. M. Feldman, R. E. Chapman and J. T. Hannibal (eds), *Paleotechniques*. The Paleontological Society Special Publication **4**. iv + 358.

FELDMAN, R. M, CHAPMAN, R. E and HANNIBAL, J. T. (eds). 1989. *Paleotechniques*. The Paleontological Society Special Publication **4**. iv + 358.

FERGUSON, L. 1988. The 'fossil cliffs' at Joggins, Nova Scotia: a Canadian case-study, pp 191-200; In: P. R. Crowther and W. A. Wimbledon (eds), *The use and conservation of palaeontological sites*. Special Papers in Palaeontology. **40**, 1-200.

FINCH, E. M. 1972. The calcareous nannofossils and their usefulness as biostratigraphic indicators. *BP Exploration and Producion Research Division, (Sunbury-on-Thames, Middlesex), Technical Note EPR/TN3107* ii + 21p.

FINGER, K. L and ARMSTRONG, G. L. 1984. A technique to obtain multiple orientations of forminifers in thin section. *Journal of Foraminiferal Research*. **14**, (4), 309-313.

FINGER, K. L. 1990. Atlas of California Neogene Foraminifera. *Cushman Foundation for Foraminiferal Research, Special Publication*, **28**. 1-271.

FIORILLO, A. R. 1998. Measuring fossil reworking within a fluvial system: an example from the Hell Creek Formation (Upper Cretaceous) of eastern Montana, pp 243-251: In; Y. Tomida, L. J. Flynn and L. L. Jacobs (eds), *Advances in Vertebrate Paleontology and Geochronology*. National Science Museum Monographs (Tokyo). xii + 292.

FISCHER, A. 1939. Rubber casts and moulds of fossils. *Journal of Paleontology*. **13**, (6), 621.

FITZPATRICK, K. 1991. Anatomical photography of the human skeleton. *Science Technology*. **24**, 12-17.

FORD, P. B and LEE, D. E. 1997. Note on a new method of using hydrofluoric acid for the study of conodonts in cherts in the Torlesse terrane, New Zealand. *Journal of Micropalaeontology*. **16**, (2), 158.

FOURCADE, E and BUTTERLIN, J. 1988. Reworked and redeposited larger foraminifers on slopes and in basins of the Bahamas, Leg 101. *Proceedings ofthe Ocean Drilling Program, Scientific Results*. **101**, 47-61.

FOURNIER, G. 1950. Photographing small foraminifera. *Micropaleontologist*. **4**, (1), 19-21.

FOURNIER, G. 1954. Use of the pinhole diaphragm in photomicrography. *Micropaleontologist*. **8**, (3), 58-62.

FOURNIER, G. 1956. New methods and techniques in the photography of microfossils. *Micropaleontology*. **2**, (1), 37-56.

FOURNIER, G. 1957. Construction of pinhole diaphragms for use in photomicrography. *Micropaleontology*. **3**, (1), 85-87.

FREEMAN, E. F. 1982. Fossil bone recovery from sediment residues by the "interfacial method". *Palaeontology*. **25**, (3), 471-484.

FREY, D. G. 1955. A differential flotation technique for recovering microfossils from inorganic sediments. *New Phytologist*. **54**, (2), 257-258.

FRICKHINGER, K. A. 1994. *The fossils of Solenhofen*. Goldschneck-Verlag (Korb). 336p.

FRIEDMAN, G. M. 1959. Identification of carbonate minerals by staining methods. *Journal of Sedimentary Petrology*. **29**, (1), 87-97.

FRIEDMAN, G. M. 1971. Staining, pp 511-530: In; R. E. Carver (ed), *Procedures in Sedimentary Petrology*. Wiley-Interscience (New York).

FRIZZELL, D. L. 1965. Otoliths, pp 125-127: In; B. Kummel and D. Raup (eds), *Handbook of Paleontological Techniques*. W. H. Freeman and Co., (San Francisco). xiii + 852p.

FROST, M. J and PETTIFER, G. S. 1975. Simulation of deformed fossils. *Geology Teaching*. **7**, 153.

FUEHRER, O. F. 1938. Liquid rubber, a new casting medium. *Museum News*. **15**, (16), 7.

FUEHRER, O. F. 1939. Liquid rubber as an enlarging medium. *Museum News*. **16**, (14), 8.

FULLER, J. A and MEISBURGER, E. P. 1982. A simple ship-based vibratory corer. *Journal of Sedimentary Petrology*. **52**, (2), 642-644.

FUNKHOUSER, J. W and EVITT, W. R. 1959. Preparation techniques for acid-insoluble microfossils. *Micropaleontology*. **5**, (3), 369-375.

GAFFEY, S. J and BRONNIMANN, C. E. 1993. Effects of bleaching on organic and mineral phases in biogenic carbonates. *Journal of Sedimentary Research*. **63**, (4), 752-754.

GAFFEY, S. J, KOLAK, J. J and BRONNIMANN, C. E. 1991. Effects of drying, heating, annealing and roasting on carbonate skeletal material, with geochemical and diagenetic implications. *Geochimica et Cosmochimica Acta*. **55**, 1627-1640.

GALIPPE, V. 1920. Recherches sur la résistance des microzymas à l'action du temps et sur leur survivance dans l'ambre. *Comptes Rendus de l'Académe des Sciences, Paris*. **170**, 856-868.

GARNER, M. R. 1953. The preparation of latex casts of soil cavities for the study of tunnelling activities of animals. *Science*. **118**, 380-381.

GAY, P. 1982. *An introduction to crystal optics*. Longman (London). ix + 262p.

GENTRY, A. W. 1979. Curation of fossil vertebrates. *Special Papers in Palaeontology*. **22**, 87-95.

GEOLOGISTS' ASSOCIATION. 1975. *A code for geological fieldwork*. (London).

GEOLOGISTS' ASSOCIATION. 1989. *Take care when coring*. (London).

GEOLOGISTS' ASSOCIATION. 1995. *A code for geological fieldwork*. (London).

GIBSON, T. G and WALKER, W. M. 1967. Flotation methods for obtaining foraminifera from sediment samples. *Journal of Paleontology*. **41**, (5), 1294-1297.

GILBERT, R and GLEW, J. 1985. A portable percussion coring device for lacustrine and marine sediments. *Journal of Sedimentary Petrology*. **55**, (4), 607-608.

GINSBURG, R. N and LLOYD, R. M. 1956. A manual piston coring device for use in shallow water. *Journal of Sedimentary Petrology*. **26**, (1), 64-66.

GINSBURG, R. N, BERNARD, H. A, MOODY, R. A and DAIGLE, E. E. 1966. The Shell method of impregnating cores of unconsolidated sediments. *Journal of Sedimentary Petrology*. **36**, (4), 1118-1125.

GIPSON, Jr, M. 1963. Ultrasonic disaggregation of shale. *Journal of Sedimentary Petrology*. **33**, (4), 955-958.

GLAUERT, A. M. (ed). 1974 to 1994. *Practical methods in electron microscopy* (Volumes 1-15). North Holland Publishing Company, Elsevier Science Publishers, (Hollond) and Portland (London).

GOCHT, H. 1972. "Grouping" preparation of fossil dinoflagellates. *Micropaleontology*. **18**, (2), 235-239.

GOLDRING, R. 1991. *Fossils in the Field*. Longman Scientific (Harlow, U. K.). xiv + 218p.

GOLDSTEIN, J. I, NEWBURY, D. E, ECHLIN, P, JOY, D. C, FIORI, C and LIFSHIN, E. 1981. *Scanning electron microscopy and X-ray microanalysis*. Plenum Press (New York). xiii + 673p.

GOODHEW, P. J. 1984. *Specimen preparation for transmission electron microscopy of materials*. Royal Microscopical Society, Handbook No. 3. Oxford Science Publications, Oxford University Press. iv + 41p.

GOODWIN, M. B and CHANEY, D. S. 1994. Molding and casting: techniques and materials, pp 235-271 in Chapter 10, Molding, casting, and painting: In; P. Leiggi and P. May (eds), *Vertebrate Paleontological Techniques*, Volume 1. Cambridge University Press. 368p.

GOTT, P. F. 1945. Procedure of simplified stereophotography of fossils. *Journal of Paleontology*. **19**, (4), 390-395.

GOWER, D. J and SENNIKOV, A. G. 1996. Endocranial casts of early archosaurian reptiles. *Paläontologische Zeitschrift*. **70**, (3/4), 579-589.

GRABAU, A. W and SHIMER, H. W. 1910. North American Index Fossils. Vol. 2. 818-819.

GRANT, R. E. 1989. Extraction of fossils from carbonate by acid, pp 237-243: In; R. M. Feldman, R. E. Chapman and J. T. Hannibal (eds), *Paleotechniques*. The Paleontological Society Special Publication. **4**. iv + 358.

GRAY, J. 1965a. Part III, Techniques in palynology (coordinator), pp 469-706: In; B. Kummel and D. Raup (eds), *Handbook of Paleontological Techniques*. W. H. Freeman and Co., (San Francisco). xiii + 852p.

GRAY, J. 1965b. Extraction techniques, pp 530-587: In; B. Kummel and D. Raup (eds), *Handbook of Paleontological Techniques*. W. H. Freeman and Co., (San Francisco). xiii + 852p.

GRAYSON, J. F. 1956. The conversion of calcite to fluorite. *Micropaleontology*. **2**, (1), 71-78.

GREEN, J. 1832. *A monograph of the Trilobites of North America: with coloured models of the species*. Published by the author. (Philadelphia). 94p.

GREEN, O. R. 1986. Vacuum and pressure impregnation of fine grained sediments. Proceedings of Geo-Techniques '85. Part 1. *Science Technology*. **15**, 21-24.

GREEN, O. R. 1993. The use and applications of stains and dyes in geological techniques. *Science Technology*. **28**, 2-26.

GREEN, O. R. 1995a. Risk assessment in Science and Technology, conference report. *Science Technology*. **35**, 13-15.

GREEN, O. R. 1995b. Pitfalls, problems and procedures in micropalaeontological preparation and conservation. *The Geological Curator*. **6**, (4), 157-166.

GREENWALD, M. T. 1989. Techniques for collecting large vertebrate fossils, pp 264-274: In; R. M. Feldman, R. E. Chapman and J. T. Hannibal (eds), *Paleotechniques*. The Paleontological Society Special Publication No. 4. iv + 358p.

GREY, K. 1993. Appendix F: Acritarch biostratigraphy of the Ediacarian of the Centralian Superbasin, pp 109-115: In; R. J. F. Jenkins, J. F. Lindsay and M. R. Walker (eds), *Field Guide to the Adelaide Geosynclinal and Amadeus Basin*. Record No. 1993/35. Australian Geological Survey Organisation. 1-133.

GRIFFITH, J. 1954. A technique for the removal of skeletal remains from bone-bed. *Proceedings of the Geologists' Association.* **65**, (2), 123-124.

GUNNELL, F. H and MOREY, P. S. 1932. The preparation of conodonts for study. *Micropaleontology Bulletin.* **3**, 77-78.

GUNTHER, V. G, GUILLOT, D. G and RIGBY, J. K. 1979. Enlargement of casts and molds of polybutadiene rubber. *Journal of Paleontology.* **53**, (3), 742-744.

GUNTHER, L. F, GUNTHER, V. G and RIGBY, J. K. 1979. An economic miniature sandblaster for preparation of fossils. *Journal of Paleontology.* **53**, (2), 506-507.

GUTSCHICK, R. C. 1960. Photography of Paleozoic arenaceous foraminifera. *Journal of Paleontology.* **34**, (4), 756-762.

HAGEMAN, J. 1976. The internal morphology of *Miniacina miniacea (Pallas). Proceedings Koninklijke Nederlandse Akademie van Wetenschappen* Series B, **79**, (1), 29-35.

HAGN, H. 1953. A new method for preparing orientated thin sections of foraminifera and other small paleontological specimens. *Micropaleontologist.* **7**, (1), 34-43.

HAMBLIN, W, K. 1962. X-ray radiography in the study of structures in homogenous sediments. *Journal of Sedimentary Petrology.* **32**, (2), 201-210.

HAMBLIN, W, K. 1971. X-ray photography, pp 251-284: In; R. E. Carver (ed), *Procedures in Sedimentary Petrology.* Wiley-Interscience (New York). xiii + 653p.

HANNA, G. D and CHURCH, C. C. 1928. Freezing and thawing to disintegrate shales. *Journal of Paleontology.* **2**, 131.

HANNIBAL, J. T. 1989. The air-abrasive technique, pp 227-236: In; R. M. Feldman, R. E. Chapman and J. T. Hannibal (eds), *Paleotechniques.* The Paleontological Society Special Publication. **4**. iv + 358.

HANNIBAL, J. T, WILLIAMS, M. E and JACKSON, G. L. 1988. An inexpensive source of dolomite powder for use with airbrassive units. *Journal of Paleontology.* **62**, (2), 316.

HANSEN, H. J. 1968. A technique for removing gold from plated calcareous microfossils. *Micropaleontology.* **14**, (4), 499-500.

HANSEN, H. J and LYKKE-ANDERSON, A-L. 1976. Wall structure and classification of fossil and recent elphidiid and nonionid foraminifera. *Fossils and Strata.* **10**, 1-37.

HANSEN, H. J and MARTINEZ MACCHIAVELLO, J. C. 1978. A new technique for depiction of diatoms. *Micropaleontology.* **24**, (3), 333-335.

HANSEN, H. J, SCHMIDT, R. R and MIKKELSEN, N. 1975. Convertible techniques for the study of the same nannoplankton specimen. *Proceedings of the Koninklijke Nederlandse Akademie van Wetenschappen.* **B 78**, 225-230.

HARBERSETZER, J. 1994. Radiography of fossils, pp 329-339 in Chapter 12, Methods and use of CT scan and X-ray: In; P. Leiggi and P. May (eds), *Vertebrate Paleontological Techniques*, Volume 1. Cambridge University Press. 368p.

HARDY, R. G and TUCKER, M. E. 1988. X-ray powder diffraction of sediments, pp 191-228: In; M. Tucker (ed), *Techniques in Sedimentology.* Blackwell Scientific Publications (Oxford). ix + 394.

HARGIS, T. G and TWILLEY, R. R. 1994. Improved coring device for measuring soil bulk density in a Louisiana deltaic marsh. *Journal of Sedimentary Research.* **A64**, (3), 681-683.

HARRIS, A. G and SWEET, W. C. 1989. Mechanical and chemical techniques for separating microfossils from rock, sediment and residue matrix, pp 70-86: In; R. M. Feldman, R. E. Chapman and J. T. Hannibal (eds), *Paleotechniques.* ThePaleontological Society Special Publication. **4**. iv + 358.

HARRIS, J. M. 1971. Chemical preparation of gypsum encrusted vertebrate material. *Journal of Paleontology.* **45**, (2), 350.

HARRIS, T. M. 1926. Note on a new method for the investigation of fossil plants. *New Phytologist.* **25**, 58-60.

HASLETT, S. K and KERSLEY, C. L. 1997. Mounting planktonic foraminifera for scanning electron microscopy. *Microscopy & Analysis.* No. 59, 33-34.

HAUBITZ, B, PROKOP, M, DOHRING, W, OSTROM, J. H and WELLENHOFER, P. 1988. Computer tomography of Archaeopteryx. *Paleobiology*. **14**, 206-213.
HAUFF, P. L and AIREY, J. 1980. The handling, hazards and maintenance of heavy liquids in the geologic laboratory. *U. S. Geological Survey Circular*. **827**, 24p.
HAY, W. W. 1961. Note on the preparation of samples for discoasterids. *Journal of Paleontology*. **35**, (4), 873.
HAY, W. W. 1965. Calcareous nannofossils, pp 3-7: In; B. Kummel and D. Raup (eds), *Handbook of Paleontological Techniques*. W. H. Freeman and Co., (San Francisco). xiii + 852p.
HAY, W. W. 1977. Calcareous nannofossils, pp 1055-1200 (techniques 1056-1069), In: A. T. S. Ramsay (ed), *Oceanic Micropalaeontology* (2 Volumes). Academic Press (London). xii + 1453p.
HAY, W. W and SANDBERG, P. A. 1967. The scanning electron microscope, a major break-through for micropalaeontology. *Micropaleontology*. **13**, (4), 407-418.
HAY, W. W and TOWE, K. M. 1962. Electron-microscopic examination of some coccoliths from Donzacq (France). *Eclogae Geologicae Helvetiae*. **55**, (2), 497-519.
HAYAT, M. A. 1989. *Principles and techniques of electron microscopy, Biological applications* (3rd Edition). Macmillan Press (Basingstoke). xv + 469.
HAYNES, J. R. 1981. *Foraminifera*. Macmillan Publishers Ltd., (London). x + 433p.
HEARD, A. 1927. On Old Red Sandstone plants showing structure, from Brecon (South Wales). *Quarterly Journal of the Geological Society, London*. **83**, 195-209.
HEATON, M. J. 1980. New advances in latex casting techniques. *Curator*. **23**, (2), 95-100.
HEDLEY, R. H. 1957. Microradiography applied to the study of foraminifera. *Micropaleontology*. **3**, (1), 19-28.
HEEGER, J. E. 1913. Ueber die mikrochemische Untersuchung fein verteiler Carbonate im Gesteinsschliff. *Centralblatt fur Mineralogie, Geologie und Paläontologie*. 44-51.
HEINTZ, N. 1963. Casting dinosaur footprints at Spitsbergen. *Curator*. **6**, 217-225.
HEITMANN, S. 1983. First aid for hydrofluoric acid burns. *Norwegian Petroleum Directorate-Bulletin 2 (Oljedirektoratet)*. 85-86.
HEMLEBEN, CH, SPINDLER, M and ANDERSON, O. R. 1989. Chapter 3, Collecting and Culture Methods, pp 33-55: In; *Modern Planktonic Foraminifera*. Springer-Verlag, (New York). xiii + 363p.
HENBEST, L. G. 1931. The use of selective stains in paleontology. *Journal of Paleontology*. **5**, (4), 355-364.
HENDEY, N. I. 1971. Electron microscope studies and the classification of diatoms, pp 625-631: In; B. Funnel and W. Riedel (eds), *Micropalaeontology of the Oceans*. Cambridge University Press. (Cambridge). x + 828p.
HENDEY, N. I. 1974. The permanganate method for cleaning freshly gathered diatoms. *Microscopy*. **32**, 423-426.
HENDRY, R. D, ROWELL, A. J and STANLEY, J. W. 1963. A rapid parallel grinding machine for serial sectioning of fossils. *Palaeontology*. **6**, (1), 145-147.
HENWOOD, A. 1992. Exceptional preservation of dipteran flight muscle and taphonomy of insects in amber. *Palaios*. **7**, (2), 203-212.
HERMAN, Y and METZ, M. C. 1972. Staining technique for recent and fossil calcareous invertebrates. *Journal of Paleontology*. **46**, (1), 152.
HERNGREEN, G. F. W. 1983. Palynological preparation techniques. *Norwegian Petroleum Directorate-Bulletin 2 (Oljedirektoratet)*. 13-34
HERON-ALLEN, E. 1894. *Prolegomena towards the study of the Chalk Foraminifera*. Nichols, (London). 36p.
HESSLAND, I. 1949. Investigations of the Lower Ordovician of the Siljan district, Sweden: 1. Lower Ordovician ostracods of the Siljan District, Sweden. *Bulletin of the Geological Institute of Uppsala*. **33**, 97-408.
HIBBARD, C. W. 1949. Techniques of collecting microvertebrate fossils. *Contributions from the Museum of Paleontology University of Michigan*. **8**, (2), 7-19.

HICKS, R and SCHULTZ, F. 1997. *The Black and White handbook.* David and Charles, (Newton Abbot, Devon). 224p.

HIGGINS, A. C and SPINNER, E. G. 1969. Techniques for the extraction of selected microfossils. *Geology - Journal of the Association of Teachers in Geology.* **1**, 12-28.

HILL, C. R. 1987. Jurassic *Angiopteris* (Marattiales) from north Yorkshire. *Review of Palaeobotany and Palynology.* **51**, (1/3), 65-93.

HILL, C. R. 1990. Scanning electron microscopy in palaeobotany: In; D. Claugher (ed), *Scanning Electron Microscopy in Taxonomy and Functional Morphology.* Clarendon Press, (Oxford). Systematics Association Special Volume No. 41. 193-204.

HILL, G, W, DORSEY, M. E, WOODS, J. C and MILLER, R. J. 1979. A radiographic scanning technique for cores. *Marine Geology.* **29**, 93-106.

HILLS, J. M. 1949. Sampling and examination of well cuttings. *Bulletin of the American Association of Petroleum Geologists.* **33**, (1), 73-91.

HILLS, L. V and SWEET, A. R. 1971. The use of "Quaternary O" in megaspore palynological preparations. *Review of Palaeobotany and Palynology.* **13**, 229-231.

HINCHEY, J. V and GREEN, O. R. 1994. A guide to the extraction of fossil diatoms from lithified or partially consolidated sediments. *Micropaleontology.* **40**, (4), 368-372.

HODGKINSON, R. L. 1991. Microfossil processing: a damage report. *Micropaleontology.* **37**, (3), 320-326.

HODGKINSON, R. L and MARGERUM, R. 1986. On the preparation of oriented sections of larger foraminifera occurring in hard limestones. *Micropaleontology.* **32**, (2), 158-162.

HODSON, F and HODSON, H. 1926. Shortcuts in picking out and sorting foraminifera. *Bulletin of the American Association of Petroleum Geologists.* **10**, (11), 1173-1174.

HOFFMEISTER, W. S. 1960. Sodium hypochlorite, a new oxidising agent for the preparation of microfossils. *Oklahoma Geological Notes.* **20**, 34-35.

HOFKER, J. 1927. *The foraminifera of the Siboga expedition.* Part 1 - Tinoporidae, Rotaliidae, Nummulitidae, Amphistiginidae. Uitk. Zool. Bot. Oceanogr. *Geol. Geb., monogr.* IVa, 1-78.

HOFKER, J. 1965. Methods for study and preparation of foraminifera, pp 251-256: In; B. Kummel and D. Raup (eds), *Handbook of Paleontological Techniques.* W. H. Freeman and Co., (San Francisco). xiii + 852p.

HONJO, S. 1960. A study of some primitive *Neoschwagerina* by a new serial section technique. *Journal of the Faculty of Science, Hokkaido University. Series 4, Geology and Mineralogy.* **10**, (4), 457-470.

HONJO, S. 1963. New serial micropeel technique. *Bulletin of the State Geological Survey of Kansas.* **165**, (6), 1-16.

HONJO, S and BERGGREN, W. A. 1967. Scanning electron microscope studies of planktonic foraminifera. *Micropaleontology.* **13**, (4), 393-406.

HOOPER, K. 1959. X-ray absorption techniques applied to statistical studies of foraminifera populations. *Journal of Paleontology.* **33**, (4), 631-60.

HOOPER, K. 1964. Electron probe microanalysis of foraminifera; an exploratory study. *Journal of Paleontology.* **38**, (6), 1082-1092.

HOOPER, K. 1965. X-ray microscopy in morphological studies of microfossils, pp 320-326: In; B. Kummel and D. Raup (eds), *Handbook of Paleontological Techniques.* W. H. Freeman and Co., (San Francisco). xiii + 852p.

HORIE, C. V. 1987. *Materials for conservation. Organic consolidants, adhesives and coatings.* Butterworth-Heinemann (Oxford). xi + 281p.

HOSKINGS, K. F. G. 1957. The identification - essentially by staining techniques - of white and near white mineral grains in composite samples. *The Camborne School of Mines Magazine.* **57**, 5-16.

HOTTINGER, L. 1978. Comparative anatomy of elementary shell structures in selected larger foraminifera, pp 203-266: In; R. H. Hedley and C. G. Adams (eds), *Foraminifera.* Volume 3. Academic Press (London). x + 290p.

HOTTINGER, L. 1979. Araldit als Helfer in der Mikropaläontologie. - aspekte. *Ciba-Geigy.* 3/1979, 1-10.

HOTTINGER, L and MEHL, J. 1991. X-ray microscopy in micropalaeontology. *Microscopy and Analysis.* No. 21, 21-23.

HOTTON, N. 1965. Tetrapods, pp 119-125: In; B. Kummel and D. Raup (eds), *Handbook of Paleontological Techniques.* W. H. Freeman and Co., (San Francisco). xiii + 852p.

HOVORSLEV, M. J and STETSON, H. C. 1946. Free-fall coring tube: a new type of gravity bottom sampler. *Bulletin of the Geological Society of America.* **57**, (10), 935-950.

HOWDEN, H. F and LING, L. E. C. 1974. Low-magnification study of uncoated specimens, In; M. A. Hayat (ed), *Principles and techniques of scanning electron microscopy. Volume 1. Biological Applications.* Van Nostrand Reinhold Co. (New York). 273p.

HOWE, H. J. 1969. Lighting techniques in photomacrography. *Journal of Paleontology.* **43**, (2), 589-592.

HOWE, H. V. 1941. Use of soap in the preparation of samples for micropaleontological study. *Journal of Paleontology.* **15**, (6), 690.

HOWE, S. R. 1987. Bruynzeel systems in use: 1, National Museum of Wales. *The Geological Curator.* **4**, (9), 545-547.

HOWELL, A. C. 1977. Problems with resin at Bolton. *Newsletter of the Geological Curators Group.* **1**, (9), 466.

HOWELL, J. 1977. Photography of low-contrast fossils. *Journal of Paleontology.* **51**, (2), 419-420.

HOWIE, F. M. P. 1974. Introduction of thioglycollic acid in preparation of vertebrate fossils. *Curator.* **17**, 159-166.

HOWIE, F. M. P. 1977a. Pyrite and conservation Part 1: historical aspects. *Newsletter of the Geological Curators Group.* **1**, (9), 457-465.

HOWIE, F. M. P. 1977b. Pyrite and conservation Part 2. *Newsletter of the Geological Curators Group.* **1**, (10), 497-512.

HOWIE, F. M. P. 1979a. Physical conservation of fossils in existing collections. *Newsletter of the Geological Curators Group.* **2** (5), 269-280.

HOWIE, F. M. P. 1979b. Museum climatology and the conservation of palaeontological material, pp 103-125: In; M. G. Bassett (ed), *Curation of palaeontological collections. Special Papers in Palaeontology.* **22**, 1-279.

HOWIE, F. M. P. 1984. The conservation and storage of geological materials, pp 308-322: In; J. M. A. Thompson (ed), *Manual of curatorship. A guide to Museum practice.* Museums Association/Butterworth & Co. xiv + 553p.

HOWIE, F. M. P. 1987. Safety considerations for the geological conservator, pp 379-401: In; P. R. Crowther and C. J. Collins (eds), The Conservation of Geological Material. *The Geological Curator.* **4**, (7), 375-474.

HOWIE, F. M. P. (ed). 1992. *The Care and Conservation of Geological Material.* Butterworth-Heinemann, (Oxford). xv + 138p.

HOYT, W. H and DEMAREST, II, J. M. 1981. A versatile twin-hull barge for shallow-water vibracoring. *Journal of Sedimentary Petrology.* **51**, (2), 656-657.

HUGHES, N. F, DE JEKOWSKY, B and SMITH, A. H. V. 1964. Extraction of spores and other organic microfossils from Paleozoic clastic sediments and coals. *C. R. Sième Congress of International Stratigraphy and Géologie, Carbonifère.* **3**, 1095-1109.

HÜGI, TH. 1945. Gesteinsbildend wichtige Karbonate und deren Nachweis mittels Fäbmethoden. *Schweizerische Mineralogische und Petrograhische Mitteilungen.* **25**, 114-140.

HULME, S. G. 1961. A mechanized method for breaking down and washing foraminferal rock samples. *Micropaleontology.* **7**, (1), 107-113.

HUSSEY, K. M and CAMPBELL, C. B. 1951. A new method of sample preparation. *Journal of Paleontology*. **25**, (2), 224-226.
HUTCHISON, C. S. 1974. *Laboratory handbook of Petrographic Techniques*. Wiley-Interscience (New York). xiii + 527p.
HUTT, J and RICKARDS, R. B. 1967. An improved transfer technique for the preparation and preservation of pyritised graptolites. *Geological Magazine*. **104**, (2), 180-181.
HÜTTEMANN, H and HEMLEBEN, CH. 1985. An improved rotating device for sputter preparation of SEM particulate specimens. *Journal of Foraminiferal Research*. **15**, (2), 108-110.
IBARAKI, M and SAMESHIMA, T. 1962. Fine structures of *Lepidocyclina nipponica* from Shimo-shiraiwa, Izu, Central Japan. *Reports of Liberal Arts and Science Faculty, Shizuoka University*. Section Natural Science. **3**, (3), 147-148.
IMPERATO, D. P. 1987. A modification to the vibracoring technique for sandy sediment. *Journal of Sedimentary Petrology*. **57**, (4), 788-789.
INSTITUTE OF GEOLOGISTS. 1985. *Code of practice for geological visits to quarries, mines and caves*. (London).
IRELAND, H. A. 1950. Curved surface sections for microscopic study of calcareous rocks. *Bulletin of the American Association of Petroleum Geologists*. **34**, (8), 1737-1739.
IRELAND, H. A. 1958. Insoluble residues, pp 74-94: In; J. D. Haun and W. L. LeRoy (eds), *Subsurface geology in petroleum exploration*. Colorado School of Mines, (Golden, Colorado).
IRELAND, H. A. 1967. Preparatory techniques for microfossils and inorganic insoluble residues. *Journal of Paleontology*. **41**, (2), 523.
IRWIN, J, PICKRILL, R. A and deL. MAIN, W. 1983. An outrigger for sampling from small boats. *Journal of Sedimentary Petrology*. **53**, (2), 675-676.
ISHAM, L. B. 1965. Preparation of drawings for paleontological publication, pp 459-468: In; B. Kummel and D. Raup (eds), *Handbook of Paleontological Techniques*. W. H. Freeman and Co., (San Francisco). xiii + 852p.
JACKSON, T. A, GERMS, A and MOORMAN, M. 1974. An improved method for the chemical maceration of sedimentary rocks. *Journal of Paleontology*. **48**, (4), 844-845.
JACOBSON, R. E, RAY, S. F and ATTRIDGE, G. G. 1988. *The manual of photography* (8th Edition). Focal Press, (Guildford). xi + 394p.
JACOBSON, S. R and SCHOPF, J. M. 1979. Preservation of palynological specimens examined by SEM for later microscopical study and reference. *Journal of Paleontology*. **53**, (3), 744-746.
JAESCHKE, H and JAESCHKE, R. 1987. Archaeological conservation: some useful applications for geology, pp 467-472: In; P. R. Crowther and C. J. Collins (eds), The Conservation of Geological Material. *The Geological Curator*. **4**, (7), 375-474.
JAGO, J. B. 1973. A hazard in the magnesium oxide method of whitening fossils. *Journal of Paleontology*. **47**, (3), 591-592.
JANSONIUS, J. 1970. Classification and stratigraphic application of Chitinozoa. *Proceedings of the North American Paleontological Convention, G*. 789-808.
JARRETT, J. E. 1988. Learning from developmental testing of exhibits. *Curator*. **29**, 295-306.
JASONIUS, J and JENKINS, W. A. M. 1978. Chitinozoa, pp 341-357: In; B. U. Haq and A. Boersma (eds), *Introduction to Marine Micropaleontology*. Elsevier Biomedical (New York). viii + 376p.
JEFFORDS, R. M and MILLER, T. H. 1960. Air brush whitening fossils, and notes on photography. *Journal of Paleontology*. **34**, (2), 275-276.
JENKINS, D. A and JOHNSON, D. B. 1993. Abandoned metal mines: a unique mineralogical and microbiological resource. *Journal of the Russell Society*. **5**, (1), 40-44.
JENKINS, W. A. M. 1967. Ordovician Chitinozoa from Shropshire (techniques pp 439-442). *Palaeontology*. **10**, (3), 436-488.

JENNINGS, J. R. 1972. A polyvinyl chloride peel technique for iron sulphide petrifications. *Journal of Paleontology*. **46**, (1), 70-71.
JEPPSSON, L, FREDHOLM, D and MATTIASSON, B. 1985. Acetic acid and phosphatic fossils - a warning. *Journal of Paleontology*. **59**, (4), 952-956.
JEPPSSON, L and ANEHUS, R. 1995. A buffered formic acid technique for conodont extraction. *Journal of Paleontology*. **64**, (4), 790-794.
JIM, C. Y. 1985.Impregnation of moist and dry unconsolidated clay samples using Spurr resin for microstructural studies. *Journal of Sedimentary Petrology*. **55**, (4), 597-599.
JOHNSON, B. 1983. Gearhart Geodata's stratigraphic laboratories, Aberdeen - the design concept. *Norwegian Petroleum Directorate-Bulletin 2 (Oljedirektoratet)*. 1-11.
JOHNSON, M. E and McKERROW, W. S. 1995. The Sutton Stone: an early Jurassic rocky shore deposit in South Wales. *Palaeontology*. **38**, (3), 529-541.
JOHNSTON, D and SHARPE, T. 1997. Visitor behaviour at *The Evolution of Wales* exhibition, National Museum and Gallery, Cardiff, Wales. *The Geological Curator*. **6**, (7), 255-266.
JONASSON, K. E and PATTERSON, R. T. 1992. Preservation potential of salt marsh foraminifera from the Fraser Delta, British Columbia. *Micropaleontology*. **38**, (3), 289-301.
JONES, B, PHIMESTER, K. F, HUNTER, I. G and BLANCHON, P. 1992. A quick, inexpensive, self-contained sediment coring system for use underwater. *Journal of Sedimentary Petrology*. **62**, (2), 725-728.
JONES, D. J. 1958. Displacement of microfossils. *Journal of Sedimentary Petrology*. **28**, (4), 453-467.
JONES, D. J. 1965. Chapter 2, Collection, preparation, and Preservation of Microfossils, pp 7-18: In; *Introduction to Microfossils*. Harper, New York.
JONES, M. D. 1969. A pneumatic power-tool for the palaeontologist. *Museum Assistance Group Newsletter, March*.
JOY, K. W, WILLIS, A. J and LACEY, W. S. 1956. A rapid cellulose peel technique in paleobotany. *Annals of Botany*. **20**, 635-637.
KATZ, B. J. 1978. Preparation of calcareous nannofossil assemblages for chemical examination. *Journal of Paleontology*. **52**, (2), 497-500.
KEAR, A, J, BRIGGS, D. E. G and DONOVAN, D. T. 1995. Decay and fossilization of non-mineralized tissue in coleoid cephalopods. *Palaeontology*. **38**, (1), 105-131.
KEEN, C. 1984. Visitor services and people with a disability. *Museums Journal*. **84**, (1), 33-38.
KEENE, S. 1987. Some adhesives and consolidants used in conservation, pp 421-425: In; P. R. Crowther and C. J. Collins (eds), The Conservation of Geological Material. *The Geological Curator*. **4**, (7), 375-474.
KELLY, S. R. A and MCLACHLAN, A. 1980. The use of silicone rubbers in the preparation of casts from natural fossil moulds. *Geological Magazine*. **117**, (5), 447-454.
KENNEDY, C and ZEIDLER, W. 1976. The preparation of oriented thin sections in micropaleontology: an improved method for revealing the internal morphology of foraminifera and other microfossils. *Micropaleontology*. **22**, (1), 104-107.
KERMACK, K. A. 1988. British Mesozoic mammal sites, pp 85-93: In; P. R. Crowther and W. A. Wimbledon (eds), *The use and conservation of palaeontological sites*. Special Papers in Palaeontology. **40**, 1-200.
KERR, P. F. 1977. *Optical Mineralogy*. (4th Edition). McGraw-Hill (New York). xvi + 492p.
KESLING, R. V and SOHN, I. G. 1958. The paleozoic ostracode genus *Alanella Boucek*, 1936. *Journal of Paleontology*. **32**, (3), 517-524.
KEYES, I. W. 1959. Paleontological casting and moulding techniques. *New Zealand Journal of Geology and Geophysics*. **2**, (1), 56-65.
KIDSON, E. J and WILLIAMS, G. L. 1969. Concentration of palynomorphs by use of sieves. *Oklahoma Geological Notes*. **29**, 117-119.

KIER, P. M, GRANT, R. E and YOCHELSON, E. L. 1965. Whitening fossils, pp 453-456: In; B. Kummel and D. Raup (eds), *Handbook of Paleontological Techniques*. W. H. Freeman and Co., (San Francisco). xiii + 852p.

KIRCHNER, Z. M. 1958. A new method of hard-rock maceration. *Micropaleontology*. **4**, (3), 327-328.

KITCHELL, J. A. 1990. Computer Applications in Palaeontology, pp 493-499: In; D. E. G. Briggs and P. R. Crowther (eds), *Palaeobiology - A synthesis*. Blackwell Scientific Publications (Oxford). xiii + 583p.

KITTRICK, J. A and HOPE, E. W. 1967. Organic dyes and strychnine-molybdate reagent as aids in isolating phosphate mineral grains. *American Mineralogist*. **52**, (1/2), 263-272.

KLING, S. A. 1978. Radiolaria, pp 203-244: In; B. U. Haq and A. Boersma (eds), *Introduction to Marine Micropaleontology*. Elsevier Biomedical (New York). viii + 376p.

KNILL, J. L, CRATCHLEY, C. R, EARLY, K. R, GALLOIS, R. W, HUMPHREYS, J. D, NEWBERY, J, PRICE, D. G and THURRELL, R. G. 1970. Geological Society Engineering Group Working Party Report on The logging of rock cores for engineering purposes. *Quarterly Journal of Engineering Geology*. **3**, (1), 1-24.

KNOX, A. S. 1942. The use of bromoform in the separation of noncalcareous microfossils. *Science*. 95, 307-308.

KOIZUMI, I. 1980. Neogene diatoms from the Emperor seamount chain, Leg 55, Deep Sea Drilling Project. *Initial Reports of the Deep Sea Drilling Project, 55, Washington, U.S. Government Printing Office*. 387-407.

KOIZUMI, I and TANIMURA, Y. 1985. Neogene diatom biostratigraphy of the middle latitude Western North Pacific, DSDP leg 86. *Initial Reports of the Deep Sea Drilling Project, 86, Washington, U.S. Government Printing Office*. 289-300.

KONTROVITZ, M. 1982. A new method of producing internal molds in ostracod valves. *Journal of Paleontology*. **56**, (3), 818-820.

KONTROVITZ, M, SLACK, J. M and YUHONG, Z. 1991. On the use of some phosphates in the preparation of ostracod shells. *Journal of Micropalaeontology*. **10**, (2), 121-126.

KONTROVITZ, M, SNYDER, S. W and BROWN, R. J. 1978. A flume study of the movement of foraminifera tests. *Palaeogeography, Palaeoclimatology, Palaeoecology*. **23**, 141-150.

KORNICKER, L. S. 1957. Concentration of ostracodes by alcohol flotation. *Journal of Paleontology*. **31**, (5), 1030.

KRAFT, P. 1926. Ontogenetische Entwicklung und die Biologie von *Diplograptus* und *Monograptus*. *Paläntologische Zeitschrift*. **7**, 207-249.

KREMP, G. 1953. Preparation of orientated thin sections of microfossils. *Micropaleontologist*. **7**, (1), 29-33.

KRINSLEY, D and BÉ, A. W. H. 1965. Electron microscopy of internal structures of foraminifera, pp 335-344: In; B. Kummel and D. Raup (eds), *Handbook of Paleontological Techniques*. W. H. Freeman and Co., (San Francisco). xiii + 852p.

KRINSLEY, D, MORTON, N, and CUTBILL, J. L. 1964. The application of vacuum shadowing to the examination and photography of small objects. *Geological Magazine*. **101**, (4), 357-359.

KRUKOWSKI, S. T. 1988. Sodium metatungstate: a new heavy mineral separation medium for the extraction of conodonts from insoluble residues. *Journal of Paleontology*. **62**, (2), 314-316.

KRUMBEIN, W. C. 1936. Application of logarithmic momments to size frequency distributions of sediments. *Journal of Sedimentary Petrology*. **6**, (1), 35-47.

KUCERA, L. J. 1981. Cutting wood specimens for observation in the scanning electron microscope. *Journal of Microscopy*. **124**, (3), 319-325.

KUCERA, L. J. 1986. Splitting wood specimens for observation in the scanning electron microscope. *Journal of Microscopy*. **142**, (1), 71-77.

KUEHNE, W. G. 1971. Photoelectric separation of microfossils from gangue. *Proceedings of the Geological Society (London)*. **1664**, 221-222.
KUGLER, H. G. 1957. Preparation of microfossils for photography. *Micropaleontology*. **3**, (2), 190.
KULLENBURG, B. 1947. The piston core sampler. *Svenska hydrogr-biol. Komm, Skr. (3, Hydr)*. **1**, (2), 1-46.
LAGLE, C. W. 1984. Recovery of siliceous microfossils by the disaggregation of dolomite, pp 185-194: In; R. E. Garrison, M. Kastner and D. H. Zenger (eds), *Dolomites of the Monterey Formation and other organic rich units*. Pacific Section, Society of Econmic Petrologists and Mineralogists (Los Angeles, California). 41, 1-215.
LANESKY, D. E, LOGAN, B. W, BROWN, R. G and HINE, A. C. 1979. A new approach to portable vibracoring underwater and on land. *Journal of Sedimentary Petrology*. **49**, (2), 654-657.
LANOOY, R. 1984. A new casting method with a modified mold rotator. *Der Präparator*. **30**, (4), 381-388.
LARSSON, S. G. 1978. Baltic amber - a palaeobiological study. *Entomonograph*. **1**, 1-192.
LAWS, R. A. 1983. Preparing strewn slides for quantitative microscopical analysis: a test using calibrated microspheres. *Micropalaeontology*. **29**, (1), 60-65.
LAYNE, Jr, N. M. 1950. A procedure for shale disintegration. *Micropaleontologist*. **4**, (1), 21.
LEARY, P. N and HART, M. B. 1988. X-raying planktonic foraminifera. *Journal of Micropalaeontology*. **7**, (1), 43-44.
LEFFINGWELL, H. A and HODGKIN, N. 1971. Techniques for preparing fossil palynomorphs for study with the scanning and transmission electron microscopes. *Review of Palaeobotany and Palynology*. **11**, 177-199.
LEHMANN, E. P. 1956. A technique of stereophotomicrography for illustrations in micropaleontology. *Journal of Paleontology*. **30**, (3), 757-759.
LEHMANN, G and RÖTTGER, R. 1997. Techniques for the concentration of foraminifera from coastal salt meadow sediments. *Journal of Micropalaeontology*. **16**, (1), 144.
LEIGGI, P and MAY, P. J. (eds). 1995. *Vertebrate Paleontological Techniques*, Volume 1. Cambridge University Press. 368p.
LENNIE, C. R. 1968. Palynological techniques used in New Zealand. *New Zealand Journal of Geology and Geophysics*. **11**, (5), 1211-1221.
LETHIERS, F and CRASQUIN-SOLEAU, S. 1988. Comment extraire les microfossiles a tests calcitiques des roches calcaires dures. *Revue de Micropaléontologie*. **31**, (1), 56-61.
LEVINSON, S. A. 1950. A technique for sectioning microfossils. *Science*. **111**, 60.
LEVINSON, S. A. 1951a. The Triebel technique for staining ostracodes. *Micropaleontologist*. **5**, (2), 27.
LEVINSON, S. A. 1951b. Thin sections of Palaeozoic Ostracoda and their bearing on taxonomy and morphology. *Journal of Paleontology*. **25**, (5), 553-560.
LEVINSON, S. A. 1956. The identification of ostracod genera and species in thin sections. (Abstract only). *Journal of Paleontology*. **30**, (4), 1010.
LEVINSON, S. A. 1961. Identification of fossil ostracodes in thin section, pp Q70-Q73: In; R. C. Moore (ed), *Treatise on Invertebrate Palaeontology, Part Q, Arthropoda 3*. Geological Society of America and the University of Kansas (Lawrence, Kansas). xxiii + 442.
LEWIN, J. C. 1961. The dissolution of silica from diatom walls. *Geochim et Cosmochim Acta*. **21**, 182-198.
LICHTER, G. 1993. *Fossil Collectors Handbook*. Sterling (New York), 160p.
LINCOLN, R. J and SHEALS, J. G. 1979. *Invertebrate animals. Collection and preservation*. British Museum (Natural History), (London and Cambridge University Press). vii + 150p.
LINDHOLM, R. C. 1987. *A Practical Approach to Sedimentology*. Allen and Unwin (London). xviii + 276p.

LINDHOLM, R. C and DEAN, D. A. 1973. Ultra-thin sections in carbonate petrology: a valuable tool. *Journal of Sedimentary Petrology*. **43**, (1), 295-297.
LINDSAY, W. 1987. The acid technique in vertebrate palaeontology: a review, pp 455-461; In: P. R. Crowther and C. J. Collins (eds), The Conservation of Geological Material. *The Geological Curator*. **4**, (7), 375-474.
LINDSAY, W and COMERFORD, G. M. 1996. An unusual method of mounting an Ichthyosaur. *The Geological Curator*. **6**, (6), 221-225.
LINSLEY, R. M. 1965. Collecting molds of fossils, pp 189-191: In; B. Kummel and D. Raup (eds), *Handbook of Palaeontological Techniques*. W.H.Freeman and Co., San Francisco. xiii + 852p.
LIPPS, J. H. (ed). 1993. *Fossil Prokaryotes and Protists*. Blackwell Scientific Publications (Oxford). ix + 342p.
LISTER, B. 1978. The preparation of polished sections. *Report of the Institute of Geological Sciences*, **78/27**. 20p.
LITWIN, R. J and TRAVERSE, A. 1989. Basic guidelines for palynomorph extraction and preparation from sedimentary rocks, pp 87-98: In; R. M. Feldman, R. E. Chapman and J. T. Hannibal (eds), *Paleotechniques*. The Paleontological Society Special Publication. **4**. iv + 358.
LOCKER, S. 1996. Quantitative radiolarian slides prepared from soft marine sediments. *Micropaleontology*. **42**, (4), 407-411.
LOFGREN, D. L, HOTTON, C. L and RUNKEL, A. C. 1990. Reworking of Cretaceous dinosaurs into Paleocene channel deposits, upper Hell Creek Formation, Montana. *Geology*. **18**, 874-877.
LOHMAN, K. E. 1972. A procedure for the microscopical study of diatomaceous sediments; In: R. Simonsen (ed), Symposium on recent and fossil diatoms. *Nova Hedwegia*. **39**, 267-285.
LOOMIS, R. J. 1987. *Museum visitor evaluation: new tool for management*. American Association of State and Local History. Nashville, Tennesee.
LÖRCHER, F and KELLER, T. 1985 (for 1984). Preparation techniques for material from the Posidonienschiefer (Lias Epsilon, Upper Liassic) of Germany. *The Geological Curator*. **4**, (3), 164-168.
LUND, R. 1970. A new technique for chemical preparation of fossils. *Journal of Paleontology*. **44**, (3), 578.
LUND, R. 1980. Photomacrography of fossils for publication. *Journal of Paleontology*. **54**, (1), 264-266.
LUTZE, G. F. 1968. Jahresgang der foraminiferen-fauna in der Bottsand-Lagune (westliche Ostsee). *Meyniana*. **18**, 13-30.
LUXON, S. G. 1992. *Hazards in the Chemical Laboratory* (5th Edition). Royal Society of Chemistry (London). xix + 675p.
MACEO, J and RISKIND, D. H. 1989. Field and laboratory moldmaking and casting of dinosaur tracks, pp 419-420: In; D. G. Gillette and M. G. Lockley (eds), *Dinosaur tracks and traces*. Cambridge University Press, (Cambridge). xvii + 454p.
MACINTYRE, G. T. 1976. A simplified method for x-ray stereographs of small specimens. *Journal of Paleontology*. **50**, (2), 357-358.
MACVICAR, D. G. 1951a. Heat treatment frees Pre-Cambrian fossils. *Science News Letter*. **59**, (17), 261.
MACVICAR, D. G. 1951b. Extraction of fossils by heat. *Micropaleontologist*. **5**, (3), 15.
MACVICAR, D. G. 1952. Thermal disintegration of sedimentary rocks. *American Journal of Science*. **250**, 271-274.
MAI, H. 1983. Korrelation cretazischer Coccolithen im Lichtmikroskop, Rasterelektronenmikroskop und Transmissionselektronenmikroskop. *Paleontographica Abteilung B, Palaophytologie*. **189**, 139-161.
MAI, H. 1988. A new method for SEM analysis of both proximal and distal sides of the same Coccolith. *Journal of Paleontology*. **62**, (1), 151-152.

MALYSKEK, V. T. 1940. Method of separation of microfauna from the sediments by centrifugal method on heavy liquids. *Azerbeidzhan Petroleum Industry.* **9**, 4-5.
MANDADO, J. A and TENA, J. M. 1986. A peel technique for sulfate (and carbonate) rocks. *Journal of Sedimentary Petrology.* **56**, (4), 548-549.
MANN, V. I. 1950. A spot test for phosphorus in rocks. *Journal of Sedimentary Petrology.* **20**, (2), 116-117.
MANN, V. I. 1955. A spot test for dolomite limestones. *Journal of Sedimentary Petrology.* **25**, (1), 58-59.
MAPLES, C. G and WATERS, J. A. 1990. Use of drain cleaner in processing shale samples. *Journal of Paleontology.* **64**, (3), 484-485.
MARSH, R. C and MARSH, L. F. 1975. New techniques for coating paleontological specimens prior to photography. *Journal of Paleontology.* **49**, (3), 565-566.
MARSHALL, D. J. 1988. *Cathodoluminescence of geological materials.* Unwin Hyman (Boston). xiv + 146p.
MARSHALL, J. E. A. 1990. Determination of Thermal Maturity, pp 511-515: In; D. E. G. Briggs and P. R. Crowther (eds), *Palaeobiology - A synthesis.* Blackwell Scientific Publications (Oxford). xiii + 583p.
MARTILL, D. M. 1991. Organically preserved dinosaur skin: taphonomic and biological implications. *Modern Geology.* **16**, 61-68.
MARTILL, D. M and HARPER, E. 1990. Critical point drying, a technique for palaeontologists. *Palaeontology.* **33**, (2), 423-428.
MARTILL, D. M, WILBY, P. R and WILLIAMS, N. 1992. Elemental mapping: a technique for investigating delicate phosphatized soft tissues. *Palaeontology.* **35**, (4), 869-874.
MARTIN, E. A and MILLER, R. J. 1982. A simple diver-operated coring device for collecting undisturbed shallow cores. *Journal of Sedimentary Petrology.* **52**, (2), 641-642.
MARTIN, R. E. 1988. Benthic foraminiferal zonation in deep-water carbonate platform margin environments, Northern Little Bahama Bank. *Journal of Paleontology.* **62**, (1), 1-8.
MARTIN, R. E and STEINKER, D. C. 1973. Evaluation of techniques for recognition of living foraminifera. *Compass of Sigma Gamma and Epsilon.* **50**, (4), 26-30.
MARTINSSON, A. 1965. Preparation of individual microfossils on cellulose acetatae films, pp 247-251: In; B. Kummel and D. Raup (eds), *Handbook of Paleontological Techniques.* W. H. Freeman and Co., (San Francisco). xiii + 852p.
MATTEN, L. C. 1968. *Actinoxylon banksii* gen. et. sp. nov.: A progymnosperm from the Middle Devonian of New York. *American Journal of Botany.* **55**, 773-782.
MAYBURY, C and AP GWYNN, I. 1993. Wet processing of Recent calcareous foraminifera: methods for preventing dissolution. *Journal of Micropalaeontology.* **12**, (1), 67-69.
MAYBURY, C, MORRISON, L and STEWART, V. 1991. The search for a reliable mounting medium for Recent "live" foraminifera. *Journal of Micropalaeontology.* **10**, (1), 16.
MAXIMOV, A, MILOSERDINA, G and EREMIN, N. 1973. *Short course of Geological Prospecting and Exploration.* MIR Publishers, (Moscow). 1-256.
McCAFFREY, W. D, BARRON, H. F, MOLYNEUX, S. G, KNELLER, B. C. 1992. Recycled acritarchs as provenance indicators: implications for Caledonian terrane reconstruction. *Geological Magazine.* **129**, (4), 457-464.
McGOWAN, C. 1989a. Computer tomography reveals further details of *Excalibosaurus*, A putative ancestor for the swordfish-like ichthyosaur *Eurhinosaurus. Journal of Vertebrate Paleontology.* **9**, 269-281.
McGOWAN, C. 1989b. The ichthyosaurian tailbend: a verification problem facilitated by computer tomography. *Paleobiology.* **15**, 429-436.
McGOWAN, C. 1990. Computer tomography confirms that Eurhinosaurus (Reptilia: Ichthyosauria) does have a tailbend. *Canadian Journal of Earth Science.* **27**, (11), 1541-1545.

McGOWAN, C. 1991. An ichthosaur forefin from the Triassic of British Columbia exemplifying Jurassic features. *Canadian Journal of Earth Science*. **28**, (10), 1553-1560.

McINTYRE, D. J and NORRIS, G. 1964. Effect of ultrasound on recent spores and pollen. *New Zealand Journal of Science*. **7**, (2), 242-257.

McKEE, K. 1977. A dialysis unit in palynological sample preparation. *The British Micropalaeontologist*. **5**, 12-13.

McKENNA, M. C. 1962. Collecting small fossils by washing and screening. *Curator*. **5**, (3), 221-235.

McKENNA, M. C. 1965. Collecting microvertebrate fossils by washing and screening, pp 193-203: In; B. Kummel and D. Raup (eds), *Handbook of Paleontological Techniques*. W. H. Freeman and Co., (San Francisco). xiii + 852p.

McNAIR, A. H. 1941. A method of photographing impressions of fossils. *Journal of Paleontology*. **15**, (1), 91.

MELLER, A. Einschlußmittel mit hohem Brechungsindex für Diatomeen. *Mikrokosmos*. **74**, 55-60.

MENDLESON, C. V. 1993. Acritarchs and prasinophytes, pp 77-104: In: Lipps, J. H. (eds), *Fossil Prokaryotes and Protists*. Blackwell Scientific Publications (Oxford). ix + 342p.

MERRILL, G. K. 1985. Interfacial alternatives to the use of dangerous heavy liquids in micropalaeontology. *Journal of Paleontology*. **59**, (2), 479-481.

MERRILL, G. K. 1987. Improvements in interfacial methods for conodont separation, pp 55-63: In; R. L. Austin (ed), *Conodonts: Investigative Techniques and Applications*. Ellis Horwood Ltd, (Chichester). 1-422.

MEYER, C. 1946. Notes on cutting and polishing thin sections. *Economic Geology*. **41**, 166-172.

MILES, R. S. 1990. Exhibit Strategies, pp 519-522: In; D. E. G. Briggs and P. R. Crowther (eds), *Palaeobiology - A synthesis*. Blackwell Scientific Publications (Oxford). xiii + 583p.

MILES, R. S, ALT, M. B, GOSLING, D. C, LEWIS, B. N and TOUT, A. F. 1988. *The design educational exhibits* (2nd Edition). Allen & Unwin (London).

MILES, R. S and TOUT, A. F. 1979. Outline of a technology for effective science exhibits, pp 209-224; In: M. G. Bassett (ed) *Curation of palaeontological collections*. Special Papers in Palaeontology. **22**, 1-279.

MILESON, P and JEPPSSON, L. 1983. Heavy liquid and solvent recovery, an economical method. *Journal of Sedimentary Petrology*. **53**, (2), 673-674.

MILLER, B. B. 1989. Screen-washing unconsolidated sediments for small macrofossils, pp 260-263: In; R. M. Feldman, R. E. Chapman and J. T. Hannibal (eds), *Puleotechniques*. The Paleontological Society Special Publication No. 4. iv + 358p.

MILLER, J. 1988. Microscopical techniques: 1. Slices, slides, stains and peels, pp 86-107: In; M. Tucker (eds), *Techniques in Sedimentology*. Blackwell Scientific Publications. ix + 394p.

MILLER, K. R, PRESCOTT, C. S, JACOBS, T. L and LASSIGNOL, N. L. 1983. Artifacts associated with quick-freezing and freeze-drying. *Journal of Ultrasturctural Research*. **82**, (2), 123-133.

MILLER, III, W. (ed). 1990. *Paleocommunity temporal dynamics: the long-term development of multispecies assemblies*. The Paleontological Society Special Publication. **5**. vii + 421.

MILNER, H. B. 1962. *Sedimentary Petrography. Volume 1, methods in sedimentary petrography* (4th Edition). Allen and Unwin (London). 643p.

MOORE, D. C. (ed). 1993. *Laboratory handbook*. Institute of Science Technology (Lichfield). 68p.

MOORE, Jr., T. C. 1973. Method of randomly distributing grains for microscopic examination. *Journal of Sedimentary Petrology*. **43**, (3), 904-906.

MOORE, P. D, WEBB, J. A and COLLINSON, M. E. 1991. *Pollen Analysis*. (2nd Edition). Blackwell Scientific Publications (Oxford). viii + 216p.

MOREMAN, W. L. 1930. Arenaceous foraminifera from Ordovician and Silurian limestones of Oklahoma. *Journal of Paleontology*. **4**, (1), 43-59.

MOSHKOWITZ, S. 1974. A new method for observing the same nannofossil specimens both by light microscope and scanning electron microsope and preservation of types. *Israel Journal of Earth Science*. **23**, 145-147.

MUIR, M. D. 1970. Scanning electron microscopy in palynology. *Review of Palaeobotany and Palynology*. **10**, 85-97.

MUIR, M. D, SPICER, R. A, GRANT, P. R and GIDDENS, R. 1974. X-ray microanalysis in the SEM for the determination of elements in modern and fossil micro-organisms. *Proceedings of the Eighth International Congress of Electron Microscopy, Canberra*. Volume 2, 104-105.

MUIR-WOOD, H. M and OWEN, E. F. 1952. The use of acetic acid in developing brachiopod shells. *Proceedings of the Geologists' Association*. **63**, (4), 313-315.

MÜLLER, K. J. 1978. Conodonts and other phosphatic microfossils, pp 277-291: In; B. U. Haq and A. Boersma (eds) *Introduction to Marine Micropaleontology*. Elsevier (New York). viii + 376.

MURRY, P. A and LEZAK, J. 1977. Recovery of vertebrate microfossils with tetrabromoethane. *Curator*. **20**, (1), 15-22.

MURRAY, J. W. 1979. *British nearshore foraminiferids*. The Linnean Society (London) and Estuarine and Brackish-Water Sciences Association, *Synopses of the British Fauna* (New Series), D. M. Kermack and R. S. K. Barnes (eds). No. 16. v + 68p.

MURRAY, J. W. 1991. *Ecology and Palaeoecology of Benthic Foraminifera*. Longman Scientific. x + 397p.

MURRAY, J. W and ALVE, E. 1994. High diversity agglutinated foraminiferal assemblages from the NE Atlantic: dissolution experiments. *Cushman Foundation Special Publication*. **32**. 33-51.

MYERS, E. H. 1942. A quantitative study of the productivity of the foraminifera in the sea. *Proceedings of the American Philosophical Society*. **85**, 325-342.

NATION, J. L. 1983. A new method using hexamethyldisilazane for preparation of soft insect tissues for scanning electron microscopy. *Stain Technology*. **58**, (6), 347-351.

NATURAL ENVIRONMENT RESEARCH COUNCIL. 1990. *Guidance note safety in fieldwork*. (Swindon). 32p.

NATURE CONSERVANCY COUNCIL. 1990. *Earth science conservation in Britain - A strategy*. Nature Conservancy Council, (Peterborough). 1-84 + Appendices - a handbook of earth science conservation techniques (compiled by A. P. McKirdy).

NEVES, R and DALE, B. 1963. A modified filtration system for palynological reparations. *Nature*. 198, 775-776.

NICHOLS, J. A and EICHENLAUB, Jr, C. M. 1991. A simple, diver-operated pneumatic corer. *Journal of Sedimentary Petrology*. **61**, (4), 634-635.

NIGAM, R and SETTY, M. G. A. P. 1980. Paleogene reworked foraminifera in Recent sediments off Daman, Western India. *Proceedings of the 3rd Indian Geological Congress, Poona*. 273-280.

NORBY, R. D. 1972. Techniques for the study of conodont assemblages. *Geological Society of America, Abstracts with Programs*. **4**, 340-341.

NOREM, W. L. 1956. An improved method for separating fossil spores and pollen from siliceous rocks. *Journal of Paleontology*. **30**, (5), 1258-1274.

NØRGAARD, I, RASMUSSEN, A-M, SCHIØLER, P and STOUGE, S. 1991. A new method for bulk palynological processing of oil-contaminated chalks. *Journal of Micropalaeontology*. **10**, (2), 202.

NORRIS, R. D. 1989. Cnidarian taphonomy and affinities of the Ediacara biota. *Lethaia*. **22**, (4), 381-393.

NOVACEK, M and CLEMENS, W. A. 1977. Aspects of intrageneric variation and evolution of *Mesodma* (multituberculata, Mammalia). *Journal of Paleontology.* **51**, (4), 701-717.

NYE, Jr., O. B, DEAN, D. A and HINDS, R. W. 1972. Improved thin section techniques for fossil and recent organisms. *Journal of Paleontology.* **46**, (2), 271-275.

OBATA, I, MARUO, H, TERAKADO, H, MURAKAMI, T, TANAKA, T and MATSUKAWA, M. 1989. Replicas of dinosaur tracks, using silicone rubber and fibreglass-reinforced plastics, pp 413-415: In; D. G. Gillette and M. G. Lockley (eds), *Dinosaur tracks and traces.* Cambridge University Press, (Cambridge). xvii + 454p.

O'CONNOR, B. 1996. Confocal laser scanning microscopy: a new technique for investigating and illustrating fossil radiolaria. *Micropaleontology.* **42**, (4), 395-402.

O'HALLORAN, D. A, MOAT, T, MATHEWS, R. E and McKIRDY, A. P. 1993. Mineralogical conservation in Great Britain - a governmental conservation agency approach. *Journal of the Russell Society.* **5**, (1), 61-63.

PÄÄBO, S. 1993. Ancient DNA. *Scientific American.* (November), 60-66.

PALMER, A. R. 1965. Preparation of plates for paleontological publication, pp 456-459: In; B. Kummel and D. Raup (eds), *Handbook of Paleontological Techniques.* W. H. Freeman and Co., (San Francisco). xiii + 852p.

PALMER, A. R. 1989. Techniques for mechanical extraction of smaller fossils, pp 208-212: In; R. M. Feldman, R. E. Chapman and J. T. Hannibal (eds), *Paleotechniques.* The Paleontological Society Special Publication. **4**. iv + 358.

PANTIN, C. F. A. 1969. *Notes on microscopical techniques for zoologists.* Cambridge University Press. 77p.

PARKER, A and McCANN, C. 1988. Percussion corer and extension device for unconsolidated sediments. *Journal of Sedimentary Petrology.* **58**, (4), 752-753.

PARSELY, R. L. 1989. Latex casting of microinvertebrate fossils, pp 275-281: In; R. M. Feldman, R. E. Chapman and J. T. Hannibal (eds), *Paleotechniques.* The Paleontological Society Special Publication. **4**. iv + 358.

PATRICK, R and REIMER, C. W. 1966. The diatoms of the United States. Volume 1. *Monographs of the Academy of Natural Sciences of Philadelphia.* 688p.

PEPPERS, R. A. 1970. Correlation and palynology of coals in the Carbondale and Spoon Formations (Pennsylvanian) of the Northeastern Part of the Illinois Basin. *Bulletin of the Illinois State Geological Survey.* **93**, 1-173.

PERCH-NIELSEN, K. 1967. Eine Präparationstechnik zur Untersuchung von Nannoplankton im Lichtmikroskop und im Elektronenmikroskop. *Dansk Geologisk Forening Meddelelser.* **17**, 129-130.

PERERA, F and PETITO, C. 1982. Formaldehyde: a question of cancer policy? *Science.* **216**, 1285-1291.

PERILLO, G. M. E, ALBERO, M. C, ANGIOLINI, F and CODIGNOTTO, J. O. 1984. An inexpensive, portable coring device for intertidal sediments. *Journal of Sedimentary Petrology.* **54**, (2), 654-655.

PESSAGNO, E. A. 1976. Radiolarian zonation and stratigraphy of the Upper Cretaceous portion of the Great Valley Sequence, California Coast Ranges. *Micropaleontology, Special Publication.* **2**. 1-95.

PESSAGNO, Jr, E. A. 1977a. Upper Jurassic Radiolaria and radiolarian biostratigraphy of the California Coast Ranges. *Micropalaeontology.* **23**, (1), 56-113.

PESSAGNO, Jr, E. A. 1977b. Lower Cretaceous radiolaria biostratigraphy of the Great Valley Sequence and Franciscan Complex, California Coast Ranges. *Cushman Foundation for Foraminiferal Research.* Special Publication. **15**. 1-87.

PESSAGNO, Jr, E. A. 1977c. Radiolaria in Mesozoic stratigraphy, pp 913-950 (techniques 918-920): In; A. T. S. Ramsay (ed), *Oceanic Micropalaeontology* (2 Volumes). Academic Press (London). xii + 1453p.

PESSAGNO, Jr, E. A and NEWPORT, R. L. 1972. A technique for extracting Radiolaria from radiolarian cherts. *Micropaleontology.* **18**, (2), 231-234.

PETTIJOHN, F. J, POTTER, P. E and SIEVER, R. 1972. *Sand and Sandstone*. Springer-Verlag (New York). xvi - 618p.

PFANNKUCHE, O and THIEL, H. 1988. Sample Processing, pp 134-145: In, R. P. Higgins and H. Thiel (eds), *Introduction to the study of meiofauna*. Smithsonian Institution Press (Washington).

PHILLIPS, L. 1972. An application of fluorescence microscopy to the problem of derived pollen in British Pleistocene deposits. *New Phytologist*. **71**, 755-762.

PHIPPS, D and PLAYFORD, G. 1984. Laboratory techniques for extraction of palynomorphs from sediments. *Papers, Department of Geology, Univesity of Queensland*. **11**, 1-23.

PHLEGER, F. B. 1945. Vertical distribution of pelagic foraminifera. *American Journal of Science*. **243**, 377-383.

PHLEGER, F. B. 1951. Ecology of Foraminifera, northwest Gulf of Mexico. Part 1, Foraminifera Distribution. *Memoir of the Geological Society of America*. **46**, 1-88.

PHLEGER, F. B. 1952. Foraminifera ecology off Portsmouth, New Hampshire. *Bulletin of the Museum of Comparative Zoology (Harvard)*. **106**, (8), 315-390.

PHLEGER, F. B. 1960. *Ecology and Distribution of Recent Foraminifera*. John Hopkins Press, (Baltimore). vii + 297p.

PIERCE, J. W and GRAUS, R. R. 1981. Use and misuses of the phi-scale: discussion. *Journal of Sedimentary Petrology*. **51**, (4), 1348-1350.

PINGITORE, Jr, N. E, FRETZDORFF, S. B, SEITZ, B. P, ESTRADA, L. Y, BORREGO, P. M, CRAWFORD, G. M and LOVE, K. M. 1993. Dissolution kinetics of $CaCO_3$ in common laboratory solvents. *Journal of Sedimentary Research*. **63**, (4), 641-645.

PLAYFORD, G and DRING, R. S. 1981. Late Devonian Acritarchs from the Carnarvon Basin, Western Australia. *Special Papers in Palaeontology*. **27**, 1-78.

PLOEM, J. S and TANKE, H. J. 1987. *Introduction to fluorescence microscopy*. Microscopy Handbooks 10. Royal Microscopical Society, Oxford University Press (Oxford). 56p.

PLOTNICK, R. E. 1986. Taphonomy of a modern shrimp: Implications for the arthropod fossil record. *Palaios*. **1**, (3), 286-293.

PLOTNICK, R. E and HARRIS, G. 1989. X-ray analysis using Energy Dispersive and Wavelength Dispersive Spectroscopy, pp 179-185: In; R. M. Feldman, R. E. Chapman and J. T. Hannibal (eds), *Paleotechniques*. The Paleontographical Society Special Publication. **4**. iv + 358p.

POHL, E. R and BROWNE, R. G. 1973. A thin section method for ultrasmall paleontological specimens. *Journal of Paleontology*. **47**, (2), 221-225.

POINAR, Jr., G. O. 1982. Amber - true or false? *Gems and Minerals*. (April, No. 534), 80-84.

POINAR, Jr., G. O. 1992. *Life in Amber*. Stanford University Press (California). xiii + 350p.

POINAR, Jr., G. O and HESS, R. 1982. Ultrastructure of 40-million-year-old insect tissue. *Science*. **215**. 1241-1242.

POINAR, Jr., G. O and POINAR, R. 1985. Preservative qualities of recent and fossil resins: electron micrograph studies on tissue in Baltic amber. *Journal for Baltic Studies*. **16**, 222-230.

POINAR, Jr., G. O and POINAR, R. 1994. *The quest for life in Amber*. Addison-Wesley (Reading, Massachusetts). xiii + 219p.

POINAR, Jr., G. O, POINAR, H. N and CANNO, R. 1994. In; B Herrmann and S. Hummel (eds), *Ancient DNA*. Springer-Verlag (New York).

POINAR, H. N, POINAR, Jr., G. O, and CANNO, R. 1993. DNA from an extinct plant. *Nature*. **363**, 677.

POJETA, Jr., J and BALANC, M. 1989a. Uses of ultrasonic cleaners in Paleontological laboratories, pp 213-217: In; R. M. Feldman, R. E. Chapman and J. T. Hannibal (eds), *Paleotechniques*. The Paleontological Society Special Publication. **4**. iv + 358.

POJETA, Jr., J and BALANC, M. 1989b. Heating and quenching fossils, pp 218-222: In; R. M. Feldman, R. E. Chapman and J. T. Hannibal (eds), *Paleotechniques*. The Paleontological Society Special Publication. **4**. iv + 358.

POJETA, Jr., J and BALANC, M. 1989c. Freezing and thawing of fossils, pp 223-226: In; R. M. Feldman, R. E. Chapman and J. T. Hannibal (eds), *Paleotechniques*. The Paleontological Society Special Publication. **4**. iv + 358.

POULSEN, C. 1957. Improved method for whitening fossils for study. *Journal of Paleontology*. **31**, (5), 1029.

POWELL, H. P. 1987. Bruynzeel systems in use: 3, Oxford University Museum. *The Geological Curator*. **4**, (9), 551.

PURVIS, O. W. 1993. The botanical interest of mine spoil heaps - the lichen story. *Journal of the Russell Society*. **5**, (1), 45-48.

QUIGLEY, M. N. 1995. Preparation of standard solutions. *Science Technology*. **36**, 15-18.

QUINN, J. H. 1940. Rubber moulds and plaster casts in the paleontological laboratory. *Fieldiana; Field Museum of Natural History, (Chicago), Field Technique Series*. **6**, 3-21.

RAINE, J. I and TREMAIN, R. 1992. Apparatus for rapid sieving of palynological residues. *New Zealand Geological Survey*. Report **PAL 151**. 14p.

RAISTRICK, A. 1934-1935. The correlation of coal-seams by microspore-content: Part 1. The seams of Northumberland. *Transactions of the Institute of Mining Engineers (London)*. **88**, 142-153; 259-264.

RASETTI, F. 1946. Optimum conditions for the photography of fossils. *Journal of Paleontology*. **20**, (5), 514-516.

RASETTI, F. 1947. Notes on techniques in invertebrate paleontology. *Journal of Paleontology*. **21**, (4), 397-399.

RASETTI, F. 1965. Photography of fossils, pp 423-430: In; B. Kummel and D. Raup (eds), *Handbook of Paleontological Techniques*. W. H. Freeman and Co., (San Francisco). xiii + 852p.

REID, W. P. 1969. Mineral staining tests. *Mineral Industrial Bulletin, Colorado School of Mines*. **12**, (3), 1-20.

REISSINGER, A. 1950. Die "Pollenanalyse" ausgedehnt auf alle Sedimentgesteine der geologischen Vergangenheit. Part II: *Palaeontographica Abteilung B*. Band **90**, 99-126.

RHUMBLER, L. 1935. Rhizopoden der Kieler Bucht, gesammelt durch R. Remane 1 Teil. *Schriften des naturwissenschaftlichen Vereins fur Schleswig-Holstein*. Band **21**, H2, 143-195.

RICE, P. C. 1980. *Amber: the golden gem of the ages*. The Kosciuszko Foundation Inc. (New York). x + 289p.

RIEDEL, W. R and FOREMAN, H. P. 1961. Type specimens of North American Paleozoic radiolaria. *Journal of Paleontology*. **35**, (3), 628-632.

RIEDEL, W. R and SANFILIPPO, A. 1974. Radiolaria from the southern Indian Ocean, DSDP Leg 26. *Initial Reports of Deep Sea Drilling Project, Leg 26*. 771-813.

RIEDEL, W. R and SANFILIPPO, A. 1977. Cainozoic Radiolaria, pp 847-912 (techniques 852-858): In; A. T. S. Ramsay (ed), *Oceanic Micropalaeontology* (2 Volumes). Academic Press, (London). xii + 1453p.

RIGBY, J. K and CLARK, D. L. 1965. Casting and molding, pp 389-413: In; B. Kummel and D. Raup (eds), *Handbook of Palaeontological Techniques*. W.H.Freeman and Co., (San Francisco). xiii + 852p.

RILEY, L. A and KING, A. D. 1973. Causes of unreliability in microfossil samples. *The Mercian Geologist*. **4**, (3), 197-203.

RIXON, A. E. 1949. The use of acetic and formic acids in the preparation of fossil vertebrates. *Museum Journal*. **49**, 116-117.

RIXON, A. E. 1976. *Fossil animal remains: their preparation and conservation*. Athlone Press, (London). viii + 304p.

RIXON, A. E and MEADE, M. J. 1960. Glass fibre resin casts of fossils. *Palaeontology*. **3**, (1), 124-126.

ROBERTS, S and MURRAY, J. W. 1995. Characterization of cement mineralogy in agglutinated foraminifera (Protista) by Raman spectroscopy. *Journal of the Geological Society, London*. **152**, (1), 7-9.

ROBISON, R. A. 1965. Use of the Vibro-tool for mechanical preparation of fossils, pp 267-270: In; B. Kummel and D. Raup (eds), *Handbook of Paleontological Techniques*. W. H. Freeman and Co., (San Francisco). xiii + 852p.

ROTTGER, R. 1984. Summaries of three scientific films on larger foraminifera. *Benthos '83; 2nd International Symposium of Benthic Foraminifera*. Pau and Bordeaux. 535.

ROWLAND, E. O. 1961. A rapid method for making plaster casts for geological work. *Proceedings of the Geologists' Association*. **72**, (4), 461-466.

RUSBRIDGE, M. M. 1989. Photomicrography without a microscope. *Proceedings of the Geologists' Association*. **100**, (2), 209-213.

RUSHTON, A. W. A. 1979. The fossil collections of the Institute of Geological Sciences, pp 57-66: In; M. G. Bassett (ed), *Curation of palaeontological collections*. Special Papers in Palaeontology. **22**, 1-279.

SACHS, Jr., K, N. 1965a. Preparing paired thin sections from single larger foraminifera, pp 238-241: In; B. Kummel and D. Raup (eds), *Handbook of Paleontological Techniques*. W. H. Freeman and Co., (San Francisco). xiii + 852p.

SACHS, Jr., K. N. 1965b. Removal of ash from plankton samples concentrated by ignition. *Deep Sea Research*. **12**, 697.

SACHS, Jr., K. N, CIFELLI, R and BOWEN, V. T. 1964. Ignition to concentrate shelled organisms in plankton samples. *Deep Sea Research*. **11**, 621-622.

SANDBERG, P and HAY, W. W. 1967. Study of microfossils by means of the scanning electron microscope. *Journal of Paleontology*. **41**, (4), 999-1001.

SANDER, N. J. 1955. An apparatus for photographing foraminifera and other small objects. *Micropaleontology*. **1**, (3), 251-256.

SANDON, P. T. S, PEACHEY, D and VICKERS, B. P. 1989. Field Methods Project 10. Assessment of methods for the determination of the chemical reactivity, and hence the agronomic potential, of African phosphate rocks. *British Geological Survey, Technical Report WI/89/5*. Analytical and Applied Geochemistry Report Series. 22p.

SANFILIPPO, A and RIEDEL, W. R. 1985. Cretaceous radiolaria, pp 573-630: In; H. M. Bolli, J. B. Saunders and K. Perch-Nielsen, (eds), *Plankton Stratigraphy*. Cambridge University Press, (Cambridge). viii + 1032p.

SANFILIPPO, A, WESTBERG-SMITH, M. J and RIEDEL, W. R. 1985. Cenozoic radiolaria, pp 631-712: In; H. M. Bolli, J. B. Saunders and K. Perch-Nielsen, (eds), *Plankton Stratigraphy*. Cambridge University Press, (Cambridge). viii + 1032p.

SARJEANT, W. A. S. 1961. The hystrichospheres - a review and discussion. *Grana Palynologica*. **2**, 101-111.

SARJEANT, W. A. S. 1974. *Fossil and living Dinoflagellates*, (techniques, appendix A, pp 128-133). Academic Press, (London). vii + 182.

SASS, D. B. 1962. Improved techniques for the photographing of fossils. *Journal of Paleontology*. **36**, (1), 171-172.

SASS, D. B. 1963. Controlled acid preparation of fossils. *Journal of Paleontology*. **37**, (5), 1129.

SAVAGE, N. M. 1988. The use of sodium polytungstate for conodont separations. *Journal of Micropalaeontology*. **7**, (1), 39-40.

SAVRDA, C. E, BOTTJER, D. J and GORSLINE, D. S. 1985. An image-enhancing ooil technique for friable, diatomaceous rocks. *Journal of Sedimentary Petrology*. **55**, (4), 604-605.

SCHAEFFER, B. 1965. Fishes, pp 110-118: In; B. Kummel and D. Raup (eds), *Handbook of Paleontological Techniques*. W. H. Freeman and Co., (San Francisco). xiii + 852p.

SCHALLREUTER, R. 1982. Extraction of ostracods from siliceous rocks, Chapter 10, pp 169-176: In; R. H. Bate, E. Robinson and L. M. Sheppard (eds), *Fossil and Recent Ostracods*. Ellis Horwood Ltd., (Chichester). 493p.

SCHLÜTER, T. 1990. Baltic amber, pp 294-297: In; D. E. G. Briggs and P. R. Crowther (eds), *Palaeobiology - A synthesis*. Blackwell Scientific Publications, (Oxford). xii + 583p.

SCHMIDT, R. A. M. 1948. Magnetic separation of microfossils. *Journal of Paleontology*. **22**, (4), 530-531.

SCHMIDT, R. A. M. 1952. Microradiography of microfossils. *Science*. **115**, (2978), 91.

SCHRADER, H-J. 1973. Proposal for a standardized method of cleaning diatom-bearing deep-sea and land-exposed marine sediments. *Nova Hedwegia*. **45**, 403-409.

SCHRADER, H-J. 1976. Cenozoic planktonic diatom biostratigraphy of the Southern Pacific Ocean. *Initial Reports of the Deep Sea Drilling Project, 35*. Washington, U.S. Government Printing Office. 605-672.

SCHRADER, H-J and GERSONDE, R. 1978. Diatoms and silicoflagellates. *Utrecht Micropalaeontological Bulletins*. **17**, 129-176.

SCHRIMPER, G. D. 1973. Hollow casting of fossil skulls in epoxy plastic. *Curator*. **16**, (4), 286-305.

SCHULZE, F. 1855. Über des vorkommen wohlerhaltener cellulose in braunkohle und steinkohle. *Bericht uber die zur Bekanntmachung geeigneten Verhandlungen Koniglich Preussischen Akademie der Wissenschaften zu Berlin*. **21**, 676-678.

SCOTT, A. C. 1989a. Geological applications of laser scanning microscopy. *Microscopy and Analysis*. No. 10, 17-19.

SCOTT, A. C. 1989b. Geological applications of scanning acoustic microscopy. *Microscopy and Analysis*. No. 11, 39-41.

SCOTT, H. W. 1944. Muscle scar patterns on some upper Paleozoic ostracodes. *Journal of Paleontology*. **18**, (2), 162-171.

SCOTT, D. B and VILKS, G. 1991. Benthic foraminifera in the surface sediments of the deep-sea Arctic Ocean. *Journal of Foraminiferal Research*. **21**, (1), 20-38.

SCOVIL, J. A. 1996. *Photographing minerals, fossils and lapidary materials*. Geoscience Press, Inc. Tucson, Arizona. xvi + 224p.

SECRIST, M. H. 1934. Technique for the recovery of Paleozoic arenaceous foraminifera. *Journal of Paleontology*. **8**, (2), 245-246.

SELA, J and BOYDE, A. 1977. Cyanide removal of gold from SEM specimens. *Journal of Microscopy*. **111**, 229-231.

SERANDERI BARERO, R and TOFFOLETTO, D. 1996. Dissolution of calcareous foraminifera in microfossil slides. *Micropaleontology*. **42**, (4), 412-414.

SETTY, M. G. P. A. 1966. Preparation and method of study of fossil diatoms. *Micropaleontology*. **12**, (4), 511-514.

SHARPE, T. 1983. *Geology in museums: a bibliography and index*. National Museum of Wales (Cardiff), Geological Series No. 6. 128p.

SHARPE, M. 1990. *Plaster waste-moulding, casting and life casting*. Alec Tiranti Ltd. (Theale) Technical Booklet. 52p.

SHAPIRO, L. 1952. Simple field method for the determination of phosphate in phosphatic rocks. *American Mineralogist*. **37**, (3/4), 341-342.

SIMES, J. E. 1973. Report on conodont techniques and research. *New Zealand Geological Survey Paleontological Reports*. Report **Rp1**. 14p.

SIMES, J. E. 1977. Extraction of insoluble microfossils from limestone. *New Zealand Geological Survey, Department of Scientific and Industrial Research*. Report **NZGS R82**. 18p.

SITTLER, C. 1955. Méthodes et techniues physicochimiques de préparation sédiments envue de leur analyse pollinique. *Revue de l'Institut Français du Pétrole et Annales des Combustibles Liquides*. **10**, (2), 103-114.

SIVETER, D. J. 1982. Casts illustrating fine ornament of a Silurian ostracod. Chapter 6, pp 105-122: In; R. H. Bate, E. Robinson and L. M. Sheppard (eds), *Fossil and Recent Ostracods.* Ellis Horwood Ltd, (Chichester). 493p.

SIVETER, D. J. 1990. Photography, pp 505-508: In; D. E. G. Briggs and P. R. Crowther (eds), *Palaeobiology - A synthesis.* Blackwell Scientific Publications (Oxford). xiii + 583p.

SKEVINGTON, D. 1963. Graptolites from the Ontikan Limestones (Ordovician) of Öland, Sweden. I. Dendroidea, Tuboidea, Camaroidea and Stolonoidea. *Bulletin of the Geological Institute, University of Uppsala.* **42**, (6), 1-62.

SKOGLUND, R. 1961. *Kinnegraptus,* a new graptolite genus from the Lower *Didymograptus* Shale of Vätergötland, Central Sweden. *Bulletin of the Geological Institute, University of Uppsala.* **40**, (33), 389-400.

SKVARLA, J. J, FARABEE, M. J and CHISSOE, W. F. 1987/1988a. Fossil pollen ultrastructure by transmission imaging in a scanning electron microscope (TSEM). *Journal of Palynology.* **23/24**, 205-211.

SKVARLA, J. J, ROWLEY, J. R and CHISSOE, W. F. 1988b. Adaptability of scanning electron microscopy to studies of pollen morphology. *Aliso, Journal of the Rancho Santa Ana Botanic Garden, Claremont, California.* **12**, 119-208.

SLAUGHTER, B. H. 1968. The neglected angle in descriptive and interpretive vertebrate paleontology. *Journal of Paleontology.* **42**, (5), 1311-1312.

SMITH, A. H. V and BUTTERWORTH, M. A. 1967. Miospores in the coal seams of the Carboniferous of Great Britain. *Special Papers in Palaeontology.* **1**, x + 324p.

SMITH, C. C. 1975. A new method for studying calcareous nannofossils using scanning and transmitted light optics. *Revista Española de Micropaleontologia.* **7**, 43-48.

SMITH, D. G. 1984. Vibracoring fluvial and deltaic sediments: tips on improving penetration and recovery. *Journal of Sedimentary Petrology.* **54**, (2), 660-663.

SMITH, D. G. 1987. A mini-vibracoring system. *Journal of Sedimentary Petrology.* **57**, (4), 757-758.

SMITH, J. A and LATIMER, B. 1985. The joy of casting. *Explorer.* **27**, (4), 4-5 & 22-24.

SMITH, J and LATIMER, B. 1989a. Making a multiple-piece mold, pp 305-319: In; R. M. Feldman, R. E. Chapman and J. T. Hannibal (eds), *Paleotechniques.* The Paleontological Society Special Publication. **4**. iv + 358.

SMITH, J and LATIMER, B. 1989b. Casting with resins, pp 326-330: In; R. M. Feldman, R. E. Chapman and J. T. Hannibal (eds), *Paleotechniques.* The Paleontological Society Special Publication. **4**. iv + 358.

SMITH, J. and LATIMER, B. 1989c. Preparation of hollow casts, pp 331-335: In; R. M. Feldman, R. E. Chapman and J. T. Hannibal (eds), *Paleotechniques.* The Paleontological Society Special Publication. **4**. iv + 358.

SMITH, P. M. R. 1984. The use of fluorescence microscopy in the characterisation of amorphous organic matter. *Organic Geochemistry.* **6**, 839-845.

SMITH, R. K. 1967. Ignition and filter methods of concentrating shelled organisms. *Journal of Paleontology.* **41**, (5), 1288-1291.

SMITH, W and McINTYRE, A. D. 1954. A spring loaded bottom sampler. *Journal of the Marine Biological Association (UK).* **33**, 257-264.

SOHL, N. F. 1989. Mechanical extraction of fossils extraction by hammers, chisels, and picks, pp 204-207: In; R. M. Feldman, R. E. Chapman and J. T. Hannibal (eds), *Paleotechniques.* The Paleontological Society Special Publication. **4**. iv + 358.

SOHN, I. G. 1956. The transformation of opaque calcium carbonate to translucent calcium fluoride in fossil ostracoda. *Journal of Paleontology.* **30**, (1), 113-114.

SOHN, I. G. 1960. Cleaning ostracode valves with ultrasonic vibrations. (Abstract only). *Bulletin of the Geological Society of America.* **71**, 1982.

SOHN, I. G. 1961. Techniques for preparation and study of fossil ostracodes, pp Q64-Q70: In; R. C. Moore (ed), *Treatise on Invertebrate Paleontology. Part Q; Arthropods 3.* Geological Society of America and the University of Kansas (Lawrence, Kansas). xxiii + 442p.

SOHN, I. G, BERDAN, J. M and PECK, R. E. 1965. Ostracods, pp 75-89: In; B. Kummel and D. Raup (eds), *Handbook of Paleontological Techniques.* W. H. Freeman and Co., (San Francisco). xiii + 852p.

SOMEKH, M. G. 1988. Scanning acoustic microscopy. *Microscopy and Analysis.* No. 6, 7-9.

SORAUF, J. E and TUTTLE, D. W. 1988. Dark-field illumination in photography of acetate peels. *Journal of Paleontology.* **62**, (1), 153-156.

SORSBY, B. D and HORNE, S. D. 1980. The readability of museum labels. *Museums Journal.* **80**, (3), 157-159.

SPENCE, T. F. 1960. A technique of plaster-casting for museum specimens. *Science Technology.* **6**.

SPICER, R. A, GRANT, P. R and MUIR, M. D. 1974. An inexpensive portable freeze-drying unit for SEM specimen preparation. In: O. Johari and I Corvin (eds), *Proceedings of the Seventh Annual Scanning Electron Microscopy Symposium.* IITRI Chicago. 290 & 299-304.

SPICER, R. A, SENGUPTA, S and PHILLIPS, L. 1976. Low temperature preparation of pteridophyte spores and sporangia for SEM. *Journal of Microscopy.* **106**, (3), 351-354.

SPIELHOLTZ, G, THOMAS, L. A and DIEHL, H. 1962. Isolation of spores by wet oxidation. *Micropaleontology.* **8**, 109-110.

SPRECHMANN, P. 1981. Paleocommunities and paleobathymetry of Maastrichtian sublittoral benthonic foraminifera from western Europe. *Neues Jahrbuch für Geologie und Paläontologie, Abhandlungen.* Band **162**, Heft 2, 188-230.

STANLEY, E. A. 1966. The problems of reworked pollen and spores in marine sediments. *Marine Geology.* **4**, 397-408.

STANLEY, J. W. 1975. Alginate impression compound:- A comparative assessment of a cheap moulding material with potential applications in palaeontology. *Newsletter of the Geological Curators Group.* **1**, (4), 191-193.

STAPLIN, F. L. 1977. Interpretation of thermal history from color of particulate organic matter - a review. *Palynology.* **1**, 9-18.

STEVENS, C. H, JONES, D. H and TOOD, R. G. 1960. Ultrasonic vibrations as a cleaning agent for fossils. *Journal of Paleontology.* **34**, (4), 727-730.

STEWART, W. N and TAYLOR, T. N. 1965. The peel technique, pp 224-232: In; B. Kummel and D. Raup (eds), *Handbook of Paleontological Techniques.* W. H. Freeman and Co., (San Francisco). xiii + 852p.

STONE, G. W and MORGAN, J. P. 1992. Jack-up pontoon barge for vibrocoring in shallow water. *Journal of Sedimentary Petrology.* 62, (4), 739-741.

STONE, J. 1987. Review of investigative techniques used in the study of conodonts, pp17-34: In; R. L. Austin (ed), *Conodonts: Investigative Techniques and Applications.* Ellis Horwood Ltd, (Chichester). 1-422.

STRACHAN, I, WHITE, D, RUSHTON, A and RICKARDS, R. B. 1991. How they are collected and prepared. Chapter 11, pp 59-68: In; D. Palmer and R. B. Rickards (eds), *Graptolites.* Writing in the rocks. Fossils Illustrated. Vol. 1. The Boydell Press, (Woodbridge). 166p + 138pl.

STRADNER, H and PAPP, A. 1961. Tertiaere Discoasteriden aus Oesterreich und deren stratigraphische Bedeutung. *Jahrbuch der Geologischen Bundesanstalt* Sonderband (Wien). **7**, 1-160.

STUBBLEFIELD, S and BANKS, H. P. 1978. The cuticle of *Drepanophycu spinaeformis*, a long-ranging Devonian lycopod from New York and eastern Canada. *American Journal of Botany.* **65**, (1), 110-118.

STUCKER, G. F. 1961. Salvaging fossils by jet. *Curator.* **4**, 332-340.

STUCKER, G. F, GALUSHA, M. J and MCKENNA, M. C. 1965. Removing matrix from fossils by miniature sandblasting, pp 273-274: In; B. Kummel and D. Raup (eds), *Handbook of Paleontological Techniques*. W. H. Freeman and Co., (San Francisco). xiii + 852p.

SURLYK, F. 1972. Morphological adaptations and population structures of Danish chalk brachiopods. *Kogelike Dansk Videnskabernes Selskab, Biologiske Skrifter*. **19**, (2), 1-57.

SWANSON, R. G. 1981. *Sample examination manual*. American Association of Petroleum Geologists, Methods in Exploration Series. **1**, 117p.

SWIFT, A and ALDRIDGE, R. J. 1984. Laboratory techniques in micropalaeontology. *Science Technology*. **7**, 28-30.

SWITZER, G and BOUCOT, A. J. 1955. The mineral composition of some microfossils. *Journal of Paleontology*. **29**, (3), 525-533.

SYLVESTER-BRADLEY, P. C. 1941. The shell structure of the Ostracoda and its application to their palaeontological investigation. *Annals and Magazine of Natural History*. Series 11. **8**, (43), 1-33.

TAPP, J. B and PREZBINDOWSKI, D. R. 1990. Application of contrast control filters in black and white photomicrography. *Journal of Sedimentary Petrology*. **60**, (4), 617-619.

TAPPAN, H and LOEBLICH, Jr., A. R. 1965. Foraminiferal remains in palynological preparations. *Revue de Micropaléontologie*. **8**, (2), 61-63.

TASCH, P and SHAFFER, B. L. 1961. Study of scolecodonts by transmitted light. *Micropaleontology*. **7**, (3), 369-371.

TAUGOURDEAU, P. 1971. Scolecodontes du Frasnien du Boulonnais. *Revista Española de Micropaleontologia*. **3**, (1), 71-108.

TAYLOR, M. A and HARTE, J. D. C. 1988. Palaeontological site conservation and the law in Britain, pp 21-39; In: P. R Crowther and W. A. Wimbledon (eds), *The use and conservation of palaeontological sites*. Special Papers in Palaeontology. **40**, 1-200.

TAYLOR, P. D. 1986. Scanning electron microscopy of uncoated fossils. *Palaeontology*. **29**, (4), 685-690.

TAYLOR, P. D and JONES, C. G. 1996. Use of the environmental chamber in uncoated SEM of Recent and fossil bryozoans. *Microscopy and Analysis*. No. 52, 27-29.

TAYLOR, R. J and HAMILTON, G. B. 1982. Techniques, pp 11-15: In; A. R. Lord (ed), *A Stratigraphical Index of Calcareous Nannofossils*. Ellis Horwood Ltd, (Chichester). 1-192.

TEICHERT, C. 1948. A simple device for coating fossils with ammonium chloride. *Journal of Paleontology*. **22**, (1), 102-104.

THEN, D. R and DOUGHERTY, B. J. 1983. A new procedure for extracting foraminifera from indurated organic shale. *Geological Survey of Canada*, Paper **83-1B**, 413-414.

THIERSTEIN, H. R, FRANZ, H. E and ROTH, P. H. 1971. Scanning electron microscopy of the same small objects. *Micropaleontology*. **17**, 501-502.

THOMAS, B. R. 1969. Kauri resins - modern and fossil, pp 599-618: In; G. Eglinton and M. T. J. Murphy (eds), *Organic Geochemistry*. Springer-Verlag (New York). 828p.

THOMAS, K. M. 1989. Health hazards in the Paleontology laboratory, pp 30-36; In: R. M. Feldman, R. E. Chapman and J. T. Hannibal (eds), *Paleotechniques*. The Paleontological Society Special Publication. **4**. iv + 358.

THOMSON, D. J and BRADBURY, S. 1987. *An introduction to photomicrography*. Microscopy Handbooks 13. Royal Microscopical Society, Oxford University Press, (Oxford). 74p.

THOMPSON, J. M. A. (eds). 1992. *Manual of Curatorship* (2nd Edition). Butterworth-Heinemann/Museums Association. 756p.

TIFFANY, M and IWAMA, B. 1994. Cast painting, pp 271-284 in Chapter 10, Molding, casting, and painting: In; P. Leiggi and P. May (eds), *Vertebrate Paleontological Techniques*, Volume 1. Cambridge University Press. 368p.

TODD, R, LOW, D and M ELLO, J. F. 1965. Smaller foraminifera, pp 14-20: In; B. Kummel and D. Raup (eds), *Handbook of Paleontological Techniques*. W. H. Freeman and Co., (San Francisco). xiii + 852p.

TOOMBS, H. A. 1948. The use of acetic acid in the development of vertebrate fossils. *Museum Journal*. **48**, 54-55.

TOOMBS, H. A and RIXON, A. E. 1950. The use of plastics in the transfer method of preparing fossils. *Museum Journal*. **50**, 105-107.

TOOMBS, H. A and RIXON, A. E. 1959. The use of acids in the preparation of vertebrate fossils. *Curator*. **2**, 304-312.

TOOMEY, D. F. 1974. The biota of the Pennsylvanian (Virgilian) Leavenworth limestone midcontinent region. Part 4: Distribution of agglutinated and silicified foraminifera. *Journal of Paleontology*. **48**, (2), 326-343.

TRAVERSE, A. 1988. *Paleopalynology*. Unwin Hyman (London). xxiii + 600p.

TRIEBEL, E. 1947. Methodische und Technische Fragen der mikropalaontologie. *Senckenberg Buch, 19 W Kramer, (Frankfurt)*. 47p.

TUCKER, M. E. 1990. *Techniques in Sedimentology*. Blackwell (Oxford). ix + 394p.

TUCKER, M. E. 1996. *Sedimentary rocks in the field*. Wiley and Sons, Inc, (Chichester). viii + 153p.

TUNNICLIFF, S. P. 1983. *I am beginning my research: What shall I do with my specimens? A note of advice*. Natural Environmental Research Council (Swindon). 16p.

TURNER, R. E. 1982. Reworked acritarchs from the type section of the Ordovician Caradoc Series, Shropshire. *Palaeontology*. **25**, (1), 119-143.

TURNER, W. M. 1966. An improved method for recovery of bromoform used in mineral separation. *U. S. Geological Survey Professional Paper*. **550-C**, 224-227.

ULRICH, E. O and BASSLER, R. S. 1923. Paleozoic Ostracoda: their morphology, classification and occurrence. *Maryland Geological Survey*, Silurian. 281-283.

UPTON, J. 1987, (for 1986). The Bruynzeel approach to geological storage. *The Geological Curator*. **4**, (9), 541-544.

VAN MORKHOVEN, F. P. C. M. 1958. A simplified method for grinding foraminifera. *Micropaleontology*. **4**, (2), 209-210.

VARKER, W. J. 1967. Conodonts of the genus *Apatognathus* Branson and Mehl from the Yoredale Series of the North of England. *Palaeontology*. **10**, (1), 124-141.

VERNON, R. O. 1957. New technique for casting fossils and forming molds. *Journal of Paleontology*. **31**, (2), 461-463.

VIDAL, G. 1988. A palynological preparation method. *Palynology*. 12, 215-220.

VON BITTER, P. H and MILLAR-CAMPBELL, C. 1984. The use of "spikes" in monitoring the effectiveness of phosphatic microfossil recovery *Journal of Paleontology*. **58**, (5), 1193-1195.

VON HOFE GRADY, F. 1979. Some new approaches to microvertebrate collecting and processing. *Newsletter of the Geological Curators Group*. **2**, (7), 439-442.

WAGNER, C. W. 1957. *Sur les ostracodes du Quaternaire Récent des Pays-bas et leur utilisation dans l'étude géologique des dépôts Holocenes*. Mouton and Co., ('s-Gravenhage). 1-250.

WALKER, D. A, LINETON, A. E and SCHAFER, C. T. 1974. Sudan Black B: a superior stain to rose Bengal for distinguishing living from non-living foraminifera. *Journal of Foraminiferal Research*. **4**, (4), 205-215.

WALKER, D. A and SCHAFER, C. T. 1973. An investigation of staining methods to differentiate living and non-living foraminifera. (Abstract only). *Geological Survey of Canada, Part B*. Paper **73-1B**, 131.

WALKER, D. A and SCHAFER, C. T. 1974. Sudan Black-B: a stain for quantitative determination of living foraminifera. (Abstract only). *Geological Survey of Canada, Part B*. Paper **74-1B**, 131.

WALLER, R. 1987. A view of ethical conservation and mineral specimen falsification, pp 433-437: In; P. R. Crowther and C. J. Collins (eds), The Conservation of Geological Material. *The Geological Curator.* **4**, (7), 375-474.

WALTON, J. 1928. A method of preparing sections of fossil plants contained in coal-balls or in other types of petrifications. *Nature.* **122**, 571-572.

WALTON, J. 1928. Improvements in the peel method. *Nature.* **125**, 413-414.

WALTON, W. R. 1952. Techniques for the recognition of living foraminifera. *Contributions from the Cushman Foundation for Foraminiferal Research.* **3**, (2), 56-64.

WARD, H. A. 1866. *Catalogue of casts of fossils from the principal museums of Europe and America, with short descriptions and illustrations.* Benton and Andrews (Rochester). 228p.

WARNE, S. St. J. 1962. A quick field or laboratory staining scheme for the differentiation of the major carbonate minerals. *Journal of Sedimentary Petrology.* **32**, (1), 29-38.

WATSON, J. 1977. Some Lower Cretaceous conifers of the Cheirolepidiaceae from the USA and England. *Palaeontology.* **20**, (4), 715-749.

WATSON, J. and ALVIN, K. L. 1976. Silicone rubber casts of silicified plants from the Cretaceous of the Sudan. *Palaeontology.* **19**, (4), 641-650.

WATSON, J and SINCOCK, C. A. 1992. *Bennettitales of the English Wealden.* Monograph of the Palaeontolographical Society (London). 1-228, pls 1-23. (Publ. No. 588, part of Vol. **145** for 1991).

WEAVER, P. P. E and SCHULTHEISS, P. J. 1983. Detection of repenetration and sediment disturbance in open-barrel gravity cores. *Journal of Sedimentary Petrology.* **53**, (2), 649-654.

WEIDE, D. L and WEBSTER, G. D. 1967. Ammonium chloride powder used in photography of artefacts. *American Antiquity.* **32**, 104-105.

WESLEY, A. 1954. A short synopsis of some microscopical methods in Palaeobotoany. *Proceedings of the Leeds Philosophical and Literary Society.* Science Section. **6**, 168-179.

WEST, R. G. 1977. *Pleistocene geology and biology* (2nd Edition). Longman, (London). xi + 440p.

WETMORE, K. L. 1988. Burrowing and sediment movement by benthic foraminifera, as shown by time-lapse cinematography. *Revue de Paléobiologie. Benthos '86.* 921-927.

WHATLEY, R. C, TRIER, K and DINGWALL, P. M. 1982. Some preliminary observations on certain mechanical and biophysical properties of the ostracod carapace. Chapter 5, pp 76-104: In; R. H. Bate, E. Robinson and L. M. Sheppard (eds), *Fossil and Recent Ostracods.* Ellis Horwood Ltd, (Chichester). 493p.

WHITCOMB, M. J. 1981. The suitability of various adhesives as mounting media for scanning electron microscopy. *Journal of Microscopy.* **121**, (3), 289-308.

WHITCOMB, M. J. 1985. The suitability of various adhesives as mounting media for scanning electron microscopy. II. General purpose glues. *Journal of Microscopy.* **139**, (1), 75-114.

WHITE, D. E, BARRON, H. F, BARNES, R. P and LINTERN, B. C. 1991. Biostratigraphy of late Llandovery (Telychian) and Wenlock turbidite sequences in the SW Southern Uplands, Scotland. *Transactions of the Royal Society of Edinburgh: Earth Sciences.* (at press).

WHITTAKER, J. E and HODGKINSON, R. L. 1991. On the preparation of specimens for scanning electron microscopy and a simple technique for plate making, using a black background. *Journal of Micropalaeontology.* **9**, (2), 219-220.

WHITTINGTON, H. B. 1956. Photographing small fossils. *Journal Paleontology.* **30**, (3), 756-757.

WHITTINGTON, H. B. 1965. Photography of small fossils, pp 430-433: In; B. Kummel and D. Raup (eds), *Handbook of Paleontological Techniques.* W. H. Freeman and Co., (San Francisco). xiii + 852p.

WHITTINGTON, H. B and CONWAY-MORRIS, S. 1985. Extraordinary fossil biotas: their ecological and evolutionary significance. *Philosophical Transactions of the Royal Society of London.* **B311**, 1-192.

WHYBROW, P. J. 1978. Fibre optics and foot operated microscope control equipment to assist in the preparation of fossils. *Newsletter of the Geological Curators Group.* **2**, (3), 141-143.

WHYBROW, P. J. 1982. Preparation of the cranium of the holotype of *Archaeopteryx lithographica* from the collections of the British Museum (Natural History). *Neues Jahrbuch für Mineralogie, Geologie und Paläontologie, Monatshefte.* **H3**, 184-192.

WHYBROW, P. J and LINDSAY, W. 1990. Practical Techniques, pp 499-502: In; D. E. G. Briggs and P. R. Crowther (eds), *Palaeobiology - A synthesis.* Blackwell Scientific Publications (Oxford). xiii + 583p.

WICHER, C. A. 1942. *Praktikum der angewandten mikropalntologie.* Berlin: Borntraeger. 143p.

WILD, R. 1988. The protection of fossils and palaeontological sites in the Federal Republic of Germany, pp 181-189: In; P. R. Crowther and W. A. Wimbledon (eds), *The use and conservation of palaeontological sites.* Special Papers in Palaeontology. **40**, 1-200.

WILLIAMS, A. 1956. The calcareous shell of the brachiopod and its importance to their classification. *Biological Reviews.* **31**, 243-287.

WILLIAMS, A. 1965. Techniques for the preparation of fossil and living brachiopods, pp H251-H256: In; R. C. Moore (ed), *Treatise on Invertebrate Palaeontology, Part H, Brachiopoda (1).* Geological Society of America and the University of Kansas (Lawrence, Kansas). xxxii + 521.

WILLIAMS, G. L. 1978. Dinoflagellates, Acritarchs and Tasmanitids, pp 293-326: In; B. U. Haq and A. Boersma (eds), *Introduction to Marine Micropaleontology.* Elsevier Biomedical, (New York). viii + 376p.

WILLIAMS-MITCHELL, E. 1948. The zonal value of foraminifera in the Chalk of England. *Proceedings of the Geologists' Association.* **59**, (2), 91-112.

WILSON, G. J. 1971. A chemical method for the palynological processing of chalk. *The Mercian Geologist.* **4**, (1), 29-36.

WILSON, G. J. 1972. A method for the recovery of mounted palynological residues. *The Mercian Geologist.* **4**, (2), 139-141.

WILSON, J. 1987. Approaches to the preparation and development of calcareous fossils from calcareous matrices, pp 447-450: In; P. R. Crowther and C. J. Collins (eds), The Conservation of Geological Material. *The Geological Curator.* **4**, (7), 375-474.

WILSON, L. R. 1964. Recycling, stratigraphic leakage, and faulty techniques in palynology. *Grana Palynologica.* **5**, (3), 425-436.

WILSON, L. R and HOFFMEISTER, W. S. 1952. Small foraminifera. *Micropaleontologist.* **6**, (2), 26-28.

WILSON, M. A. 1989. Enlarging latex molds and casts, pp 282-283: In; R. M. Feldman, R. E. Chapman and J. T. Hannibal (eds), *Paleotechniques.* The Paleontological Society Special Publication. **4**. iv + 358.

WILSON, M. A and PALMER, T. J. 1989. Preparation of acetate peels, pp 142-145: In; R. M. Feldman, R. E. Chapman and J. T. Hannibal (eds), *Paleotechniques.* The Paleontological Society Special Publication. **4**. iv + 358p.

WIMAN, C. 1895. Über *Dictyonema cavernosum* n.sp. *Bulletin of the Geological Institute, University of Uppsala.* **3**, 1-13.

WIMBLEDON, W. A. 1988. Palaeontological site conservation in Britain: facts form, function, and efficacy, pp 41-55: In; P. R. Crowther and W. A. Wimbledon (eds), *The use and conservation of palaeontological sites.* Special Papers in Palaeontology. **40**, 1-200.

WINLAND, H. D. 1971. Non-skeletal deposition of high-Mg calcite in the marine environment and its role in the retention of textures: In; O. P. Bricker (ed), *Carbonate Cements*. John Hopkins University Studies in Geology No. **19**, (Baltimore). 278-284.

WITTARD, W. F and SISSON, J. E. 1940. Phenolformaldehyde resin as a casting material. *Geological Magazine*. **77**, (6), 478-481.

WOLBERG, D. L. 1989a. Stabilizing pyritic material , pp 244-248: In; R. M. Feldman, R. E. Chapman and J. T. Hannibal (eds), *Paleotechniques*. The Paleontological Society Special Publication. **4**. iv + 358.

WOLBERG, D. L. 1989b. Glues and other stick'ums and patch'ums stabilizing compounds for strengthening fossils, pp 249-259: In; R. M. Feldman, R. E. Chapman and J. T. Hannibal (eds), *Paleotechniques*. The Paleontological Society Special Publication. **4**. iv + 358.

WOLF, K. H, EASTON, A. J and WARNE, S. 1960. Techniques of examination and analysis sedimentary carbonate skeletons, minerals and rocks. Chapter 8, pp 253-341: In; G. V. Chilingar, H. J. Bissell and R. W. Fairbridge (eds), *Carbonate rocks: physical and chemical aspects*. Developments in sedimentology 9B. Elsevier Publishing Co. 413p. (Staining techniques pp 259-272).

WOOD, A. 1948. Sectioning smaller foraminifera. *Journal of Paleontology*. **22**, (4), 530.

WOOD, A. 1963. Wall structure of foraminifera in polarised light. *Micropaleontology*. **9**, (4), 432.

WOODWARD, H. 1865. How the skull of a mammoth was got out of the Brickearth at Ilford. *Geological Magazine*. **2**, 93-94.

WORNARDT, W. W. 1965. Mechanical finger for preparation of diatom slides, pp 283-287: In; B. Kummel and D. Raup (eds), *Handbook of Paleontological Techniques*. W. H. Freeman & Co., (San Francisco). xiii + 852p.

WRAY, J. L. 1952. Endothyrid foraminifera from the Greenbrier Series of northern West Virginia. *Journal of Paleontology*. **26**, (6), 946-952.

WRIGHT, R. C and HAY, W. W. 1971. The abundance and distribution of foraminifers in a back-reef environment, Molasses Reef, Florida. *Miami Geological Society*. Memoir **1**, 121-174.

WRIGLEY, A. 1934. A Lutetian fauna of Southampton Docks. *Proceedings of the Geologists' Association*. **45**, (1), 1-16.

YAKOVLEY, N. V and SHTERENBERG, L. YE. 1967. Use of detachable centrifuge test tube for conodont separation. *Paleontological Journal*. **1**, (3), 93-94.

YOUNG, B. 1993. The mineral collector as a conservationist - the right to collect? *Journal of the Russell Society*. **5**, (1), 35-36.

ZANGERL, R. 1965. Radiographic techniques, pp 305-320: In; B. Kummel and D. Raup (eds), *Handbook of Paleontological Techniques*. W. H. Freeman and Co., (San Francisco). xiii + 852p.

ZANGERL, R and SCHULTZE, H-P. 1989. X-radiographic techniques and applications, pp 165-178: In; R. M. Feldman, R. E. Chapman and J. T. Hannibal (eds), *Paleotechniques*. The Paleontographical Society Special Publication. **4**. iv + 358p.

ZETZSCHE, F and KÄLIN, O. 1932a. Untersuchungen über die membranen der sporen und pollen VII. Eine methode zur isolierung des polymerbitumens (sporenmembranen, kutikulen usw) aus kohlen. *Braunkohle*. **31**, 345-351 and 363-366.

ZETZSCHE, F and KÄLIN, O. 1932b. Untersuchungen über die membranen der sporen und pollen VIII. *Helv. chim. acta*. **15**.

ZIEGLER, W, LINDSTRÖM and McTAVISH, R. 1971. Monochloracetic acids and conodonts - a warning. *Nature*. **230**, 584-585.

APPENDIX

1. Equipment and chemical suppliers.

2. Describing sedimentary rocks and fossils.

3. Centrifuge times for bromoform separation and water/acetone residue washing.

4. Hazard symbols.

5. Respirator and filter cartridge colour coding.

6. Fire extinguisher classification.

7. Laboratory glove resistance properties.

8. Care and handling of hydrofluoric acid.

9. Spillage and disposal of unwanted chemicals.

10. Conversion data.

11. Formulae for the preparation of standard solutions.

APPENDIX 1.

EQUIPMENT AND CHEMICAL SUPPLIERS

United Kingdom suppliers only, unless otherwise indicated. Any list of suppliers is subject to change, as companies close, merge or move. Palaeobiological preparators are well aware of this. Note that many techniques use products adapted for palaeontological application, and primary commercial needs govern their production and availability. Modify the following list accordingly.

Laboratory equipment suppliers

FISHER SCIENTIFIC UK
Bishop Meadow Road
Loughborough
Leicestershire LE11 5RG
Tel: (01509) 231166
Fax: (01509) 231893
http://www.fisher.co.uk

PHILIP HARRIS EDUCATION
Lynn Lane
Shenstone
Lichfield
Staffordshire WS14 0EE
Tel: (01543) 482202
Fax: (01543) 483056

MERCK Ltd
Merck House
Poole
Dorset BH15 1TD
Tel: (0800) 223344
Tel (Tech Services): (01202) 664778
Fax: (01455) 558586

Chemical Suppliers (see also lab suppliers above)

ALDRICH CHEMICALS Ltd
The Old Brickyard
New Road
Gillingham
Dorset SP8 4JL
Tel: (0800) 717181
Fax: (0800) 378538

SIGMA CHEMICAL Co. Ltd
Fancy Road
Poole
Dorset BH17 7NH
Tel: (0800) 373731
Fax: (0800) 378785

Conservation Products

CONSERVATION RESOURCES (UK) Ltd
Unit 1
Pony Road
Horspath Industrial Estate
Cowley
Oxford OX4 2RD
(*Paraloid* products)
Tel: (01865) 747755
Fax: (01865) 747035

Fume Cupboards

BIGNEAT Ltd	
Pipers Wood Industrial Park	Tel: (023) 9226 6400
Waterberry Drive	Fax: (023) 9226 3373
Waterlooville	
Hampshire PO7 7XU	
(Portable fume cabinet)	

LABORATORY ENGINEERING TECHNICAL SEVICES Ltd	
46 Central Road	Tel: (020) 8330 4408
Worcester Park	Fax: (020) 8330 6458
Surrey KT4 8HY	

PROTEX DEVELOPMENTS Ltd	
Wheatley	Tel: (01422) 322623
Halifax	Fax: (01422) 355924
West Yorkshire HX3 5AF	

WALDNER MSA Ltd	
Ratcliffe House	Tel: (01784) 465896
Leacroft	Fax: (01784) 458185
Kingston Road	
Staines	
Middlesex TW18 4NN	

Field Equipment and Sample Bags

GEO SUPPLIES Ltd	
16 Station Road	Tel: (0114) 2455746
Chapeltown	Fax: (0114) 2403405
Sheffield S30 4XH	

MINERS Inc	
PO Box 1301	Tel: 800-824-7452
Riggins	Fax: 208-628-3749
Idaho 83549-1301	
USA	

Graticules and finder slides

Finder graticules are available through most laboratory microscopy suppliers, as well as the primary manufacturer:

GRATICULES Ltd	
Fir Croft Way	Tel: (01732) 864863
Eden Bridge	Fax: (01732) 865544
Tonbridge	
Kent TN8 6HA.	

Hand Preparation Tools

Periosonic ultrasonic scaler

BONDENT ELECTRONIC DIVISION (U.K.) Ltd
15 Teddington Business Park
Station Road
Teddington
Middlesex TW11 9BQ

Tel: (020) 8614 6979
Fax: (020) 8614 6989

Elco S engraver (same as Brooks Walker version) Prophy-Jet 30 (airpolishing unit) Nouvag Micromotors (rotary engravers and grinders)

CLAUDIUS ASH
Unit 9
The Brickyard Business Park
Western Avenue
Cardiff CF4 3XA

Tel: (029) 2061 9277
Fax: (029) 2052 0304

VP2-X pneumatic engraving pen

DESOUTTER Ltd
319 Edgware Road
Colindale
London NW9 6ND

Tel: (020) 8205 7050
Fax: (020) 8205 5167

Dremel *Multi*

DREMEL
P.O. Box 1468
Racine
Wisconsin 53401-1468
USA

Burgess "Handy Engraver"

RECORD POWER Ltd
Parkway Works
Sheffield S9 3BL

Tel: (01742) 434370/756385
Fax: (01742) 434302

S.S. White Airbrasive 6500 system 2 plus (U.K. Distributors)

REG ABRASONICS Ltd
599-613 Princes Road
Dartford
Kent DA2 6HH

Tel: (01322) 228227
Fax: (01322) 228112

Sculptor's mallets chisels and knives

ALEC TIRANTI Ltd
70 High Street
Theale
Reading
Berkshire RG7 5AR

Tel: (01734) 302775
Fax: (01734) 323487

REWARD-CLAYGLAZE Ltd
Units A-C
Brookhouse Industrial Estate
Cheadle
Stoke-on-Trent ST10 1PW
Tel: (01538) 750052
Fax: (01538) 756892

Microscopy

LEICA UK Ltd
Davy Avenue
Knowlhill
Milton Keynes
Buckinghamshire MK5 8LB
Tel: (01908) 666663
Fax: (01908) 609992

NIKON U. K. Ltd
Instruments Division
Nikon House
380 Richmond Road
Kingston
Surrey KT2 5PR
Tel: (020) 8541 4440
Fax: (020) 8541 4584

OLYMPUS
2-8 Honduras Street
London EC1Y OTX
Tel: (020) 7253 2772
Fax: (020) 7490 7880

Micropalaeontological slides

BIOTEC
Little Lower Ease
Cuckfield Road
Ansty
West Sussex RH17 5AL
Tel: (01444) 452282
Fax: (01444) 452282

GREEN GEOLOGICAL
Paleontological Services
6727 Greenleaf Avenue
Whittier
California 90601
USA
Tel: (562) 698-5338
Fax: (562) 698- 6538

PANGEA UK
185 Oxford Road
Calne
Wiltshire SN11 8AL
Tel: (01249) 816010
Fax: (01249) 816037

Mounting media

Elvacite 2044

Charles Tennant & Co. (London) Ltd.,
Denny Avenue
Waltham Abbey
Essex EN9 1NS
Tel: (01992) 715777
Fax: (01992) 700449

Naphrax

Northern Biological Supplies Ltd
3 Betts Avenue — Tel: (01473) 623995
Martlesham Heath — Fax: (01473) 612148
Ipswich IP5 7RH

Rock Crushing, Cutting & Polishing Equipment

AGATE & GENERAL STONECUTTERS Ltd
32 Waterside — Tel: (01923) 267678
Kings Langley — Fax: (01923) 265050
Hertfordshire WD4 8HH

BEULER
Milburn Hill Road — Tel: (024) 7669 2242
University of Warwick Science Park — Fax: (024) 7669 2074
Coventry CV4 7HS

CHRISTISON Scientific Equipment Ltd
Albany Road — Tel: (0191) 477 4261
East Gateshead Industrial Estate — Fax: (0191) 490 0549
Gateshead NE8 3AT

ENGIS (UK) Ltd
5 Ambley Green — Tel: (01634) 261797
Gillingham Business Park — Fax: (01634) 261815
Gillingham
Kent ME8 0NJ

PASCALL ENGINEERING Co. Ltd
Gatwick Road — Tel: (01293) 525166
Crawley — Fax: (01293)536214
West Sussex RH10 2RS

Scanning Electron Microscopy Accessories

AGAR Ltd
66a Cambridge Road — Tel: (01279) 813519
Stansted — Fax: (01279) 815106
Essex CM24 8DA

ELECTRON OPTICAL SERVICES
52 Higher Road — Tel: (0161) 748 8448
Urmston — Fax: (0161) 7468048
Manchester M41 9AP

GISBOURNE MICROSCOPY SERVICES
Oakwood House — Tel: (01785) 665858
Brook Lane — Fax: (01785) 223334
Brocton
Stafford
Staffordshire ST17 OTZ

TAAB LABORATORIES EQUIPMENT Ltd
3 Minerva House
Calleva Industrial Park
Aldermaston
Berkshire RG7 8NA
Tel: (01734) 817775
Fax: (01734) 817881

VG MICROTECH
Bellbrook Busieness Park
Bolton Close
Uckfield
East Sussex TN22 1QZ
Tel: (01825) 761077
Fax: (01825) 768343

Sieving & Particle Size Analysis

CHRISTISON Scientific Equipment Ltd
(See above)

GLEN CRESTON Ltd
16 Dalston Gardens
Stanmore
Middlesex HA7 1BU
Tel: (020) 8206 0123
Fax: (020) 8206 2452

ENDECOTTS Ltd
9 Lombard Road
London SW19 3BR
Tel: (020) 8542 8121
Fax: (020) 8543 6629

IMPACT TEST EQUIPMENT Ltd
Block 4, Unit 1
Moorpark Industrial Estate
Moorpark Place
Stevenston
Ayrshire
Scotland KA20 3JT
Tel: (01294) 602626
Tel (Tech Services): (01294) 471117
Fax: (01294) 605199

MALVERN INSTRUMENTS Ltd
Spring Lane South
Malvern
Worcestershire WR14 1AT
Tel: (01684) 892456
Fax: (01684) 892789

RETSCH U.K. Ltd
Unit 550 Metroplex Business Park
Broadway
Salford
Manchester M5 2VW
Tel: (0161) 877 7464
Fax: (0161) 877 6771

Plastic mesh holder for palynological sieving

PLASTOK ASSOCIATES Ltd
79 Market Street
Birkenhead
Wirral
L41 6AN
Tel: (0151) 6662056
Fax:(0151) 6500073

Nylon monofilament mesh

A. J. POLAK Ltd
Unit 6
Chapman Park Industrial Estate
Willesden High Road
London NW10 2DY
Tel: (0181) 451 4141
Fax: (0181) 451 6838

Silicone rubbers

Dow Corning Silicone rubbers

MERCK Ltd.,
Merck House
Poole
Dorset BH15 ITD
Tel: (01202) 669700
Fax: (01202) 665599
http://www.merck-ltd.co.uk

Gelflex & General Electric Silicone Rubbers

ALEC TIRANTI Ltd
70 High Street
Theale
Reading
Berkshire RG7 5AR
Tel: (01743) 302775
Fax: (01734) 323487

Rhodorsil (Rhone-Poulenc)

BENTLEY CHEMICALS Ltd
Rowland Way
Hoo Farm Industrial Estate
Kidderminster
Worcestershire DY11 7RA
Tel: (01562) 515121
Fax: (01562) 515847

Personal Protective Equipment

See also general equipment suppliers

BROSCH DIRECT Ltd
Harrier Park
Southgate Way
Orton
Southgate
Peterborough PE2 6YQ
Tel: (01733) 361 351
Fax: (01733) 230 030

CHAPMAN & SMITH Ltd
Safir Works
East Hoathly
Lewes
East Sussex BN8 6EW
Tel: (01825) 840323
Fax: (01825) 840827

SHIELD MEDICARE Ltd
5 Hurlands Business Park
Hurlands Close
Farnham
Surrey GU9 9JE
Tel: (01252) 717616
Fax: (01252) 715269

Photographic slide and print archival storage

NICHOLAS HUNTER Ltd
Unit 8
Oxford Business Centre
Osney Lane
Oxford OX1 1TB
Tel: (01865) 727292
Fax: (01865) 200051

DW VIEWPACKS Ltd
Units 7/8
Peverel Drive
Granby
Milton Keynes MK1 1NL
Tel: (01908) 642323
Fax: (01908) 640164

Safety Signs and Reagent Bottle Labels

See also general equipment suppliers

SIGNS & LABELS Ltd
Kenyon House
Latham Close
Bredbury Industrial Park
Stockport
Cheshire SK6 2YF
Tel: (0161) 4946125
Fax: (0161) 4308514

Water Purification

PURITE Ltd
Unit E, Bandet Way
Chinner Road
Thame
Oxfordshire OX9 3JS
Tel: (01844) 217141
Fax: (01844) 218098

APPENDIX 2.

DESCRIBING SEDIMENTARY ROCKS AND FOSSILS IN THE FIELD

APPENDIX 2.1. Describing sedimentary rocks.

The following points should be covered in an idealised description of a sedimentary rock.

1 MEGASCOPIC DESCRIPTION (Hand specimen properties)
Record the following properties with the aid of a hand lens or stereozoom binocular microscope:

- State of induration
- Colour
- General sorting and grain size shape (roundness and sphericity)
- General mineralogical composition
- Alteration
- Sedimentary structures
- Fossils and organic remains

2 MICROSCOPIC DESCRIPTION
The following general considerations should be undertaken:

(i) Hold the slide up and with the aid of a hand lens, study the major structures and colour. Examine particles too large to be seen in their entirety under low power

(ii) Begin by looking at the slide with the lowest power objective to observe the broad framework of the rock, and to get an idea of the component percentages

(iii) Use reflected light to examine the opaque minerals

(iv) Use diagrams to illustrate significant features only (not total microscopic appearance). Features necessitating sketches might include (NB Always insert a scale on each diagram):

- different types of rock fragment
- grain or mud support
- authigenic overgrowths of quartz
- compaction effects (sutured grain contacts)
- replacement fabrics

A. Composition
Deal with the gravel fraction, then sand, and finally the mud grade, estimating the percentage of each.

1. Particle types (total percentages)

(i) Terrigenous grains (non-carbonates)

Record the following information:

- Mineralogy
- Texture
- Size
- Shape (roundness only in thin section)
- Percentages of each type
- Grain to grain relationship
- Corrosion

The following minerals (if present) should be commented on:

- Quartz (separate the different types)

- Chert
- Feldspar (separate the different types)
- Mica and other phyllosilicates
- Heavy and opaque minerals

(ii) Allochems (carbonates)

Record the following information:

- Mineralogy
- Texture
- Size
- Shape
- Percentages of each type

The following components (if present) should be commented on:

- Interclasts
- Oolites
- Bioclasts (fossil fragments)
- Pellets

2. Matrix and cement types (total percentages)

(i) Depositional

(a) Mud in sandstones

(b) Micritic mud in limestones

(ii) Post-depositional

(a) Sandstones: cements of calcite, silica or limonite

(b) Limestones: (i) primary sparite cement

(ii) replacement pseudosparite, dolomite, silica

B. Sorting and fabric

(i) General homogeneity: is specimen a single rock type, or are two or more inter-layered

(ii) Sorting of individual particle types: estimate relative sorting of various size fractions, is sediment unimodal or markedly bimodal

(iii) Sorting of whole rock, or just within a layer (well, moderately or poorly sorted, graded bedding)

(iv) Packing: do grains touch, interpenetrate, or are they separated by matrix or cement, grain or mud supported. In limestones watch for geopetal cavities and floored interstices

(v) Porosity if determinable: high or low, before or after cementation

(vi) Grain orientation structures if present (mica or elongated grains with constant orientation), also imbricate structures if present

(vii) Textural maturity, use the following guide:

Clay content

>5%	<5%→ Sorting		
IMMATURE	>710μm (poor) **SUBMATURE**	<710μm→Roundness (good)	
		(angular - sub-angular) **MATURE**	
			(sub-well rounded) **SUPERMATURE**

C. Diagenetic fabrics
Those not already described under matrix or packing may be included here, *e.g.* silicification, calcitisation, dolomitisation or partial replacement of quartz by calcite or overgrowths.

D. Petrogenesis

1. Source Area (Provenance)
   - (a) **Geology**: based on quartz fragments and heavy mineral types
   - (b) **Relief and tectonic setting**: based on mineralogy, grain size and textural maturity
   - (c) **Climate**: based on feldspar content
   - (d) **Distance from source area**: based on size of largest grains, shape and maturity
2. Depositional Environment
   - Energy of environment
   - Transport medium
   - Depth of water
   - Strength and persistence of currents
   - Effect of organisms
   - Wave action
   - Type of environment (beach, dune, river, shallow or deep sea)
3. Diagenetic History
   - Compaction
   - Cementation
   - Replacement (placed in order of occurrence)
4. Classification of Rock
   Examples of form of name:
   - A fine grained supermature orthoquartzite - quartzarenite
   - A medium grained submature subgreywacke - sublitharenite
   - Biomicrite
   - Unsorted biospar-rudite

The name is expressed in the order: cement, matrix, grains, *e.g.* calcite cemented silty sandstone, although other adjectives reflecting the energy and environment can also be incorporated within the name, *e.g.* well sorted silica cemented quart beach conglomerate.

The field description of sedimentary rocks and field logging of sections are expertly covered in Tucker (1996) and Compton (1961), and require no further expansion. The criteria applied to describing and logging of rock cores and other borehole data, was in part formalised by the Geological Society of London Engineering Group Working Party Report (see Knill *et al.* 1970). Subsequent work by Swanson (1981) provides a more detailed account of the equipment and techniques used in describing samples, and illustrates symbols and legends applicable to graphic logs. This latter work also provides an excellent reference list on the examination of samples and well cuttings.

APPENDIX 2.2. Describing fossils in the field.

The following points should be covered in an idealized description of a sedimentary rocks containing fossils.

1 FIELD DISTRIBUTION (Hand specimen properties)

When working at sites containing vertebrate fossils construct a field map showing the location of finds. Construct a log of the section when finds occur at different stratigraphic levels.

Record the following properties:

A. Fossils in growth positions

1. Does the fossil assemblage constitute a living assemblage?
This would be characterised by:
- the presence of colonial organisms, and observations relating to the interactions between different organisms (*e.g.* the presence of encrusting organisms).
- the presence of original cavities (sediment or cement infilled, are they geopetal structures?).
- unbedded or massive in appearance, is the structure part of a reef?

2. For individual fossils note and record:
- the growth form - does this change in a vertical or horizontal direction?
- are fossils infaunal or epifaunal? Has the epifauna undergone rapid sediment burial? Do fossils have exhibit a preferred orientation?
- are fossils encrusting the substrate, or boring into a hard surface? Is there evidence for the formation of a hardground?
- if terrestrial plant fossils are observed, are rootlets present?

3. For reef assemblages map out the geometry, describing:
- growth forms in the colonial organisms,
- are the skeletons of the assemblage providing the framework, or are they matrix supported?

B. Fossils not in growth positions

1. How are fossils distributed?:
- even or uneven throughout the sediment.

2. How are fossils concentrated?:
- in lenses,
- in topographic pockets,
- or form laterally persistent beds?

3. Are fossils confined/controlled by lithofacies?
4. Where evidence of fossil concentrations occurs, record:
- proportion of broken and disarticulated elements,
- preservation of delicate parts (*e.g.* spines etc)
- sorting and degree of rounding,
- evidence of imbrication (and graded or cross bedding and bedding surface structures: *e.g.* scour and sole structures).

5. Do fossils show a preferred orientation? If so measure.
6. Have fossils been bored or encrustered?
7. Note evidence of sediment bioturbation, and any biogenic or non-biogenic trace fossils.
8. Are fossils providing an indication of the top or base of a bed:
- evidence of tracks or resting positions on bedding planes.
- upper surfaces of beds indicated by encrusting animals (*e.g.* bryozoa, barnacles, formainifera) exposed on the tops of shells and hard attaching surfaces.
- rudist and boring bivalves orientated in "upright" life positions, bored holes usually, although not exclusively, open upwards.
- inequivalve molluscs, positioned with more convex valve on sediment surface.
- echinoderms orientated with their ventral (oral) side down.
- root like structures branch downwards.
- solitary corals attach at their base.

- on "level" surfaces colonial corals and coralline algae grow outward and upward.

2 FOSSIL ASSEMBLAGE AND DIVERSITY

1. Determine the fossil assemblage composition, estimate:
- relative abundance of different fossil groups within the bed and on the bedding plane (carefully examine bedding surfaces for the presence of microfauna).
2. Is the assemblage identical in all beds? Are several different assemblages present? Correlate assemblages with lithofacies variations.
3. Consider degree of reworking and transportation. Is the assemblage reflecting one or more communities.
4. Consider the composition of the fossil assemblage:
- dominated by one or more species,
- are organisms infaunal, epifaunal or nektonic? Which group dominates?
- are organisms euryhaline or stenohaline?
- is the assemblage dominated by near shore (photic zone) or distal organisms?
- are certain fossil groups conspicuous by their absence?
- Do all organisms present have a similar mode of life?

3. SKELETAL DIAGENESIS

1. Is the original mineralogy preserved? Has replacement occurred: dolomitized, silicified, pyritized, phosphatized, hematized etc.
2. Have the fossils been dissolved, leaving moulds?
3. Do the fossils occur in nodules?
4. Are the fossils preserved three-dimensionally, or have they been compressed or distorted? Distroted fossils should be examined *in-situ*, axes of deformation should be recorded before the specimen is removed.

4. MICROFOSSILS

Microfossils can provide the best means of correlating sedimentary units. Rocks should be examined for microfossils during reconnaissance mapping. The following indicators will assist in determining the presence and type of microfossils which may be present:

- view bedding planes at an inclined angle, phosphatic remains will have a high lustre, while siliceous grains may also be evident. A sugary texture may indicate calcareous microfossils. Pyritised microfossils may be evident on the surfaces of dark organic rich shales.
- target observations (and collecting) to horizons most likely to contain fossils. In an area lacking macrofossils, collect from the finest grained sediments.
- calcareous fossils are more likely to be preserved in fresh samples, although moderately weathered sediments may be easier to process.

REFERENCES

COMPTON, R. R. 1962. *Manual of Field Geology*. Wiley and Sons, Inc, (New York). x + 378p.
GOLDRING, R. 1991. *Fossils in the Field*. Longman Scientific (Harlow, U. K.). xiv + 218p.
KNILL, J. L, CRATCHLEY, C. R, EARLY, K. R, GALLOIS, R. W, HUMPHREYS, J. D, NEWBERY, J, PRICE, D. G and THURRELL, R. G. 1970. Geological Society Engineering Group Working Party Report on The logging of rock cores for engineering purposes. *Quarterly Journal of Engineering Geology*. 3, (1), 1-24.
SWANSON, R. G. 1981. *Sample examination manual*. American Association of Petroleum Geologists, Methods in Exploration Series, No. 1, 117p.
TUCKER, M. E. 1996. *Sedimentary rocks in the field*. Wiley and Sons, Inc, (Chichester). viii + 153p.

APPENDIX 3.
(All from Allman and Lawrence 1972)

CENTRIFUGE TIMES FOR BROMOFORM SEPARATION AND WATER/ACETONE RESIDUE WASHING

APPENDIX 3.1. Centrifuge times for bromoform separations.

Particle size (microns)	1000 r.p.m.			2000 r.p.m.			3000 r.p.m.			4000 r.p.m.			5000 r.p.m.		
	hr	min	sec	hr	min	sec	hr	min	sec	hr	min	sec	hr	min	sec
50·00	—	1	16·7	—	—	—	—	—	—	—	—	—	—	—	—
40·00	—	2	0	—	—	30	—	—	—	—	—	—	—	—	—
30·00	—	3	33	—	—	53·3	—	—	23·7	—	—	—	—	—	—
20·00	—	8	0	—	2	0	—	—	53·3	—	—	30	—	—	—
10·00	—	32	0	—	8	0	—	3	33	—	2	0	—	1	16·7
9·00	—	39	30	—	9	52	—	4	23	—	2	28	—	1	34·7
8·00	—	50	0	—	12	29	—	5	33	—	3	7	—	2	0
7·00	1	5	20	—	16	19	—	7	15	—	4	5	—	2	37
6·00	1	28	50	—	22	10	—	9	52	—	5	33	—	3	33
5·00	2	7	50	—	32	0	—	14	13	—	8	0	—	5	7
4·00	3	20	0	—	50	0	—	22	10	—	12	29	—	12	29
3·00	5	55	0	1	28	50	—	38	30	—	22	10	—	14	13
2·00	13	20	0	3	20	0	1	28	50	—	50	0	—	32	0
1·00	53	20	0	13	20	0	5	55	0	3	20	0	2	7	50
0·90	65	50	0	16	25	40	7	18	20	4	6	40	2	37	50
0·80	83	20	0	20	48	20	9	15	0	5	11	40	3	20	0
0·70	108	53	20	27	11	40	12	5	0	6	48	0	4	21	40
0·60	148	3	20	35	16	40	16	26	40	9	15	0	5	55	0
0·50	213	3	20	53	20	0	23	41	40	13	20	0	8	31	40
0·40	333	20	0	83	20	0	35	16	40	20	48	20	13	20	0
0·30	591	—	—	148	3	20	65	50	0	35	16	40	23	41	40
0·20	1333	—	—	333	20	0	148	3	20	83	20	0	53	20	0
0·10	—	—	—	1333	—	—	591	—	—	333	20	0	213	3	30
0·09	—	—	—	—	—	—	731	—	—	411	6	40	263	3	20
0·08	—	—	—	—	—	—	925	—	—	519	—	—	333	20	0
0·07	—	—	—	—	—	—	1208	—	—	681	—	—	436	6	40
0·06	—	—	—	—	—	—	—	—	—	925	—	—	592	—	—
0·05	—	—	—	—	—	—	—	—	—	1333	—	—	853	—	—
0·04	—	—	—	—	—	—	—	—	—	—	—	—	1333	—	—
0·03	—	—	—	—	—	—	—	—	—	—	—	—	—	—	—
0·02	—	—	—	—	—	—	—	—	—	—	—	—	—	—	—
0·01	—	—	—	—	—	—	—	—	—	—	—	—	—	—	—

Particle size (microns)	6000 r.p.m.			7000 r.p.m.			8000 r.p.m.			9000 r.p.m.			10000 r.p.m.			12000 r.p.m.			25000 r.p.m.			50000 r.p.m.			100000 r.p.m.		
	hr	min	sec	hr	min	sec	hr	min	sec	hr	min	sec	hr	min	sec	hr	min	sec	hr	min	sec	hr	min	sec	hr	min	sec
50·00	—	—	—	—	—	—	—	—	—	—	—	—	—	—	—	—	—	—	—	—	—	—	—	—	—	—	—
40·00	—	—	—	—	—	—	—	—	—	—	—	—	—	—	—	—	—	—	—	—	—	—	—	—	—	—	—
30·00	—	—	—	—	—	—	—	—	—	—	—	—	—	—	—	—	—	—	—	—	—	—	—	—	—	—	—
20·00	—	—	—	—	—	—	—	—	—	—	—	—	—	—	—	—	—	—	—	—	—	—	—	—	—	—	—
10·00	—	—	53·3	—	—	39·2	—	—	30	—	—	23·7	—	—	—	—	—	—	—	—	—	—	—	—	—	—	—
9·00	—	1	6·8	—	—	48·3	—	—	37	—	—	29·2	—	—	23·7	—	—	—	—	—	—	—	—	—	—	—	—
8·00	—	1	23·3	—	1	1	—	—	46·8	—	—	37·0	—	—	30	—	—	20·8	—	—	—	—	—	—	—	—	—
7·00	—	1	49	—	1	19·9	—	1	12	—	—	46·3	—	—	38·2	—	—	27·2	—	—	—	—	—	—	—	—	—
6·00	—	2	28	—	1	49	—	1	23·3	—	1	56	—	—	53·3	—	—	37	—	—	—	—	—	—	—	—	—
5·00	—	3	33	—	2	37	—	2	0	—	1	34·7	—	1	16·7	—	—	53·3	—	—	—	—	—	—	—	—	—
4·00	—	5	33	—	4	5	—	3	7	—	2	28	—	2	0	—	1	23·3	—	—	—	—	—	—	—	—	—
3·00	—	9	52	—	7	15	—	5	33	—	4	23	—	3	33	—	2	28	—	—	34·1	—	—	—	—	—	—
2·00	—	22	10	—	16	19	—	12	29	—	9	52	—	6	0	—	5	33	—	1	16·7	—	—	—	—	—	—
1·00	1	28	50	1	5	20	—	50	0	—	38	10	—	32	0	—	22	10	—	5	7	—	1	16·7	—	—	—
0·90	1	49	40	1	20	30	1	1	40	—	48	40	—	38	10	—	27	20	—	6	19	—	1	34·7	—	—	23·7
0·80	2	18	50	1	42	0	1	18	0	1	1	40	—	50	0	—	34	40	—	8	0	—	2	0	—	—	30
0·70	3	1	40	2	13	10	1	42	0	1	20	30	1	5	20	—	45	20	—	10	26	—	2	37	—	—	39·2
0·60	4	6	40	3	1	40	2	18	50	1	49	40	1	28	50	—	61	40	—	14	10	—	3	33	—	—	53·3
0·50	5	55	0	4	21	40	3	20	0	2	37	50	2	7	50	1	28	50	—	20	30	—	5	7	—	1	16·7
0·40	9	15	0	3	28	20	5	11	40	4	6	40	3	20	0	2	18	50	—	32	0	—	8	0	—	2	0
0·30	16	26	40	12	5	0	9	15	0	7	18	20	5	55	0	4	6	40	—	56	50	—	14	10	—	3	33
0·20	36	56	40	27	11	40	20	48	20	16	26	40	13	20	0	9	15	0	2	7	50	—	32	0	—	8	0
0·10	148	3	20	108	53	20	83	20	0	65	50	0	53	20	0	36	56	40	8	31	40	2	7	50	—	32	0
0·09	182	46	40	134	10	0	102	46	40	81	6	40	65	50	0	45	33	20	10	31	40	2	37	50	—	36	10
0·08	231	23	0	170	0	0	130	0	0	102	46	40	83	20	0	57	46	40	13	20	0	3	20	0	—	50	0
0·07	302	46	40	221	56	40	170	0	0	134	10	0	108	53	20	75	33	20	17	23	20	4	21	40	1	5	20
0·06	411	6	40	302	46	40	231	23	20	182	46	40	148	3	20	102	46	40	23	41	40	5	55	0	1	28	50
0·05	592	—	—	436	6	40	333	20	0	263	3	20	216	3	20	148	3	20	34	10	0	8	31	40	2	7	50
0·04	925	—	—	681	—	—	519	—	—	411	6	40	333	20	0	231	23	20	53	20	0	13	20	0	3	20	0
0·03	1644	—	—	1208	—	—	925	—	—	731	—	—	591	40	0	411	6	40	94	43	20	23	41	40	5	55	0
0·02	—	—	—	—	—	—	2081	—	—	1644	—	—	1333	—	—	925	—	—	216	3	20	53	20	0	13	20	0
0·01	—	—	—	—	—	—	—	—	—	—	—	—	—	—	—	3694	—	—	853	—	—	216	3	20	53	20	0

APPENDIX 3.2. Centrifuge times for washing samples in water.

Notes:
1. For particle sizes >1µm use the standard time of 2 minutes at 3000rpm.
2. Multiply all the times given in Appendix 3.2 by the factor K (see page 157).
3. It is assumed that the minerals have a S.G. >2.0.

Particle size (microns)	1000 r.p.m.			2000 r.p.m.			3000 r.p.m.			4000 r.p.m.			5000 r.p.m.		
	hr	min	sec	hr	min	sec	hr	min	sec	hr	min	sec	hr	min	sec
2·00	—	4	0	—	—	—	—	—	—	—	—	—	—	—	—
1·00	—	15	59	—	4	0	—	1	47	—	1	0	—	—	38
0·90	—	19	40	—	4	56	—	2	12	—	1	14	—	—	47
0·80	—	25	0	—	6	15	—	2	47	—	1	34	—	1	0
0·70	—	32	40	—	8	9	—	3	38	—	2	2	—	1	18
0·60	—	44	20	—	11	6	—	4	56	—	2	47	—	1	47
0·50	1	4	0	—	15	59	—	7	6	—	4	0	—	2	33
0·40	1	40	0	—	25	0	—	11	6	—	6	15	—	4	0
0·30	2	58	20	—	44	20	—	19	40	—	11	6	—	7	6
0·20	6	40	0	1	40	0	—	44	20	—	25	0	—	15	59
0·10	26	38	20	6	40	0	2	58	20	1	40	0	1	4	0
0·09	32	46	40	8	13	20	3	40	0	2	3	20	1	19	0
0·08	41	40	0	10	25	0	4	38	20	2	36	10	1	40	0
0·07	54	26	40	13	35	0	6	3	20	3	23	20	2	10	30
0·06	73	53	20	18	30	0	8	13	20	4	38	20	2	58	20
0·05	106	40	0	26	38	20	11	50	0	6	40	0	4	15	0
0·04	166	40	0	41	40	0	18	30	0	10	25	0	6	40	0
0·03	297	13	20	73	53	20	32	46	40	18	30	0	11	50	0
0·02	667	—	—	166	40	0	73	53	20	41	40	0	26	38	20
0·01	—	—	—	667	—	—	297	13	20	166	40	0	106	40	0

Particle size (microns)	6000 r.p.m.			7000 r.p.m.			8000 r.p.m.			9000 r.p.m.			10000 r.p.m.		
	hr	min	sec	hr	min	sec	hr	min	sec	hr	min	sec	hr	min	sec
2·00	—	—	—	—	—	—	—	—	—	—	—	—	—	—	—
1·00	—	—	26	—	—	—	—	—	—	—	—	—	—	—	—
0·90	—	—	33	—	—	24	—	—	—	—	—	—	—	—	—
0·80	—	—	42	—	—	31	—	—	23	—	—	—	—	—	—
0·70	—	—	54	—	—	40	—	—	31	—	—	24	—	—	—
0·60	—	1	14	—	—	54	—	—	42	—	—	33	—	—	27
0·50	—	1	47	—	1	18	—	1	0	—	—	47	—	—	38
0·40	—	2	47	—	2	2	—	1	34	—	1	14	—	1	0
0·30	—	4	56	—	3	38	—	2	47	—	2	12	—	1	47
0·20	—	11	6	—	8	9	—	6	15	—	4	56	—	4	0
0·10	—	44	20	—	32	40	—	25	0	—	19	40	—	15	59
0·09	—	54	50	—	40	20	—	30	50	—	24	20	—	19	40
0·08	1	9	20	—	51	0	—	39	0	—	30	50	—	25	0
0·07	1	30	40	1	6	40	—	51	0	—	40	20	—	32	40
0·06	2	3	20	1	30	40	1	9	20	—	54	50	—	44	20
0·05	2	58	20	2	10	30	1	40	0	1	19	0	1	4	0
0·04	4	38	20	3	23	20	2	36	10	2	3	20	1	40	0
0·03	8	13	20	6	3	20	4	38	20	3	40	0	2	58	20
0·02	18	30	0	13	35	0	10	25	0	—	—	—	6	40	10
0·01	73	53	20	54	26	40	41	40	0	32	46	40	26	38	20

APPENDIX 3.3. Centrifuge times for washing samples in acetone.

Notes:
1. For particle sizes >1µm use the standard time of 1 minute at 2000rpm.
2. Multiply all the times given in Appendix 3.3 by the factor K (see page 157).
3. It is assumed that the minerals have a S.G. >2.0.

Particle size (microns)	1000 r.p.m.			2000 r.p.m.			3000 r.p.m.			4000 r.p.m.			5000 r.p.m.		
	hr	min	sec	hr	min	sec	hr	min	sec	hr	min	sec	hr	min	sec
2·00	—	1	0	—	—	—	—	—	—	—	—	—	—	—	—
1·00	—	4	0	—	1	0	—	—	27	—	—	—	—	—	—
0·90	—	4	56	—	1	14	—	—	33	—	—	—	—	—	—
0·80	—	6	15	—	1	34	—	—	42	—	—	23	—	—	—
0·70	—	8	9	—	2	2	—	—	54	—	—	31	—	—	—
0·60	—	11	6	—	2	47	—	1	14	—	—	42	—	—	27
0·50	—	15	59	—	4	0	—	1	47	—	1	0	—	—	38
0·40	—	25	0	—	6	15	—	2	47	—	1	34	—	1	0
0·30	—	44	20	—	11	6	—	4	56	—	2	47	—	1	47
0·20	1	40	0	—	25	0	—	11	6	—	6	15	—	4	0
0·10	6	40	0	1	40	0	—	44	20	—	25	0	—	15	59
0·09	8	13	20	2	3	20	—	54	50	—	30	50	—	19	40
0·08	10	25	0	2	36	10	1	9	20	—	39	0	—	25	0
0·07	13	35	0	3	23	20	1	30	40	—	51	0	—	32	40
0·06	18	30	0	4	38	20	2	3	20	1	9	20	—	44	20
0·05	26	38	20	6	40	0	2	58	20	1	40	0	1	4	0
0·04	41	40	0	10	25	0	4	38	20	2	36	10	1	40	0
0·03	73	53	20	18	30	0	8	13	20	4	38	20	2	58	20
0·02	166	40	0	41	40	0	18	30	0	10	25	0	6	40	0
0·01	667	—	—	166	40	0	73	53	20	41	40	0	26	38	20

Particle size (microns)	6000 r.p.m.			7000 r.p.m.			8000 r.p.m.			9000 r.p.m.			10000 r.p.m.		
	hr	min	sec	hr	min	sec	hr	min	sec	hr	min	sec	hr	min	sec
2·00	—	—	—	—	—	—	—	—	—	—	—	—	—	—	—
1·00	—	—	—	—	—	—	—	—	—	—	—	—	—	—	—
0·90	—	—	—	—	—	—	—	—	—	—	—	—	—	—	—
0·80	—	—	—	—	—	—	—	—	—	—	—	—	—	—	—
0·70	—	—	—	—	—	—	—	—	—	—	—	—	—	—	—
0·60	—	—	—	—	—	—	—	—	—	—	—	—	—	—	—
0·50	—	—	27	—	—	20	—	—	—	—	—	—	—	—	—
0·40	—	—	42	—	—	31	—	—	23	—	—	19	—	—	—
0·30	—	1	14	—	—	54	—	—	42	—	—	33	—	—	27
0·20	—	2	47	—	2	2	—	1	34	—	1	14	—	1	0
0·10	—	11	6	—	8	9	—	6	15	—	4	56	—	4	0
0·09	—	13	42	—	10	4	—	7	43	—	6	5	—	4	56
0·08	—	17	20	—	12	45	—	9	45	—	7	43	—	6	15
0·07	—	22	40	—	16	39	—	12	45	—	10	4	—	8	9
0·06	—	30	50	—	22	40	—	17	20	—	13	42	—	11	6
0·05	—	44	20	—	32	40	—	25	0	—	19	40	—	15	59
0·04	1	9	35	—	51	0	—	39	0	—	30	50	—	25	0
0·03	2	3	20	1	30	40	1	9	20	—	54	50	—	44	20
0·02	4	38	20	3	23	20	2	36	10	2	3	20	1	40	0
0·01	18	30	0	13	35	0	10	25	0	8	13	20	6	40	0

APPENDIX 4.

HAZARD SYMBOLS

TOXIC, VERY TOXIC, POISON

Present serious risk of acute or chronic poisoning by inhalation, ingestion or skin absorption

HARMFUL

Present moderate risk to healthby inhalation, ingestion or skin absorption

IRRITANT

Non-corrosive, but liable to cause inflammation through immediate prolonged or repeated contact with skin or mucous membranes

EXPLOSIVE

May explode under the effect of flame, heat, friction or sensitive to sudden shock

CORROSIVE

Destroys living tissue

RADIOACTIVE

DANGEROUS TO THE ENVIRONMENT

On entering the environment may present an immediate or delayed danger for one or more components

FLAMMABLE

Extremely & highly flammable substances (flash point <0°C, boiling point <35°C, continue to burn when removed from heat source), flammable liquids (flash point equal to or greater than 21°C and less than or equal to 55°C)

OXIDIZING

Give rise to highly exothermic reactions in contact with other substances, particularly flammables

APPENDIX 5.

RESPIRATOR & FILTER CARTRIDGE COLOUR CODING

TYPE OF HAZARD	MAXIMUM USAGE CONCENTRATION	COLOUR CODE
Organic vapours (boiling point >65°C), Solvents (in general)	1000ppm or 10 x OEL, whichever is the lower	Brown
Acid gases	1000ppm or 10 x OEL, whichever is the lower	Yellow
Organic vapours & acid gases (combined)	1000ppm or 10 x OEL, whichever is the lower	Brown/Yellow banded label
Ammonia	1000ppm or 10 x OEL, whichever is the lower	Green
Organic vapours combined with particulates	Gases: 1000ppm or 10 x OEL, whichever is the lower Particulates: 4 x OEL	Brown/White
Particulates only	4 x OEL	White

APPENDIX 6.

FIRE EXTINGUISHER CLASSIFICATION

FIRE CLASSIFICATION		FOAM	STANDARD DRY POWDER	TYPE A, B, C POWDER
TYPE	CHARACTERISTICS			
A	Fires involving solid materials, e.g. wood, paper, textiles	YES Smothering & cooling effect	YES Use only on small surface fires	YES rapid knockdown & smothering effect, prevents reignition
B	Fires involving liquids, e.g. petrol, oil, paint, fat, solvents & grease	YES Blankets fire & prevents reignition	YES Rapid knockdown	YES Rapid knockdown
C	Fires involving flammable gasses, e.g. propane, butane, natural gas	NO	YES	YES
ELECTRICAL	Fires involving electrical equipment	NO	YES	YES
EXTINGUISHER COLOUR CODING		CREAM	BLUE	BLUE

APPENDIX 7.

LABORATORY GLOVE RESISTANCE PROPERTIES

APPENDIX 7.1. General properties of main gloove compounds.

GENERAL PROPERTIES	NATURAL RUBBER	NITRILE	NEOPRENE	VINYL
Tensile strength	G	E	G	F
Puncture resistance	G	E	F	F
Abrasion resistance	G	E	G	F
Tear resistance	G	G	F	F
Flexibility	E	F	G	G
Thermal resistance	P	F	F	P
Ozone resistance	F	G	G	G
UV resistance	F	G	G	F

CHEMICAL RESISTANCE	NATURAL RUBBER	NITRILE	NEOPRENE	VINYL
Acids & Alkalines	G*	E*	G*	G*
Hydrocarbon & chlorinated solvents	P	G	F	P
Ketones	G	P	G	P
Alcohols	G	G	G	G
Oils	P	E	G	G
Detergents	F	G	G	G
Miscellaneous chemicals	V	G*	G*	V

Key: P = Poor; F = Fair; G = Good; E = Excellent; V = Varies.
* - Some important exceptions, seek further technical advice from supplier or manufacturer.

APPENDIX 7.2. Gloove properties in relation to chemical solutions.

GENERAL PROPERTIES	NATURAL RUBBER	NITRILE	NEOPRENE	VINYL
Acetic acid (Glacial)	E	P	E	NR
Acetone	E	NR	G	F
Ammonium Hydroxide	E	G	E	F
Amyl Acetate	P	F	NR	P
Anionic detergent (5%)	E	E	E	E
Biological detergent	E	E	E	E
Carbon Tetrachloride	NR	G	NR	NR
Cationic detergent	E	E	E	E
Chloroform	NR	NR	NR	NR
Chromic acid	NR	F	NR	F
Citric acid (sat. sol)	E	E	E	E
Ethanol	E	G	E	G
Ethylene Glycol	E	E	E	E
Formaldehyde	E	E	E	F
Formic acid (90%)	E	F	E	E
Glycerol	E	E	E	E
Hydrochloric acid (10%)	E	E	E	E
Hydrochloric acid (40%)	G	G	G	G
Hydroflouric acid (50%)	G	F	F	F
Hydrogen Peroxide (30%)	G	F	E	G
Isobutyl alcohol	E	E	E	F
Kerosene	P	E	E	F
Lubricating oil	G	E	E	G
Methanol	E	G	E	G

GENERAL PROPERTIES	NATURAL RUBBER	NITRILE	NEOPRENE	VINYL
Methyl Ethyl Ketone	G	NR	F	NR
Methyl Isobutyl Ketone	F	P	NR	NR
Morpholine	G	NR	P	NR
Naptha	NR	G	NR	F
Nitric acid (10%)	G	G	E	G
Nitric acid (70%)	NR	NR	G	P
Oxalic acid (sat. sol)	E	E	E	E
Perchloric acid (60%)	F	G	E	E
Petrol	NR	G	P	F
Petroleum Ether	NR	G	NR	NR
Phenol	E	NR	E	G
Phosphoric acid	E	E	E	G
Potassium Hydroxide (50%)	E	E	E	F
Sodium Carbonate (sat. sol)	E	E	E	E
Sodium Chloride (sat. sol)	E	E	E	E
Sodium Hydroxide (50%)	E	E	E	F
Sodium Hypochlorite (sat. sol)	E	E	G	G
Styrene	NR	NR	NR	NR
Sulphuric acid (25%)	E	E	E	E
Sulphuric acid (50%)	E	P	F	F
Sulphuric acid (90%)	NR	NR	F	F
Toluene	NR	P	NR	NR
Turpentine	NR	E	NR	P
Xylene	NR	F	NR	NR

Key: NR = Not recommended; P = Poor; F = Fair; G = Good; E = Excellent

This list is only a guide, and is not exhaustive. Many glove manufacturers and suppliers provide more extensive lists, and some *e.g. Merck* provide semi-quantitative data i.e. breakthrough time, permeation rate and physical degredation for selected brands.

APPENDIX 8.

CARE, HANDLING AND DISPOSAL OF HYDROFLUORIC (HF) ACID

Hydrofluoric acid (HF) is an extremely dangerous chemical, and one of the strongest inorganic acids. It should only be used when absolutely necessary. If there is an alternative to using HF it should be used. It should only be used in a specialised laboratory containing an HF fume cupboard (containing integral washing facilities, and not constructed using glass or porcelain tiles). Never work in a laboratory without informing at least one other person of the procedures and chemical strengths used and location of the work.

Hydrofluoric acid must be stored in polythene, gutta-percha or plastic bottles. When using this acid all work must be done in polythene containers within a fume cupboard. It cannot be used or stored in any other type of material.

Hydrofluoric acid is a highly corrosive and colourless acid, however, it possesses a distinct pungent odour. Both the solution and its vapour (it fumes in normal commercial concentrations at room temperature) are extremely toxic and corrosive, and can cause serious injury by direct contact and inhalation. The fluoride ions of the acid readily penetrate the skin, causing deep tissue and bone destruction. Burns are not easily neutralised, particularly if the symptoms are not detected immediately.

1. PROTECTIVE CLOTHING

The operator should always wear a laboratory coat to protect clothing, together with specialised acid-resistant protective clothing. This should consist of a full-length rubber apron and long (gauntlet type) rubber gloves (which should be checked before use for holes and wear. If there are any signs that holes may develop the gloves should be discarded). Polythene sleeve protectors should be used if wrist length gloves are used. Safety goggles, or preferably a full-face mask, should be worn to protect against acid splashes. This will also offer some protection from the fumes and vapour.

If any of the clothing or equipment becomes contaminated it should be neutralised with calcium carbonate ($CaCO_3$) or sodium bicarbonate ($NaCO_3$) and thoroughly washed before re-use.

2. PRECAUTIONS WHEN USING HF IN THE LABORATORY

If HF is regularly used in a laboratory, advance arrangements with a local hospital should be made for the treatment and admission of casualties. Ensure that the telephone number for the local Burns and Poison Unit or Accident and Emergency Department is available, and known to laboratory workers, together with the position of the first aid box, safety shower and eye wash station, spillage granules and neutralising agents (which should be in the fume cupboard) for small spills which might occur during procedures. Many institutions issue HF users with basic information cards which can be presented to first aider's and clinical personnel on arrival at hospital. These outline initial first aid treatment and subsequent hospital procedures in dealing with affected tissue burns.

When using HF in the laboratory, all work must be carried out in an acid proofed fume cupboard. HF should be stored near the fume cupboard, preferably in a cupboard, away from other commonly used reagents, so that transportation within the laboratory is kept to a minimum. The plastic bottles containing HF should themselves be stored in screw top polythene containers. A neutralizing agent should be stored close by for quick use in the event of spillage. Take special care not to leave part used or empty HF containers in the fume cupboard. When not in use all acid bottles should be placed in storage cupboards.

3. MEDICAL TREATMENT

HOSPITAL TREATMENT MUST BE ADMINISTERED FOR ALL BURNS.

Symptoms may start immediately contact has been made (concentrated solutions, >50%), or may be delayed (solutions <20%), and commence as a dull throbbing that builds up to become an acutely severe with persistent pain due to death of the underlying tissues. If not treated the conditions can result in extensive and even permanent damage to both tissue and bone. Accompanying the pain there may be a visible reddening of the skin.

If treating someone who has been contaminated remember to ensure that you are adequately protected by wearing disposable gloves. HF gives rise to insidious burns which may manifest themselves several hours after contact with the acid has been received, and are extremely painful. If there has been any suspicion of direct skin contact with HF wash the area immediately for several minutes.

Eyes :The eyes must be irrigated with cold running water or an isotonic saline solution until medical aid arrives. **Do not use calcium gluconate gel in the eyes.** Calcium gluconate drops should only be administered by a medically qualified person at a hospital. Pontocaine, cortisporin, or a similar agent, can also be used following consultation and recommendations from an ophthalmologist.

Lungs :Remove from exposure, evacuate to fresh air, rest and keep warm. Oxygen may be given by a medically qualified person to alleviate or prevent pulmonary oedema. If ingested, the mouth should be washed out with plenty of water, then large quantities of water or milk given to drink. Two calcium lactate (0.5g) tablets should be given. Try not to induce vomiting.

Skin :Remove **all** contaminated clothing. Irrigate the skin immediately with cold running water until medical attention is obtained. Pay particular attention to skin under the fingernails. If medical attention is delayed apply one of the following:

(a) A bandage soaked in a 2% iced solution of Diethylene triamine (Deta, $(NH_2.CH_2.CH_2)_2NH$), and changed every 5-10 minutes. Like HF this solution readily penetrates unbroken skin and can neutralise the effect of the acid within the tissue. A 2% aqueous solution of *Hyamine* (a benzethonium chloride cationic surfactant) can be applied instead of Diethylene triamine. A stop bottle of either solution can be maintained (mix a fresh stock before HF procedures begin, or regularly replace in laboratories were HF is in regular use) in the HF first aid box.

(b) A cream or gel of 25% calcium gluconate directly on the burn, and massaged into the area. Re-apply and massage in until the pain is elieved, and for a further 15 minutes after the pain has ceased.

(c) A paste of magnesium oxide or glycerol, or a diluted solution of ammonia in water.

(d) Soak the burned area in a solution of magnesium sulphate.

All injuries **must** be referred for clinical assessment and treatment. Inform the hospital of the intended arrival of the injured person, and continue first aid treatment on the journey to the hospital. Fortunately, fatal accidents in laboratories are extremely rare, however the danger should never be under estimated (Anon 1995a).

4. DISPOSING OF HF

All used solutions containing HF should be neutralised carefully with the addition of an alkali of powdered calcium carbonate ($CaCO_3$), sodium bicarbonate ($NaCO_3$) or orthoboric acid (H_3BO_3). Test the acidity by using pH indicator paper. Leave to stand for several hours, then flush down the drain with plenty of running water. **N.B. ALL ACIDIC SOLUTIONS MUST BE NEUTRALISED BEFORE THEY ARE DISPOSED OF.** Polluting water supplies is a criminal offence, environmental, health and safety laws are constantly being reviewed and altered. It is neglect on behalf of laboratory workers not to be aware of those that affect their working area.

BIBLIOGRAPHY ON HF SAFETY

ANONYMOUS. 1995a. HF fatality. *The American Association of Stratigraphic Palynologists Newsletter.* 28, (1), 14-15.

BROWNE, T. D. 1982. The treatment of hydrofluoric acid burns. *Journal of the Society of Occupational Medicine.* 24, 80-88.

COSTA, L. I. 1983. Safety in palynology laboratories. *Norwegian Petroleum Directorate-Bulletin 2 (Oljedirektoratet).* 75-84.

HEITMANN, S. 1983. First aid for hydrofluoric acid burns. *Norwegian Petroleum Directorate-Bulletin 2 (Oljedirektoratet).* 85-86.

LUXON, S. G. 1992. *Hazards in the Chemical Laboratory* (5th Ed). Royal Society of Chemistry (London). xix + 675p.

APPENDIX 9.

SPILLAGE AND DISPOSAL OF UNWANTED CHEMICALS

There is no single comprehensive procedure dealing with laboratory spillage's and disposal of chemicals and contaminated cleaning materials. Procedures employed depend on the hazards posed by the chemical, and in particular its flammability and toxicity. Small spills of most solutions, in fume cupboards and on floors, can be dealt with in the following way:

1 Shut off all possible sources of ignition.

2 Instruct others to keep at a safe distance, and get additional help or inform others of the spillage. For large spills, seal off the contaminated area by means of barrier tape and hazard notices.

3 Check yourself, and others in the immediate area of the spillage, for injuries. Treat as appropriate. Wear breathing apparatus, or goggles/face shield, laboratory coat, gloves, sleeve protectors, plastic/rubber apron and suitable protective footwear if required. If called to deal with an unknown spillage, identify the chemicals and potential hazards to aid in formulating a plan of action.

4 Contain the spillage by surrounding it with universal absorbent minibooms. Apply neutralising compound, a little at a time, to acid spills. Absorb the spillage using inert pillows, cloth or granules. Transfer contaminated waste to suitable leak proof bags or containers for transportation/disposal or atmospheric evaporation (if volatile) before disposal of the solid waste.

5 Ventilate the area to dispel residual vapour. Ensure all recesses are clear of all spilt solutions, and all damp areas are neutralised.

6 Thoroughly wash the site of the spillage with water and detergent.

Ensure the disposal of contaminated waste is in accordance with any local regulations, particularly if general flushing of the area results in waste water entering the local water system. Sealed containers should be used to dispose of solid or soiled waste cleaning products in accordance with the Chemicals (Hazard Information and Packaging) Regulations (CHIP, 1994). Large spillage's, or those involving extremely toxic substances, should be dealt with by professional firemen. However, following the evacuation of the area, remain at a safe distance to the scene to advise on the chemicals spilt and the geography of the laboratory.

REFERENCE

ANONYMOUS. 1996. *BDH Spillage Chart* (10th Edition). BDH Ltd., (Poole).

APPENDIX 10.

CONVERSION DATA

(i) Temperature

Fahrenhite to celsius (centigrade):
Subtract 32 and multiply by 0.5555,
or subtract 32, multiply by 5, and divide by 9.

Celsius to fahrenhite:
Multiply by 1.8, and add 32,
or multiply by 9, divide by 5 and add 32.
to convert degrees celsius (°C) to degrees kelvin (°K) add 273.2.

(ii) Weight

To convert:	to:	multiply by:
Grams	Ounces	0.03527
Ounces	Grams	28.3495
Kilograms	Pounds	2.20462
Pounds	Grams	453.59237
Pounds	Kilograms	0.45359

(iii) Volume

To convert:	to:	multiply by:
Cubic inches	Cubic cms	16.3871
Cubic feet	Litres	28.317
Cubic feet	Cubic metres	0.0283
U.K pints	Litres	0.5683
U.K gallons	Litres	4.546
U.S gallons	Litres	3.785
U.K gallons	U.S gallons	1.20095
U.S gallons	U.K gallons	0.832674

(iv) Length

To convert:	to:	multiply by:
Inches	Centimetres	2.54
Feet	Centimetres	30.48
Feet	Metres	0.3048
Yards	Metres	0.9144
Miles	Kilometres	1.6093

APPENDIX 11.

PREPARATION OF STANDARD SOLUTIONS AND SOLUTIONS FROM COMMERCIAL REAGENTS

$$\text{Weight \% (w/w)} = \left(\frac{\text{Mass of a substance in solution}}{\text{Mass of solution}}\right) \times 100$$

$$\text{Volume \% (v/v)} = \left(\frac{\text{Volume of a substance in solution}}{\text{Volume of solution}}\right) \times 100$$

$$\text{Weight/Volume \% (w/v)} = \left(\frac{\text{Mass of a substance in solution}}{\text{Volume of solution}}\right) \times 100$$

$$\text{Mole fraction} = \left(\frac{\text{Moles of component}}{\text{Total moles of all components}}\right)$$

ml of commercial reagent diluted to 1lt with deionized/distilled water

$$= \left(\frac{\left(\frac{100}{\text{\% Assay*}}\right) \text{ x formula mass* x M}}{\text{Specific Gravity*}}\right)$$

* = values indicated on reagent bottle lable
M = required mole value

REFERENCE

QUIGLEY, M. N. 1995. Preparation of standard solutions. *Science Technology*. **36**, 15-18.

INDEX

Author Index

Page numbers in **bold** refer to a figure or table.

Abbott, 281, 282, **283**, 284
Abelmann, 123, 302
Abineri, 434
Adams, 43, 45, 48, 49, 50, 51
Adamczak, 120
Agnew, 25
Airey, 173
Akpati, 189
Aldridge, 258, 322, 334, 335, 338, 341
Allen, 400
Allison, 98, 106
Allman, 84, 86, 127, 154, **155**, **156**, **157**, **159**, **161**, 168, 169, 173, 182, 216, 217, 322, 334, 335, 347, 360, 400, 401, 403, 503
Alt, 447
Alve, 352, 354
Alvin, 406
Anderson, L, 165, 166
Anderson, T.F, 103
Andrews, C. W, 48
Andrews, H. N, 279
Andruleit, 204, 372
Anehus, 129, 130, 322
Anima, 34
Ap Gwynn, 35, 36, 37, 38
Armstrong, 191, 192, 193
Arnold, 34, 154, 189, 340
Artusy, J. C, 217
Artusy, R. L, 217
Ashley Smith, 82, 89, 95
Assarsson, 256
Athersuch, 183, 418
Auffenberg, 45
Austin, 138, 139, 144, 325
Awadallah, 84
Azar, 237, 238, 239

Baas, 99, 108
Babcock, 411, 412, **414**
Baird, D, 44
Baird, G. C, 98, 124
Balanc, 115, 116, 117, 122, 123, 124, 126
Baldauf, 202
Banks, 280
Banner, 143
Baird, 407
Barnes, B, 370
Barnes, S. H, 387
Barry, 23
Barss, 201, 202, 256, 264, 312, 366, 367, 368, 369
Bartlett, 393
Bassett, 3
Bassler, 115, 195, 425
Bates, 269, 271
Bathel, 445
Battarbee, 204
Batten, 260
Baumgartner, 29, 32, 33, 142, 183, 200, 289, 293, 295, 385
Baumgartner-Mora, 183, 200
Be, 170, 172, 363, 396, 433
Beckmann, 318, 323
Belka, 173, 179, 180
Bell, 130, 318, 322
Bellemo, 199, 200
Bengtson, 318, 437
Benjamin, 173
Benson, 34, 425, 429
Benton, 3, 25, 406
Berdan, 110
Berggren, 379
Bernhard, 215
Berry, 331
Besenbacher, 378
Bessom, 445
Besterman, 20, 25

Billman, 358, 359
Bishop, 67
Bitter, 323
Black, G. P, 25
Black, M, 370
Blackie, 184
Blackmon, 364
Bodén, 203, 204
Bolli, 335, 337, 338, 343, 344
Boltovskoy, 166, 172, 197, 198
Bonem, 34
Booth, 95
Bothner, 34
Boucot, 364
Bouma, 84, 362, 391
Bourdon, 346, 347
Bowen, 166
Boyde, 387
Bracegirdle, 418, 419
Bradbury, 185, 422
Bradley, 84
Bramlette, 369, 370, 372, **373**
Branson, 430
Brasier, 143, 168, 171, 289, 295, 308, 322, 335, 337, 340, 341, 349
Brett, 98
Briggs, A, 185
Briggs, D. E. G, 98, 99, 100, 105, 106, **107**, 108, 251, 445
Bromley, 33
Bronnimann, 14, 134, 135, 347
Brooks, 163, 169, 347
Brotzen, 216
Broussard, 34
Brown, R. J, 144
Browne, R. G, 191, 192
Browne, T. D, 513
Brunton, 3, **9**, 15, 20, 21, 60, 61, 75, 77, 78, 82, 96, 112, 441, 442, 443, 444, 445, 446, 448
Bulman, 331, 332
Burckle, 297
Burden, 263
Burke, 413
Burma, 200, 289
Burmann, 308
Burnham, 34
Bushinsky, 33
Butterlin, 143, 144
Butterworth, 263, 274275, 276, 277
Buzas, 171

Calder, 418
Callahan, 173
Camp, 25, 43, 45, 59, 123
Campbell, 128
Canno, 240
Caratini, 263
Carpenter, K, 445
Carpenter, W. B, 357
Carreck, 43, 45, 48, 49, 50, 51
Carson, 173
Carver, 84
Chaloner, 402
Chaney, 28, 112, 118, 400, 401, 406, 408, 409
Chapman, 378
Charlton, 174, 175, **176**, 327
Chase, 413, 441, 445, 446
Chescoe, 376, 378, 379
Chiovetti, 102
Choquette, 228, 231
Church, 123, 124
Ciesielski, 301
Cifelli, 166
Clark, D. F, 370
Clark, D. L, 400, 401, 403, 415
Clark, S, 392, 393, 397
Clay-Poole, 268, 269
Cleal, 25
Clemens, 25, 242, 245
Cohee, 173
Cohen, 259
Colbath, 256
Colenutt, 334
Collins, 82
Collinson, C, 29, 30, 173, 318, 321, 322, 325
Collinson, M. E, 280
Colthurst, 144

Comerford, 448
Compton, 3, 6, 21, 59, 502
Conrad, 401
Conroy, 392
Converse, 15, 20, 43, 44, 45, 46, 47, 48, 242, 245, 246, 247, 250, 318, 319, 320, 400, 406, 445
Conway-Morris, 98
Cook, 392
Cooper, C. F, 48
Cooper, C. L, 425, 428
Cooper, G. A, 318
Cooper, J. A, 15
Cope, 25
Copestake, 289, 292
Copley, 273
Corliss, 136
Cornish, 92, 93
Costa, 67, 68, 513
Cox, 331
Crasquin-Soleau, 129, 343, 346, 347
Cratchley, 502
Crelling, 73
Crevello, 84, 85, 86
Crilley, 201, 202, 264
Croucher, 413
Crowley, 335
Crowther, 3, 82, 98, 331
Culver, 143
Cummings, 188, 432
Cushman, 34, 350
Cwynar, 263

Dale, 263
Davies, 45
Davis, B. L, 362, 391
Davis, L, 175
Dean, A, 262, 265
Dean, B, 98
Dean, D. A, 186, 193
Demarest, 34
DeSalle, 240
De Wever, 289, 385
Dickson, 224, 227, **229**, 231
Dillé, 185
DiMichele, 256
Doher, 256
Donovan, 98
Doran, 280
Dougherty, 338
Doughty, 15
Douglass, 357, 361, 434
Dow, 160, 162, 163, **164**, 327, 328
Doyle, 82, 92, 93, 252, 253
Dring, 307
Driver, 122
Duffield, 335, 338
Dumican, 333
Dumitrica, 294
Dunn, 350
Dutcher, 73
Dyer, 289, 292

Eagar, 263
Early, 502
Easton, 422, 434
Eaton, 145
Eckert, 160, 162, 165, 349
Ediger, 263
Edwards, A. R, 154, 368, 369
Edwards, L. E, 311
Eichenlaub, 34
Eisenack, 350
Elder, 82, 84
Emerson, 136
Emery, 34
Emiliani, 188, 189, 190, 191, 364
Emo, 144
Empson, 183
Erdtman, 272
Ethington, 325
Everett, 65, 66, 68
Evitt, 155, 203, 277, 291, 305, 312, 313, 314, 431
Ewing, 318

Faegri, 264
Farley, 256
Faruqi, 273
Feigl, 223

Feldman, 118, 420, 424, 425, 426, 427, 430, 431, 437, 439, 445
Ferguson, 25
Finch, 367, 368
Finger, 191, 192, 193
Fiorillo, 145
Fischer, 400, 402
Fitzpatrick, 420
Ford, 325, 326
Foreman, 206, 208
Fourcade, 143, 144
Fournier, 419, 422, 426, 430, 432, 433
Franz, 372, **373**
Freeman, 173, 177, 178, 179
Frey, 173
Frickhinger, 98
Friedman, 56, 224, **225**, 228
Frizzell, 242, 245, 246
Frost, 400, 403
Fuehrer, 400, 415
Fuller, 34
Funkhouser, 155, 277, 291, 313

Gaffey, 14, 134, 135, 347
Galippe, 238
Gallois, 502
Garner, 52, 400
Garrett, 418
Gay, M. M, 402
Gay, P, 185
Gersonde, 204
Gibson, 171, 173
Gilbert, 34
Ginsburg, 34, 84
Gipson, 116
Glauert, 376
Glew, 34
Gocht, 203
Goodhew, 376, 378, 379
Goodwin, 400, 406
Goldring, 3, 502
Goldstein, 376
Gott, 430
Gower, 243
Grabau, 425
Granlund, 256
Grant, 130
Graus, 147
Gray, 116, 127, 256, 312
Grayson, 196
Green, J, 410
Green, O. R, **4**, **8**, 14, 38, 55, 65, 69, 84, 86, **87**, 123, 124, 136, 211, **230**, 231, 301, **306**, 347, 355, 357, 358
Greenwald, 43, 44, 242
Grey, 307, 308
Griffith, 319
Guber, 259
Gunnell, 322
Gunther, L. F, 115
Gunther, V. G, 402, 415
Gutschick, 419, 432, 433

Haedrich, 393
Hageman, 357
Hagn, 188
Hamblin, 362, 391, 394
Hamilton, 367, 368
Hanna, G. D, 25, 43, 45, 59, 123, 124
Hannibal, 113
Hansen, 199, 203, 358, 372, **373**, 383, 387
Harbersetzer, 392, 393-396
Hardy, 364
Hargis, 34
Harper, 98, 100, 103, 104, 106
Harris, A. G, 127
Harris, G, 364, 382
Harris, J. M, 248, 251, 252
Harris, T. M, 279
Hart, 363, 364, 396
Harte, 20
Haslett, 383, 384, 385
Haubitz, 392
Hauff, 173
Hay, 139, 171, 366, 367, 369, 370, 371, 379
Hayat, 100, 102, 103, 376, 378
Haynes, 34, 36, 188

Heard, 284
Heaton, 402
Hedley, 363, 396
Heeger, 224
Heintz, 410
Heitmann, 513
Hemleben, 27, 34, 36, 383, 386, 387
Henbest, 217
Hendey, 297, 298, 388
Hendry, 189
Henwood, 239
Herman, 217
Herngreen, 260, 262, 278
Heron-Allen, 342
Hess, 240
Hessland, 195, 427
Hibbard, 245
Hicks, 418
Higgins, **151**, 168, 275, 308, 322, 334, 335, 349
Hill, C. R, 103, 382, 384
Hill, G. W, 391
Hills, J. M, 31
Hills, L. V, 257
Hinchey, 123, 124, 301, **306**
Hinds, 193
Hodgkin, 383, 388
Hodgkinson, 14, 65, 115, 127, 135, 187, 188, 383, 384, 385, 437, 439
Hodson, F, 173
Hodson, H, 173
Hoffman, 165, 166
Hoffmeister, 259, 277
Hofker, 357
Honjo, 189, 361, 379
Hooper, 363, 364, 391, 396
Hope, 58
Horie, 14, 82, 92
Horne, 448
Hoskings, 58
Hottinger, 183, 358, 363, 364, 396
Hotton, 43
Hovorslev, 34
Howden, 387
Howe, H. J, 420, 421
Howe, H. V, 169
Howe, S. R, 67
Howell, A. C, 83
Howell, J, 421
Howie, 70, 73, 75, **76**, 77, 78, 82, 91, 92, 95, 96, 112, 248, 251, 413
Hoyt, 34
Hughes, 262
Hügi, 224
Hulme, 153, 340
Humphreys, 502
Hussey, 128
Hutchison, 84, 154, 182, 193
Hutt, 331, 333
Hüttmann, 383, 386, 387

Ibaraki, 197
Imperato, 34
Ireland, 194, 350, 351
Irwin, 34
Isham, 445, 446
Iversen, 264
Iwama, 412

Jackson, 260
Jacobson, R. E, 418, 419
Jacobson, S. R, 280, 281
Jago, 429
Jansonius, 204, 310
Jarrett, 446
Jeffords, 421, 425, 430
Jenkins, D. A, 25
Jenkins, W. A. M, 203, 310, 311
Jennings, 284, 285
Jepson, 23
Jeppsson, 129, 130, 173, 195, 322, 323
Jim, 84
Johnson, B, 64, 65, 66, 83
Johnson, D. B, 25
Johnson, M. E, 51, 410
Johnston, 443, 445, 446, 447
Jonasson, 136
Jones, B, 34
Jones, C, 183, 418

Jones, C. G, 387
Jones, D. J, **27**, 33, 138, 141, 142, 150, **152**
Jones, M. D, 112
Jones, T. R, 357
Joy, 284

Kaesler, 34
Kälin, 276
Katz, 154, 368, 369
Kear, 98, 99, 100, 105, 106, **107**, 108
Keen, 448
Keene, 83
Keller, 110, 111, 250, 251
Kellett, 195, 425
Kelly, 360, 400, 402, 406
Kennedy, 191, 192
Kermack, 25
Kerr, 185
Kersley, 383, 384, 385
Kesling, 193, 199
Keyes, 403, 404
Kidson, 263
Kier, 425, 429
King, 138, 140
Kirchner, 342, 343, 344, 345
Kittrick, 58
Kling, 289, 290
Knill, 502
Knox, 156, 157, 173
Koizumi, 202, 301
Kontrovitz, 14, 144, 359
Kornicker, 172, 349
Kraft, 331
Kremp, 188
Krinsley, 433
Krukowski, 173
Krumbein, 146
Kucera, 282
Kuehne, 165
Kugler, 432
Kullenburg, 34

Lagle, 300, 301, 304, 305
Lanesky, 34
Lanooy, 413
Larsson, 238
Latimer, 410, 412, 413, **414**
Lawrence, 84, 86, 127, 154, **155**, **156**, **157**, **159**, **161**, 168, 169, 173, 182, 216, 217, 322, 334, 335, 347, 360, 400, 401, 403, 503
Laws, 203
Layne, 335
Leary, 363, 364, 396
Lee, 325, 326
Leffingwell, 383, 388
Leggi, 242, 320
Lehmann, E. P, 430, 431
Lehmann, G, 169, 171
Lennie, 256, 267
Lethiers, 129, 343, 346, 347
Levinson, 188, 190, 194, 199, 217
Lewin, 300
Lezak, 173
Lichter, 237, 238
Lincoln, 35, 36, **40**, **41**, 67, 99, 100, **101**
Lindholm, 147, 186, 364
Lindsay, 83, 91, 130, 248, 249, 251, 448
Ling, 387
Linsley, 400
Lipps, 342
Lister, 86, 193
Litwin, 256
Lloyd, 34
Locker, 203, 204
Loeblich, 259
Lofgren, 145
Lohman, 303
Loomis, **442**, 447
Lörcher, 110, 111, 250, 251
Lund, 128, 421
Lutze, 170
Luxon, 74, 513
Lykke-Anderson, 199, 358

Maceo, 51, 52, 53, 406
MacIntyre, 394

MacVicar, 120, 122, 123
Mai, 371, 372, **373**
Malyskek, 156
Mandado, 220
Mann, 55, 56, 57
Maples, 335, 338
Margerum, 65, 187, 188
Marsh, L. F, 425, 428
Marsh, R. C, 425, 428
Marshall, 376
Martill, 98, 100, 103, 104, 106, 251, 379, 380
Martin, E. A, 34
Martin, R. E, 143, 144, 213
Martinez Macchiavello, 203
Martinsson, 110, 355, **356**
Matten, 284
Maximov, **30**, **31**
May, 242, 320
Maybury, 35, 36, 37, 38
McAndrews, 263
McCaffrey, 144
McCann, 34
McGowan, 392
McIntyre, 34, 263
McKee, 261
McKenna, 245
McKerrow, 51, 410
McLachlan, 360, 400, 402, 406
McNair, 419
Meade, 412
Mehl, 183, 363, 364, 396
Meisburger, 34
Meller, 202
Mendleson, 305
Merrill, 173, 179
Metz, 217
Meyer, 188
Miles, 441, 442, 443, 445, 446, 447
Mileson, 173
Millar-Campbell, 173, 323
Miller, B. B, 147, 245
Miller, J, 182
Miller, K. R, 102
Miller, R. J, 34
Miller, T. H, 421, 425, 430
Miller, W, 98
Milner, 185
Moore, D. C, 65
Moore, P. D, 256
Moore, T. C, 203, 204, 205
Morey, 322
Morgan, 34
Moreman, 350
Morrison, I, 392, 393, 397
Morrison, L, 260
Moshkowitz, 372, **373**
Muir, 376, 382
Muir-Wood, 318, 319
Müller, 288
Murry, 173
Murray, J. W, 170, 213, 350, 352, 354
Murray, P. A,
Myers, 213

Nation, 100, 105
Neves, 263
Newbery, 502
Newport, 294
Nichols, 34
Nigam, 143
Norby, 324
Norem, 256, 260, 262, 264
Nørgaard, 266
Norris, 98, 263
Novacek, 242, 245
Nye, 193, 194

Obata, 51, 52, 53, 406, 407, 412, 441
O'Connor, 183, 295
O'Halloran, 25
Owen, 318, 319

Pääbo, 240
Palmer, A. R, 110, 111, 112, 437, 445
Palmer, T. J, 220, 222
Pantin, 35
Papp, 369, 372, **373**
Parker, A, 34
Parker, W. K, 357

Parsley, 51, 402
Patrick, 122
Patterson, 136
Peachey, 57
Peppers, 263, 264, 274, 275
Perch-Nielsen, 372, **373**
Perera, 70
Perillo, 34
Pershouse, 34
Pessagno, 289, 291, 292, 293, 294, 357, 385
Petito, 70
Pettifer, 400, 403
Pettijohn, **148**
Phillips, 185
Phipps, 256
Phleger, 34, 211, 212
Pierce, 147
Pingitore, 14, 128, 135, 335, **336**, 337, 338, 343, 347
Playford, 256, 307
Ploem, 185
Plotnick, 98, 361, 364, 382
Pohl, 191, 192
Poinar, G. O, 234, **235**, 236, 237, 238, 239, 240, 241
Poinar, H. N, 240
Poinar, R, 234, 236, 239, 240
Pojeta, 115, 116, 117, 122, 123, 124, 126
Potter, **148**
Poulsen, 425
Powell, 67
Prezbindowski, 435, **436**
Price, 502
Purvis, 25
Pye, 98

Quigley, 516
Quinn, 400, 413

Raine, 263
Raistrick, 274, 276
Rasetti, 115, 130, 360, 361, 419, 420, 421, 422, 429, 431
Reid, 58
Reimer, 122
Reissinger, 259
Rhumbler, 213
Rice, 236, 237, 238
Rickards, 331, 333
Riedel, 206, 208, 289, 292, 294, 295
Rigby, 400, 401, 403, 415
Riley, 138, 140
Riskind, 51, 52, 53, 406
Rixon, 5, 15, 43, 44, 48, 50, 83, 90, 91, 92, 111, 112, 113, 126, 243, 249, 252, 253, 318, 320, 400, 401, 412, 415, 445
Roberts, 350
Robison, 112
Roth, 372, **373**
Röttger, 169, 171, 418
Rowland, 403, 404, 412
Rusbridge, 433
Rushton, **3**, 59

Sachs, 166, 172, 189
Sameshima, 197
Sandberg, 379
Sander, 419, 432
Sandon, 57
Sanfilippo, 289, 292, 294, 295
Sarjeant, 206, 263, 305, 312, 313, 314, 315
Sass, 68, 319, 320, 420, 425
Savage, 173, 174, 327
Savrda, 33, 196, 391
Schaffer, 214
Schaeffer, 249, 250, 251
Schallreuter, 197
Schlüter, 234
Schmidt, R. A. M, 160, 396
Schmidt, R. R, 372, **373**
Schopf, 280, 281
Schrader, 202, 204, 298, 300, 301
Schrimper, 413
Schultheiss, 34
Schultz, 418

Schultze, 362, 391, 393, 394, 395, 396, 397
Schulze, 274
Scott, A. C, 183, 184, 185
Scott, D. B, 183
Scott, H. W, 196
Scovil, 418, 419, 421, **423**, 424, 435
Secrist, 350, 352
Sefton, 95
Sela, 387
Sennikov, 243
Serandrei Barbero, 136
Setty, 143, 208, 298, 300
Shaffer, 315
Shapiro, 57
Sharpe, M, 401
Sharpe, T, 441, 443, 445, 446, 447
Sheals, 35, 36, **40**, **41**, 67, 99, 100, **101**
Shimer, 425
Shterenberg, 154, 155, 173
Siever, **148**
Simes, 322
Sincock, 264, 265, 279, 280
Sisson, 400
Sittler, 256
Siveter, David, J, 360
Sieveter, Derek, J, 419, 425, 428, 430, 431
Skevington, 331
Skoglund, 331, 332
Skvarla, 376
Smith, A. H. V, 263, 274, 275, 276, 277
Smith, C. C, 372, **373**
Smith, D. G, 34, 144
Smith, J, 410, 412, 413, **414**
Smith, P. M. R, 185
Smith, R. K, 166, 167, 172
Smith, W, 34
Snyder, 143
Sohl, 110, 111, 126
Sohn, 116, 123, 124, 193, 194, 196, 199, 355, 356
Somekh, 185
Sorauf, 422, 434
Sorsby, 448
Spechmann, 150, 151, **152**, 349
Spence, 411
Spicer, 103
Spielholtz, 277
Spinner, **151**, 168, 275, 308, 322, 334, 335, 349
Stanley, E. A, 143, **143**
Stanley, J. W, 401, 402, 403
Stanley, M. F, 67
Staplin, 308
Steinker, 213
Stetson, 34
Stevens, 115, 116
Stewart, 284
Stone, J, 318
Stone, G. W, 34
Stradner, 369, 372, **373**
Stubblefield, 280
Stucker, 113
Sullivan, 369, 370, 372, **373**
Surlyk, 125
Swanson, 58, 502
Sweet, A. R, 257
Sweet, W. C, 127
Swift, 341
Switzer, 364
Sylvester-Bradley, 193, 199

Tanimura, 202, 303
Tanke, 185
Tapp, 435, **436**
Tappan, 259
Tasch, 315
Taugourdeau, 315
Taylor, M. A, 20
Taylor, P. D, 387, 388, **389**
Taylor, R. J, 367, 368
Taylor, T. N, 284
Teichert, 425, 427, 428
Tena, 220
Then, 338
Thierstein, 372, **373**
Thomas, B. R, 234

Thomas, K. M, 69
Thompson, 15
Thomson, 422
Thurrell, 502
Tiffany, 412
Todd, 29, 33, 340, 364
Toffoletto, 136
Toombs, 249, 318, 320
Toomey, 350
Tout, 447
Towe, 370
Traverse, 156, 173, 208, 256, **258**, 264
Tremain, 263
Triebel, 217, 432
Trusell, 228, 231
Tucker, 173, 364, 502
Tunnicliff, 25
Turner, R. E, 144
Turner, W. M, 173
Tuttle, 422, 434
Twilley, 34

Ulrich, 425
Upton, 67

Valentine, 34
Van Morkhoven, 188, 189
Vannier, 392
Varker, 29, 321
Vernon, 403
Vickers, 57
Vidal, 263
Vilks, 183
Von Bitter, 173
Von Hofe Grady, 173, 242, 245

Wagner, 196
Walawender, 362, 391
Walker, A. D, 406
Walker, D. A, 36, 214, 215
Walker, W. M, 171, 173
Walton, 211, 213, 284
Ward, 410
Warne, 56, 224, **226**
Warshauer, 335, 338
Waters, 335, 338
Watson, 264, 265, 279, 280, 406
Weaver, 34
Webster, 175, 425
Weide, 425
Wesley, 279
West, 202, 269, 270, 271, 297, 298
Westberg-Smith, 289
Wetmore, 361, 362, 418
Whatley, 361
Whitcomb, 383, 385
White, 144
Whittaker, 383, 384, 385, 437, 439
Whittington, 98, 318, 419
Whybrow, 14, 83, 91, 111, 112, 185, 243, 244, 249, 392
Wicher, 344
Wild, 25
Wilson, G. J, 203, 265, 266, 267
Wilson, J, 112
Wilson, L. R, 139, 141, **141**, 142, 143, 259
Wilson, M. A, 220, 222, 415
Williams, A, 193
Williams, G. L, 201, 202, 256, 263, 264, 305, 308, 312, 366, 367, 368, 369
Williams-Mitchell, 140, 342, 345
Wiman, 331
Wimbledon, 3, 25
Wing, 256
Winland, 228
Wittard, 400
Wolberg, 82, 89, 91, 92
Wornardt, 203
Wood, 189, 199
Woodward, 45
Wray, 350
Wright, 171
Wrigley, 198

Yakovley, 154, 155, 173
Young, 20, 21

Zangerl, 362, 391, 393, 394, 395, 396, 397
Zeidler, 191, 192
Zetzsche, 276
Ziegler, 130, 324

Subject Index

Page numbers in **bold** refer to a figure or table.

Acetylation, 132, 267, 269, 272, 314
Acid-hydrolysis, 132
Acritarchs, 305, **309**
Air-abrasive, 113-115
Airbrush,
  latex application, 402
  magnesium oxide application, 430
Amber, 96, 234-241, **423**
  colour alterations, 238
  cutting, polishing, 236
  extracting organisms, 238
  imitations, 240
  surface degradation, 237

Bacterial activity, 95
Bitumen maceration, 273
Byne's disease, 95

Calcareous microfossil separation from
  glauconite, 161
  quartz, 162
Calcite (differentiating from dolomite through staining), 56, 223-227
Carbonised plants, 93, 279, **423**
Casting, *see* replication
Cathodoluminescence, 183, 192
Cell autolysis, 35
Cellulose, removal of, 267, 269, 272
Centrifuge, 154-159, 503
  tubes, 155, 156
Chemical Hazardous Information & Packaging (CHIP), 65, 72
Chemicals
  disposing of, 78, 514
  handling, 69
  spillages, 71, 514
  storing, 70
Chitinozoa, 308, **309**
Coal maceration, 274
Collecting
  equipment, 23, 28
  field, 20
  insurance, 22
  live material, 34
  micropalaeontological, 34
  parties, 24
  recent marine, 34
  strategy, 21, 22
Compression experiments, 361
Conodont
  extraction from
  argillaceous, 324
  calcareous, 322
  siliceous, 325
  processing, 321
  separation by
  density liquid, 173, 327
  magnetic, 162, 327
  sieving, 327
Conservation
  site, 3, 25
  specimen, 6, **8**
Copal, 234, **235**
Copy stand, 420
Cores
  field, 24, 31
Corrosion, 94
Contrast enhancing, 194-197
Control of Substances Hazardous to Health (COSHH), 65, 79
Conversion data, 515
Critical point drying, 103
Curating, 9-11

Deflocculation, 135, 269
Dehydration, 94, 271
Density liquids, 173-176, 261, 267, 313
Diatoms, 295, **306**
  extraction from,
  lithified carbonates, 304
  lithified clastics, 301
  partially consolidated sediments, 298

unconsolidated sediments, 297
Dinoflagellates, 311
Disaggregation, 127-136
carbonate removal, 129, 259, 266, 268, 269, 302
elcetrolyte removal, 135
fluro-carbonate removal, 261, 267
organic matter removal, 132, 262, 267, 270, 302
silicate removal, 131, 260, 267, 269, 271
sulphate removal, 131, 268
sulphide removal, 131
Dispersion
with peptizers, 135, 269, 303
Documentation
field samples, 59, 60
laboratory samples, 14
safety, 14
storage, 15, **16**
photographs, **16**, 422, 435
practical procedures, 14
Dolomite (determination through staining), 55
Dust (generation in work area), 73

Ebridians, 305
Electromagnetic spectrum, **377**
Electron microscopy, 376
image types, 379, **389**
Electron probe, 364
Enamel, 93
Enzymatic hydrolysis, 265
Ethics
collector, 20
field work, 21, 33
laboratory practice, 13
Exhibiting, 441
maintenance, 447
media selection, 445
planning, 441, **442**
specimen selection, 443
testing, 446
Expanded polystrene, 48
Expanded polyurethane, 49

Eye protection, 74

Filtration, 172
Finder graticules, 206, **207**
First aid, 80
Fixatives
see specimen fixation
Flotation, 169, 347
Foraminifera (agglutinated)
extraction from,
consolidated, 351
unconsolidated, 352
Foraminifera (calcareous)
extraction from,
lithified argillaceous, 335
lithified carbonates, 343
unlithified argillaceous, 334
unlithified carbonates, 342
sectioning, 188
Franz isodynamic separator, 160, 162, **164**, 349
Freeze drying, 100
Freeze substitution, 103
Freeze-thaw fracturing, 123-126
Furnace, 121, 166

Glassfibre reinforced plastic (GRP), 53, 83, 400, 405, 412-414
Graptolites, 331-333
extraction from,
carbonate, 332
siliceous, 331
mounting and storage, 333
optical enhancement, 332

Handling heavy loads, 79
Heavy liquids, *see* density liquids
Holographic microscopy, 183
Humic acid, removal of, 268
Hydration, 94
Hydrolysis, 94
Hystrichospheres, 305

Ignition techniques, 166
Illustrating, 441, **447**

Impregnation
surface, 83
thorough (vacuum), 84, 86, **87**
Interfacial techniques, 177-180
Iron sulphides
treatment of, 89-93
Ivory, 93

Jacketing
see specimen jacketing
removal of, 246

Laboratory
design, 64-67
documentation, 12, **12**
glove resistence properties, 508
practice (GLP), 13, 140
safety, 14, 65, 68-81, 506, 507
security, 81
Laser Scanning Microscopy (LSM), 183
Latex rubber, 51, 402, 415
Light
damage to specimens, 96
Light Microscopy (LM), 182-185, 280
Lignin, removal of, 262, 267, 270, 274
Liquid separation, 168-180
Locality
preparing a visit, 22
arrival, 23
departing, 25

Maceration, 279-281
Magnetic separation, 160, 327, 349
Microfossil
field contamination, 138
horizontal displacement, 142
laboratory contamination, 139
peels, 361
preparation of single specimens, 355
staining residues, 216-218
sample contamination, 138-145
stratigraphic contamination, 141
vertical displacement, 141
Micropalaeontology
collecting, 27
processing, **4**
Microtome, 193, 281
Microvertebrates, 320
Mechanical cleaning, 110-118
Mechanical separation, 160-167
Moulding, *see* also replication
enlargement, 413
multiple piece mould, 408
one piece mould, 407

Nannofossil
carbon replication, 370
collecting, 366
preparation, 366-369
slides, 369
Noise reduction, 75, **76**

Occupational hygiene, 78
Oleoresins, 234
Optical observations
field, 32
laboratory, 182-185
Optical microscopes, 183, **380**
Organic acid vapours (OAV), 95
Ostracods, 120
Oxidation, 89
of organic matter, 132, 262, 267, 270, 274, 302, 313
Ozone depleting solvents, **72**

Palaeobotany, 281
cuticle preparation, 279
- recovering material from damage slides, 265
Palynology - pollen & spore extraction from:
argillaeous, 257
bitumen, 273
carbonates, 266
coal, 274
evaporites, 268
peat, 269
siliceous, 257

- recovering assemblages from damage slides, 265
Panning, 168
Particle separation, 154
  >50μm, 154
  <50μm, 156
Peels
  microfossils, 361
  stained acetate, 219-227
  stained fossil wood, 284, 285
Percussion tools, 110, 246
  chisels, 111
  dental picks, 111
  engravers, 112
  hammers, 110
  mallets, 110
  mounted needles, 111
Personal Protective Equipment (PPE), 14, 70
Phosphate (determination through staining), 56, 319
Phosphatic fossils, 288, 318
  chemical extraction, 319
Photoelectric separation, 165
Photographic plate preparation, 437, **438**
Photomacrography, 419-424
  dark field illumination, 422
  lighting, 420, **423**, **424**
  coating, 425-430
  orientation, 422
Photomicrography, 431
  coating, 432
  filters, 434, **436**
  immersion, 433
Picking, 197, 328, 341
Pneumatic tools, 28, 112
Polished sections, 192-193
Preservatives
  see specimen preservatives
Pyritized fossils, 291, 333

Radiation, 73
Radiolaria, 289, **296**
  extraction from,
  argillaceous, 290
  calcareous, 293
  siliceous, 294
Random settling, 204
Recent material, 99-108
  dehydration, 100
  experimental studies, 106-108
  killing, 99
  preservation, 99
  staining, 211-215
Record keeping, *see* documentation
Reflected light microscopy, 194, 197
Relative humidity, 89
Replication,
  field application, 405, 410
  macrofossil, 400-417
  metal casting, 360
  microfossil, 357
  nannofossil, 370
Resins
  epoxy, 86, 186, 192, 193, 244, 358, 384
  polyester, 85, 333, 412, **414**
Respirators, 77
Reworked assemblages, 141
  recognition of, 143-145
Risk Assessment, 69
Rock crushing, 257
Rock saw
  field use, 28
RTV, *see* silicone rubber

Safety signs, 79, 506
  fire extinguisher coding, 507
  respirator coding, 507
Sample bags, 5
  boxes, 6, 60
  canvas, 5, 61
  cloth, 5, 60
  hessian, 5
  polythene, 5, 60
Sample
  describing, 498
  packing, 5, **7**, 60, 139
  transportation, 6, **7**, 61, 139

washing, 257, 277
Sample collecting, 2, **7**, 20, 27, 28
bulk, 30
continuous (channel), 29
labelling, 5, **7**, 60
microvertebrates, 244-246
reconnaissance, 30
selective, 30
spot, 29
vertebrate bones, 242
Scanning Acustic Microscope (SAM), 184
Scanning Electron Microscopy (SEM), *see also* electron microscopy, 280, 371, 378, **381**
sample preparation, 382
stub preparation, 383
Scanning Light Microscope (SLM), 183
Scolecodonts, **309**, 315
Shellac, 45
Sieving, 146-153
cleaning, 147
dry, 150
shakers, 147
size, 146, **148**
wet, 152, 263, 326, 340
Siliceous fossils, 289-305
Silicoflagellates, 305
Silicone rubber (RTV), 52, 244, 360, 400-402, 406
Skin protection, 74
Slide preparation
see thin sectioning
palynological, 264
recovering palynological assemblages, 265
Small Shelly Fossils (SSF's), 321
Soil colloids, removal of, 270
Specimen
cataloguing, **9**
consolidation, 43, 82-97, 243, 247, **248**
curating, 9, **10**
documentation, 14
fixation, 35-38
jacketing, 44-51, **46**,
orientation in thin section, 185
preservation, 9, **10**, 82-97
preservatives, 35-38, **283**
repair, 82
replication, 43, 51-53
stabilization, 43
staining, 37, 55-58, 282-284
storage, **9**, **10**, 238, 277
surface enhancement, 33
Sputter coating, 385
gold coating removal, 387
uncoated specimens, 387
Staining, 55-58, 195, 211-233, 263, 282, 315
Stereophotomicrography, 430
Stratigraphic leakage, 142
Strew slide preparation, 200-206
Stub production
palynological, 264
Sub-fossil bone, 93, 252
Suppliers, 490-497
Swirling, 261
Syntron, 165

Tasmanitids, 308, **309**
Temperature
damage to specimens, 95
Thermal extraction, 120-126
heating and quenching, 122, 198
high temperature fracturing, 120
Thermo setting vinyl rubber, 360, 403
Thin sectioning, 182-208
photography, 433
polished, 192
(double sided), 192, 200
stained, 221, 231
Toxic minerals, 78
Transfer method, 249
Transmitted light microscopy, 196
see also Optical observations
Transmission electron microscope (TEM), 378, **380**
Trilobites, 120

Ultrasonic tools, 115-118
    descalers, 118, 410
    precautions with, 115, 117
    tanks, 115-118, 198, 263
Ultrastructure studies, 199

Vertebrate fossils, 242-254
    acid development, 248
Vibration
    damage to specimens, 96
    reduction, 77

Wood preservation, 282, **283**

X-radioraphy, 244, 391
    CAT-scanning, 397
    contact radiography, 392
    diffraction, 364, 382
    images, 394
    interpretation, 395
    microscopy, 362
    photography, 392, 395
    photomicroscopy, 396
    specimen preparation, 393

Chemical Index

Page numbers in **bold** refer to a figure or table.

Acetic acid, 35, **39**, **40**, 130, 133, 165, 195, 232, 249, 251, 269-272, 273, **278**, **283**, 300, 314, 319, 323, 332, 346, 354, 357, 509
Acetic anhydride, 132, 214, 267, 269, 272, 314
Acetone, 43, 104, 132, 159, 173, 174, 179, 190, 222, 239, 243, **248**, 250, 271, **336**, 406, 425, 505, 509
Acridine orange, 282
Acrylic polymer, 83, 84, 96, 111
Acrylic plaster polymer, 410
Adenosine-5'-triphosphate (ATP), 215
Alcohol
see isopropyl alcohol
Alcoholic Bouin (Dubosq-Brasil fluid), **39**
Alginate, 401, 405
Alizarin red-S, 56, 217, 224-227, 231
Aluminium oxide, 199
Aluminium potassium sulphate (alum), 411
Alconox, **336**, 367
Alvar 1570, 83
Ambersil DP 100, 407
Ammonia, 50, 91, 195, 297, 300, 303, 319, 371, 425, 428
Ammonium acetate, 319
Ammonium chloride, 195, 251, 252, 425-429
Ammonium hydrogen carbonate, 37
Ammonium hydroxide, 91, 132, 133, 135, 251, 276, 279, 285, 509
Ammonium iron (III) sulphate *12-hydrate*, 282
Ammonium metavanadate, 57
Ammonium molybdate, 57
Ammonium oxalate, 294
Ammonium polymethacrylate, 51
Amyl acetate, 189, 280, 509
Amyl chlorohydrin
see stovaine
Antimony oxide, 430
Aqua regia, 277
Araldite, 86, 358
Argon, 386
Aroclor, 202

Barium iodide, 173
Basic fuchsin, 218, 315
Beeswax, 401
Benzamine hydrochloride, **101**
Benzene, 132
Bioden, 361
Bismarck brown, 263, 315
Biuret reagent, 212
Bouin's fluid (Picro-fomol), **39**
Bromoform, 173, 262, 327, 341, 503
Butan-1-ol, 239, 299
Butanone, 285
Butvar (B76, B98), 83, 247, **248**
Byprox, 335, 343

Cadmium iodide, 173
Caedax, 369
Calcium acetate, 323
Calcium carbonate, **41**, 260, 322, 367
Calcium chloride 6-hydate, 36
Calcium hydrogen phosphate, 251
tri-calcium phosphate, 322
Calgon, *see* sodium hexametaphosphate
Canada Balsam, 189, 190, 191, 196, 200, 205, 231, 265, 357
Carbon, 432
Carbon dioxide, 104
Carbon tetrachloride, 170, 171, 293, 308, 509
Castor oil, 196
Cellosize, 201, 203, 264, 369
Cellosolve, **101**, 190
1-(3-chlorallyl)5,7-triaza-1-azoniaadamantane, **41**
Chloral hydrate, **101**

Chloroform, 132, 190, 239, 509
Chloro-zinc-iodine
see Schulze's solution
Chromic acid
see chromium (VI) oxide
Chromium (VI) oxide, 35, **40**, 259, 279, 299, 509
Chromium sulphate, 200
Chromium trioxide
see chromium (VI) oxide
Ciment Fondu, 410
Citric acid, 324, 509
Clayton yellow
see Titan yellow
Clearcol, 264
Clorox, 315, 316, 370
Cobalt Napthenate, 85
Copper sulphate, 212
Corrosive acetic, **40**
Corrosive sublimate
see mercury (II) chloride
Cyanoacrylate, 192

Darvan No. 4, 315
Dibromoethane, 173
Diethyl ether, 214, 238
Dimethylformamide, 132
Dirax, 202
Durofix, 189

Elvacite, 202, 203, 205, 264, 369
Ethanol, 35, **41**, 104, 131, 173, 193, 214, 215, 239, 282-284, 300, **336**, 378, 509
Ethanolamine thioglycollate. 92, 93
2-Ethoxyethane, 190
2-Ethoxyethyl acetate, 44
Ethyl acetate, 252
Ethyl alcohol
see ethanol
Ethyl m-aminobenzoate, **101**
Ethyl methyl ketone
see butanone
Ethylenediamine, **283**
Ethylenediaminetetra-acetic acid (EDTA), 199, 220, 268, **278**
Ethylenediaminetetra-acetic acid disodium salt (diNaEDTA), 228, 232, 357
Ethylene glycol, **41**, 509
Ethylene glycol mono-methylether, 201
Eucaine, **101**

Fast green FCF, 284
Ferric chloride, 162
Feigl's solution, 223
Flemming's solution, **40**
Formaldehyde, 35-37, **39**, **41**, 166, 301, 304, 509
Formalin, 36, **39**, **41**, 136, 170, **283**, 298, 341
Formic acid, 130, 166, 249, 293, 322, 509

Gelatine, 401
Gelflex, 403
Gentiana violet, 197
Glutaraldehyde, 36, 105, 199
Glycerol, 38, **39**, 94, 191, 196, 202, 203, 205, 237, 251, 252, 264, 269, **283**, 333, 358, 370, 433, 509
Glyceel, 38
Grren Mold Plastiflex, 403
Gum tragacanth, 136, 191, 357

Haematoxylin, 282
Heavy distillate, 179
Heidenhain's Susa mixture, **39**
Hexamethyldisilazane, 105
Hexamine, 36
Hydrochloric (HCl) acid, 32, 55, 56, 129, 131, 132, 134, 135, 187, 195, 220, 221, 223, 231, 257, 258, 261-263, 266, 267, 269-271, 275, 277-279, 285, 290, 291, 293, 294, 297-299, 301, 302, 304, 307, 310, 312-315, 331, 350, 351, 357-359, 361, 370, 425, 426, 509

Hydofluoric (HF) acid, 32, 129, 131, 196, 197, 260, 263, 267, 269, 271, 279-280, **283**, 290-292, 294, 295, 307, 312, 315, 325, 326, 331, 357, 370, 371, 509
disposal of, 511
safety note, 32, 196, 259, 291, 511
Hydrogen peroxide, 128, 134, 135, 166, 179, 180, 276, **283**, 290, 293, 294, 297, 298, 300, 302, 304, 311, 324, **336**, 338, 347, 357, 367, 428, 509
Hydrogen sulphide, 371
Hyrax, 202
Hystomount, 203

Immersion oil, 196
Iodine, **40**
Iodized alcohol, **39**, **40**
iso-amyl acetate, 104
Isocyanates, 49
Isopropyl alcohol, **39**, **40**, 43, 167, 170, 171, 194, 197, 213, 247, **248**, 259, 262, 271, 406, 425, 430

Kerosene, 177, 178, 179, 180, 266, 415, 509

Lakeside 70C, 231, 359
Lissapol, 267
Lithium chloride, 136
Lithium hydroxide, 136
Lubricating oil, 196

Malachite green, 263
Magnesium chloride, **101**
Magnesium oxide, 195, 429, 430
Magnesium sulphate, **101**
Manganese sulphate, 223
Mercaptoacetic, 251
Mercury, 211
Mercury (II) chloride, 35, **39**, **40**
Mercuric chloride
see mercury (II) chloride
Methanol, **101**, 132, 273, 293, 509
Methyl Amyl alcohol, 129
Methyl Ethyl Ketone, 509
Methyl Ethyl Ketone Peroxide, 86
Methyl green-eosin, 213
Methyl Isobutyl Ketone, 509
Methyl violet, 218
Methylated spirits (IMS), 50, 92, 93, 189
Methylene blue, 218
Mikrops, 202, 298
Millon's reagent, 211
Monochloracetic acid, 130, 323
Morpholine, 92, 509
MP-10, 266, 290, 292

Naphrax, 202, 205
Naphtha, 273, 509
Nitric acid, 130, 131, 132, 133, 134, 211, 257, 262, 263, 267, 274-276, 277, 279, 285, 299, 302, 307, 311, 314, 332, 509
p-Nitrobenzene-azoresourcinol, 55
Nitro-cellulose, 280
Nitrogen, 428

Orthoboric acid, 260, 292, 295, 307, 312, 326
Osmic acid, 35, **40**
Osmium tetroxide, 102, 103
Oudemans' fluid, **39**
Oxalic acid, 130, 131, 285, 294, 297, 509

Paraformaldehyde, 37
Paraloid (B67, B72), 83, 247
Paraffin oil, *see* kerosene
Paraffin wax, 357, 401
Paratoluenesolphonic acid, **283**
Parmetol K40, 253
Perchloric acid, 277, 509
Periodic acid, 277
Petroleum ether, 509
Petroleum jelly, 177, 178, 407
Petroleum spirit, 128, 132, 291, 324, 337, 343, 509
Phenol, 509

Phenosulphonic acid, **283**
Phenoxetols, 36, **39**, **41**
Phenoxetol BPC, **41**
Phosphoric acid, 509
Picric acid, 35, **39**
Plaster (of Paris), 45, 410-412, **414**
Pleurax, 202
Poly(butadiene), 415
Poly(butyl methacrylate), 92, 250, 252, 319, 320
Poly(ethylene glycol) (PEG), 84, 91, **283**, 407
Poly(vinyl acetal), 83, 92
Poly(vinyl acetate) (PVA), 43, 44, 84, 112, 121, 252, 406
Poly(vinyl butyral) (PVB), 43, 44, 45, 83, 84, 92, 95, 111, 243, 247, **248**
Poly(vinyl chloride) (PVC), 284, 413
Polystyrene, 177
Polyurethane, 92
Potassium carbonate, 276
Potassium chlorate, 134, 273, 274, 275, 277, 279, 322
Potassium dichromate, 35, **39**, 277, 297
Potassium ferricyanide, 56, 224-227, 231
Potassium hydroxide, 133, 135, 212, 262, 263, 270, 275, 277, **278**, 307, 509
Potassium iodide, **40**
Potassium permanganate, 297, 303
Propylene glycol, **39**
Propylene phenoxetol, **39**, **41**, **101**
Pyridine, 277

Quaternary-O, 191, 259, 335, 338
Quebracho, 367

Rhodorsil, 402
Rose bengal, 213, 352
Ruthenium red, 218

Safranine O, 263, 273, 280, 282, 284, 308, 315
Schaudinn's solution, **40**
Schultze's solution, 134, 262, 274, 275, 279, 311, 314
Shellac, 240
Silcoset, 360
Silver nitrate, 195, 217, 432
Silver sulphate, 223
Sodium acetate, 162, 323
Sodium acetate *3-hydrate*, 124, 125
Sodium bicarbonate, 245, 251, 367
Sodium cacodylate, 37
Sodium carbonate, 128, 132, 135, 245, 303, 307, 325, 326, 335, 337, 343, 509
Sodium carbonate *10-hydrate*, 50
Sodium chlorate, 134, 270
Sodium chloride, **39**, 170, 172, 509
tri-Sodium citrate, 126
Sodium cyanide, 387
di-Sodium disulphite, 303
Sodium hexametaphosphate, 136, 298, 299, 303, 367
di-Sodium hydrogen orthophosphate, 345
Sodium hydroxide, 37, 55, 56, 133, 135, 199, 212, 220, 223, 228, 232, 262, 268, 270, 271, 272, **278**, 279, 285, 290, 291, 307, 314, 335-338, 345, 347, 387, 509
Sodium hypochlorite, 128, 134, 166, 199, 277, 279, 294, 307, 311, 314, 324, **336**-338, 357, 509
Sodium nitrite, 303
Sodium oxalate, 136
Sodium polytungstate, 129, 173, 322, 327
tetra-Sodium pyrophosphate, 135, 268, 271, 293, 301, 304
Sodium sulphate, **39**, 344
Sodium sulphate *10-hydrate*, 125
Sodium sulphite, 311
di-Sodium tetraborate, 37
di-Sodium tetraborate *10-hydrate*, 411
Sodium thiosulphate, 131, 338
penta-Sodium triphosphate, 136

Stearic acid, 128
Steedman's solution, **39**
Stovaine, **101**
Styrax, 202
Styrene, 86, 509
Sudan black B, 214
Sulphuric acid, 132, 267, 269, 270, 272, **278**, 297, 303, 314, 509
Super Glue
see cyanoacrylate

T.A.F., **40**
Tartaric acid, 324
Teepol, 335, 343
Tetrabromoethane, 173, 327, 341
Tetrachloroethylene, 177
Thioglycollic
see mercaptoacetic
Thoulet solution, 262
Titan Yellow, 56, 232
Toluene, 43, 202, 239, 415, 509
Toluidine Blue O, 194
Trichloracetic acid, **39**
1,1,1-trichloroethane, 239
Trichloroethylene, 354
Triethylene glycol, **283**
Triton X-100, 266
Turpentine, 509

Viets' solution, **39**
Vinagel, 401
Vinamold, 360, 403

Xylene, 131, 189, 197, 202, 204, 282, 333, 357, 509

Zenker's fluid, **39**
Zinc bromide, 173, 263, 262, 268, 313, 341
Zinc chloride, 173, 262

GPSR Compliance

The European Union's (EU) General Product Safety Regulation (GPSR) is a set of rules that requires consumer products to be safe and our obligations to ensure this.

If you have any concerns about our products, you can contact us on ProductSafety@springernature.com

In case Publisher is established outside the EU, the EU authorized representative is:

Springer Nature Customer Service Center GmbH
Europaplatz 3
69115 Heidelberg, Germany

Batch number: 10809436

Printed by Printforce, the Netherlands